Planning and Design of Airports

Robert Horonjeff

Francis X. McKelvey

William J. Sproule

Seth B. Young

Fifth Edition

New York Chicago San Francisco
Lisbon London Madrid Mexico City
Milan New Delhi San Juan
Seoul Singapore Sydney Toronto

The *McGraw·Hill* Companies

Cataloging-in-Publication Data is on file with the Library of Congress

1 2 3 4 5 6 7 8 9 0   DOC/DOC   1 6 5 4 3 2 1 0

ISBN 978-0-07-144641-9
MHID 0-07-144641-9

Sponsoring Editor
Larry S. Hager

Editing Supervisor
Stephen M. Smith

Production Supervisor
Richard C. Ruzycka

Acquisitions Coordinator
Alexis Richard

Project Manager
Vastavikta Sharma,
Glyph International

Copy Editor
Surendra Nath,
Glyph International

Proofreader
Bhavna Gupta,
Glyph International

Indexer
Arc Films, Inc.

Art Director, Cover
Jeff Weeks

Composition
Glyph International

Printed and bound by RR Donnelley.

McGraw-Hill books are available at special quantity discounts to use as premiums and sales promotions, or for use in corporate training programs. To contact a representative, please e-mail us at bulksales@mcgraw-hill.com.

This book is printed on acid-free paper.

About the Authors

Robert Horonjeff (deceased) was an internationally known consultant on airport design and professor of transportation engineering at the University of California, Berkeley.

Francis X. McKelvey (deceased) was a professor of civil engineering at Michigan State University. He served as a consultant on transportation and airport planning to federal, state, and local agencies, as well as to private firms in the United States and abroad.

William J. Sproule is a professor of civil and environmental engineering at Michigan Technological University. He has many years of experience in government service, consulting, and university teaching and research in transportation planning, traffic engineering, airport planning and design, and automated people movers in Canada and the United States. He is active in several professional societies and is the 2008 recipient of the ASCE Robert Horonjeff Award for his work in airport engineering.

Seth B. Young is an associate professor in the Department of Aviation at The Ohio State University and president of the International Aviation Management Group, Inc. He serves as a consultant and instructor to airports around the world on issues of airport management, planning, and design. Dr. Young is the chair of the National Academies Transportation Research Board Committee on Aviation System Planning. He is a certified member of the American Association of Airport Executives, a licensed airplane and seaplane commercial pilot, and a certified flight instructor.

Contents

Preface

In the preface to the fourth edition of this text, the late Dr. Francis McKelvey remarked that the technological and legislative developments related to the air transportation industry in the 1980s and early 1990s were of such significance that an updating of the book was needed. The fourth edition, published in 1994, enhanced previous editions, the first of which was published in 1962.

In the 16 years since this last update, it may be said that the changes to the practice of airport planning and design have been more significant than in any other era in the history of aviation. Implementation of twenty-first-century technologies has resulted in the first major overhaul to aircraft and air navigation systems in generations, computer-based analytical and design models have replaced antiquated monographs and estimation tables, and highly significant geopolitical events have all but rewritten the rules of planning, designing, and operating civil-use airports.

These significant enhancements to the aviation system have resulted in unique challenges in creating an updated fifth edition of this important and highly accepted text. While every attempt was made to keep to the traditional structure of the book and to preserve the theoretical strengths for which it is most well known, much of the material in the previous edition required more replacement than simply being made current. Within this latest edition the reader will find, for example, new and entirely different strategies to estimate required runway lengths and their associated required pavement thicknesses. This text attempts to maintain the flavor of previous editions while understanding, for example, that airport navigational aids of the previous century are becoming all but obsolete, in favor of a digital, satellite-based communication and navigational system, and that airport financing strategies are in a revolutionary state, given anticipated changes to federal aviation funding mechanisms. Updating this edition has, in fact, been a continuous "race against time," as important changes to the aviation system were constantly occurring during the process.

In light of these challenges, this fifth edition is hoped to again be the standard text for those interested in the fundamentals of airport planning and design. The information located within these chapters is applicable both for academic coursework and as a reference on the desks of airport planning and design professionals. As the industry continues to move forward, it is of course recommended that the latest design standards published by the Federal Aviation Administration, the International Civil Aviation Organization, and local, state, and other federal agencies be consulted.

Seth B. Young, Ph.D.

Acknowledgments

The fifth edition of this historic text could never have been created without the career efforts of its original author, the late Dr. Robert Horonjeff, and his coauthor on later editions, the late Dr. Francis McKelvey. Their authorship will always be first credited. Updating this book without their personal guidance was immensely challenging. It is only hoped that they would be satisfied with knowing their original philosophies still form the basis for this text. Contributing to this update have been the fine efforts of Dr. William Sproule, who studied under Dr. McKelvey and helped bring his goal of maintaining the currency of this text to fruition.

Many thanks go out to the institutions at which the original authors were, and those who helped update this latest edition have been, lucky enough to be employed: the University of California at Berkeley, Michigan State University, Embry-Riddle Aeronautical University, Michigan Technological University, Jacobs Consultancy, and The Ohio State University. It is hoped that the students, faculty, and professionals of these and all such institutions continue to find this text a valuable resource.

This book is dedicated to the memories of Dr. Horonjeff and Dr. McKelvey, who have helped to immortalize the formal practice of planning and designing the world's airports. Their life's efforts have resulted in bettering the lives of countless students, professionals, and users of civil aviation.

Seth B. Young, Ph.D.

PART 1

Airport Planning

CHAPTER 1

The Nature of Civil Aviation and Airports

Introduction

Since its beginning in the early twentieth century, civil aviation has become one of the most fascinating, important, and complex industries in the world. The civil aviation system, particularly its airports, has come to be the backbone of world transport and a necessity to twenty-first-century trade and commerce.

In 2008, the commercial service segment of civil aviation, consisting of more than 900 airlines and 22,000 aircraft, carried more than 2 billion passengers and 85 million tons of cargo on more than 74 million flights to more than 1700 airports in more than 180 countries worldwide. Millions more private, corporate, and charter "general aviation" operations were conducted at thousands of commercial and general aviation airports throughout the world. In many parts of the world, commercial service and general aviation serve as the primary, if not the only method of transportation between communities.

The magnitude of the impact of the commercial air transportation industry on the world economy is tremendous, contributing more than $2.6 trillion in economic activity, equivalent to 8 percent of the world gross domestic product, and supporting 29 million jobs. In the United States alone civil aviation is responsible for $900 billon in economic activity and 11 million jobs. General aviation serves an equally important role in the world's economy, providing charter, cargo, corporate, medical, and private transport, as well as such services as aerial photography, firefighting, surveillance, and recreation. In the United States alone, there are more than 225,000 registered general aviation aircraft and more than 600,000 registered pilots.

The presence of civil aviation has affected our economic way of life, it has made changes in our social and cultural viewpoints, and has had a hand in shaping the course of political history.

The sociological changes brought about by air transportation are perhaps as important as those it has brought about in the economy. People have been brought closer together and so have reached a better understanding of interregional problems. Industry has found new ways to do business. The opportunity for more frequent exchanges of information has been facilitated, and air transport is enabling more people to enjoy the cultures and traditions of distant lands.

In recent years, profound changes in technology and policy have had significant impacts on civil aviation and its supporting airport infrastructure. The industry continues to grow in numbers of aircraft, passengers and cargo carried, and markets served, from nonstop service on superjumbo aircraft between cities half-way across the planet, to privately operated "very light jets" between any of thousands of small airports domestically. Growth encouraged from technological advancements countered with increased constraints on the civil aviation system due to increased capacity limitations, security regulations, and financial constraints have resulted in ever increasing challenges to airport planning and design.

Civil aviation is typically considered in three sectors, commercial service aviation (more commonly known as air carriers or airlines), air cargo, and general aviation. Although the lines between these traditional sectors are becoming increasingly blurred, the regulations and characteristics regarding their individual operations are often mutually exclusive, and as such, those involved in airport planning and design should have an understanding of each sector.

Commercial Service Aviation

Commercial service aviation, supported by the world's airlines, is by far the most well known, most utilized, and most highly regulated segment of civil aviation. It is the segment of the industry responsible for providing public air transportation between the world's cities.

In the United States, domestic commercial air service accommodated nearly 650 million enplaning passengers in 2008, flying approximately 570 billion passenger-miles, reflecting a slight decline following the most recent surge in the growth of air transportation since the mid-twentieth century, and forecasted to carry more than 1 billion passengers by 2020, as illustrated in Fig. 1-1.

Intercity travel, of course, is not solely available through commercial service aviation. Intercity travel may be accommodated using either private modes of transportation, most commonly via private automobile travel, or through other modes of public transportation, such as bus, rail, or ship. Private automobile travel, accounts for

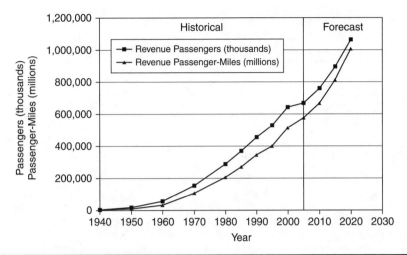

Figure 1-1 Total scheduled U.S. domestic passenger traffic: 1940 to 2020 (*U.S. Bureau of Transportation Statistics*).

nearly 90 percent of the total intercity (defined as trips greater than 50 mi in distance) travel in the United States, and public transportation or common-carrier travel (bus, rail, and air) accounts for the remaining portion.

Since the later half of the twentieth century, there has been a steady increase in overall travel, by private automobile and public transportation. Air transportation has had the greatest increase in overall passengers served. In the United States, this period has also witnessed a dramatic reduction in rail travel except for the rail markets in the Northeast United States. These relationships are shown in Table 1-1.

As illustrated in Table 1-2, air transportation in the United States accounts for the vast majority of domestic travel for trips exceeding 750 mi, and approximately one third of trips 500 to 750 mi in length. In all, air transportation accounts for approximately 70 percent of the United States' public intercity transportation. With the exception of travel to Canada and Mexico, air transportation serves nearly 100 percent of travel between the United States and international destinations.

In many parts of the world the use of private automobile is much less significant, and the use of rail transportation is much more prevalent. However, growth in commercial aviation in markets such as Europe and India are forecast to take greater numbers of passengers off the rails and onto airlines.

Much of the historical growth in air carrier transportation has been largely credited to the 1978 Federal Airline Deregulation Act, which allowed air carriers to freely enter and compete in domestic markets in

Year	Highway	Annual Growth, %	Civil Aviation	Annual Growth, %	Rail	Annual Growth, %
1960	1,272,078		33,399		21,261	
1965	1,555,237	22	57,626	73	17,388	−18
1970	2,042,002	31	117,542	104	10,771	−38
1975	2,404,954	18	147,400	25	8,444	−22
1980	2,653,510	10	219,068	49	11,019	30
1985	3,012,953	14	290,136	32	11,359	3
1990	3,561,209	18	358,873	24	13,139	16
1995	3,868,070	9	414,688	16	13,789	5
2000	4,390,076	13	531,329	28	14,900	8
2005	4,884,557	11	603,689	14	15,381	3

Source: U.S. Bureau of Transportation Statistics.

TABLE 1-1   U.S. Passenger Travel by Mode

the United States, and "open skies" agreements throughout the 1990s between nations to allow for more service between international destinations. The most recent growth in air transportation is attributable to changing airline business models, such as the emergence of the "low-cost carrier" (LCC), as well as increasing numbers of international open skies agreements that have proliferated since 2000.

Mode	One-Way Distance				
	50–499 Miles	500–749 Miles	750–999 Miles	1000–1499 Miles	1500 + Miles
Personal vehicle	95.4	61.8	42.3	31.5	14.8
Air	1.6	33.7	55.2	65.6	82.1
Bus	2.1	3.3	1.5	1.5	1.4
Train	0.8	1.0	0.9	0.7	0.8
Other	0.2	0.1	0.1	0.7	1.0
Total	89.8	3.1	2.0	2.3	2.8

TABLE 1-2   Percent of Trips by Mode for One-Way Travel Distance

Passenger Air Carriers

Commercial air carriers are defined in the United States as those that operate under Title 14 Part 121 of the U.S. Code of Federal Regulations to provide scheduled air transportation to the public. In the United States, these airlines are categorized by their annual revenues. Major airlines are those that generate at least $1 billion in annual revenues. National carriers generate between $100 million and $1 billion in annual revenues, and regional carriers generate between $25 million and $100 million in annual revenues.

International air carriers receive operating certificates as prescribed by standards set by the International Civil Aviation Organization (ICAO) and defined by the country in which the airline is based. Historically, international air carriers were owned and operated by their nations, hence the term "flag" carriers. In recent years, most of the traditional international carriers have been transferred to private ownership. In addition, there has been an emergence of new international air carriers, most following the LCC model of serving point-to-point markets for fares that are on the whole far lower than their historical airline counterparts. The emergence of the LCC models in Europe and more recently in the Far East and India are resulting in a tremendous growth in aviation activity in these regions.

Air carriers using aircraft with less than 75 seats providing scheduled or unscheduled air charter services operating under Title 14 Part 121 of the U.S. Code of Federal Regulations are known as regional air carriers. Those carriers operating aircraft with less than 30 seats and those that operate under Title 14 Part 135 of the U.S. Code of Federal Regulations are known as commuter air carriers. If service frequency between city pairs is provided less than 5 times weekly, these carriers are known as air taxi operators.

In 2008, there were over 700 air taxis, commuter and small regional air carriers operating more than 2750 aircraft, over 50 percent of which were regional jet aircraft. Six hundred and forty two airports in the United States received service by small regional and commuter airlines. Regional and commuter air service is the sole provider of public air transportation to 492 airports in the United States (source: RAA). Throughout the 1980s and 1990s, commuter airline growth was encouraged by their increasing roles as code-share partners with major air carriers. In 2006, over 95 percent of all passengers traveling on commuter and regional air carriers purchased their tickets through these code-share partnerships. Table 1-3 illustrates the growth of the commuter and regional carriers since 1970.

International Air Transportation

Although international air transport was inaugurated in the mid-1930s, rapid growth did not begin until 1950. Since that time the average annual growth rate in the number of worldwide passengers

Year	Enplaned Passengers (thousands)	Passenger Miles (millions)	Average Trip Length (miles)	Number of Aircraft	Average Seats per Aircraft
1970	4,270	399	98	741	11
1975	7,243	689	110	948	13
1980	14,810	1,920	129	1,339	14
1985	26,000	4,410	173	1,745	19
1990	42,099	7,610	183	1,917	22
1995	55,800	11,461	213	2,109	30
2000	82,800	23,638	285	2,275	39
2003	112,120	43,100	384	2,189	45
2007	161,390	73,690	457	2,579	51

TABLE 1-3   Commuter and Regional Airline Statistics 1970 to 2007

was nearly 14 percent in the 1960s, slightly less than 7 percent in the 1970s, and slightly less than 5 percent in the 1980s. As illustrated in Fig. 1-2, worldwide growth in air transportation has increased by more than 60 percent between 1990 and 2005.

Air Cargo

Originating as the transport of mail by air in the early part of the twentieth century, air cargo has come to be defined as a $40 billion

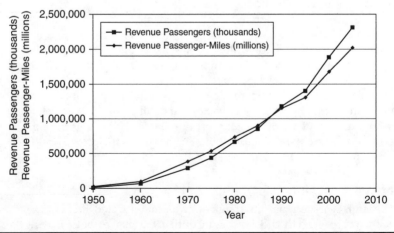

FIGURE 1-2   Worldwide growth in civil air transport passenger traffic.

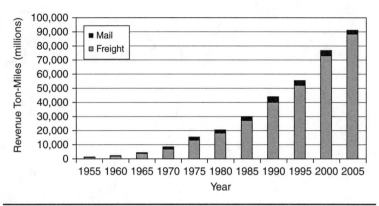

FIGURE 1-3   Worldwide air cargo.

industry focused on the air transport of mail, bulk freight, high-value goods, and all other revenue generating payload other than passengers and their luggage. As illustrated in Fig. 1-3, the transport of air cargo has increased tremendously since the mid-twentieth century, with its greatest rate of growth occurring since the late 1980s.

The top 50 carriers of air cargo in the global air cargo industry carried nearly one-hundred billion freight-ton-miles of cargo in 2008. Approximately 15 percent of the air cargo transported globally is performed by industry leaders and exclusive cargo carriers FedEx and UPS. The majority of air cargo is transported by air carriers, using aircraft designed exclusively for air cargo carriage, as well as on commercial passenger aircraft. Cargo carried on commercial passenger aircraft is often referred to as "belly cargo" as the cargo is stowed in the belly of the passenger aircraft. Cargo carried on aircraft designed exclusively for the carriage of cargo is often referred to as "palette" or "containerized" cargo, describing the containers within which cargo is stowed and the palettes used to load and unload cargo. Cargo operations using each type aircraft pose unique challenges for airport planning and design.

The geographic distribution of world air transport is also of interest. For statistical purposes ICAO has divided the world into six regions: Asia and Pacific, Europe, North America, Latin American, Caribbean, and the Middle East.

While slightly more than 60 percent of all traffic is generated in North America and Europe, the relative growth rates of traffic in the Asian and Pacific region, as well as in the Middle East, is expected to dominate worldwide air transportation growth, reflecting the growth of importance of this area in the political, social, and economic sectors.

Year	Asia and Pacific	Europe	North America	Latin America and Caribbean	Africa	Middle East
1972	8.4	35.9	47.6	4.4	2	1.7
1976	12.4	36.5	41	4.8	2.5	2.8
1980	15.5	35	38.6	5.5	2.6	2.8
1984	17.6	33	38	4.9	2.8	3.8
1988	19.8	31	39.3	4.6	2.2	3
1990	19.8	31.9	38.5	4.7	2.2	2.9
1992	21.2	27.3	41.8	4.7	2.2	2.8
2002	26.7	26.2	36.8	4.5	2.2	3.6
2005	33.2	25.4	30.3	4.3	2.1	4.7

TABLE 1-4   Percentage of Worldwide Distribution of Air Cargo Traffic

The air cargo market is forecast to triple between 2008 and 2030, led by the growth of air freight demand to China, as illustrated by the forecast percentage distribution of worldwide air cargo activity in Table 1-4. It is forecast that increasing percentages of air cargo would be shipped on dedicated cargo aircraft, requiring the need for expanded exclusive air cargo facilities at airports throughout the world.

General Aviation

General aviation is the term used to designate all flying done other than by the commercial air service carriers. General aviation operations range from local recreational flying to global business transport, performed on aircraft not operating under the federal aviation regulations for commercial air carriers.

While, by definition, general aviation operations carry no "commercial" passengers, it is estimated that more than 166 million people traveled by general aviation on nearly 20 million flights in 2008. During 2007, general aviation accounted for nearly 75 percent of all aircraft operations in the United States (source: FAA TAF). General aviation supports more than 1.3 million jobs and contributes more than $103 billion annually to the United States economy.

As of 2008, there were approximately 225,000 general aviation aircraft registered in the United States and an estimated 340,000 aircraft worldwide (source: GAMA). These aircraft range in type and size from small single-engine propeller aircraft to large jet aircraft, to

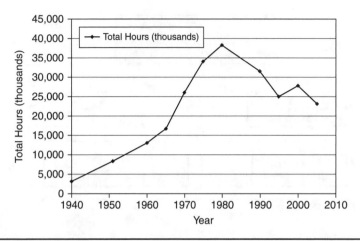

FIGURE 1-4   Total flight hours in general aviation aircraft in the United States.

"ultralight" aircraft, to helicopters. General aviation aircraft are served by nearly 20,000 landing facilities in the United States alone.

General aviation activity has experienced a decline in activity between 1980 and 2008, as illustrated in Fig. 1-4. Despite this recent historical decline, general aviation activity is forecast to increase with the proliferation of new aircraft technology which is expected to reduce the cost of general aviation operations. This forecast growth in general aviation, combined with new technologies, will pose interesting challenges for airport planning and design.

Civil Aviation Airports

Airports serving civil aviation range from private nonpaved strips that serve less than one privately operated aircraft per day to major international airports covering tens of thousands of acres, serving hundreds of thousands of flights and hundreds of millions of passengers annually. In the United States there are approximately 20,000 recognized civil airports, most of which are privately owned and closed to general public use. Of the approximately 5200 airports open to the public, approximately 700 are certified to accommodate commercial air service, with the remaining serving general aviation exclusively.

Airports currently serving at least 2500 enplaned passengers using commercial air service are known as *commercial service airports*. *Primary airports* are designated as those commercial airports serving at least 10,000 annual enplaned passengers. Airports serving less than 2500 annual enplaned passengers are considered *general aviation airports*. General aviation airports designed to accommodate smaller

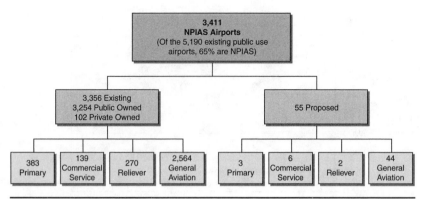

FIGURE 1-5 NPIAS categories.

single and twin-engine aircraft are considered basic utility airports. Those general aviation airports that accommodate larger aircraft are considered general utility airports.

The United States' Federal Aviation Administration (FAA), the governmental body with administrative oversight to the nation's civil aviation system, categorizes airports through its National Plan of Integrated Airport Systems (NPIAS). As illustrated in Fig. 1-5 the NPIAS recognizes approximately 3400 airports considered by the FAA to be essential to civil aviation and classifies these airports by the levels of commercial service activity within their respective standard metropolitan statistical areas (SMSAs).

Primary airports are further classified into what are known as "hub classifications" (not to be confused with the airline "hub and spoke" route models). The hub classifications used by the FAA are large hub primary, medium hub primary, small hub primary, and nonhub primary airports. Large hubs are those airports that account for at least 1 percent of the total annual passenger enplanements in the United States. Medium hubs account for at least 0.25 but less than 1 percent of the total passenger enplanements. Small hubs account for at least 0.05 percent but less than 0.25 percent, and nonhubs account for less than 0.05 percent but at least 10,000 annual enplaned passengers. The number of airports, by hub classification, is illustrated in Table 1-5.

Reliever airports are airports not currently serving regular commercial service but have been designated by the FAA as "general aviation-type airports that provide relief" when necessary to commercial service airports, typically by accommodating high volumes of general aviation activity within a metropolitan area and accommodating commercial service operations when the nearby commercial service airport is closed or otherwise cannot accommodate normal operations. Airports are typically given "reliever" status if they

Number of Airports	Airport Type	Percentage of 2006 Total Enplanements	Percentage of All Based Aircraft
30	Large hub primary	68.7	0.9
37	Medium hub primary	20.0	2.6
72	Small hub primary	8.1	4.3
244	Nonhub primary	3.0	10.9
139	Nonprimary commercial service	0.1	2.4
270	Relievers	0.0	28.2
2,564	General aviation	0.0	40.8
3,356	*Existing NPIAS airports*	99.9	89.8
16,459	*Low-activity landing areas (Non-NPIAS)*	0.1	10.2

TABLE 1-5   Number of NPIAS Airports by Hub Classification

are located within an SMSA of population of at least 5,000,000 or where passenger enplanements exceed 250,000 annually. In addition, the airport must have at least 100 aircraft based at the field or handle at least 25,000 itinerant operations annually. Reliever airports, although not serving regular commercial service operations, are among the busiest airports in the United States.

While most of the airports in the United States are privately owned and operated, the majority of public use airports are in fact publicly owned. Public use airports, and commercial service airports in particular, are typically owned and operated by local municipalities, counties, states, or some public "authority" typically overseen by representatives from a combination of local and regional jurisdictions. There are a few public use airports that are operated by private airport management companies but rarely do private firms actually own the property on which the airport is located. As such, in the United States, most planning and design programs at civil public use airports must go through extensive governmental processes for ultimate approval and often funding support.

While the United States has by far the greatest number of commercial service and general aviation airports in the world, many of the world's largest and most important airports are located all over the globe. Table 1-6 lists the world's busiest airports.

Total Passenger Traffic 2008

Rank	City	Airport	Passengers
1	Atlanta, GA	ATL	90,039,280
2	Chicago, IL	ORD	69,353,654
3	London, GB	LHR	67,056,228
4	Tokyo, JP	HND	66,735,587
5	Paris, FR	CDG	60,851,998
6	Los Angeles, CA	LAX	59,542,151
7	Dallas/Ft Worth, TX	DFW	57,069,331
8	Beijing, CN	PEK	55,662,256
9	Frankfurt, DE	FRA	53,467,450
10	Denver, CO	DEN	51,435,575
11	Madrid, ES	MAD	50,823,105
12	Hong Kong, CN	HKG	47,898,000

Total Operations (Movements) 2008

Rank	City	Airport	Movements
1	Atlanta, GA	ATL	978,824
2	Chicago, IL	ORD	881,566
3	Dallas/Ft Worth, TX	DFW	656,310
4	Denver, CO	DEN	615,573
5	Los Angeles, CA	LAX	615,525
6	Las Vegas, NV	LAS	579,949
7	Houston, TX	IAH	576,062
8	Paris, FR	CDG	559,812
9	Charlotte, NC	CLT	536,253
10	Phoenix, AZ	PHX	502,499
11	Philadelphia, PA	PHL	492,010
12	Frankfurt, DE	FRA	485,783

Total Cargo (Metric Tons) 2008

Rank	City	Airport	Metic Tons
1	Memphis, TN	MEM	3,695,561
2	Hong Kong, CN	HKG	3,656,724
3	Shanghai, CN	PVG	2,698,795
4	Seoul, KR	ICN	2,423,717
5	Anchorage, AK	ANC	2,361,088
6	Paris, FR	CDG	2,280,049
7	Frankfurt, DE	FRA	2,111,116
8	Tokyo, JP	NRT	2,099,349
9	Louisville, KY	SDF	1,973,965
10	Singapore, SG	SIN	1,883,894
11	Dubai, AE	DXB	1,824,992
12	Miami, FL	MIA	1,806,769

Rank	City	Code	Passengers	Rank	City	Code	Passengers	Rank	City	Code	Passengers
13	New York, NY	JFK	47,790,485	13	London, GB	LHR	478,569	13	Los Angeles, CA	LAX	1,630,385
14	Amsterdam, NL	AMS	47,429,741	14	Madrid, ES	MAD	469,740	14	Amsterdam, NL	AMS	1,602,584
15	Las Vegas, NV	LAS	44,074,707	15	Detroit, MI	DTW	462,284	15	Taipei, TW	TPE	1,493,120
16	Houston, TX	IAH	41,698,832	16	Minneapolis/St Paul, MN	MSP	446,840	16	London, GB	LHR	1,486,260
17	Phoenix, AZ	PHX	39,890,896	17	Amsterdam, NL	AMS	446,626	17	New York, NY	JFK	1,446,491
18	Bangkok, TH	BKK	38,604,009	18	New York, NY	JFK	435,750	18	Chicago, IL	ORD	1,324,820
19	Singapore, SG	SIN	37,694,824	19	Newark, NJ	EWR	433,463	19	Beijing, CN	PEK	1,303,258
20	Dubai, AE	DXB	37,441,440	20	Munich, DE	MUC	432,296	20	Bangkok, TH	BKK	1,173,131
21	San Francisco, CA	SFO	37,405,467	21	Beijing, CN	PEK	431,675	21	Indianapolis, IN	IND	1,025,895
22	Orlando, FL	MCO	35,622,252	22	Toronto, ON, CA	YYZ	429,829	22	Newark, NJ	EWR	889,121
23	Newark, NJ	EWR	35,299,719	23	San Francisco, CA	SFO	387,710	23	Tokyo, JP	HND	849,378

TABLE 1-6 The World's Busiest Airports

Total Passenger Traffic 2008				Total Operations (Movements) 2008				Total Cargo (Metric Tons) 2008			
Rank	City	Airport	Passengers	Rank	City	Airport	Movements	Rank	City	Airport	Metric Tons
24	Detroit, MI	DTW	35,144,841	24	Salt Lake City, UT	SLC	387,695	24	Osaka, JP	KIX	845,496
25	Rome, IT	FCO	35,132,879	25	Los Angeles, CA	VNY	386,706	25	Luxembourg, LU	LUX	788,223
26	Charlotte, NC	CLT	34,732,584	26	New York, NY	LGA	377,940	26	Guangzhou, CN	CAN	685,866
27	Munich, DE	MUC	34,530,593	27	Phoenix, AZ	DVT	376,210	27	Kuala lumpur, MY	KUL	661,212
28	London, GB	LGW	34,214,474	28	Miami, FL	MIA	372,635	28	Dallas/Pt Worth, TX	DFW	660,465
29	Miami, FL	MIA	34,063,531	29	Boston, MA	BOS	371,604	29	Atlanta, GA	ATL	655,277
30	Minneapolis/St Paul, MN	MSP	34,032,710	30	Mexico City, MX	MEX	366,561	30	Brussels, BE	BRU	616,423

Source: Airports Council International.

TABLE 1-6 The World's Busiest Airports (*Continued*)

Contrary to the ownership structures in the United States, airports worldwide have traditionally been owned and operated by their respective federal governments through their ministries of transport, however many are increasingly becoming privatized to operate as for-profit business entities.

Historical Review of the Legislative Role in Aviation

Both in the United States and internationally, legislative actions taken by federal and state governments have had a profound impact on the growth of civil aviation and the planning and design of its airports.

As early as 1911, the Post Office Department showed an interest in civil aviation, particularly the transportation of mail by air, and from then on the department did much to encourage civil aviation. Attempts to obtain federal appropriations for air-mail began in 1912 but met with little success until 1916, when an appropriation for experimental purposes was made. In 1918, the first air-mail route in the United States was established between Washington, D.C. and New York City. At the start of this service the flying operations were conducted by the War Department but later that year the Post Office Department took over the entire operation with its own equipment and pilots. Service was inaugurated between New York City and Chicago in 1919 and was extended to San Francisco in 1920.

The Post Office Department, having demonstrated the practicality of moving mail by air, desired to turn over the operation to private enterprise. By 1925, the development work of the government had reached a stage where private operation seemed feasible. Accordingly, legislation permitting the Post Office Department to contract with private operators for the carriage of mail by air was provided by the Air Mail Act of 1925 (Kelly Act). However, it was not until 1926 that a number of contract routes were opened. Some of the early contractors were the Ford Motor Company, Boeing Air Transport (predecessor of United Airlines), and National Air Transport.

Air Commerce Act of 1926

The first year of the carriage of mail also saw the passage of the first federal law dealing with air commerce, the Air Commerce Act of 1926 (Public Law 64-254). Although this law provided regulatory measures, it did more to aid and encourage civil aviation than to regulate. The principal provisions of this act were as follows:

1. All aircraft owned by United States citizens operating in common-carrier service or in connection with any business must be registered.

2. All aircraft must be certificated and operated by certified airmen.

3. Authority was given to the Secretary of Commerce to establish air traffic rules.

4. The Secretary of Commerce was authorized to establish, operate, and maintain lighted civil airways.

In drafting the legislation, Congress relied considerably upon the precedents in maritime law. An analogy was utilized between the role of the government in meeting water navigation needs and the role of the government in air navigation. In water navigation these services included the signing, lighting, and marking of channels, safety inspection of ships and operating personnel, assistance in the development and improvement of ports and waterways, and laws concerning the operations of the industry. The provision of docks and terminal facilities were the responsibility of local government or the private sector of the economy. Therefore, the legislation was adopted in such a framework which held that airports were analogous to the docks of waterborne transportation [30].

Under the Air Commerce Act, control of air transportation was divided among several government agencies. The air-mail contracts were let by the Post Office Department, air-mail rates were subject to regulation by the Interstate Commerce Commission, and matters having to do with registration, certification, and airways were vested in the Bureau of Air Commerce in the Department of Commerce. The result of this divided jurisdiction was a lack of coordination in the efforts of government to develop the air transportation industry. In addition, the Act specifically prohibited any direct federal funding for airport development.

Civil Aeronautics Act of 1938

The Air Commerce Act of 1926 had been passed before the carriage of mail and passengers had developed into a substantial business enterprise. The failure of this legislation to provide adequate economic control led to wasteful and destructive competitive practices. The carriers had little security in their routes and therefore could not attract private investors and develop traffic volumes sufficient to achieve economic stability. These particular weaknesses in the existing legislation led to the enactment of the Civil Aeronautics Act of 1938 (Public Law 76-706). This act defined in a precise manner the role of the federal government in respect to the economic phases of air transport. It created one independent agency to foster and regulate air transport in lieu of the three agencies operating under the Air Commerce Act. This new agency was called the Civil Aeronautics Authority (not to be confused with the Civil Aeronautics Administration). It consisted of a five-member authority, a three-member air safety board, and an administrator. The five-member authority was principally concerned with the economic regulation of air carriers;

the safety board was an independent body for the investigation of accidents; and the concerns of the administrator dealt primarily with construction, operation, and maintenance of the airways.

During the first year and a half of its existence a number of organizational difficulties arose within the Civil Aeronautics Authority. As a result, President Franklin D. Roosevelt, acting within the authority granted to him in the Reorganization Act of 1939 (55 Stat. 561), reorganized the Civil Aeronautics Authority and created two separate agencies, the Civil Aeronautics Board and the Civil Aeronautics Administration. The five-member authority remained as an independent agency and became known as the Civil Aeronautics Board; the Air Safety Board was abolished and its functions given to the Civil Aeronautics Board; and the administrator became the head of an agency within the Department of Commerce known as the Civil Aeronautics Administration (CAA). The duties of the original five-member authority were unchanged, except that certain responsibilities, such as accident investigation, were added because of the abolition of the Air Safety Board. The administrator, in addition to retaining the functions of supervising construction, maintenance, and operation of the airways, was required to undertake the administration and enforcement of safety regulations and the administration of the laws with regard to aircraft operation. Subsequently, the administrator became directly responsible to the Secretary of Commerce (1950).

The Civil Aeronautics Act, like its predecessor the Air Commerce Act, authorized the federal government to establish, operate, and maintain the airways, but again, authorization for actively aiding airport development was lacking. The act, however, authorized the expenditure of federal funds for the construction of landing areas provided the administrator certified "that such landing area was reasonably necessary for use in air commerce or in the interests of national defense." The act also directed the administrator to make a survey of airport needs in the United States and report to Congress about the desirability of federal participation and the extent to which the federal government should participate.

In accordance with the requirements of the Act, the Civil Aeronautics Authority conducted a detailed survey of the airport needs of the nation. An advisory committee was appointed, composed of representatives of interested federal agencies (both military and civil), state aviation officials, airport managers, airline representatives, and others. A report was submitted to Congress on March 23, 1939 (House Document 245, 76th Congress, 1st Session). Some of the more important recommendations in this report were as follows:

1. Development and maintenance of an adequate system of airports and seaplane bases should be recognized in principle as a matter of national concern.

2. Such a system should be regarded, under certain conditions, as a proper object of federal expenditure.

3. In passing upon applications for federal expenditure on airport development or improvement, the highest preference should be given to airports which are important to the maintenance of safe and efficient operation of air transportation along the major trade routes of the nation; and to those rendering special service to the national defense.

4. At such times as the national policy includes the making of grants to local units of government for public-works purposes, or any work-relief activity, a proportion of the funds involved should be allocated to airport purposes. Such purposes should be given preference as rendering an important service to the localities concerned and at the same time being of particular importance to the nation's commerce and defense.

5. Whenever emergency public-works programs may be terminated, or when such programs may be curtailed to a degree not enabling adequate airport development to continue, or when the Congress for other reasons may determine federal assistance for airports should be continued through annual appropriations for that purpose, based upon annual reports which should include a review of the general status of the nation's airport system and of the work recently done or currently in course of being done, and recommendations for future work in the interest of developing and maintaining a system adequate to national needs, expenditures at these periods should be limited to projects of exceptional national interest.

6. In connection with such public-works or work-relief programs as normally involve joint contributions by the federal government and by local government, there should be a provision of supplementary funds to enable the federal government to increase its share of the total expense, in any proportion justified by the importance of the project.

7. All applications for federal airport grants from such a supplementary appropriation should be presented through agencies of state government.

8. In deciding upon the wisdom and propriety of granting any such applications, and the priority that should be given to them, consideration should be given to the aeronautical policy of the state in question, with reference to such matters as the state's policy in protecting the approaches to airports; the state's policy in respect to the employment of any taxes collected on the fuel used in aircraft; and any measures taken by the state to insure the proper maintenance of airports and the maintenance of reasonable charges for the services given them.

9. The detailed plans for the location and development of any airport with respect to which there is federal contribution of any kind, should be subject to the approval of the federal agency charged with the establishment of civil airways, landing areas, and necessary air navigation facilities.

10. There should be no direct federal contribution to the cost of maintaining airports, other than federal airports; except that the administrator of the Civil Aeronautics Authority may, in accordance with the Civil Aeronautics Act and so far as available funds permit, assume the cost of operating any lighting equipment and other air navigation facility as a part of the cost of operation of the federal airways system.

The airport survey submitted in 1939 was updated with new studies completed in 1940. Continuing studies were made through the war years. While first importance was attached to the military requirements, care was taken whenever possible to anticipate the needs of postwar civil aviation. During the war years the federal government, through the CAA, spent $353 million for the development of military landing areas in the continental United States. This does not include funds spent by the military agencies. During the same period the CAA spent $9.5 million for the development of landing areas in the United States solely for civil purposes.

Federal Airport Act of 1946

At the end of World War II, over 500 airports constructed for the military by the CAA were declared surplus and were turned over to cities, counties, and states for airport use. The interest in adequate airport facilities by various political subdivisions of government continued. The needs were made known to Congress by various interests. As a result, the House of Representatives passed a resolution (H.R. 598, 78th Congress) directing the CAA to make a survey of "need for a system of airports and landing areas throughout the United States" and report back to Congress.

The results of this survey were completed in 1944 (House Document 807, 78th Congress, 2nd Session) and contained the following principal recommendations:

1. That Congress authorize an appropriation to the Office of the Administrator of Civil Aeronautics not to exceed $100 million annually to be used in a program of federal aid to public agencies for the development of a nationwide system of public airports adequate to meet the present and immediate future needs of civil aeronautics. The administrator be authorized to allocate such funds for any construction work involved in constructing, improving, or repairing an airport, including the construction, alteration, and repair of airport

buildings other than hangars, and the removal, lowering, marking, and lighting of airport obstructions; for the acquisition of any lands or property interest necessary either for any such construction or to protect airport approaches; for making field and specifications; supervising and inspecting construction work, and for any necessary federal expenses in the administration of this program.

2. That such a program can be conducted in cooperation with the state and other nonfederal public agencies on a basis to be determined by the Congress. That the federal contribution be determined by the Congress in passing the necessary enabling legislation. A good precedent for the proportionate sharing of costs exists in the public-roads program which has operated satisfactorily for many years on a 50-50 basis.

3. That any project for which federal aid is requested must meet with the approval of the administrator of Civil Aeronautics as to scope of development and cost, conform to Civil Aeronautics Administration standards for location, layout, grading, drainage, paving, and lighting and all work thereon be subject to the inspection and approval of the Civil Aeronautics Administration.

4. In order to participate in the federal-aid program, a state shall

   a. Establish and empower an official body equipped to conduct its share of the program.

   b. Have legislation adequate for the clearing and protection of airport approaches, and such other legislation as may be necessary to vest in its political subdivisions all powers necessary to enable them to participate through the state as sponsors of airport projects.

   c. Have no special tax on aviation facilities, fuel, operations, or businesses, the proceeds of which are not used entirely for aviation purposes.

   d. Ensure the operation of all public airports public interest, without unjust discrimination or unreasonable charges.

   e. Ensure the proper operation and maintenance of all public airports within its jurisdiction.

   f. Make airports developed with federal aid available for unrestricted use by United States government aircraft without charge other than an amount sufficient to cover the cost of repairing damage done by such aircraft.

   g. Require the installation at all airports for which federal funds have been provided for a standard accounting and fiscal reporting system satisfactory to the administrator.

5. That sponsors of projects be required to enter into contracts with the Civil Aeronautics Administration ensuring the proper maintenance and protection of airports developed with federal aid and their operation in the public interest.

The recommendations contained in the airport needs survey report were written into an airport development bill, introduced into the House of Representatives (H.R. 5024) but no action was taken on it. After extensive hearings in both houses of Congress, the Senate passed an airport bill (S. 2) in 1945 and later that year the House passed a bill (H.R. 3615). The language in these two bills differed in several respects. One of the principal differences was the method employed in channeling funds to the municipalities. The Senate bill provided that funds be channeled to the municipalities through appropriate state aviation organizations unless a state did not have an appropriate agency to handle the matter. The House bill permitted channeling of funds either through the state or directly to a municipality or other political subdivision of government. The substitute bill agreed to in conference conformed more nearly to the House language. Another difference had to do with the size of the discretionary fund, which, instead of being apportioned among the states by a fixed formula, would be available for use by the administrator at his sole determination. The House bill provided 25 percent of the total appropriation for airport development as a discretionary fund, the Senate bill 35 percent. The compromise reached in conference retained the House version. Other differences which were worked out in conference concerned whether or not the costs of the acquisition of land and interest in airspace should be eligible for federal aid, project sponsorship requirements, and the reimbursement for damage to public airports caused by federal agencies.

The conference report was approved by the Congress and the Federal Airport Act of 1946 was enacted (Public Law 79-377). Known as the Federal Airport-Aid Program, appropriations of $500 million over a 7-year period were authorized for projects within the United States plus $20 million for projects in Alaska, Hawaii, Puerto Rico, and the Virgin Islands. In 1950, the 7-year period was extended an additional 5 years (Public Law 81-846). However, annual appropriations approved by Congress were much less than the amounts authorized by the act.

The original act provided that a project shall not be approved for federal aid unless "sufficient funds are available for that portion of the project which is not to be paid by the United States."

Local governments often required 2 to 3 years to make arrangements for raising funds. Most of the larger projects are financed locally through the sale of bonds. This method of financing requires legislation at the local level and, in some cases, also at the state level. General obligation bonds normally require approval by the electorate.

Programs to inform the public on the needs for airport improvement must be carefully planned and executed. Thus, after the completion of these events, local governments frequently found that sufficient federal funds were not appropriated to match local funds, and the projects were delayed. Another complaint of local governments had been that Congress failed to fulfill its obligation, since the amount appropriated by Congress fell far short of the amount authorized by the Federal Airport Act. These deficiencies as well as other matters were incorporated in a bill (S. 1855). Hearings on the bill were held before the subcommittee of the Committee on Interstate and Foreign Commerce of the United States Senate in 1955. Representatives of the Council of State Governments, the American Municipal Association, the National Association of State Aviation Officials, airport and industry trade associations, and individuals were unanimous in the feeling that air transportation had reached a stage of maturity where many airports were woefully inadequate and greater financial assistance from the federal government would be required to meet the current needs of aviation. After much debate, the bill was approved by the President (Public Law 84-211).

This amending act made no change in the basic policies and purposes expressed in the original act. There were no changes in the requirements with respect to the administration of the grants authorized, such as the distribution and apportionment of funds, the eligibility of the various types of airport construction, sponsorship requirements, etc. The primary purpose of the act was to substitute for the procedure of authorizing annual appropriations for airport projects, provisions granting substantial annual contract authorization in specific amounts over a period of four fiscal years. Airport sponsors were thus furnished assurance that federal funds would be available at the time projects were to be undertaken.

This law provided $40 million for fiscal year 1956 and $60 million for each of fiscal years 1957, 1958, and 1959 for airport construction in the continental United States. It also provided $2.5 million in fiscal year 1956 and $3 million for the three succeeding fiscal years for airport construction in Alaska, Hawaii, Puerto Rico, and the Virgin Islands. Besides the $42.5 million made available in fiscal year 1956 by Public Law 84-211, Congress approved an additional appropriation of $20 million for airport projects.

Federal Aviation Act of 1958

For a number of years there had been a growing concern about the division of responsibility in aviation matters among different agencies of the federal government. Unlike highway or other forms of transport, aviation is unique in its relation to the federal government. It was historically the only mode whose operations are conducted almost wholly within federal jurisdiction, and one subject to little or no regulation by states or local authorities. Thus, the federal government bears virtually complete responsibility for the promotion and supervision of the industry in the public interest. The military interest and the entire national defense concept are also intimately related to aviation.

Recognizing that the demands on the federal government in the years ahead would be substantial, the director of the Bureau of the Budget requested a review of aviation-facilities problems in 1955. A report was issued later that year recommending that a study of "long-range needs for aviation facilities and aids be undertaken" and that such a study be made under the direction of an individual of national reputation.

President Dwight D. Eisenhower accepted these recommendations and appointed Edward P. Curtis as his Special Assistant for Aviation Matters in 1957. Curtis was charged with the responsibility of preparing a comprehensive aviation-facilities plan which would "provide the basis for the timely installation of technically adequate aids, for optimum coordination of the efforts of the civil and military departments, and for effective participation by state and local authorities and the aircraft operators in meeting facilities requirements." Curtis completed his report and submitted it to the President. In this report Curtis stated that "it has become evident that the fundamental reason for our previous failures lies with the inability of our governmental organizations to keep pace with the tremendous growth in private, commercial, and military aviation which has occurred in the last 20 years." Curtis recommended the consolidation of all aviation functions, other than military, into one independent agency responsible only to the President. However, the report recognized that to "develop new management structures and policy, to coordinate proposals within the executive branch and to obtain legislation implementing a new permanent organization might be as long as 2 or 3 years." The most urgent matter requiring attention was in the area of air traffic control. The collision of two aircraft over the Grand Canyon in 1956 provided the impetus for rapid legislative action for remedying midair collisions. Curtis recommended that, as an interim measure, there be created an Airways Modernization Board whose function was to "develop, modify, test, and evaluate systems, procedures, facilities, and devices, as well as define the performance characteristics thereof, to meet the needs for safe and efficient navigation and traffic control of all civil and military aviation except for those needs of military agencies which are peculiar to air warfare and primarily of military concern, and select such systems, procedures, facilities, and devices which will best serve such needs and will promote maximum coordination of air traffic control and air defense systems." The board was to consist of the Secretary of Commerce, the Secretary of Defense, and an independent chairman.

Congress was receptive to this recommendation and passed the Airways Modernization Act of 1957 (Public Law 85-133) establishing the board for a 3-year term.

In the meantime, there were more midair collisions and reports of near misses were given wide circulation. Costly disagreements between the CAA and the military on the type of navigational aids to be used on the airways no doubt also spurred Congressional action. As a result, instead of taking 2 or 3 years to create a single aviation agency as was predicted, the Congress acted favorably on the legislation within a year of the passage of the Airways Modernization Act. This legislation is known as the Federal Aviation Act of 1958 (Public Law 85-726). This law superseded the Civil Aeronautics Act of 1938 but not the Federal Airport Act of 1946.

The principal provisions of the law insofar as organizational changes were concerned are as follows:

1. The Federal Aviation Agency was created as an independent agency with an administrator directly responsible to the President. The agency incorporated the functions of the Civil Aeronautics Administration and the Airways Modernization Board, both of which were abolished.

2. The Civil Aeronautics Board was retained as an independent agency including all its functions except its safety rule-making powers, which were transferred to the Federal Aviation Agency.

Creation of the U.S. Department of Transportation

For many years it had been argued that there had been a proliferation of federal activities with regard to transportation. For example, the Bureau of Public Roads was part of the Department of Commerce whereas the Federal Aviation Agency was an independent agency. It was felt by different transport interests that there was a lack of coordination and effective administration of the transportation programs of the federal government resulting in a lack of a sound national transportation policy. It is interesting to note that the first legislative proposal in this direction dates back to 1874. However, in recent years, the involvement of the federal government in the development of the transportation systems of the nation has been enormous, requiring much more coordination among federal transport activities than ever before. With this as a background, a Cabinet-level Department of Transportation (DOT) was created headed by the Secretary of Transportation (Public Law 89-670). The department began to function on April 1, 1967.

The agencies and functions transferred to the Department of Transportation related to air transportation included the Federal Aviation Agency in its entirety and the safety functions of the Civil

Aeronautics Board, including the responsibility for investigating and determining the probable cause of aircraft accidents, and its appellate safety functions involving review on appeal of the suspension, modification, or denial of certificates or licenses. The name of the Federal Aviation Agency was changed to the Federal Aviation Administration (FAA). The administrator is still appointed by the President but reports directly to the Secretary of Transportation.

A National Transportation Safety Board (NTSB) was established by the same act which created the Department of Transportation to determine "the cause or probable cause of transportation accidents and reporting the facts, conditions, and circumstances relating to such accidents" for all modes of transportation. Although created by the act which established the DOT, the board in carrying out its functions is "independent of the secretary and other offices and officers of the department." The board consists of five members appointed by the President and annually reports directly to Congress.

The creation of the DOT did not alter any legislation in the Federal Aviation Act of 1958, with the exception of the transfer of the safety functions from the Civil Aeronautics Board to the National Transportation Safety Board. In the act of establishing the new department, however, there was a statutory requirement to establish an Office of Noise Abatement to provide policy guidance with respect to interagency activities related to the reduction of transportation noise. With the introduction of jet aircraft in 1958 the complaints against aircraft noise increased significantly. As a result, in 1968 the Federal Aviation Act was amended by Congress (Public Law 90-411) to require noise abatement regulations. Its principal purpose was to establish noise levels which aircraft manufacturers cannot exceed in the development of new aircraft.

Airport and Airway Development Act of 1970

In the mid-1960s, as air traffic was expanding at a fairly rapid pace, air traffic delays getting into and out of major airports began to increase rapidly. Along with the delays in the air, congestion was also taking place on the ground in parking areas, on access roads, and in terminal buildings. It was evident that to reduce congestion substantial financial resources would be required for investment in airway and airport improvements. For airports alone it was estimated that $13 billion in new capital improvements would be required for public airports in the period 1970 to 1980. The amount of money authorized by the Federal Airport Act of 1946 was insufficient to assist in financing such a vast program. The normal and anticipated sources of revenue available to public airports were also not sufficient to raise the required funds for capital expenditures. It was argued that much of the congestion in the air at major airports was due to a lack of funds to modernize the airways system. Funds for airport development came from the budget of the FAA authorized by Congress each year and not from the Federal Airport Act. The deficiencies in airport and airway development were documented in several reports. It was the consensus of industry and government that the only way to provide the funds needed for airports and airways was through increased or new taxes imposed upon the users of the air transport system. It was also argued that the revenues from these taxes should be specifically earmarked for aviation and not go to the general fund. The concept of establishing a trust fund similar to that of the national highways program was agreed upon. Finally after much debate in Congress, the Airport and Airway Development Act of 1970 and the Airport and Airway Revenue Act of 1970 were enacted (Public Law 91-258).

As finally passed, the act was divided into two sections: Title I detailed the airport assistance programs and established a financing program for airport grants, airways hardware acquisition, and research and development; Title II created the Airport and Airway Trust Fund and established the pattern of aviation excise taxes which would provide the resources upon which the Title I capital programs would depend through 1980. The excise taxes adopted consisted of a tax on domestic passenger tickets, a head tax on international passenger departures, a flowage tax on all fuel used by general aviation, and tax on all air cargo waybills. In addition, an annual aircraft registration tax was levied on all aircraft (commercial and general aviation) plus an annual weight surtax for all aircraft weighing in excess of 2500 lb. Finally, revenues from existing taxes on aircraft tires and tubes were transferred from the Highway Trust Fund to the Airport and Airway Trust Fund.

The amount of these excise taxes were changed in the Omnibus Budget Reconciliation Act of 1990 (Public Law 101-508) and currently consists of a 10 percent tax on domestic passenger tickets, a $6 head tax on international passenger departures, a 17.5 cents a gallon flowage tax on all fuel used by general aviation, and a 6.25 percent tax on all air cargo waybills.

Significant changes from the Federal Airport Act were as follows:

1. The provision of funds to local agencies for airport system planning and master planning

2. The emphasis on airports served by air carriers and general aviation airports to relieve congested air carrier airports

3. The provision of funds for commuter service airports

4. The requirement that the FAA issue airport operating certificates to ensure that airports were adequately equipped for safe operations

5. Provision of requirements to ensure that airport projects did not adversely affect the environment and were consistent with long-range development plans of the area in which the project was proposed

6. Provision for terminal facility development in non-revenue producing public areas

7. The requirement that the FAA develop a National Airport System Plan (NASP)

To be eligible for federal aid, airport ownership was required to be vested in a public agency and the airport must be included in the NASP. This plan was reviewed and revised as necessary to keep it current. It was prepared by the FAA and submitted to Congress by the Secretary of Transportation. The plan specified, in terms of general location and type of development, the projects considered necessary to provide a system of public airports adequate to anticipate and meet the needs of civil aeronautics. These projects included all types of airport development eligible for federal aid under the act and were not limited to any classes or categories of public airports. The plan was based on projected needs over 5- and 10-year periods.

Because of the mounting public concern for the enhancement of the environment, the act specifically stated that authorized projects provide for the protection and enhancement of the natural resources and the quality of the environment of the nation. The Secretary of Transportation was required to consult with the Secretary of the Interior and the Secretary of Health, Education and Welfare regarding the effect of certain projects on natural resources and whether "all possible steps have been taken to minimize such adverse effects." The Act required that airport sponsors provide the "opportunity for public hearings for the purpose of considering the social, economic, and environmental effect on any project involving the location of an airport, the location of a runway, and a runway extension." In addition, the National Environmental Policy Act of 1969 (Public Law 91-190), supported by a Presidential Executive Order (11514, March 5, 1970), required the preparation of detailed environmental impact statements for all major airport development actions significantly affecting the quality of the environment. The environmental impact statement was required to include the probable impact of the proposed project on both the human and natural environment, including impact on ecological systems such as wildlife, fish, and marine life, and any probable adverse environmental effects which could not be avoided if the project was implemented. The Act also stipulated that no airport project involving "airport location, a major runway extension, or runway location" could be approved for federal funding unless the governor of the state in which the project was located certified to the Secretary of Transportation that there was reasonable assurance that the project would comply with applicable air and water quality standards. Finally the project had to be consistent with the plans of other agencies for development of the area, and the airport sponsor had to assure the government that adequate housing was available for any displaced people. The Aviation Safety and Noise Abatement

Act of 1979 (Public Law 96-193) amended this legislation to place increased emphasis on reducing the noise impacts of airports. Thus one of the principal differences between the Federal Airport Aid Program and the Airport Development Aid Program was the emphasis on environmental protection in the latter.

The Airport and Airway Development Act made no mention concerning specific standards for determining airports to be included in the National Airport System Plan. It did state, however, that

> The Plan shall set forth, for at least a ten-year period, the type and estimated cost of an airport development considered by the Secretary to be necessary to provide a system of public airports adequate to anticipate and meet the needs of civil aeronautics, to meet the requirements in support of the national defense as determined by the Secretary of Defense, and to meet the special needs of the Postal Service. In formulating and revising the plan, the Secretary shall take into consideration, among other things, the relationship of each airport to the rest of the transportation system in the particular area, to the forecasted technological developments in aeronautics, and to developments forecasted in the other modes of transportation.

With this and other policy guidelines, the FAA developed entry criteria which described a broad and balanced airport system. The 1980 NASP, for example, included about 3600 airport locations, indicating that the federal interest in developing a basic airport system extended well beyond the major airports with scheduled airline service. In an effort to provide a safe and adequate airport for as many communities as possible, NASP criteria were developed to include the general aviation airports which serve smaller cities and towns.

The NASP airport entry criteria evolved from both policy and legislative considerations and focused on two broad categories of airports, those with scheduled service and those without significant scheduled service in the general aviation and reliever category. Airports with scheduled service were included in NASP because of their use by the general public, legislative provisions which specifically designated airports to receive development funds, and their use by CAB certified carriers. Commuter airports were identified in NASP starting in 1976, when legislation was enacted which designated them as a type of air carrier airport and provided them with special development funds. About 70 percent of the airports in the NASP were general aviation locations which met the criteria of viability because of the number of based aircraft or aircraft activity, and which provided reasonable access for aircraft owners and users to their community. Reliever airports have been included as a separate NASP category since the 1960s when Congress designated special funding for the purpose of relieving congestion in large metropolitan areas by providing additional general aviation capacity.

Airline Deregulation Act of 1978

The Airline Deregulation Act (Public Law 95-504) was passed by Congress in October 1978. This legislation eliminated the statutory authority for the economic regulation of the passenger airline industry in the United States. It provided that the Civil Aeronautics Board would be abolished in 1985. The legislation was intended to increase competition in the passenger airline industry by phasing out federal authority to exercise regulatory controls during the period of time between 1978 and 1985. The principal provisions of this legislation:

1. Required the CAB to place maximum reliance on competition in its regulation of interstate airline service, while continuing to ensure the safety of air transportation; to maintain service to small communities; and to prevent practices which were deemed anticompetitive in nature.

2. Required CAB approval of airline acquisitions, consolidations, mergers, purchases, and operating contracts; the burden to prove that an action was anticompetitive in nature was placed upon the party challenging that action.

3. Permitted carriers to change rates within a range of reasonableness from the standard industry fare without prior CAB approval; the CAB was authorized to disallow a fare change if it considered the change predatory.

4. Provided interstate carriers an exemption from state regulation of rates and routes.

5. Required the CAB to authorize new routes and services that were consistent with the public convenience and necessity.

6. Allowed carriers to be granted operating rights to any route on which only one other carrier was providing service and on which other airlines were authorized to provide service but were not actually providing a minimum level of service. If more than one airline was providing service on this route, the CAB was required to determine if the granting of additional route authority was consistent with the public convenience and necessity before allowing additional carriers to service the route. An airline not providing the specified minimum level of service on a route (dormant authority) could begin providing such service and retain its authority. Otherwise, the CAB was required to revoke the unused authority.

7. Provided for an automatic market entry program, whereby airlines could begin service on one additional route each year during the period 1979 to 1981 without formal CAB approval. Each carrier was also permitted to protect one of its existing routes each year by declaring it as ineligible for automatic market entry by another airline.

8. Authorized the CAB to order an airline to continue to provide "essential air transportation service" and, for a 10-year period, to provide subsidies or seek other willing carriers to ensure the continuation of essential service.

9. Required the CAB to determine within 1 year of the enactment of the legislation what it considered to be "essential air transportation service" for each point being serviced at the time of enactment of the legislation.

10. Required the CAB and the Department of Transportation to determine mechanism by which the state and local governments should share the cost of subsidies from the federal government to preserve small community air service and to make policy recommendations to Congress on this matter.

11. Exempted from most CAB regulation commuter aircraft weighing less than 18,000 lb and carrying fewer than 56 passengers.

12. Made commuter and intrastate air carriers eligible for the federal loan guarantee program.

13. Provided that the domestic route authority of the board would cease in 1981; its authority over domestic fares, acquisitions, and mergers would cease in 1983; and the board would be abolished (sunset) in 1985.

14. Provided that after the board was abolished, the local service carrier subsidy program was to be transferred to the Department of Transportation; the foreign air transportation authority of the board was to be transferred to the Transportation and Justice Departments, in consultation with the State Department; and the mail subsidy program was to be transferred to the U.S. Postal Service.

Impact of Airline Deregulation

In the United States prior to 1978, air carriers applied to the Civil Aeronautics Board (CAB) for permission to serve markets. The CAB granted air carriers service to markets with a determined operating and fare schedule. Upon deregulation, air carriers freely entered new markets, increasing the number of markets served, increasing competition and lowering overall airfares. To maximize their market share in the industry, several air carriers concentrated their route structures on one or more "hub" airports. The early years of this hub and spoke route system resulted in the greatest growth in commercial aviation in its history. The beginning of the twenty-first century exposed many of the weaknesses in the airline hub and spoke model, including the increased costs of operating through congested hub airports, increasing fuel and other operating expenses, combined with the ability for the public to use the Internet to avoid extraordinarily high full fares thereby reducing the overall revenue stream of the airlines. The air carriers dependent on their traditional "legacy" business models fell into serious financial distress, and combined with the short-term drop in air travel demand following the terrorist attacks of September 11, 2001, many fell into bankruptcy to emerge under more efficient business models years later.

These more efficient business models were based on the emergence of "low-cost carriers" or LCCs, that operate primarily on a market-based origin to destination route network and price fares relative to operating costs, rather than solely by passenger demand. The LCC airline model has been the largest growth segment of the domestic airline industry in the United States.

The Airport and Airway Improvement Act of 1982

In 1982, Congress enacted the Airport and Airway Improvement Act (Title V of the Tax Equity and Fiscal Responsibility Act of 1982, Public Law 97-248). This act continued to provide funding for airport planning and development under a single program called the Airport Improvement Program (AIP). The Act also authorized funding for noise compatibility planning and implementing noise compatibility programs contained in the Noise Abatement Act of 1979 (Public Law 96-193). It required that to be eligible for a grant the airport must be included in the National Plan of Integrated Airport Systems (NPIAS). The NPIAS, the successor to the National Airport System Plan (NASP), is prepared by the FAA and published every 2 years and identifies public use airports considered necessary to provide a safe, efficient, and integrated system of airports to meet the needs of civil aviation, national defense, and the postal service.

The Airport and Airway Improvement Act has been amended several times resulting in significant changes in the provisions of the act and in the appropriations authorized under the Act. These amendments are included in the Continuing Appropriations Act of 1982 (Public Law 97-276), the Surface Transportation Assistance Act (Public Law 97-424), the Airport and Airway Safety and Capacity Expansion Act of 1987 (Public Law 100-223), the Airway Safety and Capacity Expansion Act of 1990 (Public Law 101-508), and the Airport and Airway Safety, Capacity, Noise Improvement and Intermodal Transportation Act of 1992 (102nd Congress H.R. 6168).

This legislation, as amended, significantly increased the level of federal funding for airports to an aggregate total of more than $14 billion for the period from 1982 through 1992.

The Aviation Safety and Capacity Act of 1990

In response to issues concerning the provision of limited AIP funds to the largest airports, thereby leaving the smaller airports with little in capital improvement funding support, in 1990, Congress passed the Aviation Safety and Capacity Expansion Act. This act established the policy of allowing airports to impose a passenger facility charge (PFC) to supplement their capital improvement programs, while allowing greater amounts of AIP funding to be allocated to smaller airports with capital improvement needs. Under this Act, an airport applied to collect a $1, $2, or $3 charge, on any passenger enplaning at the airport. The fee would be collected by the air carriers, upon purchase of a ticket. Revenues generated by PFCs would then be spent by the airport that generated the revenue on allowable costs associated with certain capital improvement projects approved by the FAA that enhance safety, security, or capacity, or increase air carrier competition. In 2001, the maximum allowable PFC was raised to $4.50. As of June 2007, approximately $58.6 billion in PFCs have been collected at 367 airports nationwide. More than 1500 projects utilizing PFC revenues have been approved since the 1990 Aviation Safety and Capacity Expansion Act introduced the PFC program.

AIR-21: The Wendell Ford Aviation Investment Act for the 21st Century

In April 2000, funding for airport planning and design through the AIP and PFC programs was increased with the Wendell H. Ford Aviation Investment and Reform Act for the Twenty-First Century, known as AIR-21 (Public Law 106-181). This funding increase was designed to assist larger airports which have become highly congested, as well as smaller airports struggling to preserve commercial air service.

The AIR-21 Act was introduced at a time when the nation's air carriers were coming off record profits and growth in air transportation was at its highest in history. As part of the act, AIP funding was increased, on the order of 300 percent to many airports to allow for capital improvement projects designed to relieve the increased congestion and delays encountered at the nation's largest airports at the end of the 1990s.

The Aviation and Transportation Security Act of 2001

In response to the terrorist attacks involving the hijacking of four U.S. airliners used in suicide attack missions on Washington, D.C. and New York City, on September 11, 2001, The Aviation and Transportation Security Act (Public Law 107-071) was signed into law. This Act created the Transportation Security Administration (TSA), which took authority over aviation security and imposed a series of requirements for screening air carrier passengers and luggage including mandatory electronic inspection of all checked luggage. This has had profound effects on airport terminal planning and design. To fund these policies, the Act authorized a passenger surcharge of $2.50 per flight segment and a fee imposed to air carriers equivalent to each airline's costs of providing passenger security screening in the year 2000. These funds are collected by air carriers through ticket purchases and are used to fund the operation of the TSA and to contribute to airport development to accommodate enhanced security policies and procedures.

Vision 100 Century of Aviation Act of 2003

The AIR-21 Act authorized AIP funding through 2003 at which time reauthorization legislation was to occur. This reauthorization of AIP funding was accomplished with the Vision 100 Century of Aviation Act (Public Law 108-176) in December 2003. The purpose of the Vision 100 Act was to further increase, yet diversify, federal funding for airport and airspace improvements as the commercial air carrier industry recovers, and restructures, from the severe economic industry downturns following the September 11, 2001 attacks, and other economic and geopolitical issues. The act increased annual AIP authorizations to approximately $3.4 billion in 2004, up to $3.7 billion in 2007, the last year of the act's term and broadened the use of AIP and PFC funds to include airport improvements that have certain environmental benefits, investments to attract air service to underutilized airports, and to fund debt-service for projects previously funded through bond issuances.

NextGen Financing Reform Act of 2007/ FAA Reauthorization Act of 2009

The financial recovery the nation's airlines combined with increases in general aviation activity have begun to put increased strains on an aging air traffic control system and debates in Congress ensued regarding how to appropriately reauthorize funding for civil aviation as the terms of Vision 100 were due to expire in 2007. The Congress debate focused around a complete restructuring of the current funding programs, including major AIP reform. Rather than a system of funding airport and air traffic management through airline passenger, cargo, and fuel taxes, an aircraft-based user fee system was introduced to Congress for debate. This new legislation will be the first to implement fees directly on commercial and general aviation operations using the busiest areas of the national airspace system.

The NextGen Financing Reform Act focused funding on creating the Next Generation Air Traffic Management System to replace nearly 50-year-old air traffic control technology.

As of the end of 2007, the NextGen Financing Reform Act of 2007 had yet to be signed into law. Two versions of the act are being debated in Congress. The version supported by the House of Representatives (H.R. 2881) supports reauthorizing funding by increasing fuel taxes on general aviation fuel to between 24.1 and 35.9 cents per gallon, while maintaining the remaining tax structure implemented in Vision 100. The version of the act in the Senate (S. 1300) proposes a user-based approach, including a per-flight surcharge of $25 for "air traffic control costs," for general aviation jet and turbo-prop aircraft operating inside of controlled airspace.

The NextGen Financing Reform Act of 2007 was ultimately discarded with the new presidential administration in 2009. In March 2009, a new reauthorization bill, the FAA Reauthorization Act of 2009, was introduced as H.R. 915. As of publication of this text there is continued debate on how the federal government of the United States will fund the modernization of the national airspace system, particularly in the face of the economic downturn of beginning in late 2007. The focus of debate has continued to be around the potential implementation of user-based fees and other such taxes that may prohibit growth. The funding of the nation's aviation system continues to be a critical and highly debated topic.

State Roles in Aviation and Airports

State interest in aviation began as early as 1921, when the state of Oregon established an agency to handle matters concerning aviation. Since that time virtually all states have established aeronautical agencies either as commissions, departments, bureaus, boards, or divisions. Their responsibilities vary considerably and include channeling federal aid funds, planning state airport systems, providing state aid to local airport authorities, constructing and maintaining navigational equipment, investigating small aircraft accidents, enforcing safety regulations, and licensing airports.

Despite the growing concern of the states in airport development and aviation planning, their participation in the past has had little resemblance to the pattern associated with highway development. The states have always played a leading role in the development of roads and streets within their boundaries, whereas in airport development this has not been true. The reason for this pattern can best be explained by looking into the background of the entry of the states into aviation.

The majority of airports for civil aviation served by air carriers are municipally owned and operated. In a large number of states these airports were in operation prior to the formation of a state aeronautical agency. From its inception air transportation, because it is inherently of an interstate nature, became a matter of federal concern. The federal government provided much technical and financial assistance to municipalities. During World War II a great number of municipalities were the recipients of federal aid from the Civil Aeronautics Administration through the Defense Landing Area Program. Thus, in the early stages of airport development in this country, a fairly close relationship was established between the federal government and the municipalities. This relationship was furthered when the Federal Airport Act of 1946 authorized the CAA to issue grants directly to municipalities, as long as such a procedure was not opposed to state policy. In the meantime, the majority of states were doing very little in the way of providing funds to municipalities for airports. While significant increases in state aid for airport development have occurred in the last several years, the amount of federal aid has been substantially higher.

States such as Alaska, Rhode Island, and Hawaii directly own and operate many of the airports within their respective boundaries. Other states support municipality-owned airports through state block grant funding programs. In those states where monetary aid is made available for airport development, the plans, specifications, and design for airport construction are generally reviewed by the state aeronautical agency.

There is no doubt that participation by the states in airport development is assuming a more significant role with the emergence of recent legislation in Congress to channel funds directly to state aeronautical agencies through block grant programs. The public concern for environmental control has resulted in legislation being passed at the state level, in addition to federal statutes, aimed at the control of aircraft noise and pollution. As general aviation and commuter activities continue to grow, the states will have to share the burden with the federal government in providing facilities for these activities, enforcing safety regulations, and other matters.

Aviation Organizations and Their Functions

The organizations directly involved in United States and international air carrier transportation and general aviation activity have an important influence on airport development as well as aircraft operations. These organizations and their functions can be classified into four groups, namely, federal agencies, state agencies, international government agencies, and industry or trade organizations.

Federal Agencies of the United States Government

There are several agencies at the federal level which dictate policy of direct and indirect effects on air transportation. The Federal Aviation Administration (FAA), the Transportation Security Administration (TSA), and the Environmental Protection Agency (EPA) are those agencies with the most direct influence on civil aviation policy, and airport planning and design.

Federal Aviation Administration

The Federal Aviation Administration is the agency within the U.S. Department of Transportation responsible for the safe and efficient operation of the nation's civil aviation system. The FAA is headed by the chief executive known as the administrator who is appointed by the President. The FAA performs the following functions:

1. Encourages the establishment of civil airways, landing areas, and other air facilities

2. Designates federal airways and acquires, establishes, operates, and conducts research and development and maintains air navigation facilities along such civil airways

3. Makes provision for the control and protection of air traffic moving in air commerce

4. Undertakes or supervises technical development work in the field of aeronautics and the development of aeronautical facilities

5. Prescribes and enforces the civil air regulations for safety standards, including:
   a. Effectuation of safety standards, rules, and regulations
   b. Examination, inspection, or rating of airmen, aircraft engines, air navigation facilities, aircraft, and air agencies
   c. Issuance of various types of safety certificates

6. Provides for aircraft registration

7. Requires notice and issues orders with respect to hazards to air commerce

8. Issues airport operating certificates to airports serving air carriers

The FAA develops, directs, and fosters the coordination of a national system of airports, the National Plan of Integrated Airport Systems [22], an aviation system capacity enhancement plan, the Aviation System Capacity Plan [9], a plan to modernize and significantly upgrade the air traffic control system, the National Airspace System Plan [21], and oversees funding for airports through the Airport Improvement Program and PFC Program. In this connection it performs the following functions:

1. Provides consultation and advisory assistance on airport planning, design, construction, management, operation, and maintenance to governmental, professional, industrial, and other individuals and agencies.

2. Develops and establishes standards, government planning methods and procedures; airport and seaplane base design and construction; and airport management, operation, and maintenance.

3. Collects and maintains an accurate record of all available airport facilities in the United States.

4. Directs, formulates, and keeps current a national plan (NPIAS) for the development of an adequate system of airports in cooperation with federal, state, and local agencies, and determines and recommends the extent to which portions or units of that system should be developed or improved.

5. Develops and recommends principles, for incorporation in state and local legislation, to permit or facilitate airport development, regulation, and protection of approaches through zoning or property acquisition.

6. Secures compliance with statutory and contractual requirements relative to airport operation practices, conditions, and arrangements.

7. Develops and recommends policies, requirements, and procedures governing the participation of states, municipalities, and other public agencies in federal-aid airport projects and secures adherence to such policies, requirements, and procedures.

As illustrated in Fig. 1-6, the FAA is organized into a number of offices within its headquarters in Washington, D.C. The offices most directly related to airport planning and development are the Office of Airport Planning and Programming, the Office of Airport Safety and Standards. In addition to Headquarters Offices, the FAA is divided into nine Airports Regional Offices, as illustrated in Fig. 1-7. Within these Regional Offices are Airports District Offices (ADOs). It's within these ADOs where specific consultation between the FAA and airport planners on airport planning and design programs are primarily discussed, analyzed, and ultimately approved.

The FAA publishes most of the regulations of concern to civil aviation and these—are found in Title 14—"Aeronautics and Space" of the United States Code of Federal Regulations. The Federal Aviation Regulations (FARs) are made up of more than 100 chapters, known as "parts," regulating various aspects of the civil aviation system, including pilots, aircraft, the airspace system, and airports. The FARs of most concern to airport planning and design that will be further discussed in this text include

FAR Part 1: Definitions and Abbreviations

FAR Part 11: General Rule Making Procedures

PAR Part 36: Noise Standards: Aircraft Type and Airworthiness Certification

FAR Part 71: Designation of Class A, Class B, Class C, Class D, and Class E Airspace Areas, Airways, Routes, and Reporting Points

FAR Part 73: Special Use Airspace

FAR Part 77: Objects Affecting Navigable Airspace

FAR Part 91: General Operating and Flight Rules

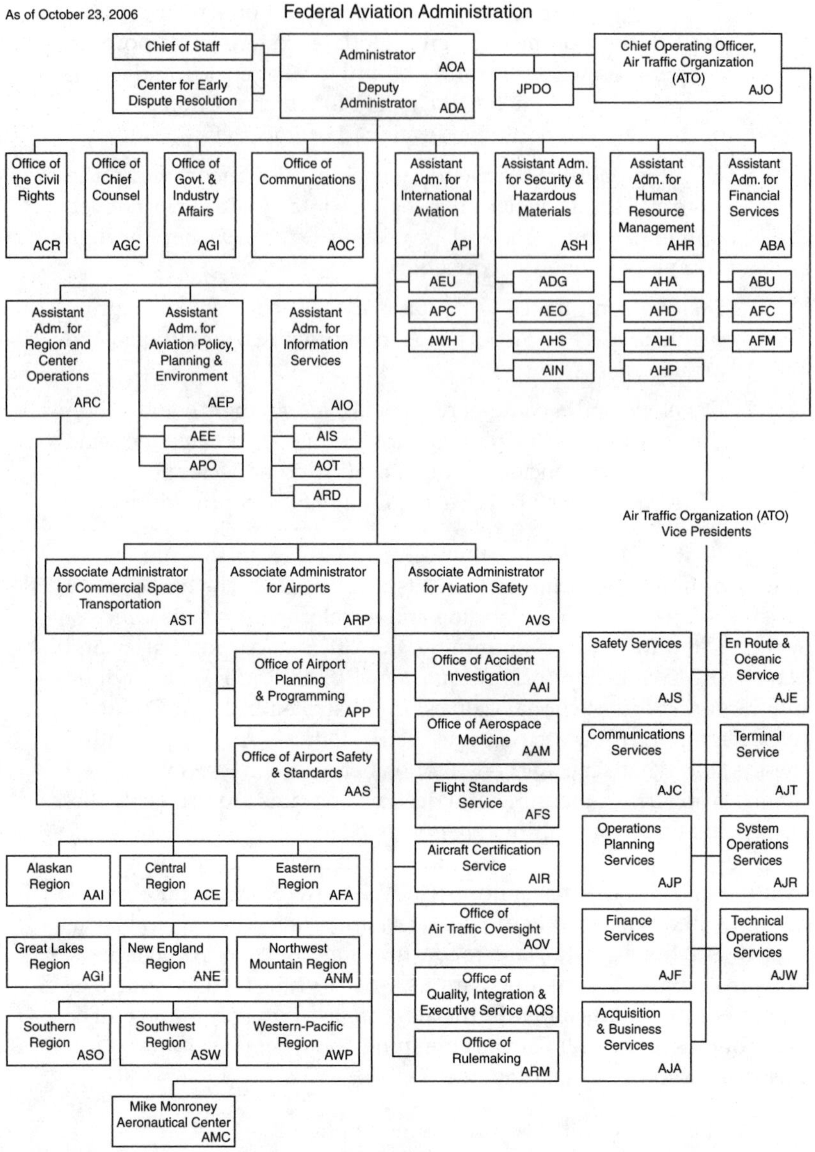
As of October 23, 2006
Federal Aviation Administration
Chief of Staff
Center for Early Dispute Resolution
Administrator AOA
Deputy Administrator ADA
JPDO
Chief Operating Officer, Air Traffic Organization (ATO) AJO
Office of the Civil Rights ACR
Office of Chief Counsel AGC
Office of Govt. & Industry Affairs AGI
Office of Communications AOC
Assistant Adm. for International Aviation API
AEU
APC
AWH
Assistant Adm. for Security & Hazardous Materials ASH
ADG
AEO
AHS
AIN
Assistant Adm. for Human Resource Management AHR
AHA
AHD
AHL
AHP
Assistant Adm. for Financial Services ABA
ABU
AFC
AFM
Assistant Adm. for Region and Center Operations ARC
Assistant Adm. for Aviation Policy, Planning & Environment AEP
AEE
APO
Assistant Adm. for Information Services AIO
AIS
AOT
ARD
Air Traffic Organization (ATO) Vice Presidents
Associate Administrator for Commercial Space Transportation AST
Associate Administrator for Airports ARP
Office of Airport Planning & Programming APP
Office of Airport Safety & Standards AAS
Associate Administrator for Aviation Safety AVS
Office of Accident Investigation AAI
Office of Aerospace Medicine AAM
Flight Standards Service AFS
Aircraft Certification Service AIR
Office of Air Traffic Oversight AOV
Office of Quality, Integration & Executive Service AQS
Office of Rulemaking ARM
Safety Services AJS
En Route & Oceanic Service AJE
Communications Services AJC
Terminal Service AJT
Operations Planning Services AJP
System Operations Services AJR
Finance Services AJF
Technical Operations Services AJW
Acquisition & Business Services AJA
Alaskan Region AAI
Central Region ACE
Eastern Region AFA
Great Lakes Region AGI
New England Region ANE
Northwest Mountain Region ANM
Southern Region ASO
Southwest Region ASW
Western-Pacific Region AWP
Mike Monroney Aeronautical Center AMC

FIGURE 1-6 FAA headquarters organizational chart.

FAR Part 121: Operating Requirements: Domestic, Flag, and Supplemental Air Carrier Operations

FAR Part 139: Certification of Airports

FAR Part 150: Airport Noise Compatibility Planning

FAR Part 151: Federal Aid to Airports

FIGURE 1-7   FAA regions.

FAR Part 152: Airport Aid Program

FAR Part 156: State Block Grant Pilot Program

FAR Part 157: Notice of Construction, Alteration, Activation, and Deactivation of Airports

FAR Part 158: Passenger Facility Charges

FAR Part 161: Notice and Approval of Airport Noise and Access Restrictions

A complete list of FARs may be found at the Federal Aviation Administration website http://www.faa.gov.

In addition to federal regulations, the FAA publishes a series of Advisory Circulars (ACs) to provide guidance into the application of the regulations. The "150 series" of Advisory Circulars are focused on guiding airport managers and planners. There are more than 100 current and historical Advisory Circulars within the 150 series. Those of most direct application to airport planning and design include

AC 150/5020-1: Noise Control and Compatibility Planning for Airports

AC 150/5060-5: Airport Capacity and Delay

AC 150/5070-6B: Airport Master Plans

AC 150/5070-7: The Airport System Planning Process

AC 150/5300-13: Airport Design

AC 150/5325-4B: Runway Length Requirements for Airport Design

AC 150/5340-1J: Standards for Airport Markings

AC 150/5360-12D: Airport Signing and Graphics

AC 150/5360-13: Planning and Design Guidelines for Airport Terminal Facilities

AC 150/5360-14: Access to Airports by Individuals with Disabilities

Advisory Circulars are in a constant state of updating. The latest available ACs may be found at the FAA website at http://www.faa.gov.

Transportation Security Administration

The Transportation Security Administration (TSA) is the agency within the U.S. Department of Homeland Security responsible for the security nation's transportation systems, including civil aviation. The TSA was formed in 2001 in response to the terrorist attacks of September 11, 2001. In 2003, the TSA moved from the Department of Transportation to become part of the newly formed Department of Homeland Security. The TSA is led by an administrator appointed by the President.

With the formation of the TSA, all federal regulations pertaining to the security of the civil aviation system were moved from Title 14 of the Code of Federal Regulations to Title 49—Transportation, and have become commonly known as Transportation Security Regulations (TSRs). TSRs of specific importance to airport planners are those within Subsection C of the TSRs including

49 CFR Part 1500: Applicability, Terms, and Abbreviations

49 CFR Part 1510: Passenger Civil Aviation Security Service Fees

49 CFR Part 1540: Civil Aviation Security: General Rules

49 CFR Part 1542: Airport Security

49 CFR Part 1544: Aircraft Operator Security: Air Carriers and Commercial Operators

49 CFR Part 1546: Foreign Air Carrier Security

49 CFR Part 1550: Aircraft Security under General Operating and Flight Rules

Since 2001, the focus of the TSA has been on airport and commercial aviation security. In the future, it is expected that the TSA will further expand its regulatory role in air cargo and general aviation security, as well as into other modes of transportation, such as transit, rail, shipping, and the nation's highways. More information on the TSA may be found at its website at http://www.tsa.gov.

Environmental Protection Agency

Established in 1970 as part of the National Environmental Policy Act (NEPA), the Environmental Protection Agency (EPA) is responsible for preserving the environment with the goal of protecting human health.

The EPA has directed many of its efforts to minimizing environmental damage resulting from civil aviation activities, with focus on aircraft noise levels, emissions, air quality, and water runoff.

Most of the EPA requirements pertaining to civil aviation are incorporated into the FAA's Federal Aviation Regulations and policies regarding mandatory environmental impact evaluation of any proposed airport planning projects. More information on the EPA may be found at its website at http://www.epa.gov.

National Transportation Safety Board

The National Transportation Safety Board consists of five members appointed by the President. The NTSB performs the following functions:

1. Investigates certain aviation, highway, marine, pipeline, and railroad accidents, and reports publicly on the facts, conditions and circumstances, and the cause or probable cause of such accidents.

2. Recommends to Congress and federal, state, and local agencies measures to reduce the incidence of transportation accidents.

3. Initiates and conducts transportation safety studies and investigations.

4. Establishes procedures for reporting accidents to the board.

5. Assesses accident investigation techniques and issues recommendations for improving accident investigation procedures.

6. Evaluates the adequacy of the procedures and safeguards used for the transportation of hazardous materials.

7. Reviews, on appeal, the suspension, amendment, modification, revocation, or denial of certain operating certificates, documents, or licenses issued by the Federal Aviation Administration or the U.S. Coast Guard.

Information about the activities of the NTSB, including all records of civil aviation accidents and incidents, many of which impact airport planning and design, may be found on the NTSB website at http://www.ntsb.org.

State Agencies

As mentioned earlier, the states are involved in varying degrees in the many aspects of aviation including airport financial assistance, flight safety, enforcement, aviation education, airport licensing, accident investigation, zoning, and environmental control. Because of the interstate nature of air transportation, the federal government has preempted the legislative and administrative controls since the early days of aviation. However, for those aviation activities which occur wholly within the borders of a state, there have been formed regulatory agencies at the state level to oversee that these activities are operated in the best interests of the state.

Many state aviation agencies are participant members in the National Association of State Aviation Officials (NASAO). The mission of NASAO is to provide representation in Washington, D.C. on behalf of state aviation departments. Links to individual state aviation departments as well as a host of informational materials may be found on the NASAO website at http://www.nasao.org.

The International Civil Aviation Organization

Perhaps the most important international agency concerned with airport development is the International Civil Aviation Organization (ICAO), which is now a specialized agency of the United Nations with headquarters in Montreal, Canada. One hundred and eighty-eight nations were members of ICAO in 2009.

The ICAO concept was formed during a conference of 52 nations held in Chicago in 1944. This conference was by the invitation of the United States to consider matters of mutual interest in the field of air transportation. The objectives of ICAO as stated in its charter are to develop the principles and techniques of international air transportation so as to

1. Ensure the safe and orderly growth of international civil aviation throughout the world

2. Encourage the arts of aircraft design and operation for peaceful purposes

3. Encourage the development of airways, airports, and air navigation facilities for international aviation

4. Meet the needs of the peoples of the world for safe, regular, efficient, and economical air transport

5. Prevent economic waste by unreasonable competition

6. Ensure that the rights of contracting states are fully respected and that every contracting state has a fair opportunity to operate international airlines

7. Avoid discrimination between contracting states

8. Promote safety of flight in international air navigation

9. Promote generally the development of all aspects of international civil aeronautics

The ICAO has two governing bodies, the Assembly and the Council. The Council is a permanent body responsible to the Assembly and is composed of representatives from 30 countries. The Council is the working group for the organization. It carries out the directives of the Assembly and discharges the duties and obligations specified in the ICAO charter.

To the airport planner and designer perhaps the most important document issued by ICAO is "Aerodromes," Annex 14 to the Convention on International Civil Aviation [1]. Annex 14 contains the international design standards and recommended practices which are applicable to nearly all airports serving international air commerce. In addition to Annex 14, ICAO publishes a great deal of technical and statistical information relative to international air transport [19] and is available on its website at http://www.icao.org.

Industry and Trade Organizations

There are many groups involved in the technical and promotional aspects of aviation. The following is a partial list of those groups which are primarily concerned with the airport aspects of aviation, most of these professional organizations serve as lobbying groups promoting the perspectives of the industry groups they represent.

1. *Aerospace Industries Association of America (AIA).* The national trade association of companies in the United States engaged in research, development, and manufacture of aerospace systems.

2. *Aircraft Owners and Pilots Association (AOPA).* An association of owners and pilots of general aviation aircraft. It is headquartered in Frederick, MD, a suburb of Washington, D.C.

3. *Air Line Pilots Association, International (ALPA).* An association of airline pilots. It is headquartered in Herndon, VA, a suburb of Washington, D.C.

4. *Airports Council International (ACI).* An association of over 400 large airports and airport authorities throughout the world. It is based in Geneva, Switzerland. The North American region of this organization (ACI-NA) is headquartered in Washington, D.C. Other regions include Europe, Africa, Asia/Pacific, and Latin America/Caribbean.

5. *Air Transport Association of America (ATA).* An association of scheduled domestic and international airlines in the United States. The headquarters are in Washington, D.C.

6. *American Association of Airport Executives (AAAE).* An association of the managers of public and private airports. It is located in Alexandria, VA, a suburb of Washington, D.C.

7. *General Aviation Manufacturers Association (GAMA).* An association promoting the interests of general aviation. It is located in Washington, D.C.

8. *Helicopter Association International (HAI).* An association which represents the interests of manufacturers and users of helicopters and promotes the use of helicopters. It is located in Alexandria, VA, a suburb of Washington, D.C.

9. *International Air Transport Association (IATA).* An association of scheduled carriers in international air transportation. This organization is headquartered in Montreal, Canada.

10. *Regional Airline Association (RAA).* An association of small regional and commuter aircraft operators promoting the needs of this segment of the air transportation industry. It was formerly called the Commuter Airline Association of America (CAAA). It is located in Washington, D.C.

References

1. *Aerodromes, Annex 14 to the Convention on International Civil Aviation,* vol. 1, *Aerodrome Design and Operations,* 5th ed., International Civil Aviation Organization, Montreal, Canada, November 2009.
2. *Aerospace Facts and Figures,* Aerospace Industries Association of America, Inc., Washington, D.C., 1980.
3. *Airline Capital Requirements in the 1980s,* Economics and Finance Department, Air Transportation Association of America, Washington, D.C., September 1979.
4. *Airline Deregulation,* M. A. Brenner, J. O. Leet, and E. Schott, Eno Foundation for Transportation, Inc, Westport, Conn., 1985.
5. *Air Taxi Operators and Commercial Operators, Federal Aviation Regulations,* Part 135, Federal Aviation Administration, Washington, D.C., 1978.
6. *Air Transport Facts and Figures,* Air Transportation Association of America, Washington, D.C., annual.
7. *Air Transportation,* 10th ed., Robert M. Kane, Kendall/Hunt Publishing Company, Dubuque, Iowa, 1990.
8. *Annual Report of the Regional Airline Association,* Regional Airline Association, Washington, D.C., 1991.
9. *Aviation System Capacity Plan 1991–1992,* Report No. DOT/FAA/ASC-91-1, U.S. Department of Transportation, Federal Aviation Administration, Washington, D.C., 1991.
10. *Certification and Operations: Domestic, Flag and Supplemental Air Carriers and Commercial Operators of Large Aircraft,* Part 121, Federal Aviation Regulations, Federal Aviation Administration, Washington, D.C., 1980.
11. *Commuter Air,* 1981 yearbook edition, Commuter Airline Association of America, Washington, D.C., April 1981.
12. *Commuter Air Carrier Traffic Statistics,* 12 months ended June 30, 1980, Civil Aeronautics Board, Washington, D.C.
13. *Current Market Outlook,* Boeing Commercial Airplane Group, Seattle, Wash., March 1992.
14. *Developments in the Deregulated Airline Industry,* D. R. Graham and D. P. Kaplan, Office of Economic Analysis, Civil Aeronautics Board, Washington, D.C., June, 1981.
15. *FAA Aviation Forecasts, Fiscal Years 1992–2003,* Federal Aviation Administration, Washington, D.C., February 1992.
16. *FAA Statistical Handbook of Civil Aviation,* Federal Aviation Administration, Washington, D.C., 1990.
17. *Hearings before Subcommittee on Aviation of the Committee on Commerce, Science, and Transportation,* U.S. Senate, Washington, D.C., August 1980.

18. *ICAO Journal*, International Civil Aviation Organization, Montreal, Quebec, Canada, monthly.
19. *ICAO Publications and Audio Visual Training Aids*, Catalogue, International Civil Aviation Organization, Montreal, Quebec, Canada, 1992.
20. *National Airport System Plan, 1978–1987*, Federal Aviation Administration, Department of Transportation, Washington, D.C.
21. *National Airspace System Plan*, Federal Aviation Administration, Washington, D.C., 1989.
22. *National Plan of Integrated Airport Systems (NPIAS) 1990–1999*, Federal Aviation Administration, U.S. Department of Transportation, Washington, D.C., 1991
23. *National Transportation Statistics*, annual report, Research and Special Programs Administration, Department of Transportation, Washington, D.C., July 1990.
24. *National Transportation Strategic Planning Study*, U.S. Department of Transportation, Washington, D.C., 1990.
25. *Report on Airline Service, Fares, Traffic, Load Factors, and Market Share*, a staff study, fourteenth in a series, Civil Aeronautics Board, Washington, D.C., 1981.
26. *Secretary's Task Force on Competition in the U.S. Domestic Airline Industry*, U.S. Department of Transportation, Washington, D.C., February 1990.
27. *Terminal Area Air Traffic Relationships*, fiscal year 1980, Federal Aviation Administration, Washington, D.C.
28. *Terminal Area Forecasts*, Federal Aviation Administration, Washington, D.C., annual.
29. *The Changing Airline Industry: A Status Report through 1979*, Comptroller General of the United States, General Accounting Office, Washington, D.C., September 1980.
30. *The Federal Turnaround on Aid to Airports 1926–38*, The Federal Aviation Administration, Department of Transportation, Washington, D.C., 1973.
31. *Transportation in America*, Frank A. Smith, Eno Foundation for Transportation, Inc., Waldorf, Md., annual.
32. *Travel Market Closeup 1989*, National Travel Survey Tabulations, US Travel Data Center, Washington, D.C., 1990.
33. *Winds of Change, Domestic Air Transport Since Deregulation*, Special Report 230, Transportation Research Board, National Research Council, Washington, D.C., 1991.
34. *Worldwide Airport Traffic Report*, Airports Association Council International, Inc., Washington, D.C., annual.
35. *National Plan of Integrated Airport Systems (NPIAS) 2009–2013*, Federal Aviation Administration, Washington D.C., 2008.
36. *Airport Planning and Management*, Alex T. Wells and Seth B. Young, McGraw Hill, 2003.

Web References

Bureau of Transportation Statistics: http://www.bts.gov
Federal Aviation Administration: http://www.faa.gov
International Civil Aviation Organization: http://www.icao.org

CHAPTER 2

Aircraft Characteristics Related to Airport Design

One of the great challenges for airport planning and design is creating facilities that accommodate a very wide variety of aircraft. Aircraft vary widely in terms of their physical dimensions and performance characteristics, whether they be operated for commercial air service, cargo, or general aviation activities.

There are a large number of specifications for which aircraft may be categorized. Depending on the portion of the area of the airport, certain aircraft specifications become more critical. For example, aircraft weight is important for determining the thickness and strengths of the runway, taxiway, and apron pavements, and affects the takeoff and landing runway length requirements at an airport, which in turn to a large extent influences planning of the entire airport property. The wingspan and the fuselage length influence the size of parking aprons, which in turn influences the configuration of the terminal buildings. Wingspan and turning radii dictate width of runways and taxiways, the distances between these traffic ways, and affects the required turning radius on pavement curves. An aircraft's passenger capacity has an important bearing on facilities within and adjacent to the terminal building.

Since the initial success of the Wright Flyer in 1903, fixed-wing aircraft have gone through more than 100 years of design enhancements, resulting in vastly improved performance, including the ability to fly at greater speeds and higher altitudes over larger ranges with more revenue generating carrying capacity (known as *payload*) at greater operating efficiencies. These improvements are primarily the results of the implementation of new technologies into aircraft specifications, ranging from materials from which the airframes are built, to the engines that power the aircraft. Of great challenge to airport planning and design, historically has been to adapt the airport environment to accommodate changes in aircraft physical and performance specifications. For example:

- The introduction of "cabin-class" aircraft, such as the Douglas DC-3, in the mid-1930s resulted in the need for airports to construct longer, paved runways from the shorter grass strips that previously existed.

- The introduction of aircraft equipped with turbofan and turbojet engines in the late 1950s added requirements for longer and stronger runways, facilities to mitigate jet-blast, and policies to reduce the impact of aircraft noise at and around the airport.

- The introduction of "jumbo-jet" or "heavy" aircraft, such as the Boeing-747, in the late 1960s added new requirements for runway specifications, as well as terminal area design requirements for accommodating large volumes of passengers and cargo.

- The proliferation of regional jet aircraft, introduced because of more efficient engine technologies, resulted in the need for airports to modify many terminal areas that had accommodated larger jets or smaller turbo-prop aircraft.

Most recently, the introduction of the world's largest passenger aircraft, the Airbus A-380, as well as the smallest of certified general aviation jet aircraft, continues to affect design specifications of airport airfield and terminal areas.

Table 2-1 provides a summary of some of the important aircraft characteristics of some of the aircraft that make up the world's commercial airline fleet. Many regional airlines use smaller aircraft with less than 50 seats, while the world's major airlines use very large aircraft, with potential configurations for more than 800 seats.

Table 2-2 provides a summary of important aircraft characteristics for common general aviation aircraft. While it should be noted that aircraft designed primarily for air carrier purposes are also often used for general aviation activity (e.g., the Boeing 737 is often configured for personal or business use and marketed as the Boeing Business Jet), most general aviation aircraft are smaller than typical commercial airline aircraft. Some of the aircraft listed in Table 2-2 are part of the fleet of "very light jets" that have emerged into the market since 2007.

Many of the values provided in Tables 2-1 and 2-2 are only approximate and tend to vary by specific model, as well as by each individual operation. For more precise values appropriate references, such

Turboprop Aircraft

Aircraft	Wingspan	Length	MSTOW† (lb)	# Engines	Avg. # Seats	Runway Required (ft)*
Beech 1900c	54'06"	57'10"	16,600	2	19	3,300
Shorts 360	74'10"	70'10"	27,100	2	35	4,300
Dornier 328-100	68'10"	68'08"	27,557	2	30	3,300
SAAB 340B	70'04"	64'09"	28,500	2	37	4,200
AT-42-300	80'06"	74'05"	36,815	2	45	3,600
EMB 120	64'11"	65'7"	26,433	2	30	5,200

Jet Aircraft Less than 100,000 lb MSTOW† (Regional Jets)

Aircraft	Manufacturer	Wingspan	Length	MSTOW† (lb)	# Engines	Avg. # Seats	Runway Required (ft)*
ERJ 135	Embraer	65'9"	86'5"	41,887	2	35	5,800
ERJ 140	Embraer	65'9"	93'4"	44,313	2	40	6,100
ERJ 145	Embraer	65'9"	98'0"	46,275	2	50	7,500
CRJ 200	Bombardier	69'7"	87'10"	51,000	2	50	5,800
CRJ 700	Bombardier	76'3"	106'8"	72,750	2	70	5,500
CRJ 900	Bombardier	81'6"	119'4"	80,500	2	90	5,800

TABLE 2-1 Characteristics of Commercial Service Aircraft

Jet Aircraft Less than 100,000 lb MSTOW† (Regional Jets)

Aircraft	Manufacturer	Wingspan	Length	MSTOW† (lb)	Avg. # Seats	# Engines	Runway Required (ft)*
BAe-RJ70	British Aerospace	86'00"	78'9"	89,999	95	2	4,700
BAe-RJ85	British Aerospace	86'00"	86'11"	92,999	110	2	5,400
Bae-RJ100	British Aerospace	86'00"	94'10"	97,499	110	2	6,000

Jet Aircraft between 100,000 and 250,000 lb MSTOW† (Narrow Body Jets)

Aircraft	Manufacturer	Wingspan	Length	Wheel Base	Wheel Track	MSTOW† (lb)	# Engines	Avg. # Seats	Runway Required (ft)*
A-319	Airbus Industrie	111'25"	111'02"	41'33"	24'93"	141,095	2	140	5,800
MD-87	McDonnell-Douglas	107'10"	130'05"	62'11"	16'08"	149,500	2	135	7,600
MD-90-30	McDonnell-Douglas	107'10"	152'07"	77'02"	16'08"	156,000	2	165	6,800
A-320-200	Airbus Industrie	111'03"	123'03"	41'05"	24'11"	158,730	2	160	5,700
B-737-800	Boeing	112'06"	124'11"	50'09"	18'8"	172,445	2	175	
B-727-200	Boeing	108'00"	153'03"	63'03"	18'09"	184,800	3	165	8,600
B-757-200	Boeing	124'10"	155'03"	60'00"	24'00"	220,000	2	210	5,800

Jet Aircraft Greater than 250,000 lb MSTOW† (Wide Body Jets)

Model	Manufacturer								
A310-300	Airbus Industrie	144'00"	153'01"	49'11"	31'06"	330,690	2	240	7,500
B-767-300	Boeing	156'01"	180'03"	74'08"	30'06"	345,000	2	275	8,000
A-300-600	Airbus Industrie	147'01"	175'06"	61'01"	31'06"	363,765	2	310	7,600
L-1011-500	Lockheed	164'04"	164'03"	61'08"	36'00"	510,000	3	290	9,200
B-777-200	Boeing	199'11"	209'01"	84'11"	36'00"	535,000	2	375	8,700
DC-10-40	McDonnell-Douglas	165'04"	182'03"	72'05"	35'00"	555,000	3	325	9,500
A-340-200	Airbus Industrie	197'10"	195'00"	62'11"	16'09"	558,900	4	320	7,600
DC-10-30	McDonnell-Douglas	165'04"	182'03"	72'05"	35'00"	572,000	3	320	9,290
MD-11	McDonnell-Douglas	170'06"	201'04"	80'09"	35'00"	602,500	3	365	9,800
B-747SP	Boeing	195'08"	184'09"	67'04"	36'01"	630,000	4	315	7,000
B-747-400	Boeing	213'00"	231'10"	84'00"	36'01"	800,000	4	535	8,800

TABLE 2-1 Characteristics of Commercial Service Aircraft (*Continued*)

		Jet Aircraft between 100,000 and 250,000 lb MSTOW[†] (Narrow Body Jets)							
Aircraft	Manufacturer	Wingspan	Length	Wheel Base	Wheel Track	MSTOW[†] (lb)	# Engines	Avg. # Seats	Runway Required (ft)*
B-787-8 Dreamliner	Boeing	197'04"	186'02"	74'09"	32'07"	242,000	2	230	9,600
A-380	Airbus Industrie	261'08"	239'03"	99'08"	46'11"	1,235,000	4	525	10,000

*Runway lengths are takeoff runway length estimates based on sea level elevation, temperature 20°C at maximum takeoff weight. It should be noted that required runway length varies considerably based on aircraft weight and local atmospheric conditions.

[†] MSTOW is maximum structural takeoff weight.

Table 2-1 Characteristics of Commercial Service Aircraft (*Continued*)

Piston and Turbo-Prop Engine Aircraft

Aircraft	Manufacturer	Wingspan	Length	MSTOW (lb)	# Engines	Avg. # Seats	Runway Required*
PA28-Archer	Piper	35'00"	23'09"	2,550	1	4	1,660
DA-40	Diamond	39'06"	26'09"	2,645	1	4	1,198
PA28-Arrow	Piper	35'05"	24'08"	2,750	1	4	1,525
C-182 Skylane	Cessna	35'10"	28'01"	2,950	1	4	1,350
SR20-G2	Cirrus	35'07"	26'00"	3,000	1	4	1,446
SR-22	Cirrus	38'04"	26'00"	3,400	1	4	1,028
PA-32 Saratoga	Piper	36'02"	27'08"	3,600	1	6	1,760
Corvalis 400	Cessna	36'01"	25'02"	3,600	1	4	2,600
DA-42 Twin Star	Diamond	44'06"	28'01"	3,748	2	4	1,130
C-310	Cessna	37'06"	29'07"	5,500	2	6	1,790
BN2B-Islander	Britten-Norman	49'00"	35'08"	6,600	2	9	1,155
C-402c	Cessna	44'01"	36'05"	6,850	2	10	2,195
Cheyenne IIIA	Piper Aircraft	47'08"	43'05"	11,200	2	10	2,400
Super KingAir	Beechcraft	54'06"	43'09"	12,500	2	12	2,600
C-208 Grand Caravan	Cessna	52'01"	41'07"	8,750	1	14	1,500

TABLE 2-2 Characteristics of General Aviation Aircraft

Aircraft	Manufacturer	Wingspan	Length	MSTOW (lb)	# Engines	Avg. # Seats	Runway Required*
Very Light Jet Aircraft							
Mustang	Cessna	43'2"	40"7"	8,645	2	5	3,100
Eclipse 500	Eclipse	33'6"	33'6"	5,995	2	5	2,400
Hondajet	Honda	39'10"	41'8"	9,200	2	5	3,100
Business Jet Aircraft							
Citation CJ1	Cessna	46'11"	42'7"	10,800	2	5	3,300
Citation X	Cessna	56'4"	52'6"	36,400	2	10	3,560
Lear 45 XR	Bombardier	47'9"	57'6"	21,500	2	9	5,040
Lear 60 XR	Bombardier	43'9"	58'6"	23,500	2	9	3,400
Hawker 850 XP	Beechcraft	54'04"	51'02"	28,000	2	8	5,200
G-IV	Gulfstream	77'10"	88'04"	73,200	2	19	5,000
G-550	Gulfstream	93'06"	96'05"	85,100	2	19	5,150

*Runway lengths are takeoff runway length estimates based on sea level elevation, temperature 20°C at maximum takeoff weight. It should be noted that required runway length varies considerably based on aircraft weight and local atmospheric conditions.

TABLE 2-2 Characteristics of General Aviation Aircraft (*Continued*)

as an airplane's characteristics and performance handbook, should be consulted. In particular, the runway length required to operate a particular aircraft, whether it be a takeoff or a landing, can vary considerably based on aircraft engine performance and total operating weight, as well as by the local environmental and atmospheric conditions. Calculation of required runway length is often performed prior to each operation as part of aircraft flight planning, often using tables, charts, or formulas provided by the aircraft manufacturer.

While there have certainly been recent breakthroughs in the introduction of very large aircraft such as the Airbus A-380, the overall trend in aircraft manufactured for civil air transport has focused design on efficiency, rather than the historical goals of increased size. More efficient aircraft may be smaller than older generation aircraft, but their increased efficiencies allow operators to focus on increasing service frequencies. This increase in operating efficiency has also shifted the focus of increasing aircraft speeds, at least in the realm of producing supersonic aircraft (i.e., those that travel at speeds greater than the speed of sound), to more efficient subsonic aircraft. As such production and operation of supersonic aircraft, such as the Concorde, was retired in the early part of the twenty-first century.

Dimensional Standards

Figure 2-1 illustrates some of the terms related to aircraft dimensions that are important to airport planning and design.

The *length* of an aircraft is defined as the distance from the front tip of the *fuselage*, or main body of the aircraft, to the back end of the tail section, known as the *empennage*. The length of an aircraft is used to determine the length of an aircraft's parking area, hangars. In addition for a commercial service airport, the length of the largest aircraft to perform at least five departures per day determines the required amount of aircraft rescue and firefighting equipment on the airfield.

The *wingspan* of an aircraft is defined as the distance from wingtip to wingtip of the aircraft's main wings. The wingspan of an aircraft is used to determine the width of aircraft parking areas and gate spacing, as well as determining the width and separations of runways and taxiways on the airfield.

The *maximum height* of an aircraft is typically defined as the distance from the ground to the top of the aircraft's tail section. Only in rare cases is an aircraft's maximum height found elsewhere on the aircraft, for example, the Airbus Beluga's maximum height is noted as the distance from the ground to the top of the forward fuselage entry door when it is fully extended upward in the open position.

The *wheelbase* of an aircraft is defined as the distance between the center of the aircraft's *main landing gear* and the center of its *nose gear*, or *tail-wheel*, in the case of a tail-wheel aircraft. An aircraft's *wheel track* is defined as the distance between the outer wheels of an aircraft's

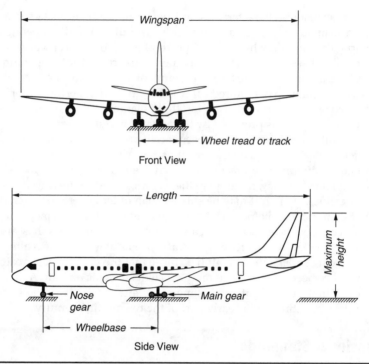

Figure 2-1 Aircraft dimensions.

main landing gear. The wheelbase and wheel track of an aircraft determine its *minimum turning radius,* which in turn plays a large role in the design of taxiway turnoffs, intersections, and other areas on an airfield which require an aircraft to turn.

Turning radii are a function of the nose gear steering angle. The larger the angle, the smaller the radii. From the center of rotation the distances to the various parts of the aircraft, such as the wingtips, the nose, or the tail, result in a number of radii. The largest radius is the most critical from the standpoint of clearance to buildings or adjacent aircraft. The minimum turning radius corresponds to the maximum nose gear steering angle specified by the aircraft manufacturer. The maximum angles vary from 60° to 80°, although for design purposes a steering angle of approximately 50° is often applied.

The turning radius of an aircraft may be expressed using the following formula:

$$R_{180° \text{ turn}} = b \tan (90 - \beta) + t/2 \qquad (2\text{-}1)$$

where b = wheelbase of an aircraft
t = wheel track of the aircraft
β = maximum steering angle

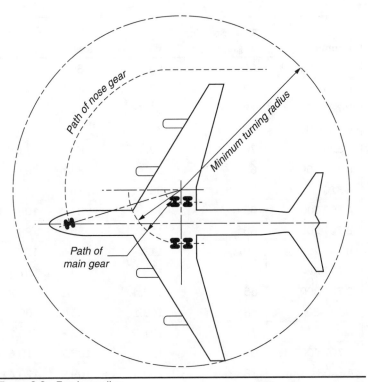

Figure 2-2   Turning radius.

The center of rotation can be easily determined by drawing a line through the axis of the nose gear at whatever steering angle is desired. The intersection of this line with a line drawn through the axes of the two main gears is the center of rotation. Some of the newer large aircraft have the capability of swiveling the main gear when making sharp turns. The effect of the swivel is to reduce the turning radius (Fig. 2-2). Minimum turning radii for some typical transport aircraft are given in Table 2-3.

Landing Gear Configurations

Aircraft currently operating in the world's civil use airports have been designed with various configurations of their landing gear. Most aircraft are designed with one of three basic landing gear configurations; the *single-wheel* configuration, defined as a main gear of having a total of two wheels, one on each strut, the *dual-wheel* configuration, defined as a main gear of having a total of four wheels, two on each strut, and the *dual-tandem* configuration, defined as two sets of wheels on each strut. These configurations are illustrated in Fig. 2-3.

Aircraft	Max. Steering Angle, deg	Radius, ft		
		Wingtips	Nose	Tail
MD-81/83/88	82	65.9	80.7	74.3
MD-90	82	66.5	85.5	74.6
B-737-800	78	69.4	65.4	73.6
B-727-200	78	71	79.5	80
A-320	70	72.2	60	71.9
B-757-200	65	92	84	91
A-310	65	98	75.6	94.9
A-300-600	65	104.9	87.7	108.4
B-767-200	65	112	85	98
B-747-200	70	113	110	125
B-747-SP	70	113	93	97
B-767-300	65	116.4	96.1	108.4
DC-10-30	68	118.1	105	100.8
MD-11	70	121.5	113.8	10.2
B-767-400	65	129.5	108.2	119.6
A-340	78	130.6	109.9	120.4
B-777-300	70	132	125	142
B-787-8	70	132	96.4	111
B-747-400	70	157	117	96

TABLE 2-3   Minimum Turning Radii for Typical Passenger Aircraft

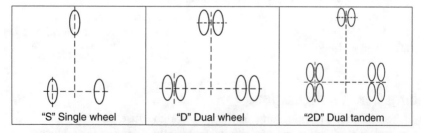

"S" Single wheel	"D" Dual wheel	"2D" Dual tandem

FIGURE 2-3   Traditional landing gear configurations (*Federal Aviation Administration*).

   The landing configurations of the largest of commercial service aircraft have become more complex than the simple configurations illustrated in Fig. 2-3. For example, the Boeing 747, Boeing 777, and Airbus A-380 landing gear configurations are illustrated in Fig. 2-4.

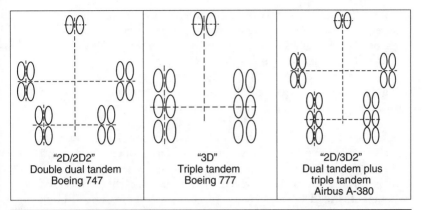

"2D/2D2"	"3D"	"2D/3D2"
Double dual tandem	Triple tandem	Dual tandem plus
Boeing 747	Boeing 777	triple tandem
		Airbus A-380

FIGURE 2-4   Complex landing gear configurations (*Federal Aviation Administration*).

The complexity of landing configurations prompted the FAA to adopt standard naming conventions for aircraft landing gear configurations [60]. Examples of this naming convention are represented in quotes in Figs. 2-3 and 2-4.

The landing gear configuration plays a critical role in distributing the weight of an aircraft on the ground it sits on, and thus in turn has a significant impact on the design of airfield pavements. Specifically, the more wheels on a landing gear, the heavier an aircraft can be and still be supported on a ramp, taxiway, or runway of a given pavement strength.

Aircraft Weight

While the concept of aircraft weight may be thought to be a simple one, the measurement of the weight of a given aircraft is actually relatively complex. An aircraft will in fact be measured with a certain number of weight measurements, depending on its level of loading with fuel, payload, and crew, and assigned maximum allowable weight values for takeoff, landing, and at rest.

These various measurements of aircraft weight are important to airport planning and design, in particular the facilities such as ramps, taxiways, and runways that are designed to support the aircraft.

While it is rare that any two aircraft, even those of the same model and configuration, have the same weight measurements (as there are almost always variations between aircraft in equipment, seating configurations, galleys, and other objects), most manufactures will assign typical weights to their aircraft for planning and design purposes. These weights are as follows.

The "lightest" measure of an aircraft's weight is known as the *operating empty weight* (OEW), the basic weight of the aircraft including crew and all the necessary gear required for flight but not including payload and fuel. The OEW of an aircraft is considered for the design of aircraft that may occupy maintenance hangars, aircraft storage facilities, or any other areas that are not intended to support the weight of an aircraft when loaded with fuel or payload.

The *zero fuel weight* (ZFW) is the OEW of an aircraft plus the weight of its payload. The ZFW is the weight of the aircraft at which all additional weight must be fuel, so that when the aircraft is in flight, the bending moments at the junction of the wing and fuselage do not become excessive. The *payload* is a term which refers to the total revenue-producing load. This includes the weight of passengers and their baggage, mail, express, and cargo. The *maximum structural payload* is the maximum load which the aircraft is certified to carry, whether this load be passengers, cargo, or a combination of both. Theoretically, the maximum structural payload is a difference between the zero fuel weight and the operating empty weight. The maximum payload actually carried is usually less than the maximum structural payload because of space limitations. This is especially true for passenger aircraft, in which seats and other items consume a considerable amount of space.

The *maximum ramp weight* is the maximum weight authorized for ground maneuver including taxi and run-up fuel. As the aircraft taxis between the apron and the end of the runway, it burns fuel and consequently loses weight.

The maximum gross takeoff weight is the maximum weight authorized at brake release for takeoff. It excludes taxi and run-up fuel and includes the operating empty weight, trip and reserve fuel, and payload. The difference between the maximum structural takeoff weight and the maximum ramp weight is very nominal, only a few thousand pounds for the heaviest aircraft. The maximum gross landing weight actually varies with certain atmospheric conditions (namely, air density, which is a function of field elevation and ambient air temperature). This is due to the fact that at times of low air density (such as at high elevations and/or high temperatures), an airplane of a given weight may simply not have the engine power to get takeoff, while at the same weight it may be able to at a higher air density, found at lower elevations and/or lower air temperatures.

The *maximum structural takeoff weight* (MSTOW), is typically designed as the maximum gross takeoff weight for an aircraft operating at sea level elevation at a temperature of 59°F (15°C). It is also the maximum weight that the aircraft's landing gear can support. The MSTOW is the standard design weight measurement used in airport planning and design.

The *maximum structural landing weight* (MLW) is the structural capability of the aircraft in landing. The main gear is structurally designed to absorb the forces encountered during landing; the larger the forces, the heavier must be the gear. Normally the main gears of transport category aircraft are structurally designed for a landing at a weight less than the maximum structural takeoff weight. This is so because an aircraft loses weight en route by burning fuel. This loss in weight is considerable if the journey is long, being in excess of 80,000 lb for large jet transports. It is therefore not economical to design the main gear of an aircraft to support the maximum structural takeoff weight during landing, since this situation will rarely occur. If it does occur, as in the case of aircraft malfunction just after takeoff, the pilot must jettison or burn off sufficient fuel prior to returning to the airport so as not to exceed the maximum landing weight. For short range aircraft, the main gear is designed to support, in a landing operation, a weight nearly equal to the maximum structural takeoff weight. This is so because the distances between stops are short, and therefore a large amount of fuel is not consumed between stops.

On landing, the weight of an aircraft is the sum of the operating weight empty, the payload, and the fuel reserve, assuming that the aircraft lands at its destination and is not diverted to an alternate airport. This landing weight cannot exceed the maximum structural landing weight of the aircraft. The takeoff weight is the sum of the landing weight and the trip fuel. This weight cannot exceed the maximum structural takeoff weight of the aircraft.

Engine Types

Perhaps the most significant contributor to increased aircraft performance has historically come from improvements in aircraft engine technology, from early twentieth-century piston engines to twenty-first-century high-performance jet engine technology.

While there are many makes and models of aircraft engines produced by a number of engine manufacturers, aircraft engine types can generally be placed into three categories, piston engines, turboprops, and turbofan (or jet) engines.

The term *piston engine* applies to all propeller-driven aircraft powered by high-octane gasoline-fed reciprocating engines. Most small general aviation aircraft are powered by piston engines. The term *turboprop* refers to propeller-driven aircraft powered by turbine engines. The term *turbofan* or *jet* has reference to those aircraft which are not dependent on propellers for thrust, but which obtain the thrust directly from a turbine engine. Jet engines are typically powered using a form of diesel fuel, known as Jet-A. While historically jet engines have been used to power larger general aviation and commercial service aircraft, jet engines recently have been increasingly produced for smaller "regional jet" commercial service aircraft, and even smaller "very light jet" general aviation aircraft.

In the early part of the twenty-first century, most of the transport category aircraft in service are equipped with jet engines, and as such, much of the planning and design of airports serving commercial service and business general aviation are based around jet engine aircraft.

Jet engines can be classified into two general categories, turbojet and turbofan. A turbojet engine consists of a compressor, a combustion chamber, and a turbine at the rear of the engine. The early jet airline aircraft, particularly the Boeing 707 and the DC-8, were powered by turbojet engines, but these were discarded in favor of turbofan engines principally because the latter are far more economical.

A turbofan is essentially a turbojet engine to which has been added large-diameter blades, usually located in front of the compressor. These blades are normally referred to as the fan. A single row of blades is referred to as single stage, two rows of blades as multistage. In dealing with turbofan engines reference is made to the bypass ratio. This is the ratio of the mass airflow through the fan to the mass airflow through the core of the engine or the turbojet portion. In a turbofan engines the air flow through the core of the engine, the inner flow, is hot and very compressed and is burned in it. The air flow through the fan, the outer flow, is compressed much less and exits from the engine without burning into an annulus around the inner core. Fan engines are quieter than turbojet engines and the development of quiet propulsive integrated power plants in modern turbofan has included extensive acoustic lining development both in the inlet and the fan exhaust [38].

Most fans are installed in front of the main engine. A fan can be thought of as a small diameter propeller driven by the turbine of the main engine. Nearly all airline transport aircraft are now powered by turbofan. Current technological advances in engines are concentrated toward the development of *propfan* engines for short and medium haul aircraft and ultrahigh bypass ratio turbofan engines for long haul aircraft. These engine technologies reduce fuel consumption by 25 to 35 percent. These engines, which are variously termed *unducted fan* (UDF) engines and *ultrahigh bypass ratio* (UHB) turbofan engines, have brought on the emergence of very light jet aircraft.

Jet engine performance is made in measured both in terms of power and efficiency. The power of an aircraft engine is typically measured in pounds of forward moving force, or "thrust." Table 2-4 lists a sample of jet engines, and their measurements of thrust installed on historical and current transport category aircraft.

Aircraft engine power efficiency is measured in terms of the thrust-to-weight ratio, defined simply as the pounds of thrust provided by the engine, divided by the weight of the engine. Early jet engines were produced with thrust-to-weight ratios of approximately 3:1. In the early part of the twenty-first century, new light but powerful jet engines with thrust to weight ratios nearing 5:1 have significantly improved the operating efficiency of air transport aircraft and have made the emergence of the very light jet market feasible.

One important measure of engine performance efficiency is that of specific fuel consumption, expressed in terms of pounds of fuel per

Engine Family	Manufacturer	Max. Thrust (lb) Aircraft
PW610F	Pratt and Whitney	900 Eclipse 500
PW615F	Pratt and Whitney	1,350 Cessna Mustang
PW617F	Pratt and Whitney	1,700 Embraer Phenom 100
JT8D	Pratt and Whitney	21,000 DC-9, MD-80, SUPER 27
PW6000	Pratt and Whitney	24,000 A318
V2500	Pratt and Whitney	32,000 A-319, A-320, A-321, MD-90
PW2000	Pratt and Whitney	43,000 B-757, C -17, IL-96
JT9D	Pratt and Whitney	56,000 B-747, B-767, A-300, A-310, DC-10
PW4000-94	Pratt and Whitney	62,000 B-747-400, B767-200/300, MD-11, A-300, A-310
PW4000-100	Pratt and Whitney	69,000 A-300-200/300
GP7000	Pratt and Whitney	70,000 A-380
PW4000-112	Pratt and Whitney	98,000 B-777-200/300
RB211-535	Rolls-Royce	43,000 B-757-200/300, Tu-204
Trent 500	Rolls-Royce	56,000 B-340-500/600
RB211-524	Rolls-Royce	61,000 L-1011, B-747-200/400/400/SP/F, B-767-300
Trent 700	Rolls-Royce	71,000 A-330

TABLE 2-4 Turbojet Aircraft Engines

Engine Family	Manufacturer	Max. Thrust (lb) Aircraft
Trent 900	Rolls-Royce	76,000 A-380
Trent 800	Rolls-Royce	95,000 B-777-200/300
CT7	General Electric	2,100 Bell-214ST, Saab 340a
CF34	General Electric	20,000 CRJ-100-200/700/900, ARJ21, EMBRAER 170,175,190,195
CF6	General Electric	72,000 A-300, A-310, A-330
Genx	General Electric	75,000 8787, B-747-800
GE90	General Electric	115,000 B-777-200/ER/LR/300ER
CFM56-5B	GE/International Aerospace	33,000 A-318, A-319, A-320, A-321
CFM56-3	GE/International Aerospace	24,000 A-737-300/400/500
CFM56-2	GE/International Aerospace	24,000 B-707, KC-135
CFM56-7B	GE/International Aerospace	27,000 B-737-600/700/800/900, BBJ
CFM56-5A	GE/International Aerospace	27,000 A-319, A-320
CFM56-5C	GE/International Aerospace	34,000 A-340-200/300
V2500	International Aero	33,000 A-319, A-320, A-321, ACJ, MD-90

TABLE 2-4 Turbojet Aircraft Engines (*Continued*)

Aircraft	Engine	Bypass Radio	Specific Fuel Consumption*
A340	CFM56-5C2	6.4	0.32
B-757	PW2037	6.0	0.33
A-330-300	CF6-80E1A2	5.1	0.33
A320	CFM56-5A1	6.0	0.33
B737-400/500	CFM56-3Ca	6.0	0.33
A-310	PW4152	4.9	0.348
B-767-200	CF6-80A2	4.7	0.35
B-747-400	PW4056	4.9	0.359
B-737-600	CFM56-7B20	5.5	0.36
A-321-200	V2533-A5	4.6	0.37
BA-146-300	LF507	5.6	0.406
MD-80	JT8D-219	1.8	0.519

*Specific Fuel Consumption is the amount of fuel required, in pounds, to create 1 lb of thrust.

TABLE 2-5   Performance Characteristics of Typical Jet Aircraft Engines

hour per pound of thrust. Fuel consumption of jet aircraft engines tends to be expressed in pounds rather than in gallons. This is because the volumetric expansion and contraction of fuel with changes in temperature can be misleading in the amount of fuel which is available. Each gallon of jet fuel weighs about 6.7 lb.

Specific fuel consumption for a particular type of aircraft, defined as the amount of fuel required (in pounds) to create 1 lb of thrust, is a function of its weight, altitude, and speed. Some typical values are given in Table 2-5 merely to illustrate the fuel economy of a turbofan engine particularly at high bypass ratios (a jet engine's *bypass ratio* is defined as the ratio between the mass flow rate of air drawn in by the fan but bypassing the engine core to the mass flow rate passing through the engine core). Significant gains in specific fuel consumption have been made with modern aircraft. Table 2-6 gives the approximate average consumption of fuel for typical aircraft.

Fuel consumption improvements in the last two decades have been significant. New engines, such as the CFM56, CF6, RB211-524D, and PW4000, as well as derivatives of current engines, have resulted in significant fuel economy gains.

An indication of the differences in fuel consumption attained by the various types of passenger aircraft in the different trip modes is given in Table 2-5. It should be pointed out, however, that the data are only indicators of fuel consumption and not productivity. Those aircraft which burn the higher rates of fuel generally are capable of greater speeds and have greater passenger capacity.

Aircraft	Engine	Fuel Consumption, lb/h	Fuel Consumption per Engine lb/h
EMB-145	AE3007A	2,253	1,127
A320-200	CFM56-5A3	4,054	2,027
A-319-100	CFM56-5A4	6,966	3,483
B-737-500	FM56-3B1R	7,879	3,940
B-737-200	JT8B-15A	8,829	4,415
B-757-200	RB211-535E4B	11,109	5,555
B-767-300	CF6-802C2B2F	11,893	5,947
A340-300	CFM-56-5C4	16,093	4,023
B-747-200	RB211-524D4	28,638	7,160

TABLE 2-6   Average Fuel Consumption of Typical Jet Aircraft

As observed in Fig. 2-5, the fuel consumption in gallons per available seat mile decreases with increasing route segment length. This ratio has become increasingly significant to aircraft operations as the price of fuel has increased dramatically in the early part of the twenty-first century. Most significantly for airport planning and

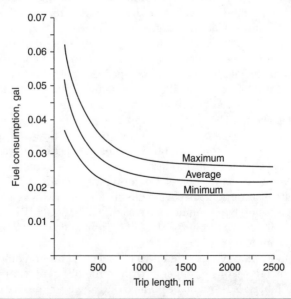

FIGURE 2-5   Fuel consumption in gallons per seat-mile as a function of route distance.

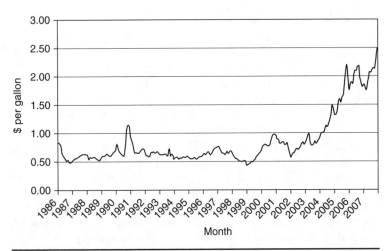

FIGURE 2-6 Jet fuel prices, 1986 to 2007 (*BTS, ATA*).

design, aircraft operators are placing increasing effort into minimizing aircraft operating time at airports, including searching for shorter taxi times between aircraft parking areas and runways, turnaround times at gate areas, and operating in areas where there is reduced congestion in the local airspace.

Recent increases in fuel costs, combined with the efforts of air carriers to reduce other operating expenses, have resulted in fuel being the greatest expense to most air carriers. The historical trends in and the projections for the price of oil and the price of jet fuel for U.S. airlines are shown in Fig. 2-6. The cost of jet fuel per gallon had increased from less than $0.50 in 1987 to nearly $3.50 in 2008 before decreasing to approximately $1.00 per gallon by the end of 2008 (source: BTS, ATA), further motivating the aircraft industry to engineer more efficient engine propulsion and aircraft technologies and for aircraft operators and airport planners to create environments that allow for more efficient operations.

Atmospheric Conditions Affecting Aircraft Performance

Just as they vary in dimensional characteristics, the current fleet of civil use aircraft varies widely in their respective abilities to fly at certain speeds and altitudes over certain distances, the runway lengths required to safely perform landing and takeoff operations, as well as in the amount of noise emissions and energy consumption. Many of these variations are not only functions of the aircraft themselves but in the varying environments at which they operate.

To fully understand the varying performance characteristics of aircraft, it is necessary to understand certain elements the environment in which they operate.

Air Pressure and Temperature

Since aircraft are designed to operate in the altitudes of the earth's atmosphere from sea level to nearly 50,000 ft above sea level, it is important to understand the characteristics of the atmosphere at these altitudes and how altitudes, as well as other atmospheric characteristics, affect aircraft performance.

The performance of all aircraft is affected significantly by the atmospheric conditions in which they operate. These conditions are constantly varying, based simply on the daily heating and cooling of the earth by the sun, and the associated winds and precipitation that occur.

In general, the performance of aircraft depends primarily on the density of the air through which it is operating. The greater the density of the air, the more air molecules flow over the wings, creating more lift, allowing the aircraft to fly. As air density decreases, aircraft require larger airspeed to maintain lift. For airport design, for example, this translates to longer runway length requirements when air is less dense. The density of the air is primarily a function of the air pressure, measured in English units as inches of mercury (inHg) and in metric units as millibars (mb) or hectopascals.

Air density is affected by air pressure and air temperature. As air pressure decreases, there are less air molecules per unit volume and thus air density decreases. As air temperature increases, the velocity and thus spacing between air molecules increases, thus reducing air density.

While these characteristics of the atmosphere vary from day to day and from place to place, for practical convenience for comparing the performance of aircraft, as well as for planning and design of airports, a *standard atmosphere* has been defined. A *standard atmosphere* represents the average conditions found in the actual atmosphere in a particular geographic region. Several different standard atmospheres are in use, but the one most commonly used is the one proposed by ICAO.

In the standard atmosphere it is assumed that from sea level to an altitude of about 36,000 ft, known as the *troposphere*, the temperature decreases linearly. Above 36,000 to about 65,000 ft, known as the *stratosphere*, the temperature remains constant; and above 65,000 ft, the temperature rises. Many conventional jet aircraft fly as high as 41,000 ft. The supersonic transports flew at altitudes on the order of 60,000 ft or more.

In the troposphere the standard atmosphere is defined as follows:

1. The temperature at sea level is 59°F or 15°C. This is known as the *standard temperature* at sea level.

2. The pressure at sea level is 29.92126 inHg or 1015 mb. This is known as the *standard pressure* at sea level.

3. The temperature gradient from sea level to the altitude at which the temperature becomes −69.7°F is 3.566°F per thousand feet. That is, for every increase in altitude of 1000 ft, the temperature decreases by approximately 3.5°F or 2°C.

Both standard pressure and standard temperature decrease with increasing altitude above sea level. The following relation establishes the standard pressure in the troposphere up to a temperature of –69.7°F.

$$\frac{P_0}{P} = \frac{T_0^{5.2561}}{T}$$ (2-2)

where P_0 = standard pressure at sea level (29.92 inHg)
  P = standard pressure at a specified altitude
  T_0 = standard temperature at sea level (59°F)
  T = standard temperature at a specified altitude

In the above formula, the temperature is expressed in "absolute" or Rankine units. Absolute zero is equal to –459.7°F, 0°F is equal to 459.7°R, and 59°F is equal to 518.7°R.

Using these criteria, the standard temperature at an altitude of 5000 ft is 41.2°F, and the standard pressure is 24.90 inHg. Table 2-7 contains a partial listing of standard temperatures and pressures. It is common to refer to *standard conditions* or *standard day*. A standard

Altitude, ft	Temperature, °F	Pressure, inHg	Speed of Sound, kn
0	59.0	29.92	661.2
1,000	55.4	28.86	658.9
2,000	51.9	27.82	656.6
3,000	48.3	26.82	654.3
4,000	44.7	25.84	652.0
5,000	41.2	24.90	649.7
6,000	37.6	23.98	647.7
7,000	34.0	23.09	645.1
8,000	30.5	22.23	642.7
9,000	26.9	21.39	640.4
10,000	23.3	20.58	638.0
20,000	–12.2	16.89	626.2
30,000	–47.8	13.76	614.1
40,000	–69.7	8.90	589.2
50,000	–69.7	7.06	576.3
60,000	–69.7	6.41	573.3

TABLE 2-7 Table of Standard Atmospheres

condition is one in which the actual temperature and pressure correspond to the standard temperature and pressure at a particular altitude. When reference is made to the temperature being "above standard" it means that the temperature is higher than the standard temperature.

As aircraft takeoff performance data is typically related to the local barometric pressure and ambient air temperature, which in turn affects the density of the air, a defined value known as *density altitude* is often used to estimate the density of the air at any given time. Density altitude is a function of the effect of barometric pressure on air density, defined through the measurement known as pressure altitude, and the ambient temperature.

Assuming that at a standard day at sea level, where the elevation above sea level is effectively 0, the density altitude on a standard day would also be 0. If the barometric pressure was less than the standard pressure of 29.92 inHg, the pressure altitude would be greater than 0. Conversely, if the barometric pressure was greater than standard pressure, the pressure altitude would be less than 0. This relates to the fact that, when the atmospheric pressure drops, the air becomes less dense, requiring a longer run on the ground to obtain the same amount of lift as on a day when the pressure is high. Thus a reduction in atmospheric pressure at an airport has the same effect on its air density as if the airport had been moved to a higher elevation. *Pressure altitude* is defined as the altitude corresponding to the pressure of the standard atmosphere. Thus if the atmospheric pressure is 29.92 inHg, the pressure altitude is 0. If the pressure drops to 28.86 inHg, the pressure altitude is 1000 ft. This can be obtained from the formula relating pressure and temperature. If this lower pressure occurred at a sea level airport, the geographic altitude would be 0, but the pressure altitude would be 1000 ft. For airport planning purposes, it is satisfactory to assume that the geographic and pressure altitudes are equal unless the barometric pressures at a particular site are unusually low a great deal of the time.

Density altitude is defined as pressure altitude adjusted for temperature. Similar to the effect of barometric pressure on aircraft performance, if the temperature of the air was greater than standard temperature, the density of the air would be lower and the density altitude would increase, and if the temperature were lower than standard, the density altitude would decrease. It is because of the effect of both barometric pressure and ambient air temperature on aircraft performance that airports located at high elevations, where air pressure is generally lower than at sea level, and in locations where the ambient air temperature often rises well above 59°F, are airports constructed with longer runways, as longer runways are required for aircraft to reach needed airspeeds to get sufficient lift for takeoff, than at sea level elevations, or when temperatures are lower.

Wind Speed and Direction

Since aircraft depend on the velocity of air flowing over their wings to achieve lift, and fly through streams of moving air, similar to ships moving along water with currents, the direction and speed of wind, both near the surface of airports and at altitudes have great effect on aircraft performance.

As winds primarily affect the speed at which aircraft operate at an airport, it is important to understand the basic difference between two ways of measuring speed in an aircraft, *groundspeed* and *airspeed*. The groundspeed is the speed of the aircraft relative to the ground. True airspeed is the speed of an aircraft relative to the air flowing over the airfoil, or wing. For example, if an aircraft is flying at a groundspeed of 500 kn in air where the wind is blowing in the opposite direction, known as a *headwind*, at a speed of 100 kn, the true airspeed is 600 kn. Likewise, if the wind is blowing in the same direction, a *tailwind*, and the aircraft maintained a groundspeed of 500 kn, the true airspeed would be 400 kn.

On the airport surface, the speed and direction of winds directly affect aircraft runway utilization. For takeoff and landings, for example, aircraft perform best when operating with the wind blowing directly toward them, that is, with a direct *headwind*. Headwinds allow an aircraft to achieve lift at slower groundspeeds, and thus allow takeoffs and landings with slower groundspeeds and shorter runway lengths. While wind blowing from behind an aircraft, that is, a *tailwind* is preferable for aircraft flying at altitude, as they achieve greater groundspeeds at a given airspeed, it is not preferable for takeoff or landing, for precisely the same reason. As such, airports tend to plan and design runways so that aircraft may operate most often with direct headwinds, and orient their primary runways in the direction of the prevailing winds.

It is not very often the case that aircraft fly into a direct *headwind* or *tailwind*. Moreover, it is quite common for an aircraft to takeoff or land from an airport at such a time when the runways are not oriented directly into the existing wind. When this situation occurs, aircraft performance takes into consideration any effect of what are known as *crosswinds*.

While operating in direct headwind, tailwind, or calm conditions, the direction toward which an aircraft is pointing, or *heading*, is the same direction as the aircraft is actually traveling, or *tracking* over the ground. However, when operating with a crosswind, the aircraft heading is different than its track. A common analogy to this situation is the swimmer swimming across a river with a swift current. Even though such a swimmer may be pointing directly to the opposite shore of the river, he or she may end up farther downstream than simply straight across the river, and to end up directly across the river, the swimmer would have to point, or head, at some angle upstream.

Aircraft navigating a route at altitude operate in precisely the same manner. A heading is calculated, based on the speed and direction of the wind, and the speed of the aircraft itself, that will give the aircraft the desired track. The angle between the desired track and the calculated heading is known as the *crab angle*. The magnitude of this angle can be obtained from the following relation:

$$\sin x = \frac{V_c}{V_h} \tag{2-3}$$

where V_c is the crosswind in miles per hour or knots and V_h is the true airspeed in miles per hour or knots.

The crosswind, V_c, is defined as the component of the wind, V_w, that is at a right angle to the track. The angle x is referred to as the *crab angle*. It will be noted that the magnitude of the angle is directly proportional to the speed of the wind and inversely proportional to the speed of the aircraft.

As an aircraft approaches a runway, its heading (direction in which the nose is pointing) is of course also dependent on the strength of the wind traveling across the path of the aircraft (crosswind). The approach flight path to the runway is an extension of the centerline of the runway. An aircraft must fly along this track to safely reach the runway. The relation between track, heading, and crosswind is illustrated in Fig. 2-7. In order not to be blown laterally off the track by the wind, the aircraft must fly at an angle x from the track. This means that when the aircraft is moving slowly, as it does when it approaches a runway, and there is a strong crosswind, the angle x will be large. The term V_t is the true airspeed along the track and is equal to $V_h \cos x$. To obtain the groundspeed along the track, the component of the wind along the track must be subtracted from V_t. In the diagram the groundspeed along the track is equal to V_t minus the wind along the track, $V_w \sin x$. For example, assume that an aircraft was approaching a runway at a speed of 135 kn and the crosswind was 25 kn. The

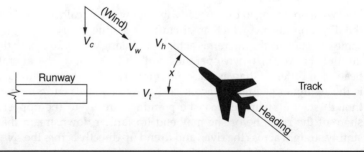

Figure 2-7   Crosswind correction.

crab angle x would be 10°10'. This crab angle is reduced to 0 just prior to touchdown, so that the aircraft is appropriately pointed straight down the center of the runway.

While aircraft operators are trained to safely operate aircraft in these crosswind conditions, it is clearly desirable to minimize this occurrence. Furthermore, the physical ability of an aircraft to properly land in crosswind conditions is limited by the aircraft's weight, landing speed, and existing winds. Often times, small aircraft cannot safely land if crosswinds on a runway are too great. For this reason, airports accommodating smaller, slower aircraft are often designed with runways in several directions, to accommodate varying wind conditions. As opposed to the *primary* runways that are oriented into the prevailing winds, *crosswind* runways are oriented into the direction of winds occurring less frequently.

The FAA categorizes aircraft by the airspeeds at which they make approaches to land at an airport, known as the Aircraft Approach Category, and provides requirements to airports that runways be provided that allow for safe operation of the aircraft that use the airport for at least 95 percent of the annual wind conditions at the airport. The design process for estimating the number and orientation of primary, as well as crosswind runways based on the approach category of selected aircraft is detailed in Chap. 6 of this book.

Aircraft Performance Characteristics

Aircraft Speed

Reference is made to aircraft speed in several ways. Aircraft performance data is typically made reference two airspeeds, namely, *true airspeed* (TAS) and *indicated airspeed* (IAS). The pilot obtains his speed from an airspeed indicator. This indicator works by comparing the dynamic air pressure due to the forward motion of the aircraft with the static atmospheric pressure. As the forward speed is increased so does the dynamic pressure. The airspeed indicator works on the principle of the pitot tube. From physics it is known that the dynamic pressure is proportional both to the square of the speed and to the density of the air. The variation with the square of the speed is taken care of by the mechanism of the airspeed indicator, but not the variation in density. The indicator is sensitive to the product of the density of the air and the square of the velocity. At high altitudes the density becomes smaller and thus the indicated airspeed is less than the true airspeed.

If the true airspeed is required, it can be found with the aid of tables. As a very rough guide, one can add 2 percent to the indicated speed for each 1000 ft above sea level to obtain true airspeed.

The indicated airspeed is of more importance to the pilot than is the true airspeed. The concern is with the generation of lift, specifically the stall speed, the speed at which there is not enough airflow over the wings to sustain lift, which is dependent on speed and air density. At high altitudes an aircraft will stall at a higher speed than it does at sea level. At higher altitudes, however, the airspeed indicator is indicating speeds lower than true speeds; consequently this is on the safe side and no corrections are necessary. Thus, an aircraft with a stalling speed of 90 kn will stall at the same indicated airspeed regardless of altitude. This is why aircraft manufacturers always report stalling speeds in terms of indicated airspeed rather than true airspeed. With the introduction of jet transports and high speed military aircraft, the reference datum for speed is often the speed of sound. The speed of sound is defined as *Mach 1* (after Ernst Mach, Austrian scientist). Thus Mach 3 means three times the speed of sound. Most of our current jet transports are *subsonic* (slower than the speed of sound) and cruise at a speed in the neighborhood of 0.8 to 0.9 Mach. Many military aircraft are *supersonic* (faster than the speed of sound). Again the reader is reminded that when the maximum speed of an aircraft is quoted as 0.9 Mach, this is in terms of true airspeed and not groundspeed. Such an aircraft can conceivably be traveling at a groundspeed higher than the speed of sound, depending on the magnitude of the tailwind.

The speed of sound is not a fixed speed; it depends on temperature and not on atmospheric pressure. As the temperature decreases, so does the speed of sound. The speed of sound at 32°F (0°C) is 742 mi/h (1090 ft/s), at −13°F (−25°C) it is 707 mi/h, and at 86°F (30°C) it is 785 mi/h. In fact, the speed of sound varies 2 ft/s for every change in temperature of 1°C above or below the speed at 0°C. The speed of sound at the altitudes at which jets normally fly is less than 700 mi/h, but at altitudes at which small aircraft normally fly (20,000 ft or less) it is greater than 700 mi/h.

The speed of sound may be computed from the formula

$$V_{sm} = 33.4T^{0.5} \tag{2-4}$$

$$V_{sf} = 49.04T^{0.5} \tag{2-5}$$

where V_{sm} = speed of sound in miles per hour at some temperature
V_{sf} = speed of sound in feet per second at some temperature
T = temperature in degrees Rankine

For convenience in navigation, aircraft distances and speeds are measured in nautical miles and knots, just like measurement on the high seas. One nautical mile (6080 ft) is practically equal to 1 min of arc of the earth's circumference. One knot is defined as 1 nmi/h. One nautical mile is approximately 1.15 land miles.

The performances of aircraft are, in part, defined by the various speeds at which they can safely liftoff, cruise, maneuver, and approach to land. These speeds are defined in aircraft performance manuals as *V-speeds*. Such V-speeds include:

V_{ne}: Do-Not-Exceed Speed, the fastest an aircraft may cruise in smooth air to maintain safe structural integrity.

V_a: Design Maneuvering Speed, the recommended speed for an aircraft performing maneuvers (such as turns) or operating in turbulent air.

V_{lo}: Liftoff Speed, the recommended speed at which the aircraft can safely liftoff.

V_r: Rotate Speed, the recommended speed at which the nose wheel may be lifted off the runway during takeoff.

V_1: Decision Speed, the speed at which, during a takeoff run, the pilot decides to continue with the takeoff, even if there might be an engine failure from this point before takeoff. If an aircraft develops an engine issue prior to reaching V_1, the pilot will abort the takeoff.

V_{so}: Stall Speed (landing configuration), the minimum possible speed for an aircraft in landing configuration (landing gear down, flaps extended) to maintain lift. If the aircraft's airspeed goes below V_{so}, the airplane loses all lift and is said to *stall*. This speed is also typically the speed at which an aircraft will touch down on a runway during landing.

V_{ref}: Reference Landing Approach Speed, the speed at which an aircraft travels when on approach to landing. V_{ref} is typically calculated as $1.3 \times V_{so}$.

For airport planning and design, many of these speeds contribute to determining required runway lengths for takeoff and landing, as well as in determining the maximum number of operations (i.e., the capacity) that can be performed on runways over a given period of time.

Payload and Range

The maximum distance that an aircraft can fly, given a certain level of fuel in the tanks is known as the aircraft's *range*. There are a number of factors that influence the range of an aircraft, among the most important is payload. Normally as the range is increased the payload is decreased, a weight trade-off occurring between fuel to fly to the destination and the payload which can be carried.

The relationship between payload and range is illustrated in Fig. 2-8. The point A, the range at maximum payload, designates the farthest distance, R_a, that an aircraft can fly with a maximum structural payload. To fly a distance of R_a and carry a payload of P_a the aircraft has to take off at its maximum structural takeoff weight; however, its fuel tanks are not completely filled. Point B, the range at maximum

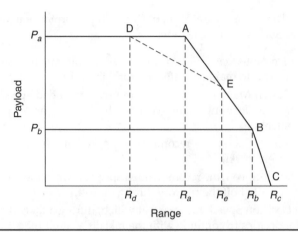

FIGURE 2-8   Typical relationship between payload and range.

fuel, represents the farthest distance, R_b, an aircraft can fly if its fuel tanks are completely filled at the start of the journey. The corresponding payload that can be carried is P_b. To travel the distance R_b, the aircraft must take off at its maximum structural takeoff weight. Therefore to extend the distance of travel from R_a to R_b the payload has to be reduced in favor of adding more fuel. Point C represents the maximum distance an aircraft can fly without any payload. Sometimes this is referred to as the *ferry range* and is used, if necessary, for delivery of aircraft. To travel this distance R_c, the maximum amount of fuel is necessary, but since there is no payload, the takeoff weight is less than maximum. In some cases the maximum structural landing weight may dictate how long an aircraft can fly with a maximum structural payload. If this is the case, the line DE represents the trade-off between payload and range which must occur since the payload is limited by the maximum structural landing weight. The shape of the payload versus range curve would then follow the line DEBC instead of ABC. Payload versus range depends on a number of factors such as meteorological conditions en route, flight altitude, speed, fuel, wind, and amount of reserve fuel. For performance comparison of different aircraft in an approximate way the payload range curves are usually shown for standard day, no wind, and long range cruise.

The actual payload, particularly in passenger aircraft, is normally less than the maximum structural payload even when the aircraft is completely full. This is due to the limitation in the use of space when passengers are carried. For computing payload, passengers and their baggage are normally considered as 200 lb units.

The aircraft manufacturers publish payload versus range diagrams in aircraft characteristics manuals for each aircraft which may be used for airport planning purposes. These diagrams are most useful in airport planning for determining the most probable weight characteristics of aircraft flying particular stage lengths between airports.

The distribution of the load between the main gears and the nose gear depends on the type of aircraft and the location of the center of gravity of the aircraft. For any gross weight there is a maximum aft and forward center of gravity position to which the aircraft can be loaded for flight in order to maintain stability. Thus the distribution of weight between the nose and main gears is not a constant. For the design of pavements it is normally assumed that 5 percent of the weight is supported on the nose gear and the remainder on the main gears. Thus if there are two main gears, each gear supports 47.5 percent of the total weight. For example, if the takeoff weight of an aircraft is 300,000 lb, each main gear is assumed to support 135,000 lb. If the main gear has four tires, it is assumed that each tire supports an equal fraction of the weight on the gear, in this example, 33,750 lb. As will be discussed in Chap. 7, pavement strengths are designed based on the maximum structural takeoff weights, as well as the landing gear and loading configurations, of the aircraft of intended use.

Runway Performance

One of the most critical elements of aircraft performance is how such characteristics, along with local atmospheric conditions, affect the runway length for an aircraft to safely takeoff and land.

For any given operation, whether it be a takeoff or landing, an aircraft will require a certain amount of runway. Required runway length may vary widely for a specific aircraft, as a result of the aircraft's weight at the time of the operation, as well as the local atmospheric conditions. For the airport planner and designer, such variations have less direct impact on the design length of runways, and more to aircraft operators who must determine whether the length of a runway at a given time is safe for a particular operation. Nevertheless, the airport planner and designer should be aware of how an aircraft's performance characteristics specifically affect its runway length requirements.

The factors which have a bearing on and aircraft's runway length requirements for a given operations may be grouped into two general categories:

1. The physical capabilities of the aircraft under given environmental conditions

2. Requirements set by the government to protect for safe operations

An aircraft's performance capabilities and hence runway length requirements are often significantly affected by certain natural environmental conditions at the airport. The more important of these conditions are temperature, surface wind, runway gradient, altitude of the airport, and condition of the runway surface.

Field Elevation

All other things being equal, the higher the field elevation of the airport, the less dense the atmosphere, requiring longer runway lengths for the aircraft to get to the appropriate groundspeed to achieve sufficient lift for takeoff. This increase is not linear but varies with the weight of the aircraft and with the ambient air temperature.

At higher altitudes the rate of increase is higher than at lower altitudes. For planning purposes, it can be estimated that between sea level and 5000 ft above sea level, runway lengths required for a given aircraft increases approximately 7 percent for every 1000 ft of increase in elevation, and greater under very hot temperatures those that experience very hot temperatures or are located at higher altitudes, the rate of increase can be as much as 10 percent. Thus, while an aircraft may require 5000 ft of runway to takeoff at an airport at sea level, the same aircraft may require 7500 ft or more at an airport 5000 ft above sea level, especially during periods of high temperatures.

Surface Wind

Wind speed and direction at an airport also have a significance on runway length requirements. Simply, the greater the headwind the shorter the runway length required, and the greater the tailwind the longer the runway required. Further, the presence of crosswinds will also increase the amount of runway required for takeoff and landing. From the perspective of the planner, it is often estimated that for every 5 kn of headwind, required runway length is reduced by approximately 3 percent and for every 7 kn of tailwind, runway length requirements increase by approximately 7 percent. For airport planning purposes runway lengths are often designed assuming calm wind conditions.

Runway Gradient

To accommodate natural topographic or other conditions, runways are often designed with some level of slope or *gradient*. As such, aircraft operating for takeoff on a runway with an uphill gradient requires more runway length than a level or downhill gradient, the specific amount depending on elevation of the airport and temperature. Conversely, landing aircraft require less runway length when landing on a runway with an uphill gradient, and more length for a downhill gradient.

Studies that have been made indicate that the relationship between uniform gradient and increase or decrease in runway length is nearly linear [55]. For turbine-powered aircraft this amounts to 7 to 10 percent for each 1 percent of uniform gradient. Airport design criteria limit the gradient to a maximum of 1½ percent. Information provided by aircraft manufacturers in flight manuals is based on uniform gradient, yet most runway profiles are not uniform. In the United States aircraft operators are allowed to substitute an average uniform gradient, which is a straight line joining the ends of the runway, as long as no intervening point along the actual path profile lies more than 5 ft above or below the average line. Fortunately most runways meet this requirement. For airport planning purposes only, the FAA uses an effective gradient. The effective gradient is defined as the difference in elevation between the highest and lowest points on the actual runway profile divided by the length of the runway. Studies indicate that within the degree of accuracy required for airport planning there is not very much difference between the use of the average uniform gradient and effective gradient.

Condition of Runway Surface

Slush or standing water on the runway has an undesirable effect on aircraft performance. Slush is equivalent to wet snow. It has a slippery texture which makes braking extremely poor. Being a fluid, it is displaced by tires rolling through it, causing a significant retarding force, especially on takeoff. The retarding forces can get so large that aircraft can no longer accelerate to takeoff speed. In the process slush is sprayed on the aircraft, which further increases the resisting forces on the vehicle and can cause damage to some parts. Considerable experimental work has been conducted by NASA and the FAA on the effect of standing water and slush. As a result of these tests, jet operations are limited to no more than ½ in of slush or water. Between ¼ and ½ in depth, the takeoff weight of an aircraft must be reduced substantially to overcome the retarding force of water or slush. It is therefore important to provide adequate drainage on the surface of the runway for removal of water and means for rapidly removing slush. Both water and slush result in a very poor coefficient of braking friction. When tires ride on the surface of the water or slush the phenomenon is known as hydroplaning. When the tires hydroplane, the coefficient of friction is on the order of wet ice and steering ability is completely lost. Hydroplaning is primarily a function of tire inflation pressure and to some extent the condition and type of grooves in the tires. According to tests made by NASA, the approximate speed at which hydroplaning develops may be determined by the following formula:

$$V_p = 10 \, p^{0.5} \tag{2-6}$$

where V_p is the speed in miles per hour at which hydroplaning develops and p is the tire inflation pressure in pounds per square inch.

The range of inflation pressures for commercial jet transports varies from 120 to over 200 lb/in². Therefore the hydroplaning speeds would range from 110 to 140 mi/h or more. The landing speeds are in the same range. Therefore hydroplaning can be a hazard to jet operations. Hydroplaning can develop when the depth of water or slush is on the order of 0.2 in or less, the exact depth depending on tire tread design, condition of the tires, and the texture of the pavement surface. Smooth tread operating on a smooth pavement surface requires the least depth of fluid for hydroplaning.

To reduce the hazard of hydroplaning and to improve the coefficient of braking friction, runway pavements have been grooved in a transverse direction. The grooves form reservoirs for the water on the surface. The FAA is conducting extensive research to establish standards for groove dimensions and shape [54]. In the past the grooves were normally ¼ in wide and deep and spaced 1 in apart [44].

Declared Distances

Transport category aircraft are licensed and operated under the code of regulations known as the Federal Aviation Regulations (FAR). This code is promulgated by the federal government in coordination with industry. The regulations govern the aircraft gross weights at takeoff and landing by specifying performance requirements, known as *declared distances* which must be met in terms related to the runway lengths available. The regulations pertaining to turbine aircraft consider three general cases in establishing the length of a runway necessary for safe operations. These three cases are

1. A normal takeoff where all engines are available and sufficient runway is required to accommodate variations in liftoff techniques and the distinctive performance characteristics of these aircraft

2. Takeoff involving an engine failure, where sufficient runway is required to allow aircraft to continue the takeoff despite the loss of power, or else brake to a stop

3. Landing, where sufficient runway is required to allow for normal variation in landing technique, overshoots, poor approaches, and the like

The regulations pertaining to piston-engine aircraft retain in principal the above criteria, but the first criterion is not used. This particular regulation is aimed toward the everyday, normal takeoff maneuver, since engine failure occurs rather infrequently with turbine-powered aircraft. The runway length needed at an airport by a particular type and weight of turbine-powered aircraft is established by one of the foregoing three cases, whichever yields the longest length.

In the regulations for both piston-engine aircraft and turbine-powered aircraft, the word runway refers to *full-strength pavement* (FS).

Thus, in the discussion which follows, the terms runway and full-strength pavement are synonymous. In any discussion of the effect of the regulations on the length of the runway, however, it is important to note that the current regulations for turbine-powered aircraft do not require a runway for the entire takeoff distance, while the regulations for piston-engine aircraft normally do.

To indicate why there is a difference in the two regulations with regard to the length of full-strength pavement, it is necessary to examine in more detail the regulations pertaining to turbine-powered transports.

These three criteria as defined by the current turbine-powered transport regulations, FAR Part 25 [12] and Part 121 [23], are illustrated in Fig. 2-9.

Figure 2-9a illustrates the required *landing distance*. The regulations state that the *landing distance* (LD) required for an aircraft landing on a given runway must be sufficient to permit the aircraft to come to a full stop, *stop distance* (SD), within 60 percent of this distance, assuming that the pilot makes an approach at the proper speed and crosses the threshold of the runway at a height of 50 ft. The landing distance must be of full-strength pavement. The landing distance for piston-engine aircraft is defined in exactly the same manner.

Figure 2-9c, illustrates the runway length requirements for a normal takeoff with all engines fully operating, Fig. 2-9c defines a *takeoff distance* (TOD), which, for a specific weight of aircraft, must be 115 percent of the actual distance the aircraft uses to reach a height of 35 ft (D35). Not all of this distance has to be of full-strength pavement. What is necessary is that all this distance be free from obstructions to protect against an overshooting takeoff. Consequently the regulations permit the use of a *clearway* (CL) for part of this distance. A clearway is defined as a rectangular area beyond the runway not less than 500 ft wide and not longer than 1000 ft in length, centrally located about the extended centerline of the runway, and under the control of the airport authorities. The clearway is expressed in terms of a clearway plane, extending from the end of the runway with an upward slope not exceeding 1.25 percent above which no object nor any portion of the terrain protrudes, except that threshold lights may protrude above the plane if their height above the end of the runway is not greater than 26 in and if they are located to each side of the runway. Up to one-half the difference between 115 percent of the distance to reach the point of liftoff, *liftoff distance* (LOD), and the takeoff distance may be clearway. The remainder of the takeoff distance must be full-strength pavement and is identified as the *takeoff run* (TOR).

Figure 2-9b illustrates the engine-failure case, described as the case where one engine fails at a critical point during an aircraft takeoff roll, and the pilot makes an immediate judgmental decision whether or not to continue with a takeoff, or perform an emergency stop.

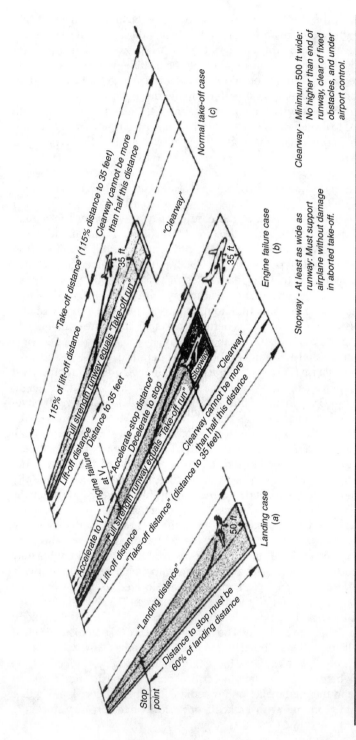

Figure 2-9 Declared distances, balanced field concept.

84

Federal regulations specify that the takeoff distance required during the engine-failure case is the actual distance to reach a height of 35 ft (D35) with no percentage applied as in the all-engine takeoff case. This recognizes the infrequency of occurrence of engine failure. The regulations again permit the use of a clearway, in this case up to one-half the difference between the liftoff distance and the takeoff distance, the remainder being full-strength pavement. The regulations for piston-engine aircraft normally require full-strength pavement for the entire takeoff distance.

The engine-failure case also requires that sufficient distance must also be available to stop the airplane rather than continue the takeoff. The speed at which engine failure is assumed to occur is selected by the aircraft manufacturer and is referred to as the *critical engine-failure speed* or *decision speed*, V_1. If the engine fails at a speed greater than this speed, the pilot has no choice but to continue the takeoff. If an engine actually fails at or prior to this selected speed, the pilot brakes to a stop. This distance required, from beginning of the takeoff roll to the emergency stop is referred to as the *accelerate-stop distance* (DAS). For piston-engine aircraft only full-strength pavement is normally used for this purpose. The regulations for turbine-powered aircraft, however, recognize that an aborted takeoff is relatively rare and permit use of lesser strength pavement, known as *stopway* (SW), for that part of the accelerate-stop distance beyond the takeoff run. The stopway is defined as an area beyond the runway, not less in width than the width of the runway, centrally located about the extended centerline of the runway, and designated by the airport authorities for use in decelerating the aircraft during an aborted takeoff. To be considered as such, the stopway must be capable of supporting the airplane during an aborted takeoff without inducing structural damage to the aircraft. Engineered material arresting systems (EMAS) are being used as for this purpose with increasing frequency.

Based on the above requirements, aircraft operators estimate a required *field length* (FL) for each operation. The field length is generally made up of three components, namely, the full-strength pavement (FS), the partial strength pavement or stopway (SW), and the clearway (CL).

The preceding regulations for turbine-powered aircraft may be summarized for each of the cases in equation form to find the required field length.

Normal takeoff case:

$$FL_1 = FS_1 + CL_{1max} \tag{2-7}$$

where

$$TOD_1 = 1.15(D35_1) \tag{2-7a}$$

$$CL_{1max} = 0.50[TOD_1 - 1.15(LOD_1)] \tag{2-7b}$$

$$TOR_1 = TOD_1 - CL_{1max} \tag{2-7c}$$

$$FS_1 = TOR_1 \tag{2-7d}$$

Engine-failure takeoff case:

$$FL_2 = FS_2 + CL_{2max} \tag{2-8}$$

where

$$TOD_2 = D35_2 \tag{2-8a}$$

$$CL_{2max} = 0.50(TOD_2 - LOD_2) \tag{2-8b}$$

$$TOR_2 = TOD_2 - CL_{2max} \tag{2-8c}$$

$$FS_2 = TOR_2 \tag{2-8d}$$

Engine-failure aborted takeoff:

$$FL_3 = FS + SW \tag{2-9}$$

where

$$FL_3 = DAS \tag{2-9a}$$

Landing case:

$$FL_4 = LD \tag{2-10}$$

where

$$LD = \frac{SD}{0.60} \tag{2-10a}$$

$$FS_4 = LD \tag{2-10b}$$

To determine the required field length and the various components of length which are made up of full-strength pavement, stopway, and clearway, the above equations must each be solved for the critical design aircraft at the airport. This will result in finding each of the following values:

$$FL = \max\,[(TOD_1), (TOD_2), (DAS), (LD)] \tag{2-11}$$

$$FS = \max\,[(TOR_1), (TOR_2), (LD)] \tag{2-12}$$

$$SW = [(DAS) - \max\,(TOR_1, TOR_2, LD)] \tag{2-13}$$

where SW_{min} is zero.

$$CL = \min [(FL - DAS), (CL_{1max}), (CL_{2max})] \qquad (2\text{-}14)$$

where CL_{min} is zero and CL_{max} is 1000 ft.

If operations are to take place on the runway in both directions, as is the usual case, the field length components must exist in each direction. Example Problem 2-1 illustrates the application of these requirements for a hypothetical aircraft.

Example Problem 2-1   Determine the runway length requirements according to the specifications of FAR 25 and FAR 121 for a turbine-powered aircraft with the following performance characteristics:

Normal takeoff:

> Liftoff distance = 7000 ft
> Distance to height of 35 ft = 8000 ft

Engine failure:

> Liftoff distance = 8200 ft
> Distance to height of 35 ft = 9100 ft

Engine-failure aborted takeoff:

> Accelerate-stop distance = 9500 ft

Normal landing:

> Stop distance = 5000 ft

From Eq. (2-4) for a normal takeoff

$$TOD_1 = 1.15 \, D35_1 = (1.15)(8000) = 9200 \text{ ft}$$

$$CL_{1max} = 0.50[TOD_1 - 1.15(LOD_1)] = (0.50)[9200 - 1.15(7000)] = 575 \text{ ft}$$

$$TOR_1 = TOD_1 - CL_{1max} = 9200 - 575 = 8625 \text{ ft}$$

From Eq. (2-5) for an engine-failure takeoff

$$TOD_2 = D35_2 = 9100 \text{ ft}$$

$$CL_{2max} = 0.50(TOD_2 - LOD_2) = 0.50(9100 - 8200) = 450 \text{ ft}$$

$$TOR_2 = TOD_2 - CL_{2max} = 9100 - 450 = 8650 \text{ ft}$$

From Eq. (2-6) for an engine-failure aborted takeoff DAS = 9500 ft

From Eq. (2-7) for a normal landing

$$LD = \frac{SD}{0.60} = \frac{5000}{0.60} = 8333 \text{ ft}$$

Using the above quantities in Eqs. (2-8) through (2-11), the actual runway component requirements become

$$FL = \max [(TOD_1), (TOD_2), (DAS), (LD)]$$

$$= \max [(9200), (9100), (9500), (8333)] = 9500 \text{ ft}$$

$$FS = \max [(TOR_1), (TOR_2), (LD)]$$

$$= \max [(8625), (8650), (8333)] = 8650 \text{ ft}$$

$$SW = [(DAS) - max (TOR_1, TOR_2, LD)]$$

$$= (9500) - max [(8625), (8650), (8333)] = (9500 - 8650) = 850 \text{ ft}$$

$$CL = min [(FL - DAS), CL_{1max}, CL_{2max}]$$

$$= min [(9500 - 9500), 575, 450] = 0 \text{ ft}$$

The above regulations, as illustrated in Example Problem 2-1, are applied at all airports, in the form of *declared distances* for each runway [1, 9]. Declared distances are the distances that are declared available and suitable for satisfying the takeoff run, takeoff distance, accelerate-stop distance and landing distance requirements of aircraft. Four declared distances are commonly reported for each runway. They are the takeoff run available (TORA), takeoff distance available (TODA), accelerate-stop distance available (ASDA), and landing distance available (LDA).

The *takeoff run available* (TORA) is the runway length declared available and suitable for the ground run of an aircraft during takeoff. For Example Problem 2-1, the TORA would be 8650 ft. The *takeoff distance available* (TODA) is the takeoff run available plus the length of any remaining runway and clearway beyond the far end of the takeoff run available. For Example Problem 2-1, the TODA would be 9500 ft. The *accelerate-stop distance available* (ASDA) is the amount of runway plus stopway declared available and suitable for the acceleration and deceleration of an aircraft during an aborted takeoff. For Example Problem 2-1, the ASDA would also be 9500 ft. The *landing distance available* (LDA) is the runway length available and suitable for landing an aircraft. For Example Problem 2-1, the LDA would be 8650 ft.

It is apparent that both the takeoff distance and accelerate-stop distance will depend on the speed the aircraft has achieved when an engine fails.

Since, for piston-engine aircraft, full-strength pavement was normally used for the entire accelerate-stop distance and the takeoff distance, it was the general practice to select V_1, so that the distance required to stop from the point where V_1 was reached was equal to the distance (from the same point) to reach a specified height above the runway. The runway length established on this basis is referred to as the balanced field concept or balanced runway and results in the shortest runway. For turbine-powered aircraft, the selection of V_1 on this basis will not necessarily result in the shortest runway if a clearway or a stopway is provided.

From an airport planning perspective, it is not typical to design a runway's full-strength pavement, stopway, and clearway based on a given aircraft. Rather, for each individual aircraft operation, a V_1 speed is selected which best accommodates the runway on which it will be operating.

For example, for an aircraft operating on a relatively short runway, a lower V_1 may be selected, which will allow for a shorter accelerate-stop distance, but would require at least some clearway to allow for the aircraft to safely climb out to 35 ft. Conversely, for relatively long runways that may have obstacles near the runway's end, or for runways with less full-strength pavement but a stopway at the runway end, a higher V_1 may be selected, to allow for steeper climb-out under engine-failure conditions, and the ability to accommodate a longer accelerate-stop distance.

Thus, one can see that the regulations pertaining to turbine-powered aircraft offer a number of alternatives to the aircraft operator. It should be emphasized that the takeoff distance and the takeoff run for the engine-failure case must be compared with the corresponding distance for the normal all engine takeoff case. The longer distance always governs. A further discussion of these concepts is presented by ICAO [1].

Both aircraft operators and airport planners are interested in clearways, because clearways will, for a fixed available length of runway, allow the operator additional gross takeoff weight with less expense to airport management than building full-strength pavement would require.

Wingtip Vortices

Whenever the wings lift an aircraft, vortices form near the ends of the wings. The vortices are made up of two counter-rotating cylindrical air masses about a wingspan apart, extending aft along the flight path. The velocity of the wind within these cylinders can be hazardous to other aircraft encountering them in flight. This is particularly true if a lighter aircraft encounters a vortex generated by a much heavier aircraft. The tangential velocities in a vortex are directly proportional to the weight of the aircraft and inversely proportional to the speed. The more intense vortices are therefore generated when the aircraft is flying slowly near an airport [52]. The winds created by vortices are often referred to as wake turbulence or wake vortex.

Once vortices are generated they move downward and drift laterally in the direction of the wind. The rate at which vortices settle toward the ground is dependent to some extent on the weight of an aircraft, the heavier the vehicle the faster the vortex will settle. About one wingspan height above the ground the vortices begin to move laterally away from the aircraft, as shown in Fig. 2-10. The duration of a vortex is dependent to a great extent on the velocity of the wind. When there is very little or no wind they can persist for longer than 2 min. As a result of these tests, the FAA and ICAO divide aircraft into three classes for the purposes wake-turbulence separation minima.

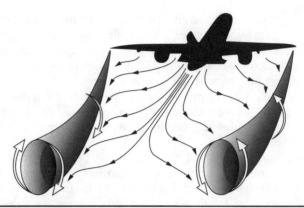

Figure 2-10   An illustration of wake turbulence.

FAA Wake Turbulence Classifications by Aircraft Weight (MSTOW)		ICAO Wake Turbulence Classifications by Aircraft Weight (MSTOM)	
Category	Weight	Category	Weight
Small	≤41,000 lb	Light	≤7,000 kg
Large	41,000–255,000 lb	Medium	7,000–136,000 kg
Heavy	>255,000 lb	Heavy	>136,000 kg

Table 2-8   FAA and ICAO Wake Turbulence Classification

For airport planning and design, as well as air traffic safety purposes, aircraft have been categorized into wake-turbulence classifications, based primarily their maximum structural takeoff weights, as illustrated in Table 2-8. Operating aircraft of varying wake-turbulence classifications in the same vicinity has significant effects on the safe and efficient operation of an airfield.

References

1. *Aerodrome Design Manual, Part 1: Runways,* 2d ed., Document 9157-AN/901, International Civil Aviation Organization, Montreal, Canada, 1984.
2. *Airbus Industrie A300 Airplane Characteristics for Airport Planning,* A.AC E00A, Airbus Industrie, Biagnac, France, October 1987.
3. *Airbus Industrie A300-600 Airplane Characteristics for Airport Planning,* D.AC E00A, Airbus Industrie, Biagnac, France, October 1990.
4. *Airbus Industrie A310 Airplane Characteristics for Airport Planning,* B.AC E00A, Airbus Industrie, Biagnac, France, December 1991.
5. *Airbus Industrie A320 Airplane Characteristics for Airport Planning,* Airbus Industrie, Biagnac, France, February 1988.
6. *Airbus Industrie A340 Airplane Characteristics Airport Planning,* Preliminary, Airbus Industrie, Biagnac, France, July 1991.

7. "Aircraft of the Future," W. E. Parsons and J. A. Stern, *International Air Transportation Meeting*, Paper 800743, Society of Automotive Engineers, Warrendale, Pa., May 1980.

8. "Aircraft Wake Turbulence Avoidance," W. A. McGowan, *12th Anglo-American Aeronautical Conference*, National Aeronautics and Space Administration, Paper 72/6, Washington, D.C., July 1971.

9. *Airport Design*, Advisory Circular, AC150/5300-13 change 13, Federal Aviation Administration, Washington, D.C., 2008.

10. *Air Taxi Operators and Commercial Operators of Small Aircraft, Federal Aviation Regulations*, Part 135, Federal Aviation Administration, Washington, D.C., 1978.

11. *Airworthiness Standards: Normal, Utility, and Acrobatic Category Airplanes, Federal Aviation Regulations*, Part 23, Federal Aviation Administration, Washington, D.C., 1974.

12. *Airworthiness Standards: Transport Category Airplanes, Federal Aviation Regulations*, Part 25, Federal Aviation Administration, Washington, D.C., 1974.

13. *Boeing 707 Airplane Characteristics—Airport Planning*, Document D6-58322, Boeing Commercial Airplane Company, Seattle, Wash., December 1968.

14. *Boeing 727 Airplane Characteristics—Airport Planning*, Document D6-58324-R2, Boeing Commercial Airplane Company, Seattle, Wash., June 1978.

15. *Boeing 737-100/200 Airplane Characteristics-Airport Planning*, Document D6-58325 Revision D, Boeing Commercial Airplane Group, Seattle, Wash., September 1988.

16. *Boeing 737-300/400/500 Airplane Characteristics—Airport Planning*, Document D6-58325-2 Revision A, Boeing Commercial Airplane Group, Seattle, Wash., July 1990.

17. *Boeing 747 Airplane Characteristics—Airport Planning*, Document D6-58326, Rev. E, Boeing Commercial Airplane Group, Seattle, Wash., May 1984.

18. *Boeing 747-400 Airplane Characteristics—Airport Planning*, Document D6-58326-1, Rev. B, Boeing Commercial Airplane Group, Seattle, Wash., March 1990.

19. *Boeing 757 Airplane Characteristics—Airport Planning*, Document D6-58327 Rev D, Boeing Commercial Airplane Group, Seattle, Wash., September 1989.

20. *Boeing 767 Airplane Characteristics—Airport Planning*, Document D6-58328 Rev F, Boeing Commercial Airplane Group, Seattle, Wash., February 1989.

21. *Boeing 777 Airplane Characteristics—Airport Planning*, Preliminary Information, Document D6-58329, Boeing Commercial Airplane Group, Seattle, Wash., February 1992.

22. *British Aerospace 146 Airplane Characteristics for Airport Planning*, APM 146.1, British Aerospace Limited, Hatfield, Hertfordshire, England, June 1984.

23. *Certification and Operations—Domestic, Flag, and Supplemental Air Carriers and Commercial Operators of Large Aircraft, Federal Aviation Regulations*, Part 121, Federal Aviation Administration, Washington, D.C. 1974.

24. *Certification and Operations of Scheduled Air Carriers with Helicopters, Federal Aviation Regulations*, Part 127, Federal Aviation Administration, Washington, D.C., 1974.

25. *Commercial Air Transportation in the Next Three Decades*, H. W. Withington, Boeing Commercial Airplane Company, Seattle, Wash., 1980.

26. *Commercial Air Transportation 1980's and Beyond*, H. W. Withington, Boeing Commercial Airplane Company, Seattle, Wash., November 1980.

27. "CTOL Concepts and Technology Development," D. William Conner, *Astronautics and Aeronautics*, American Institute of Aeronautics and Astronautics, July-August 1978.

28. *CTOL Transport Aircraft Characteristics, Trends, and Growth Projections*, Aerospace Industries Association of America, Inc., Washington, D.C., 1979.

29. *Current Market Outlook*, Boeing Commercial Airplane Group, Seattle, Wash., March 1992.

30. *DC-8 Airplane Characteristics, Airport Planning*, Report DAC-67492, Douglas Aircraft Company, McDonnell-Douglas Corporation, Long Beach, Calif., March 1969.

31. *DC-9 Airplane Characteristics, Airport Planning,* Report DAC-67264, Douglas Aircraft Company, McDonnell-Douglas Corporation, Long Beach, Calif., September 1978.

32. *DC-10 Airplane Characteristics, Airport Planning,* Report DAC-67803A, Douglas Aircraft Company, McDonnell-Douglas Corporation, Long Beach, Calif., January 1991.

33. *Dimensions of Airline Growth,* Boeing Commercial Airplane Company, Seattle, Wash., March 1980.

34. *Energy and Transportation Systems,* Final Report, California Department of Transportation, Sacramento, Calif., December 1981.

35. *Environmental Protection, Annex 16 to the Convention on International Civil Aviation,* vol. 1: Aircraft Noise, 2d ed., International Civil Aviation Organization, Montreal, Canada, 1988.

36. *High Speed Civil Transport,* Program Review, Boeing Commercial Airplane Group, Seattle, Wash., 1990.

37. *Jane's All the World's Aircraft,* Franklin Watts, Inc., New York, annual.

38. *Jet Aviation Development: One Company's Perspective,* John E. Steiner, Boeing Commercial Airplane Group, Seattle, Wash, 1989.

39. "Jet Transport Characteristics Related to Airports," R. Horonjeff and G. Ahlborn, *Journal of the Aerospace Transport Division,* vol. 91 AT1, American Society of Civil Engineers, New York, April 1965.

40. *L1011 Airplane Characteristics,* Airport Planning Document CER-12013, Lockheed California Company, Burbank, Calif., December 1972.

41. *MD-11 Airplane Characteristics for Airport Planning,* Report MDC-K0388, McDonnell-Douglas Corporation, Long Beach, Calif., October 1990.

42. *MD-80 Series Airplane Characteristics for Airport Planning,* Report MDC-J2904, McDonnell-Douglas Corporation, Long Beach, Calif., February 1992.

43. *MD-90-30 Aircraft Airport Compatibility Brochure,* Report MDC-91K0393, McDonnell-Douglas Corporation, Long Beach, Calif., February 1992.

44. *Measurement, Construction and Maintenance of Skid Resistant Airport Pavement Surfaces,* Advisory Circular, AC 150/5320-12A, Federal Aviation Administration, Washington, D.C., July 1986.

45. *Noise Standards: Aircraft Type and Airworthiness Certification, Federal Aviation Regulations,* Part 36, Federal Aviation Administration, Washington, D.C., 1974.

46. *Outlook for Commercial Aircraft 1980-1994,* Douglas Aircraft Company, McDonnell-Douglas Corporation, Long Beach, Calif., June 1980.

47. Pavement Grooving and Traction Studies, NASA SP-5073, *Proceedings of Conference at Langley Research Center,* National Aeronautics and Space Administration, Langley Field, Va., November 1968.

48. "Pneumatic Tire Hydroplaning and Some Effects on Vehicle Performance," W. B. Horne and U. T. Joyner, *Proceedings of the Society of Automotive Engineers,* Paper 97UC, New York, 1965.

49. "Runway Grooving for Increasing Traction—the Current Program and an Assessment of Available Results," W. B. Horne and G. W. Brooks, *20th Annual International Air Safety Seminar,* Williamsburg, Va., December 1967.

50. *Runway Length Requirements for Airport Design,* Advisory Circular AC 150/5325-4B, Federal Aviation Administration, Washington, D.C., January 2005.

51. *Short-Haul Transport Aircraft Future Trends,* Aerospace Industries Association of America, Inc., Washington, D.C., January 1978.

52. "Simulated Vortex Encounters by a Twin-Engine Commercial Transport Aircraft during Final Approach," E. C. Hastings, Jr. and G. L. Keyser, Jr., *International Air Transportation Meeting,* Paper 800775, Society of Automotive Engineers, Warrendale, Pa., May 1980.

53. "Technology Requirements and Readiness for Very Large Vehicles," D. William Conner, *AIAA Very Large Vehicle Conference,* American Institute of Aeronautics and Astronautics, Arlington, Va., April 1979.

54. *The Braking Performance of an Aircraft Tire on Grooved Portland Cement Concrete Surfaces,* S. K. Agrawal and H. Diautolo, Federal Aviation Administration Technical Center, Report FAA-RD-80-78, Federal Aviation Administration, Atlantic City, N.J., January 1981.

55. *The Effect of Variable Runway Slopes on Takeoff Runway Length for Transport Aeroplanes,* ICAO Circular 91-AN/75, International Civil Aviation Organization, Montreal, Canada, 1970.
56. "Trailing Vortex Hazard," W. A. McGowan, *Proceedings of the Society of Automotive Engineers,* Paper 680220, New York, April 1968.
57. *Water, Slush, and Snow on the Runway,* Advisory Circular, AC 91-6A, Federal Aviation Administration, Washington, D.C., May 1978.
58. *Aeronautical Information Manual,* Federal Aviation Administration, Washington, D.C., 2008.
59. *Aircraft Wake Turbulence,* Advisory Circular AC 190-23F, Federal Aviation Administration, Washington, D.C., 2002.
60. *Standard Naming Convention for Aircraft Landing Gear Configurations,* US DOT FAA Order 5300.7, October 6, 2005.

CHAPTER 3

Air Traffic Management

Introduction

In order that the airport planner and designer may be aware of the importance of the rules and technologies that define the aviation operations within the airspace, a very brief summary of what constitutes air traffic control, increasingly being known as air traffic management, how it is managed and operated, and the principal aids to air navigation, is presented in this chapter.

An appreciation of air traffic management and its current and future operating and technological characteristics will focus attention on the fact that any extensive reorientation of runways on existing airports or the construction of entirely new airports requires consultation with the organizations in charge of operating surrounding airspace and very often an airspace study. This is particularly true in large metropolitan areas where several airports are present and the existing airspace must be shared by several airports. In addition, the design of local airspace procedures include procedures for the departure and arrival of aircraft to airport runways requires a fundamental knowledge in current and future air traffic control technologies and policies. Conflicts in air traffic procedures can seriously affect the efficiency of any single airport or a system of airports in a region. The planning of airports must include provisions for facilities located at airports that support the air traffic management system.

As enhanced air traffic management technologies and strategic plans continue to be implemented, consideration of local air traffic procedures has become increasingly relevant and important for even the smallest of airports.

As of 2008, the air traffic management system was just beginning a complete system transformation. As such it is imperative of the airport planner to have an understanding of both the fundamentals of air traffic management and the constant enhancements to the system. This chapter is intended to only briefly introduce the system to the airport planner.

A Brief History of Air Traffic Management

The first attempt to set up rules for air traffic control was made by the International Commission for Air Navigation (ICAN), which was under the direction of the League of Nations. The procedures which the commission promulgated in July of 1922 were adopted by 14 countries. Although the United States was not a member of the League of Nations, and therefore did not officially adopt the rules, many of the procedures established by ICAN were used in the promulgation of air traffic procedures in the United States as well as in most regions of the world.

Construction and operation of the airways system in the United States prior to 1926 were controlled by the military and by the Post Office Department. The formal entry of the federal government into the regulation of air traffic came with the passage of the Air Commerce Act of 1926 (Public Law 64-254). This act directed the Bureau of Air Commerce to establish, maintain, and operate lighted civil airways. At the present time the Federal Aviation Administration maintains and operates the airways system of the United States.

The establishment of the International Civil Aviation Organization (ICAO) in 1944 helped to standardize recommended air traffic control procedures internationally. Today, air traffic control in each country is operated either by its federal government or by private corporations under governmental supervision and regulations. Examples of international air traffic control organizations include the Federal Aviation Administration in the United States, National Air Traffic Services Ltd. (NATS) serving the United Kingdom, NAV Canada in Canada, and Air Services Australia serving the Australian continent. In addition, Eurocontrol, an intergovernmental organization comprising 38 member states within the European Union, coordinates, standardizes, and assists in managing air traffic in the airspace over the European continent.

The primary mission of the Federal Aviation Administration, as well as its international counterparts, is to provide for safe and efficient movement of aircraft throughout the airspace system. The primary function of the air traffic management system is to prevent collisions between aircraft. As such, the FAA office of air traffic management is made up of and responsible for a series of hierarchical control facilities, ground and satellite based navigational aides and aircraft routing procedures, as well as a defined system of air routes and airspace classifications. While much of the current air traffic system is in many ways based on the original development of air traffic control in the early twentieth century, it should be noted that air traffic control policies are constantly changing as the most modern technologies are implemented to better manage increasing air traffic volumes.

The Organizational Hierarchy of Air Traffic Management in the United States

In general, aircraft operate in what is known as the National Airspace System (NAS). The NAS is defined by a series of air routes, airspace classifications, and navigational aids. Aircraft operate within the NAS under varying levels of air traffic control, based primarily on the weather conditions and the type and amount of flight activity within the area. In areas with very low volumes of flight activity during excellent visibility conditions, aircraft may operate in the complete absence of air traffic control, whereas in the busiest airspace or when visibility is limited, aircraft may be under full "positive" control, only being able to change speed, course, or altitude by direct orders from an air traffic controller.

The NAS is operated and managed by a hierarchical organization of air traffic control facilities. The specific purpose of the air traffic control service is to prevent collisions between aircraft and on the maneuvering area between aircraft and obstructions, to expedite and maintain an orderly flow of air traffic [3].

The Air Traffic Control System Command Center

In the United States, air traffic control is managed on a macro level at the air traffic control system command center (ATCSCC) in Herndon, Virginia. In 2007, ATCSCC monitored an average of 25,000 flights per day, with an average of 6000 flights airborne during peak periods. In addition, ATCSCC manages flights planned 6 to 12 h in the future, with the purpose of planning for limiting congestion within the nation's airspace. In doing so, ATCSCC has the authority to implement ground delay programs by dictating certain aircraft to remain at their airports of departure to prevent further congestion in points of the airspace or at airports suffering from delays due to weather or heavy traffic volumes.

Air Route Traffic Control Centers

Air route traffic control centers (ARTCCs) have the responsibility of controlling the movement of en route aircraft along the airways and jet routes, and in other parts of the airspace. Each of the 21 air traffic control centers within the United States has control of a defined geographical area which may be greater than 100,000 mi^2 in size. At the boundary point, which marks the limits of the control area of the center, control of aircraft may be transferred to an adjacent center or an approach control facility, or radar service may be terminated and VFR aircraft are free to contact the next center. Air traffic control centers are normally not located at airports. Air traffic control centers can also provide approach control service to nontowered airports and to nonterminal radar approach control airports.

Each ARTCC geographical area is divided into sectors. The configuration of each sector is based on equalizing the workload of the controllers. Control of aircraft is passed from one sector to another. The geographical area is sectored not only in the horizontal but also in the vertical plane. Thus there can be a high-altitude sector above one or more low-altitude sectors. Each sector is manned by one or more controllers, depending on the volume and complexity of traffic. The average number of aircraft that each sector can handle depends on the number of people assigned to the sector, the complexity of traffic, and the degree of automation provided.

Each sector is normally provided with one or more air route surveillance radar (ARSR) units which cover the entire sector and allow for monitoring of separation between aircraft in the sector. In addition, each sector has information on the identification of the aircraft, destination, flight plan route, estimated speed, and flight altitude, which is posted on pieces of paper called flight progress strips, and are superimposed on the radarscope adjacent to the blips which identify the position and identity of aircraft. The strips are continuously updated as the need arises.

At present, communication between the pilot and controller is by voice. Therefore each ARTCC is assigned a number of VHF and UHF radio communication frequencies. The controller in turn assigns a specific frequency to the pilot. However, modernization of air traffic control is planned to include further proliferation of digital communications, known as controller pilot data link communications (CPDLC) between controllers and pilots.

Terminal Approach Control Facilities

The terminal approach control facility (TRACON) monitors the air traffic in the airspace surrounding airports with moderate to high density traffic. It has jurisdiction in the control and separation of air traffic from the boundary area of the air traffic control tower at an airport to a distance of up to 50 mi from the airport and to an altitude ranging up to 17,000 ft. This is commonly referred to as the terminal area. Where there are several airports in an urban area, one facility may control traffic to all of these airports. In essence the facility receives aircraft from the ARTCC and guides them to one of several airports. In providing this guidance, it performs the important function of metering and sequencing aircraft to provide uniform and orderly flow to the airports.

The organizational structure of an approach control facility is very similar to the ARTCC. Like the ARTCC, the geographic area of the facility is divided into sectors to equalize the workload of the controllers. The approach control facility transfers control of an arriving aircraft to the airport control tower when it is lined up with the runway about 5 mi from the airport. Likewise, control of departing aircraft is transferred to the approach control facility by the airport control tower.

Airport Traffic Control Tower

The airport traffic control tower (ATCT) is the facility which supervises, directs, and monitors the arrival and departure traffic at the airport and in the immediate airspace within 5 mi from the airport. The tower is responsible for issuing clearances to all departing aircraft, providing pilots with information on wind, temperature, barometric pressure, and operating conditions at the airport, and for the control of all aircraft on the ground except in the maneuvering area immediately adjacent to the aircraft parking positions called the ramp area. In the United States in 2007, there were more than 550 air traffic control towers. While most towers are operated by the FAA, as of January 2007, 233 were operated by the private sector under the FAA's contract tower program. The number of operating contract towers, as they are known, has increased tremendously since the inception of the program in 1982. Figure 3-1 provides an illustration of the new ATCT, overshadowing the previously active ATCT, at the Hartsfield-Jackson Atlanta International Airport, Atlanta, Georgia.

FIGURE 3-1   The new ATCT dwarfing the old tower at Hartsfield-Jackson Atlanta International Airport (*ATCmonitor.com*).

Flight Service Stations

While not providing specific control, flight service stations (FSS) are the element of the air traffic management system that provides information and other noncontrol communications to aircraft operating in the system. Their principal functions are to accept and close flight plans, brief pilots about their routes of flight, and to provide important information, in the form of notices to airmen (NOTAMs) before flight and in flight, on such items as severe weather, the status of navigational aids, airport runway closures, and changes in published approach and departure procedures. A secondary function is to relay traffic control messages between aircraft and the appropriate control facility on the ground.

Flight service has gone through a number of changes since the early 1990s. In the 1990s the FAA consolidated more than 180 flight service stations into approximately 60 automated flight service stations (AFSS) which allow many functions, particularly with respect to disseminating weather and other NOTAMs and the filing of flight plans to be performed electronically by voicemail or computer.

In 2005, the FAA awarded a contract to operate the AFSS system to the Lockheed Martin Corporation, representing another step in the privatization of major components of the nation's air traffic control system. While the privatization of the AFSS system has caused some controversy within the aviation industry, there has been relatively little impact of this or any other FAA privatization efforts on airport planning and design.

Air Traffic Management Rules

Air traffic rules are traditionally applied based on prevailing meteorological conditions. Visual meteorological conditions (VMC) are applied when there is sufficient visibility for pilots of aircraft to be able to navigate by referencing locations on the ground, as well as to be able to see and avoid other aircraft in the area. Around airports, VMC is defined as at least 3 statute miles visibility and cloud "ceilings" (defined as at least $\frac{5}{8}$ of the sky covered by clouds) of at least 1000 ft above the ground (AGL). Conversely, instrument meteorological conditions (IMC) exist when visibilities are less than 3 statute miles and cloud ceilings are less than 1000 ft above the ground.

At its most basic level, aircraft operating in VMC tend to fly under visual flight rules (VFR). VFR flight rules depend on aircraft operators to visually maintain adequate separation from terrain, clouds, and other aircraft. Under VFR, aircraft navigation is based on visual reference to locations on the ground, including visual identification and approaches to airports.

While flying under VFR conditions, pilots may request from air traffic control to be under "flight following." Under flight following, air traffic control operators provide assistance to pilots by supervising course and altitude changes, as well as actively notifying pilots of nearby aircraft. Pilots flying under VFR conditions are required to fly under flight following in the busiest of airspace.

Aircraft flying in IMC or at altitudes over 18,000 ft above sea level (AMSL) fly under instrument flight rules (IFR). Aircraft flying under IFR navigate using ground-based and satellite-based navigation aides and are fully controlled along planned routes by air traffic control personnel. Often times, flights operating under IFR will fly defined departure and approach procedures to and from airports which depend on flying precise courses and altitudes to and from waypoints as defined by ground- and satellite-based navigation systems. These published instrument procedures provide for aircraft to safely and efficiently depart from and arrive to airport runways while avoiding collisions with terrain and other aircraft during poor visibility conditions. In many ways, IFR rules, routes, and departure and approach procedures have significant influence on the planning, design, and operation of airports.

Airspace Classifications and Airways

In the United States, domestic airspace is defined into six classes, plus areas with special operating restrictions, and a designated series routes between airports and waypoints. Aircraft are subject to different levels of air traffic control depending on which airspace classification they are currently operating in, the type of defined route they are on, and whether they are flying under VFR or IFR flight rules.

Classes of airspace in the United States are identified alphabetically, as Class A, B, C, D, E, or G airspace, as illustrated in Fig. 3-2.

Class A airspace, also known as positive control airspace, is the airspace between 18,000 ft above mean sea level (AMSL) (known as FL 180) and 60,000 ft (FL 600) AMSL over the 48 contiguous United States and Alaska, extending out to 12 nm off the coast of the United States.

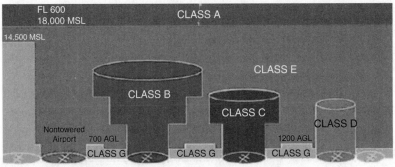

AGL-above ground level FL-flight level MSL-mean sea level   Effective September 16, 1993

Figure 3-2 Illustration of airspace classes.

Since aircraft flying in Class A airspace are generally fast moving commercial airline or general aviation aircraft, all aircraft operating in Class A airspace operate under IFR.

Class B airspace are defined areas within a 30 nm radius around the busiest airports, including areas of multiple large airports, in the United States. Class B airspace surrounds 36 of the busiest commercial service airports in the United States. Class B airspace is typically shaped in the form of what is known in the industry as an "inverted wedding cake." Nearest the busiest airports within the radius of Class B airspace, Class B airspace extends from the surface of the busiest airports in the area to generally 10,000 ft MSL. Farther away from the airport, Class B may begin at some altitude above the surface and extend to 10,000 ft MSL. The purpose of Class B airspace is to provide an area of positive air traffic control to coordinate the many high-speed aircraft transitioning from high altitudes to landing at the busiest airports, and vice versa, with local lower altitude traffic within the area, while providing airspace at lower altitudes further away from the airport to be used with lower levels of control for smaller and slower general aviation aircraft in the region. Aircraft operating within Class B airspace are under positive air traffic control, and as such must either be flying under IFR rules or, with permission from air traffic control, under VFR rules with flight following. An example depiction of Class B airspace is illustrated in Fig. 3-3. This illustration is a portion of an airspace sectional chart, provided by the U.S. Department of Defense and the Federal Aviation Administration as one standard for identifying classes of airspace, airports, navigational aids, and air routes in the NAS.

United States Airspace Class B Areas, centered around the following civil airports:

- PHX   Phoenix Sky Harbor International
- LAX   Los Angeles International
- SAN   San Diego International Lindbergh Field
- SFO   San Francisco International
- DEN   Denver International
- MIA   Miami International
- MCO   Orlando International
- TPA   Tampa International
- HNL   Honolulu International
- ORD   Chicago O'Hare International
- CVG   Cincinnati/Northern Kentucky International
- MSY   Louis Armstrong New Orleans International
- BWI   Baltimore/Washington International

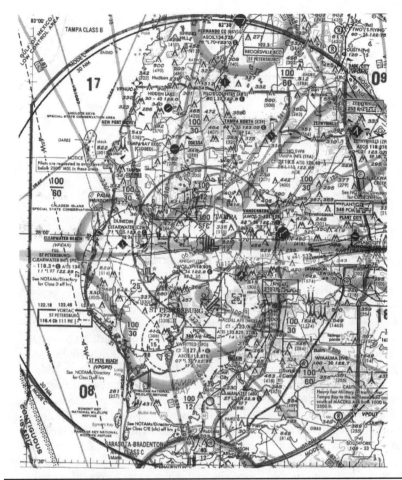

Figure 3-3 Class B airspace around Tampa International Airport, Tampa, Florida.

- BOS   General E. L. Logan International (Boston)
- DTW   Detroit Metropolitan Wayne County
- MSP   Minneapolis-St. Paul International
- MCI   Kansas City International
- STL   Lambert-St. Louis International
- LAS   Las Vegas McCarran International
- EWR   Newark Liberty International
- JFK   John F. Kennedy International
- LGA   New York LaGuardia
- CLT   Charlotte/Douglas International
- CLE   Cleveland-Hopkins International

- CVG   Cincinnati/Northern Kentucky International Airport
- PHL   Philadelphia International
- PIT   Pittsburgh International
- MEM   Memphis International
- DAL   Dallas Love Field
- DFW   Dallas-Fort Worth International
- HOU   Houston William P. Hobby
- IAH   George Bush Intercontinental (Houston)
- SLC   Salt Lake City International
- DCA   Ronald Reagan Washington National
- IAD   Washington Dulles International
- SEA   Seattle-Tacoma International

Class C airspace is found around airports without as much operating volume as those around Class B airspace, but is busy enough to warrant some active level of air traffic control within 10 mi of the airport. VFR traffic operating within Class C airspace must adhere to strict cloud separation requirements and have at least 3 mi of visibility so that they may sufficiently be able to see and avoid other traffic. In addition, all traffic operating within Class C airspace must have established radio communication with air traffic control. The shape of Class C airspace is also in the form of an upside down wedding cake, extending from the surface to typically 4000 ft AGL around the inner 5-nm radius around the airport, and from 1000–2000 ft to 4000 ft AGL from 5 to 10 nm from the airport. Figure 3-4 provides an illustration of Class C airspace surrounding the Daytona Beach International Airport, depicted by a two concentric rings of radii 5 and 10 mi around the airport.

Class D airspace is found within a 5-mi radius of an airport with an operating air traffic control tower, extending from the surface to typically 2500 ft AGL. The purpose of Class D airspace is to provide an area of air traffic control authority to controllers in the airport's control tower, who are responsible for the safe separation of arriving and departing aircraft to and from the airport. Aircraft operating under VFR flight rules are allowed to operate within Class D airspace as along as they establish communication with the air traffic controllers in the tower. When an airport's control tower is in operation, the airport is said to be a "controlled" airport. When the airport's tower is not operational, the airport is considered "uncontrolled" and Class D airspace is no longer active. Airports without a control tower are considered "uncontrolled airports," as well. Figure 3-5 illustrates Class D airspace surrounding the Southwest Georgia Regional Airport in Albany, Georgia, depicted by a dashed 5-mi radius circle around the

FIGURE 3-4   Class C airspace around Daytona Beach International Airport.

airport. The outer shaded ring depicts Class E airspace beginning at 700 ft above ground level.

Class E airspace is found in several locations with the purpose of providing areas of at least "passive" control for airplanes flying in areas of low altitude but moderate traffic activity, on defined air routes,

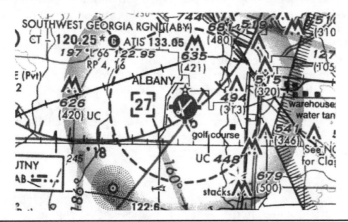

FIGURE 3-5   Class D airspace around Southwest Georgia Regional Airport, Albany, Georgia.

as well as those flying on instrument-based approach and departure procedures to or from an airport. Specifically, Class E airspace exists in most parts of the United States, from the surface, 700 ft AGL, or 1200 ft AGL to 14,500 ft AGL, and 3 nm surrounding the nation's airways.

Any airspace that does not fall within Class A, B, C, D, or E airspace is considered Class G, or uncontrolled airspace. This airspace is found only at very low altitudes (typically less than 700 or 1200 ft AGL) or in rural areas of low volume air traffic. Within Class G airspace, aircraft may move freely as long as there is sufficient visibility (1 mi during day hours, 3 mi during night hours, 5 mi day or night when above 10,000 ft AMSL) to see and avoid other air traffic.

Within the National Airspace System are a number of special use airspace classifications. Some of these define permanent location of special use or restricted activity, others define locations where flight operations are restricted for security or other reasons.

Prohibited areas are defined within the NAS as areas prohibited to any civil aviation activity. These areas are typically defined around highly sensitive locations, such as the White House in Washington, D.C.

Restricted areas are defined within the NAS as areas where regular, but not constant, sensitive operations occur, precluding the safe passage of civil aircraft. These areas, such as around the Kennedy Space Center on the east coast of Florida, will periodically restrict civilian access when sensitive activities are occurring.

Military operations areas (MOAs) are defined as areas with periodic military aviation or other activity. These areas may be entered only by permission from air traffic control, which coordinates with the military for civilian use.

Temporary flight restrictions (TFRs) are defined as areas that temporarily restrict or prohibit most civil aviation operations for reasons of national security. TFRs are implemented with little advance notice for a variety of reasons, ranging from protecting nuclear power facilities, to national sporting events, to the travels of the President of the United States. Oftentimes, the activation of a TFR will have serious impacts on the accessibility of an airport to the aviation system.

Figure 3-6 provides an illustration of multiple classes of airspace within the same region, including Class E airspace under Palatka-Larkin Airport and restricted areas within a military operations area south of the airport. Restricted use airspace presents challenges to airport planners seeking maximum efficiency of air traffic to and from the airfield.

Airways

Aircraft flying from one point to another have traditionally followed designated routes. In the United States these are referred to as victor airways and jet routes. These routes have evolved over time as discussed below.

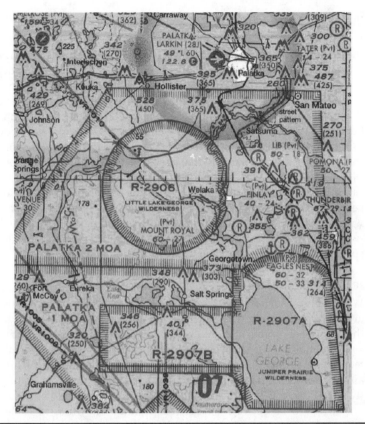

FIGURE 3-6   MOAs and restricted areas south of Kay Larkin Airport, Palatka, Florida.

Colored Airways

The earliest airways, created in the 1920s were initially given a color designation on aeronautical charts and described by their color. The trunk lines east and west were green, trunk lines north and south were amber, secondary lines east and west were red, and secondaries north and south were blue. Each of these colored airways was then given a number, such as green 3, red 4, etc. The numbering for the airways began at the Canadian border and the Pacific Coast, then progressed to the south and east. These airways were then assigned an altitude level, which for green and red was at odd-thousand feet eastbound and at even-thousand feet westbound. On the amber and blue airways northbound, odd-thousand-foot levels were assigned, and southbound even-thousand-foot levels were assigned. These airways were delineated on the ground by low-frequency medium-frequency (LF/MF) four course radio ranges. The colored airways were phased out as aircraft became equipped to use the victor airways in the late 1940s.

Victor Airways

Following the development of the LF/MF four course radio ranges the routes now known as the victor airways were established. The victor airways are delineated on the ground by very high frequency omnirange radio equipment (VORs). Each VOR station has a discrete radio frequency to which a pilot could tune a navigational radio and thus be able to maintain a course from one VOR to the next. The numbering system for these airways is even numbers east and west, odd numbers north and south. The advantages of the victor airways were that the VORs were relatively free of static and it is much easier for a pilot to determine air position relative to a VOR station than with the LF/MF four course radio range. Victor airways are designated on aeronautical charts as V-1, V-2, etc. The airway includes the airspace within parallel lines 4 mi each side of the centerline of the airway. If two VORs delineating an airway are more than 120 mi apart, the airspace included in the airway is as indicated for jet routes.

Jet Routes

With the introduction of commercial jet aircraft in 1958, the altitudes at which these aircraft flew increased significantly. At higher altitudes the number of ground stations (VORs) required to delineate a specific route is smaller than at lower altitudes because the signal is transmitted on a line of sight. Therefore there was no need to clutter the high altitude routes with all the ground stations required for low altitude flying. All the routes in the continental United States could be placed on one chart. These were established what are known as jet routes. Although in one sense these routes are airways, they are not referred to as such. Today both victor airways and jet routes exist. Thus the jet routes are delineated by the same aids to navigation on the ground (VORs) as are victor airways but fewer stations are used. Victor airways extend from 1200 ft above the terrain to, but not including, 18,000 ft AMSL. Jet routes extend from 18,000 ft to 45,000 ft AMSL. Above 45,000 ft there are no designated routes and aircraft are handled on an individual basis. The numbering system for the jet routes is the same as for the victor airways. Jet routes are designated on aeronautical charts as J-1, J-2, etc.

Area Navigation

For many years all aircraft were required to fly on designated routes, airways, or jet routes. That is, all aircraft had to fly from one VOR to the next VOR since the VORs delineate the airways and jet routes. This required the funneling of all traffic on the designated routes that resulted in congestion on certain routes. Also the designated routes were often not the shortest distance between two points, resulting in additional fuel consumption, flight time, and cost. Furthermore, if the designated route penetrates into an area of thunderstorms, aircraft have to be vectored around the storm by controllers on the ground. This imposed an extra workload on the controllers that is compensated for by the use of the severe weather avoidance program (SWAP). Despite these inefficiencies, the vast majority of transient aviation still fly along the victor airways and jet routes. This trend, however, has begun to change dramatically since the beginning of the twenty-first century, with the proliferation of GPS based navigation systems, under what is known as RNAV.

Area navigation, RNAV, is a method of aircraft navigation that permits aircraft operation on any desired course within the coverage of station-referenced navigational signals or within the limits of a self-contained system capability. Area navigation routes are direct routes, based upon the area navigation capability of aircraft, between waypoints defined in terms of latitude and longitude coordinates, degree and distance fixes, or offsets from established routes and airways.

RNAV is possible due to the proliferation of onboard aircraft technologies that take advantage of the global positioning system (GPS). GPS is based on 24 satellites located approximately 12,000 mi about the earth in a geostatic orbit. Technology that references an aircraft's position in relation to these satellites allows an aircraft to navigate by referencing its position to a detailed database that identifies airports, waypoints, terrain, and man-made infrastructure. Enhancements to the accuracy of the GPS system, with technologies such as the wide area augmentation system (WAAS) have made it possible for the air traffic control system to approve defined approaches to airport runways with far greater accuracy than with traditional radio-frequency-based systems.

Area navigation provides a more flexible routing capability that allows for better utilization of the airspace. The greater utilization reduces delays in the airspace and results in more economical operation of the aircraft. For example, routes parallel to the designated routes from one VOR to another can be established without requiring additional aids to navigation on the ground. Another example is the establishment of a more direct route from one point to another by establishing waypoints that provide for a shorter trip. Routing around a thunderstorm without continuous radar guidance from the ground is another example.

Area navigation is not limited to the horizontal plane but can also be utilized in the vertical plane, termed VNAV. It can also include a time reference capability. A properly equipped aircraft could arrive at a specified point in space, called a fix, with no need for ground vectoring or directions and could additionally be at that point at a specified altitude and time. This is a four-dimensional capability giving latitude, longitude, altitude, and time (4D RNAV). Thus area navigation

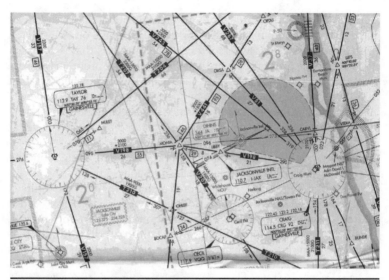

FIGURE 3-7 Airways, navigational aids, and airports as depicted on an IFR en route low altitude navigational chart.

has the potential of increasing airspace capacity, enhancing safety, and reducing the workload of the pilot and the air traffic controller.

As part of RNAV enhancements, the FAA began establishing T-routes as alternative routes to the victor airways for those aircraft equipped with GPS systems. These T-routes were established to provide aircraft with more direct routing, often around congested traffic areas such as Class B airspace and areas where victor airways intersect, often in the vicinity of a VOR station.

Figure 3-7 illustrates a portion of airspace as depicted in an IFR en route low altitude chart, published by the FAA's National Aeronautical Charting Office. This figure depicts both victor airways (identified by V followed by a route number) and T-routes (identified by a T followed by a route number), as well as the locations of airports, classes of airspace (such as the shaded area around Jacksonville International Airport), VOR stations (such as Taylor, Cecil, and Craig), and other navigational facilities. It is recommended that airport planners become familiar with understanding the information provided on this and other aeronautical charts.

Air Traffic Separation Rules

Air traffic rules governing the minimum separation of aircraft in the vertical, horizontal or longitudinal, and lateral directions are established in each country by the appropriate government authority. The current rules described in this text are those that are prescribed by the

FAA for use in the United States. The separation rules are prescribed for IFR operations and these rules apply whether or not IMC conditions prevail. Minimum separations are a function of aircraft type, aircraft speed, availability of radar facilities, navigational aids, and other factors such as the severity of wake vortices [3].

Vertical Separation in the Airspace

The minimum vertical separation of aircraft outside of the terminal area from the ground up to and including 41,000 ft AMSL is 1000 ft. In 2005, vertical separation minimums above 29,000 ft AMSL were reduced from 2000 to 1000 ft under the reduced vertical separation minima (RVSM) program. Implementation of this program allowed for additional jet routes thereby increasing the capacity within the NAS. Within a terminal area a vertical separation of 500 ft is maintained between aircraft, except that a 1000-ft vertical separation is maintained below a heavy aircraft.

Assigned Flight Altitudes

To formalize the separation of air traffic in the airspace, air traffic control assigns flight altitudes to aircraft based on their direction, or more precisely magnetic heading, of flight, and whether or not they are flying under VFR versus IFR rules.

Aircraft flying under IFR are typically assigned altitudes of odd-thousand feet (i.e., 3000 ft, 5000 ft, etc.) AMSL while on an easterly heading (magnetic compass heading of 0° to 179°) and even-thousand feet (i.e., 4000 ft, 6000 ft, etc.) while on a westerly heading (magnetic compass heading of 180° to 359°). Between 29,000 ft AMSL (FL 290) and 41,000 ft AMSL (FL 410), aircraft are assigned a flight level of either FL 290, FL 330, FL 370, or FL 410 when traveling on an easterly heading, and either FL 310, FL 350, FL 390 when traveling on a westerly heading. If an aircraft is RVSM certified, it may be assigned an RVSM altitude of FL 300, 320, 340, etc. between FL 290 and FL 410.

Aircraft flying under VFR above 3000 ft AMSL are typically assigned altitudes of odd-thousand feet plus 500 ft (i.e., 3500 ft, 5500 ft, etc.) while on an easterly heading, and even-thousand feet plus 500 feet (i.e., 4500 ft, 6500 ft, etc.) while on a westerly heading (magnetic compass heading of 180° to 359°). Above 29,000 ft (FL 290), VFR traffic is assigned every-other even or odd thousand (FL 290, FL 330, etc. if traveling on an easterly heading, and FL 320, FL 360, etc. if traveling on a westerly heading). It should be noted that above FL 180, all traffic is required to be on an IFR flight plan.

Longitudinal Separation in the Airspace

The minimum longitudinal separation depends on a number of factors. Among the most important are aircraft size, aircraft speed, and the availability of radar for the control of air traffic. For the purposes of maintaining aircraft separations aircraft are classified by the FAA as heavy, large, or small based upon their maximum gross takeoff weight (MGTOW). Heavy aircraft are classified as those aircraft which have a MGTOW of 300,000 lb or more. Large aircraft are as those aircraft which have a MGTOW of in excess of 12,500 lb but less than 300,000 lb. Small aircraft are as those aircraft which have a MGTOW of 12,500 lb or less. Aircraft size is related to wake turbulence. Heavy aircraft create trailing wake vortices which are a hazard to lighter aircraft following them.

The minimum longitudinal separations en route are expressed in terms of time or distance as follows:

1. For en route aircraft following a preceding en route aircraft, if the lead aircraft maintains a speed at least 44 kn faster than the trail aircraft, 5 mi between aircraft using distance measuring equipment (DME) or area navigation (RNAV) and 3 minutes between all other aircraft

2. For en route aircraft following a preceding en route aircraft, if the lead aircraft maintains a speed at least 22 kn faster than the trail aircraft, 10 mi between aircraft using DME or RNAV and 5 min for all other aircraft

3. For en route aircraft following a preceding en route aircraft, if both aircraft are at the same speed, 20 mi between aircraft using DME or RNAV and 10 min for all other aircraft

4. When an aircraft is climbing or descending through the altitude of another aircraft, 10 mi for aircraft using DME or RNAV if the descending aircraft is leading or the climbing aircraft is following and 5 min for all other aircraft

5. Between aircraft in which one aircraft is using DME or RNAV and the other is not, 30 mi

Minimum longitudinal separations over the oceans is normally 10 minutes for supersonic flights and 15 minutes for subsonic flights but in some locations it can be slightly more or less than this value [3].

When the aircraft mix is such that wake turbulence is not a factor and radar coverage is available, the minimum longitudinal separation for two aircraft traveling in the same direction and at the same altitude is 5 nm, except that when the aircraft are in the terminal environment within 40 nm of the radar antenna the separation can be reduced to 3 nm. For this reason the minimum spacing in the terminal area is 3 nm because the airport is almost always within 40 nm of a radar antenna. Under certain specified conditions a separation between aircraft on final approach within 10 nm of the landing runway may be reduced to 2.5 nm [3].

If wake turbulence is a factor, the minimum separation in the terminal area between a small or large aircraft and a preceding heavy

Lead Aircraft Type	VFR* Trail Aircraft Type			IFR (Wake Vortex) Trail Aircraft Type		
	Heavy	Large	Small	Heavy	Large	Small
Heavy	2.7	3.6	4.5	4.0	5.0	6.0
Large	1.9	1.9	2.7	3.0	3.0	4.0
Small	1.9	1.9	1.9	3.0	3.0	3.0

*These are shown to appropriately represent these operations and are not regulatory in nature.
Source: Federal Aviation Administration [18].

TABLE 3-1   Horizontal Separation in Landing for Arrival-Arrival Spacing of Aircraft on Same Runway Approaches in VFR and IFR Conditions (nautical miles)

aircraft is 5 nm. The spacing between two heavy aircraft following each other is 4 nm. The spacing between a heavy aircraft and a preceding large aircraft is 3 nm.

For landing aircraft when wake turbulence is a factor, the longitudinal separation is increased between a small aircraft and a preceding large aircraft to 4 mi and between a small aircraft and a preceding heavy aircraft to 6 mi.

The IFR separation rules for consecutive arrivals on the same runway which are used when wake vortices are a factor are shown in Table 3-1. The VFR and IFR separation rules for consecutive departures from the same runway are expressed in terms of time and these are shown in Table 3-2.

Lateral Separation in the Airspace

The minimum en route lateral separation below 18,000 ft MSL is 8 nm, and at and above 18,000 ft MSL the minimum en route lateral separation

Lead Aircraft Type	VFR* Trail Aircraft Type			IFR Trail Aircraft Type		
	Heavy	Large	Small	Heavy	Large	Small
Heavy	90	120	120	120	120	120
Large	60	60	50	60	60	60
Small	50	45	35	60	60	60

*These are shown to appropriately represent these operations and are not regulatory in nature.
Source: Federal Aviation Administration [18].

TABLE 3-2   Separation for Same Runway Consecutive Departures in VFR and IFR Conditions (seconds)

is 20 nm. Over the oceans the separation varies from 60 to 120 nm depending on location [3].

Navigational Aids

Aids to navigation, known as NAVAIDS, can be broadly classified into two groups, ground-based systems and satellite-based systems. Each system is complimented by systems installed in the cockpit.

Ground-Based Systems

Nondirectional Beacon

The oldest active ground-based navigational aid is the nondirectional beacon (NDB). The NDB emits radio frequency signals on frequencies between 400 and 1020 Hz modulation. NDBs are typically mounted on a pole approximately 35 ft tall. They may be located on or off airport property, at least 100 ft clear of metal buildings, power lines, or metal fences. While the NDB is quickly being phased out in the United States, it is still a very common piece of navigational equipment in other parts of the world, particularly in developing nations. Figure 3-8 provides an illustration of an NDB.

Aircraft navigate using the NDB by referencing an automatic direction finder (ADF) located on the aircraft's panel. The ADF simply points toward the location of the NDB. Figure 3-9 illustrates an ADF system.

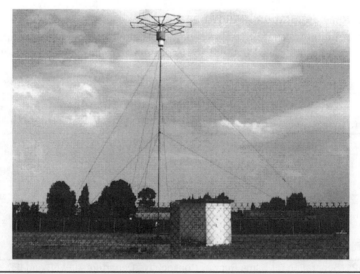

Figure 3-8 Nondirectional beacon.

FIGURE 3-9   Automatic direction finder.

Very High Frequency Omnirange Radio

The advances in radio and electronics during and after World War II led to the installation of the very high frequency omnirange (VOR) radio stations. These stations are located on the ground and send out radio signals in all directions. Each signal can be considered as a course or a route, referred to as a radial that can be followed by an aircraft. In terms of 1° intervals, there are 360 courses or routes that are radiated from a VOR station, from 0° pointing toward magnetic north increasing to 359° in a clockwise direction. The VOR transmitter station is a small square building topped with what appears to be a white derby hat. It broadcasts on a frequency just above that of FM radio stations. The very high frequencies it uses are virtually free of static. The system of VOR stations establish the network of airways and jet routes and are also essential to area navigation. The range of a VOR station varies but is usually less than 200 nm. A typical VOR beacon is illustrated in Fig. 3-10.

Aircraft equipped with a VOR receiver in the cockpit have a dial for tuning in the desired VOR frequency. A pilot can select the VOR radial or route he wishes to follow to the VOR station. In the cockpit there is also an omnibearing selector (OBS) which indicates the heading of the aircraft relative to the direction of the desired radial and whether the aircraft is to the right or left of the radial. An illustration of an OBS is provided in Fig. 3-11.

Distance Measuring Equipment

Distance measuring equipment (DME) has traditionally been installed at VOR stations in the United States. The DME shows the pilot the

FIGURE 3-10   VOR beacon on the airfield at Ronald Reagan Washington National Airport.

slant distance between the aircraft and a particular VOR station. Since it is the air distance in nautical miles that is measured, the receiving equipment in an aircraft flying at 35,000 ft directly over the DME station would read 5.8 nm.

An en route air navigation aid which best suited the tactical needs of the military was developed by the Navy in the early 1950s. This aid is known as TACAN, which stands for tactical air navigation. This aid combines azimuth and distance measuring into one unit instead of two and is operated in the ultra-high-frequency band. As a compromise between civilian and military requirements, the FAA replaced the DME

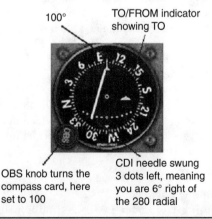

100°

TO/FROM indicator
showing TO

OBS knob turns the
compass card, here
set to 100

CDI needle swung
3 dots left, meaning
you are 6° right of
the 280 radial

FIGURE 3-11   Omnibearing selector.

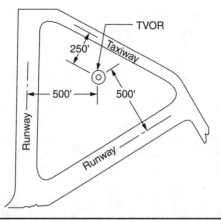

Figure 3-12   TVOR location standards.

portion of its VOR facilities with the distance measuring components of TACAN. These stations are known as VORTAC stations. If a station has full TACAN equipment, both azimuth and distance measuring equipment, and also VOR, it is designated as VORTAC.

NDB and VOR systems are often located on airport airfields. The location of these systems on airport, known as TVORs, are significant to airport planners and designers, as the location of other facilities, such as large buildings, particularly constructed of metal, may adversely affect the performance of the navaid.

As illustrated in Fig. 3-12, TVORs should be located at least 500 ft from any runways and 250 ft from any taxiways. Any structures or trees should be located at least 1000 ft from the TVOR antenna. There should also be a clearance angle of at least 2.5° for any structures and 2.0° for any trees beyond 1000 ft, as illustrated in Fig. 3-13.

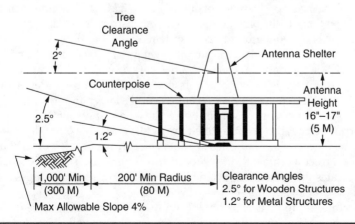

Figure 3-13   TVOR clearance requirements.

Air Route Surveillance Radar

A long-range radar for tracking en route aircraft has been established throughout the continental United States and in other parts of the world. While in the United States there is complete radar coverage in the 48 contiguous states, this is not the case elsewhere in the world. These radars have a range of about 250 nm. Strictly speaking radar is not an aid to navigation. Its principal function is to provide air traffic controllers with a visual display of the position of each aircraft so they can monitor their spacings and intervene when necessary. However, it can be and is used by air traffic controllers to guide aircraft whenever this is necessary. For this reason it has been included as an aid to navigation.

The VOR and NDB, often combined with radar-based surveillance from air traffic control, have traditionally been used in both en route navigation and for navigation on approach to landing at an airport. Navigation on approach to an airport using these ground-based systems is performed by following predetermined, published, approach procedures. These procedures are often updated and published by the FAA in the form of approach charts. Figure 3-14 provides an example of an approach chart depicting an approach procedure using an NDB as an aid to navigation, while Fig. 3-15 illustrates a similar approach using a VOR as the primary aid to navigation. It is strongly recommended that the airport planner understand the information provided in these charts. Approaches based on NDB and VOR navaids are considered "nonprecision" approaches, as they provide lateral navigation assistance but not vertical navigation. That is, these instruments may be referenced to determine which direction to fly when approaching an airport, but do not provide instrument-based guidance in determining the appropriate altitude or descent rate on approach.

Instrument Landing System

Until the recent proliferation of published navigation procedures which rely on the satellite based GPS system, the instrument landing system (ILS) was the only ground-based system certified to provide both lateral and vertical guidance to aircraft on approach to an airport, and as of 2008 is the only navigational aid certified by the FAA to provide "precision" navigation for aircraft, and is still the most widely used method of approach navigation at the world's larger airports.

An ILS system consists of two radio transmitters located on the airport. One radio beam is called the localizer and the other the glide slope. The localizer indicates to pilots whether they are left or right of the correct alignment for approach to the runway. The glide slope indicates the correct angle of descent to the runway. Glide slopes are in the order of from 2°–3° to 7.5°.

In order to further help pilots on their ILS approach, up to three low-power fan markers called ILS markers are usually installed so

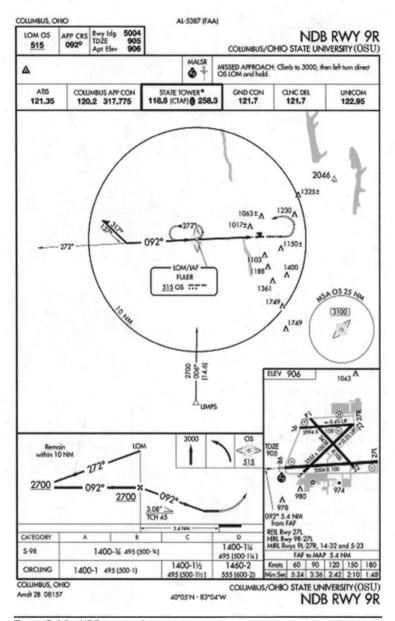

COLUMBUS, OHIO   AL-5387 (FAA)

LOM OS 515	APP CRS 092°	Rwy Idg 5004 TDZE 905 Apt Elev 906

NDB RWY 9R
COLUMBUS/OHIO STATE UNIVERSITY (OSU)

MALSR

MISSED APPROACH: Climb to 3000, then left turn direct OS LOM and hold.

ATIS 121.35	COLUMBUS APP CON 120.2 317.775	STATE TOWER* 118.8 (CTAF) 258.3	GND CON 121.7	CLNC DEL 121.7	UNICOM 122.95

CATEGORY	A	B	C	D
S-9R	1400-¾ 495 (500-¾)			1400-1¼ 495 (500-1¼)
CIRCLING	1400-1 495 (500-1)		1400-1½ 495 (500-1½)	1460-2 555 (600-2)

092° 5.4 NM from FAF
REIL Rwy 27L
HIRL Rwy 9R-27L
MIRL Rwys 9L-27R, 14-32 and 5-23
FAF to MAP 5.4 NM

Knots	60	90	120	150	180
Min:Sec	5:24	3:36	2:42	2:10	1:48

COLUMBUS, OHIO
Amdt 2B 08157

40°05'N - 83°04'W

COLUMBUS/OHIO STATE UNIVERSITY (OSU)
NDB RWY 9R

FIGURE 3-14   NDB approach.

that they may know just how far along the approach to the runway they have progressed. The first is called the outer marker (LOM) and is located about 3.5 to 5 mi from the end of the runway. The middle marker (MM) is located about 3000 ft from the end of the runway. On

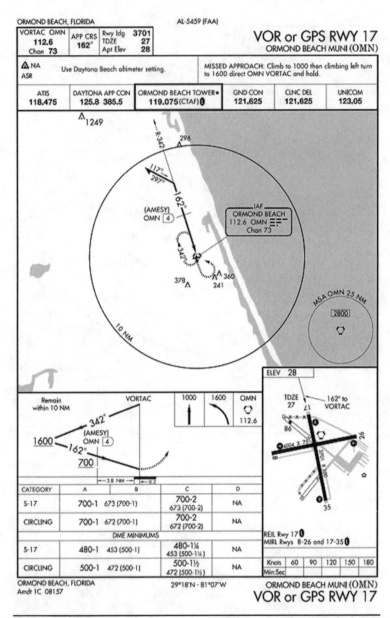

Figure 3-15 VOR approach.

some ILS systems, an additional marker called the inner marker (IM) is located 1000 ft from the end of the runway. When the plane passes over a marker, a light goes on in the cockpit and a tone sounds. The configuration of the ILS system is shown in Fig. 3-16.

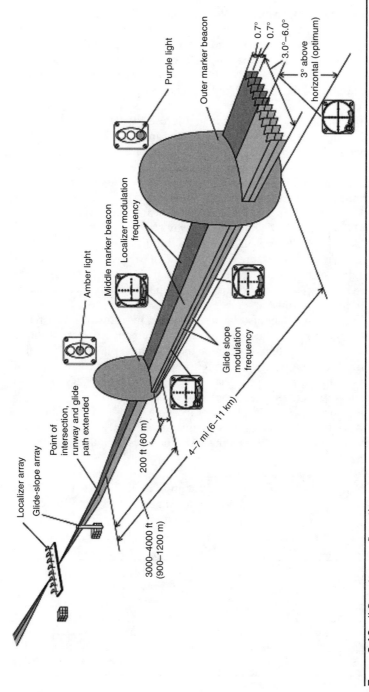

Localizer array

Glide-slope array

Point of
intersection,
runway and glide
path extended

Amber light

Purple light

Middle marker beacon

Localizer modulation
frequency

Outer marker beacon

0.7°
0.7°

3.0°–6.0°

3° above
horizontal (optimum)

Glide slope
modulation
frequency

200 ft (60 m)

4–7 mi (6–11 km)

3000–4000 ft
(900–1200 m)

FIGURE 3-16 ILS system configuration.

121

The localizer consists of an antenna, which is located on the extension of the runway centerline approximately 1000 ft from the far end of the runway, and a localizer transmitter building located about 300 ft to one side of the runway at the same distance from the end of the runway as the antenna. The glide slope facility is placed 750 to 1250 ft down the runway from the threshold and is located to one side of the runway centerline at a distance which can vary from 400 to 650 ft. The functioning of the localizer and the glide slope facility is affected by the close proximity of moving objects such as vehicular and aircraft traffic. During inclement weather the use of the ILS critical areas inhibit aircraft and vehicles from entering into areas that would impede an aircraft inside of the outer marker from receiving a clear signal. Stationary objects nearby can also cause a deterioration of signals. Abrupt changes of slope in proximity of the antennas are not permitted or the signal will not be transmitted properly. Another limitation of the ILS is that the glide slope beam is not reliable below a height of about 200 ft above the runway.

As with VOR and NDB systems, the localizer and glide slope components of the ILS are highly sensitive to their proximity to surrounding objects that may interfere with their radio signals. As such, there are specific restrictions to construction in the immediate vicinity of these systems.

ILS systems may also be accompanied by runway visual range (RVR) equipment, which provide a measurement of lateral visibility to a pilot. RVR systems determine the distance a pilot should be able to see down the runway, given current atmospheric conditions and existing lighting systems. Depending on the type of RVR system installed, pilots can safely approach to land on a runway using ILS navigation in varying levels of cloud ceiling levels and horizontal visibility. Table 3-3 provides the ceiling and visibility levels for ILS systems equipped with RVR. Figure 3-17 illustrates a published approach using an ILS.

The most critical point of approach to landing comes when the aircraft breaks through the overcast and the pilot must change from instrument to visual conditions. Sometimes, only a few seconds are

ILS Category	Cloud Ceiling	Visibility (RVR)
I	200 ft	1,800–2,400 ft
II	100 ft	1,200 ft
IIIa (auto land)	0–100 ft	700 ft
IIIb (auto rollout)	0–50 ft	150 ft
IIIc (auto taxi)	0 ft	0 ft

TABLE 3-3   ILS Capabilities

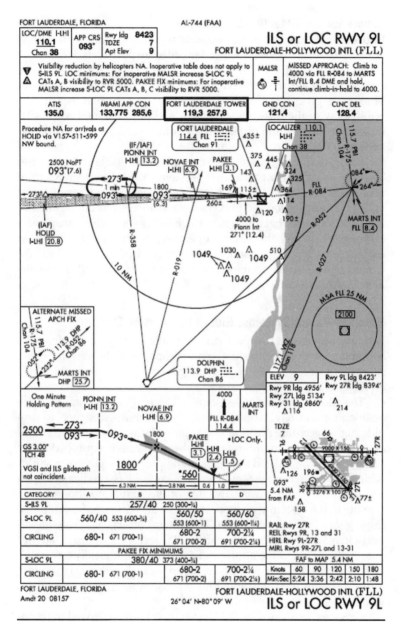

FIGURE 3-17   Published ILS approach procedure.

available for the pilot to make the transition and complete the landing. To aid in making this transition, lights are installed on the approach to and on the runways. These are generally termed approach lighting systems (ALS). More details concerning these systems are contained in the chapter on signing, marking, and lighting.

Airport Surface Detection Equipment

In use in many of the busier airports within the United States and elsewhere, specially designed radar called airport surface detection equipment (ASDE), often referred to as *ground radar*, has been developed to aid the controller in regulating traffic on the airport. The system gives the air traffic controller in the control tower a pictorial display of the runways, taxiways, and terminal area with radar indicating the position of aircraft and other vehicles moving on the surface of the airport.

ASDE technologies are most commonly available in two forms. The airport movement area safety system (AMASS) integrates third generation (ASDE-3) ground-based radar systems with audio and visual warning systems to prevent runway incursions on the airfield. ASDE-X, a system that integrates ASDE technology with transponder systems to identify the aircraft operating on the aircraft, has been employed at less busy commercial service airports throughout the United States. These systems are monitored by air traffic management personnel in the airport's ATCT.

Satellite-Based Systems: Global Positioning System

Perhaps the greatest impact on air traffic management since the beginning of the twenty-first century has been the development, acceptance, and proliferation of navigational procedures which rely on the global positioning system.

The global positioning system (GPS) is a satellite-based radio positioning and navigation system. The system is designed to provide highly accurate position and velocity information on a continuous global basis to an unlimited number of properly equipped users. The system is unaffected by weather and provides a common worldwide grid reference system. The GPS concept is predicated upon accurate and continuous knowledge of the spatial position of each satellite in the system with respect to time and distance from a transmitting satellite to the user. The GPS system consists of 24 satellites in near-circular orbit about the earth. GPS receivers onboard aircraft automatically select the appropriate signals from typically four satellites which are in view of the receiver and translate these signals into a three-dimensional position, and when the receiver is in motion, velocity.

GPS was developed by the United States military to aid in reconnaissance and strategic operations. Under military operations, signals transmitted from the orbiting GPS satellites were encrypted to reduce the accuracy of positioning for nonauthorized users. This policy became known as "selective availability." However, on May 1, 2000, President Bill Clinton ordered the removal of selective availability, allowing the civilian world to take advantage of GPS positioning accuracy on the order of 1 to 3 m. This level of accuracy has been deemed sufficient to allow aircraft to navigate using properly equipped GPS receivers for both en route navigation and approaches to airports, in both VFR and IFR conditions.

In the first decade of the twenty-first century, the proliferation of GPS-based air navigation systems has been dramatic, to the point where the use of traditional ground-based navigation aids such as the NDB and VOR is becoming obsolete. GPS navigation systems have become available as both in-panel fixed navigation systems, and portable units, and have become widely used in many areas of society outside of aviation.

As Fig. 3-18 illustrates, GPS systems, particularly enhanced with comprehensive databases of area terrain, landmarks, and airport infrastructure, provide pilots with "virtually visual conditions" and the ability to navigate from origin to destination without any reliance on traditional ground-based analog navigational aids.

GPS-based RNAV approaches have been refined with improving technology, known as the wide area augmentation system (WAAS), and training to allow aircraft to approach airports using very precise navigation procedures. These approaches, known as RNP (required navigation performance), have allowed aircraft to navigate around such obstacles as mountainous terrain and security sensitive areas, resulting in a more efficient use of airport runways. Juneau, Alaska and Washington, D.C. are airports that have benefited from these enhancements in air traffic control. An example RNP approach is illustrated in Fig. 3-19.

Figure 3-18 Aircraft GPS-based navigation equipment (*Cirrus Aircraft Inc.*).

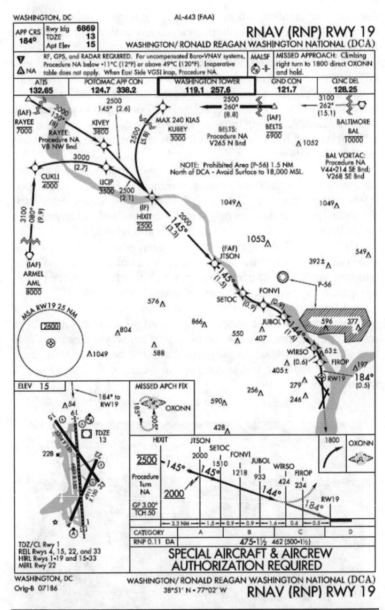

FIGURE 3-19   RNP approach into Ronald Reagan Washington National Airport, Washington, D.C.

ADS-B

Further enhancements to the air traffic management system include the use of advanced digital data-link systems, known as automated dependent surveillance. ADS-address (ADS-A) systems send digitally transmitted information between specific aircraft and ADS-broadcast

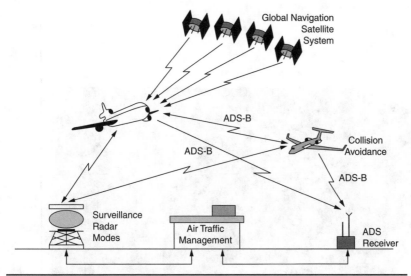

Figure 3-20   Rendering of ADS-B system (*Atcmonitor.com*).

(ADS-B) systems broadcast information to all equipped aircraft and air traffic management facilities, identifying their locations to other traffic in the system, providing the added ability to safety avoid collisions even in poor visibility conditions. Originally tested in Alaska between 2000 and 2003, ADS-B is quickly becoming a standard component of air traffic navigation systems in the United States. A rendering of the ADS-B system is Fig. 3-20.

The Modernization of Air Traffic Management

Despite the proliferation of GPS-based navigation since the beginning of the twenty-first century, the principal aids for the control of air traffic by air traffic management personnel are still voice communication and radar. Air traffic controllers monitor the spacing between aircraft on the radarscope and instruct pilots by means of voice communication.

Radar returns appear on the radarscope as small blips. These are reflections from the aircraft body. Primary radar requires the installation of rotating antennas on the ground and the range of the primary radar is a function of its frequency. Secondary radar consists of a radar receiver and transmitter on the ground that transmits a coded signal to an aircraft if that aircraft has a transponder. A transponder is an airborne receiver and transmitter which receives the signal from the ground and responds by returning a coded reply to the interrogator on the ground. The coded reply normally contains information on aircraft identity and altitude.

Information from primary and secondary radar returns are provided to air traffic controllers via an alphanumeric display on their radar "scopes," as illustrated in Fig. 3-21. The first line shows the

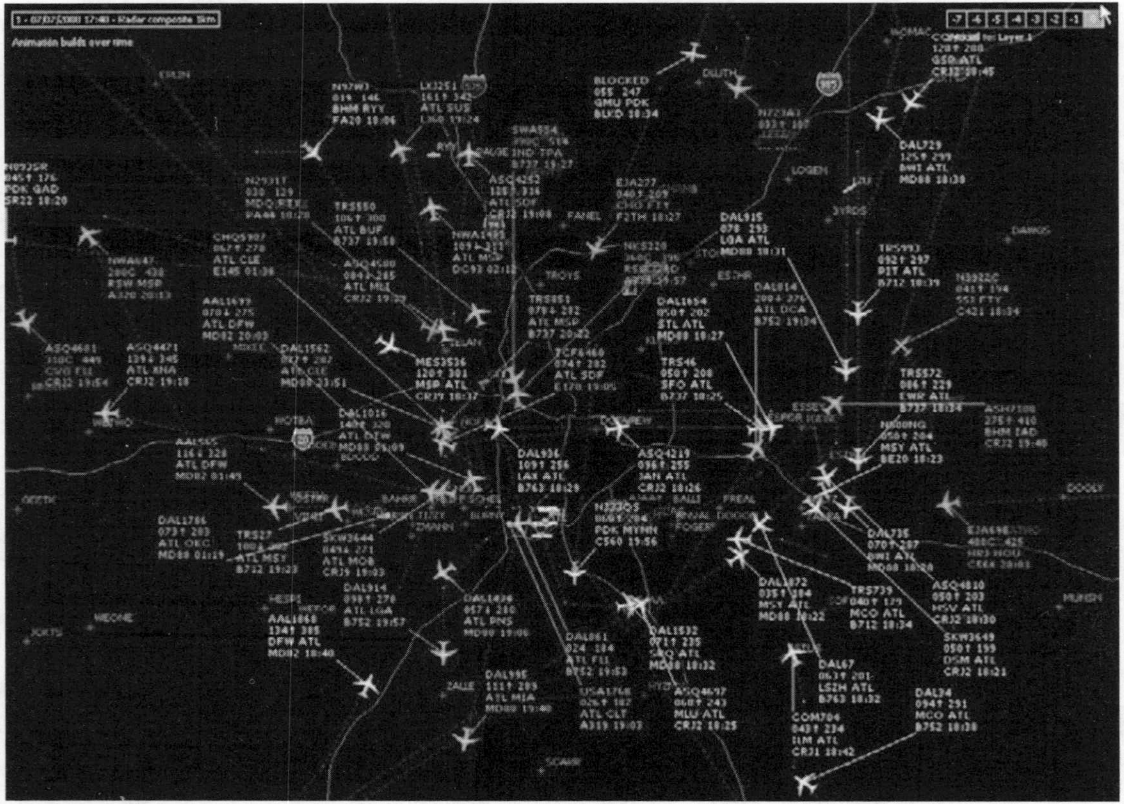

FIGURE 3-21   Air route surveillance radar, Atlanta TRACON.

identity of the aircraft, the second line its altitude and ground speed, and the third line gives the beacon code transponder number and the aircraft track number. To be able to have this information presented on the radarscope, the aircraft must carry a mode-C or mode-S transponder that has the capability of altitude reporting along with aircraft identity. All commercial airline aircraft carry a transponder, which satisfies the requirement for reporting altitude. Further, all aircraft flying in Class A, B, or C airspace are required to have an operating transponder onboard.

NextGen

For more than 50 years air traffic control systems have gone through a number of incremental technological enhancements, such as enhanced radar capabilities, automated flight service systems, and ground-based navigation systems. Despite these upgrades, it has been widely recognized that the traditional radar and analog-based communication system will not be sufficient to accommodate the increasing demands on the system in the twenty-first century.

As part of the Vision 100 Century of Aviation Reauthorization Act of 2003, the U.S. federal government called for a complete transformation of the national airspace system and a modernization of its air traffic control facilities. This modernization has come to be known as the "next generation air traffic system" or *NextGen*. Through the act, Congress directed the formation of a "Joint Planning and Development Office" (JPDO) to facilitate the mammoth task of converting the current system to a fully automated, digital, satellite-based air traffic management system. The JPDO comprises of representatives from the FAA, NASA, The U.S. Departments of Transportation, Defense, Homeland Security, Commerce, and the White House Office of Science and Technology Policy, and directed to develop a next generation air traffic system that is technologically advanced and fully integrates the interests of all who use the nation's aviation system.

NextGen will focus on making the satellite-based GPS system and digital data communications the backbone of air traffic management. Integrated into NextGen are GPS, WAAS, and ADS-B technology to allow for digital surveillance of air traffic between both ground-based air traffic management facilities as well as among aircraft themselves. In addition to ADS-B technology, NextGen features the following capabilities, as described by the FAA:

SWIM

System wide information management (SWIM) provides the infrastructure and services to deliver network-enabled information access across the NextGen air transportation operations. As an early opportunity investment, SWIM will provide high-quality, timely data to many users and applications—extending beyond the previous focus on unique, point-to-point interfaces for application-to-application data exchange. By reducing the number and types of interfaces and systems, SWIM will reduce redundancy of information and better facilitate multiagency information-sharing. SWIM will also enable new modes of decision making, as information is more easily accessed by all stakeholders affected by operational decisions.

NextGen Data Communications

NextGen data communications will provide for two-way digital communications between air traffic controllers and pilots for air traffic control clearances, instructions, and other advisories. In addition, digital communications will provide broadcast text-based and graphical advisory information such as weather reports and notices to airmen without relying on voice communications.

NextGen Enabled Weather

The NextGen network enabled weather (NNEW) will serve as the core of the NextGen weather support services and provide a common weather picture across the national airspace system. These services will, in turn, be integrated into other key components of NextGen required to enable better air transportation decision making. It is anticipated that tens of thousands of global weather observations and sensor reports from ground-, airborne-, and space-based sources would fuse into a single national weather information system, updated as needed in real time.

NextGen is due to be a phased transformation of the NAS through 2025 at an estimated cost of $20 to $25 billion. It should be noted the early stages of NextGen development have been very volatile with regard to the selection of technology platforms on which to base the future air traffic management system, and it should be expected that further developments in technology will result in variations to current system plans. As such, it is imperative that the airport planner keep up with current progress. The JPDO and the FAA frequently update their Internet sites with NextGen system progress.

References

1. *Airport Design*, Advisory Circular AC150/5300-13, Federal Aviation Administration, Washington, D.C., 2008.
2. *Air Route Traffic Control*, Airway Planning Standard Number Two, Order 7031.3, Federal Aviation Administration, Washington, D.C., September 1977.
3. *Air Traffic Control Handbook*, Order 7110.65G, Federal Aviation Administration, Washington, D.C., March 1992.
4. *Air Traffic Management Plan (ATMP) Program, Development and Control Procedures*, Order No. 7000.3, Federal Aviation Administration, Washington, D.C., February 1988.
5. *An Analysis of the Requirements for, and the Benefits and Costs of the National Microwave Landing System (MLS)*, Office of Systems Engineering Management, Report No. FAA-EM-80-7, Federal Aviation Administration, Washington, D.C., June 1980.

6. *Aviation System Capacity Plan 1991-92*, Report No. DOT/FAA/ASC-91-1, U.S. Department of Transportation, Federal Aviation Administration, Washington, D.C., 1991.
7. *Chicago Delay Task Force Technical Report*, vol. 1: Chicago Airport/Airspace Operating Environment, Landrum and Brown Aviation Consultants, Chicago, Ill., April 1991.
8. *Civil Aviation Research and Development Policy Study*, Department of Transportation and National Aeronautics and Space Administration, Washington, D.C., March 1971.
9. *Designation of Federal Airways, Area Low Routes, Controlled Airspace, Reporting Points, Jet Routes and Area High Routes*, Part 71, Federal Aviation Regulations, Federal Aviation Administration, Washington, D.C., February 1992.
10. *Enroute High Altitude—U.S.*, Flight Information Publication, National Ocean Survey, National Oceanic and Atmospheric Administration, Department of Commerce, Washington, D.C., December 1975.
11. *Establishment of Jet Routes and Area High Routes*, Part 75, Federal Aviation Regulations, Federal Aviation Administration, Washington, D.C., December 1991.
12. *FAA Long-Range Aviation Projections, Fiscal Years 2004-2015*, Report No. FAA-APO-92-4, Office of Aviation Policy, Plans, and Management Analysis, Federal Aviation Administration, Washington, D.C., May 1992.
13. *FAA Report on Airport Capacity*, The MITRE Corporation, Report FAA-EM-74-5, Federal Aviation Administration, Washington, D.C., 1974.
14. "Future ATC Technology Improvements and the Impact on Airport Capacity," R. M. Harris, The MITRE Corporation, Air Transportation Systems Division, McLean, Va.
15. "Future System Concepts for Air Traffic Management, W. E. Simpson, Office of Systems Engineering, Department of Transportation, *Presented at 19th Technical Conference of the International Air Transportation, Association*, Washington, D.C., October 1972.
16. *General Operating and Flight Rules*, Part 91, Federal Aviation Regulations, Federal Aviation Administration, Washington, D.C., February 1992.
17. *National Airspace System Plan, Facilities, Equipment and Associated Development*, Federal Aviation Administration, Washington, D.C., 1989.
18. *Parameters of Future ATC Systems Relating to Airport Capacity and Delay*, Report No. FAA-EM-78-8A, Federal Aviation Administration, Washington, D.C., 1978.
19. *Planning the Metropolitan Airport System*, Advisory Circular, AC 150/5070-5, Federal Aviation Administration, Washington, D.C., May 1970.
20. "Relative Navigation Offers Alternatives to Differential GPS,. *Aviation Week and Space Technology*, vol. 137, No. 22, New York, November 1992.
21. *Report of Department of Transportation Air Traffic Control Advisory Committee*, U.S. Department of Transportation, Washington, D.C., December 1969.
22. *Summary Report 1972*, National Aviation System Planning Review Conference, Federal Aviation Administration, Washington, D.C.
23. "The Advanced Air Traffic Management System Study," R. L. Maxwell, Office of Systems Engineering, Department of Transportation, *Presented at the 19th Technical Conference of the International Air Transportation Association*, Washington, D.C., October 1972.
24. *Traffic Management System (TMS) Air Traffic Operation Requirements*, Order No. 7032.9, Federal Aviation Administration, Washington, D.C., September 1992.
25. *United States Aeronautical Information Publication*, 12th ed., Federal Aviation Administration, Washington, D.C., October 1992.
26. *United States Standard for Terminal Instrument Procedures (TERPS)*, Order 8260.3B, Federal Aviation Administration, Washington, D.C., May 1992.

CHAPTER 4

Airport Planning Studies

Introduction

The planning of an airport is such a complex process that the analysis of one activity without regard to the effect on other activities will not provide acceptable solutions. An airport encompasses a wide range of activities which have different and often conflicting requirements. Yet they are interdependent so that a single activity may limit the capacity of the entire complex. In the past airport master plans were developed on the basis of local aviation needs. In more recent times these plans have been integrated into an airport system plan which assessed not only the needs at a specific airport site but also the overall needs of the system of airports which service an area, region, state, or country. If future airport planning efforts are to be successful, they must be founded on guidelines established on the basis of comprehensive airport system and master plans.

The elements of a large airport are shown in Fig. 4-1. It is divided into two major components, the airside and the landside. The aircraft gates at the terminal buildings form the division between the two components. Within the system, the characteristics of the vehicles, both ground and air, have a large influence on planning. The passenger and shipper of goods are interested primarily in the overall door-to-door travel time and not just the duration of the air journey. For this reason access to airports is an essential consideration in planning.

The problems resulting from the incorporation of airport operations into the web of metropolitan life are complex. In the early days of air transport, airports were located at a distance from the city, where inexpensive land and a limited number of obstructions permitted flexibility in airport operations. Because of the nature of aircraft and the infrequency of flights, noise was not a problem to the

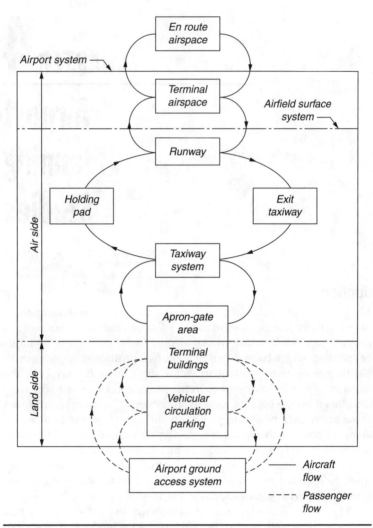

FIGURE 4-1   Components of the airport system for a large airport.

community. In many cases the arrival and departure of passenger and cargo planes was often a source of local pride. In addition, low population density in the vicinity of the airport and light air traffic prevented occasional accidents from alarming the community. In spite of early lawsuits, the relationship between airport and community was relatively free of strife resulting from problems of nuisance or hazard.

Airport operations have been increasingly hampered by obstructions resulting from industrial development related to the airport and from industry attracted by adjacent inexpensive land and access to the transportation afforded by the airfield and its associated highways.

While increasingly dense residential development has resulted from this economic stimulation, one must not overlook the effects of the unprecedented suburban spread during the post-World War II era, resulting from the backlog of housing needs and a period of economic prosperity.

Radical developments in the nature of air transport have produced new problems. The phenomenal growth of air traffic has increased the probability of unfavorable community reaction, but developments in the aircraft themselves have had the most profound effect on airport community relations. The greater size and speed of aircraft have resulted in increases in approach and runway requirements, while increases in the output of power plants have brought increases in noise. Faced with these problems the airport must cope with the problems of securing sufficient airspace for access to the airport, sufficient land for ground operations, and, at the same time, adequate access to the metropolitan area.

Types of Studies

Many different types of studies are performed in airport planning. These include studies related to facility planning, financial planning, traffic and markets, economics, and the environment. However, each of these studies can usually be classified as being performed at one of three levels: the system planning level, the master planning level, or the project planning level.

The Airport System Plan

An airport system plan is a representation of the aviation facilities required to meet the immediate and future needs of a metropolitan area, region, state, or country. The National Plan of Integrated Airport Systems (NPIAS) [11] is an example of a system plan representing the airport development needs of the United States. The Michigan Aviation System Plan [10] is an example of a system plan representing the airport development needs of the state of Michigan, and the Southeast Michigan Regional Aviation System Plan [13] is a system plan representing the airport development needs of a seven county region comprising the Detroit Metropolitan area.

The system plan presents the recommendations for the general location and characteristics of new airports and heliports and the nature of expansion for existing ones to meet forecasts of aggregate demand. It identifies the aviation role of existing and recommended new airports and facilities. It includes the timing and estimated costs of development and relates airport system planning to the policy and objectives of the relevant jurisdiction. Its overall purpose is to determine the extent, type, nature, location, and timing of airport development needed to establish a viable, balanced, and integrated system of airports [1, 8]. It also provides the basis for detailed airport planning such as that contained in the airport master plan.

The airport system plan provides both broad and specific policies, plans, and programs required to establish a viable and integrated system of airports to meet the needs of the region. The objectives of the system plan include

1. The orderly and timely development of a system of airports adequate to meet present and future aviation needs and to promote the desired pattern of regional growth relative to industrial, employment, social, environmental, and recreational goals.

2. The development of aviation to meet its role in a balanced and multimodal transportation system to foster the overall goals of the area as reflected in the transportation system plan and comprehensive development plan.

3. The protection and enhancement of the environment through the location and expansion of aviation facilities in a manner which avoids ecological and environmental impairment.

4. The provision of the framework within which specific airport programs may be developed consistent with the short- and long-range airport system requirements.

5. The implementation of land-use and airspace plans which optimize these resources in an often constrained environment.

6. The development of long-range fiscal plans and the establishment of priorities for airport financing within the governmental budgeting process.

7. The establishment of the mechanism for the implementation of the system plan through the normal political framework, including the necessary coordination between governmental agencies, the involvement of both public and private aviation and nonaviation interests, and compatibility with the content, standards, and criteria of existing legislation.

The airport system planning process must be consistent with state, regional, or national goals for transportation, land use, and the environment. The elements in a typical airport system planning process [8] include the following:

1. Exploration of issues that impact aviation in the study area

2. Inventory of the current system

3. Identification of air transportation needs

4. Forecast of system demand

5. Consideration of alternative airport systems

6. Definition of airport roles and policy strategies

7. Recommendation of system changes, funding strategies, and airport development

8. Preparation of an implementation plan

Although the process involves many varied elements, the final product will result in the identification, preservation, and enhancement of the aviation system to meet current and future demand. The ultimate result of the process will be the establishment of a viable, balanced, and integrated system of airports.

Airport Site Selection

The emphasis in airport planning is normally on the expansion and improvement of existing airports. However if an existing airport cannot be expanded to meet the future demand or the need for a new airport is identified in an airport system plan, a process to select a new airport site may be required. The scope of the site selection process will vary with size, complexity, and role of the new airport, but there are basically three steps—identification, screening, and selection.

Identification—criteria is developed that will be used to evaluate different sites and determine if a site can function as an airport and meets the needs of the community and users. One criterion will be to identify the land area and basic facility requirements for the new airport. Part of this analysis will be a definition of airport roles if more than two airports serve the region. Other criteria might be that sites are within a certain radius or distance from the existing airport or community, or that sites should be relatively flat. Several potential sites that meet the criteria are identified.

Screening—once sites are identified, a screening process can be applied to each site. An evaluation of all potential sites that meet the initial criteria should be conducted, screening out those with the most obvious shortcomings. Screening factors might include topography, natural and man-made obstructions, airspace, access, environmental impacts, and development costs. If any sites are eliminated from further consideration, thorough documentation of the reasons for that decision is recommended. The remaining potential sites should then undergo a detailed comparison using comprehensive evaluation criteria. While the criteria will vary, the following is typically considered:

Operational capability—airspace considerations, obstructions, weather

Capacity potential—available land, suitability for construction, weather

Ground access—distance from the demand for aviation services, regional highway infrastructure, available public transportation modes

Development costs—terrain, land costs, land values, soil conditions, availability of utilities

Environmental consequences—aircraft noise, air quality, groundwater runoff, impact on flora and fauna, existence of endangered species or cultural artifacts, historical features, changes in local land use, relocation of families and businesses, changes in socioeconomic characteristics

Compatibility with area-wide planning—impact on land use, effect on comprehensive land-use plans and transportation plans at the local and regional levels

Selection—the final step is selecting and recommending a preferred site. While a weighting of the evaluation criteria and weighted ratings or ranking of the alternative sites is often used in selecting a site, caution must be used in applying this technique since it introduces an element of sensitivity into the analysis. The process should focus on providing decision makers with information on the various sites in a manner that is understandable and unbiased.

The Airport Master Plan

An airport master plan is a concept of the ultimate development of a specific airport. The term development includes the entire airport area, both for aviation and nonaviation uses, and the use of land adjacent to the airport [1, 4, 9]. It presents the development concept graphically and contains the data and rationale upon which the plan is based. Figure 4-2 shows a simple flowchart of the steps for preparing an airport master plan. Master plans are prepared to support expansion and modernization of existing airports and guide the development of new airports.

The overall objective of the airport master plan is to provide guidelines for future development which will satisfy aviation demand in a financially feasible manner and be compatible with the environment, community development, and other modes of transportation. More specifically it is a guide for

1. Developing the physical facilities of an airport

2. Developing land on and adjacent to the airport

3. Determining the environmental effects of airport construction and operations

4. Establishing access requirements

5. Establishing the technical, economic and financial feasibility of proposed developments through a thorough investigation of alternative concepts

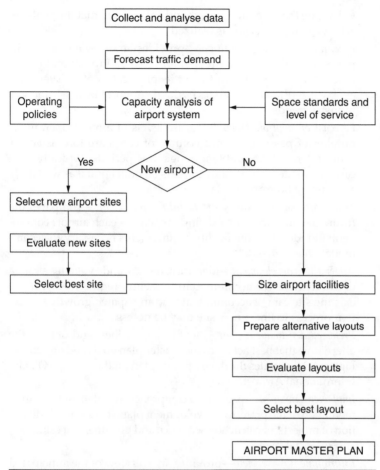

FIGURE 4-2   Flowchart for preparing an airport master plan.

6. Establishing a schedule of priorities and phasing for the improvements proposed in the plan

7. Establishing an achievable financial plan to support the implementation schedule

8. Establishing a continuing planning process which will monitor conditions and adjust plan recommendations as circumstances warrant

Guidelines for completing an airport master plan are described by ICAO [4] and in the United States by the FAA [1]. A master plan report is typically organized as follows:

Master plan vision, goals, and objectives—establishes the vision and overarching goals for the master plan as well as objectives that will guide the planning process and help ensure that the goals are achieved and the vision is realized.

Inventory of existing conditions—provides an overview of the airport's history, role in the region and nation, growth and development over time, description of its physical assets (airfield and airspace, terminal, ground access, and support facilities), and key industry trends.

Forecast of aviation demand—future levels of aircraft operations, number of passengers, and volume of cargo are forecasted for short, intermediate, and long-range time periods. Typically forecasts are made for 5, 10, and 20 years on both annual as well daily and busiest hours of the day.

Demand/capacity analysis and facility requirements—compares the future demand with the existing capacity of each airport component and identifies the facility requirements necessary to accommodate the demand.

Alternatives development—identifies, refines, and evaluates a range of alternatives for accommodating facility requirements. If the existing site cannot accommodate the anticipated growth, a selection process to find a new site may be necessary.

Preferred development plan—identifies, describes, and defines the alternative that best achieves the master plan goals and objectives. Figure 4-3 illustrates the development plan for the Chicago O'Hare International Airport.

Implementation plan—provides a comprehensive plan for the implementation of the preferred development plan, including the definition of projects, construction sequence and timeline, cost estimates, and financial plan.

Environmental overview—provides an overview of the anticipated environmental impacts associated with the preferred development plan in order to understand the severity and to help expedite subsequent environmental processing at the project specific stage.

Airport plans package—documents that show the existing as well as planned modifications are prepared and the more notable is the airport layout plan (ALP). It comprises drawings that include the airfield's physical facilities, obstruction clearance and runway approach profiles, land-use plans, terminal area and ground access plans, and a property map. Specific guidelines for the airport layout plan in the United States are identified by FAA [1].

Stakeholder and public involvement—documents the coordination efforts that occur among the stakeholders throughout the study.

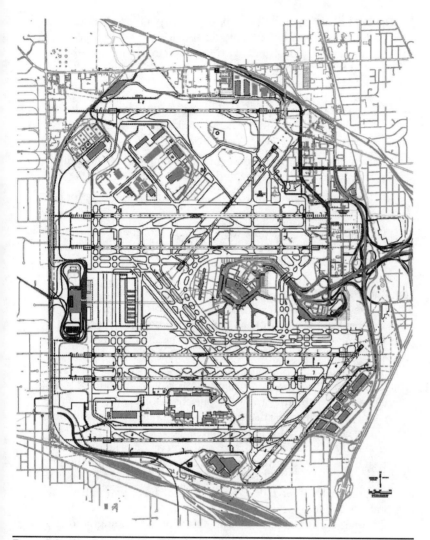

FIGURE 4-3   Development plan for Chicago O'Hare International Airport.

The Airport Project Plan

A project plan focuses on a specific element of the airport master plan which is to be implemented in the short term and may include such items as the addition of a new runway, the modification of existing of runways, the provision of taxiways or taxiway exits, the addition of gates, the addition to or the renovation of terminal building facilities, or the modification of ground access facilities.

The overall objective of the airport project plan is to provide the specific details of the development which will satisfy immediate aviation needs and be consistent with the objectives and constraints identified in the airport master plan. More specifically it is a detailed plan for

1. Developing the specific physical facilities at an airport including the architectural and engineering design for these facilities

2. Determining the environmental effects of this development through the construction and operational phases

3. Determining the detailed costs and financial planning for the development

4. Establishing a schedule for the construction and phasing of the specific items of development in the plan

Land-Use Planning

A land-use plan for property within the airport boundary and in areas adjacent to the airport is an essential part of an airport master plan. The land-use plan on and off the airport is an integral part of an area wide comprehensive planning program, and therefore it must be coordinated with the objectives, policies, and programs for the area which the airport is to serve. Incompatibility of the airport with its neighbors stems primarily from the objections of people to aircraft noise. A land-use plan must therefore project the extent of aircraft noise that will be generated by airport operations in the future. Contours of equal intensity of noise can be drawn and overlaid on a land-use map and from these contours an estimate can be made of the compatibility of existing land use with airport operations. If the land outside the airport is underdeveloped, the contours are the basis for establishing comprehensive land-use zoning requirements.

Although zoning is used as a method for controlling land use adjacent to an airport, it is not effective in areas which are already built-up because it is usually not retroactive. Furthermore jurisdictions having zoning powers may not take effective zoning action. Aircraft operations into and out of the airport may be made unnecessarily complex to minimize noise encroachment on incompatible land uses. Despite these shortcomings the planner should utilize zoning as a vehicle to achieve compatibility wherever this approach is feasible.

Airports become involved in two types of zoning. One type is height and hazard zoning, which is mainly to protect the approaches to the airport from obstructions. The other type is land-use zoning.

The extent of land use in the airport depends a great deal on the amount of acreage available. Land uses can be classified as either closely related to aviation or remotely related to aviation. Those closely related to aviation use include the runways, taxiways, aprons, terminal buildings, parking, and maintenance facilities. Nonaviation uses include space for recreational, industrial, and commercial activities. When considering commercial or industrial activities, care should be taken to ensure that they will not interfere with aircraft operations, communications equipment, and aids to navigation on the ground. Recreational facilities such as golf courses may be suitable within the immediate proximity of the airport boundary or certain agricultural uses are also appropriate as long as they do not attract birds. When there is acreage within the airport boundary in excess of aviation needs, it is sound fiscal planning to provide the greatest financial return from leases of the excess property. Thus the land-use plan within the airport is a very effective tool in helping airport management make decisions concerning requests for land use by various interests and often airports delineate areas on the airport property for the development of industrial parks.

The principal objective of the land-use plan for areas outside the airport boundary is to minimize the disturbing effects of noise. As stated earlier the delineation of noise contours is the most promising approach for establishing noise-sensitive areas. The contours define the areas which are or are not suitable for residential use or other use and, likewise, those which are suitable for light industrial, commercial, or recreational activity. Although the responsibility for developing land uses adjacent to the airport lies with the governing bodies of adjacent communities, the land-use plan provided by the airport authority will greatly influence and assist the governing bodies in their task of establishing comprehensive land-use zoning.

Environmental Impact Assessment

Environmental factors must be considered carefully in the development of a new airport or the expansion of an existing one. In the United States, this is a requirement of the Airport and Airway Improvement Act of 1982 and the Environmental Policy Act of 1969. Studies of the impact of the construction and operation of a new airport or the expansion of an existing one upon acceptable levels of air and water quality, noise levels, ecological processes, and demographic development of the region must be conducted to determine how the airport requirements can best be met with minimal adverse environmental and social consequences.

Aircraft noise is the severest environmental problem to be considered in the development of airport facilities. Much has been done to quiet engines and modify flight procedures, resulting in substantial reductions in noise. Another effective means for reducing noise is through proper planning of land use for areas adjacent to the airport. For an existing airport this may be difficult as the land may have already been built up. Every effort should be made to orient air traffic away from noise-sensitive land development.

Other important environmental factors include air and water pollution, industrial wastes and domestic sewage originating at the airport, and the disturbance of natural environmental values. In regard to air pollution, the federal government and industry have worked jointly toward alleviating the problem, and there is a reason to believe that it will probably be eliminated in the near future as an environmental factor. An airport can be a major contributor to water pollution if suitable treatment facilities for airport wastes are not provided. Chemicals used to deice aircraft are a major source of potential ground water pollution and provisions need to be made to safely dispose of this waste product. The environmental study must include a statement detailing the methods for handling sources of water pollution.

The construction of a new airport or the expansion of an existing one may have major impacts on the natural environment. This is particularly true for large developments where streams and major drainage courses may be changed, the habitats of wildlife may be disrupted, and wilderness and recreational areas may be reshaped. The environmental study should indicate how these disruptions might be alleviated.

In the preparation of an environmental study, or an environmental impact statement, the findings must include the following items [12]:

1. The environmental impact of the proposed development

2. Any adverse environmental effects which cannot be avoided should the development be implemented

3. Alternatives to the proposed development

4. The relationship between local short-term uses of the environment and the maintenance and enhancement of long-term productivity

5. Any irreversible environmental and irretrievable commitments of resources which would be involved in the proposed development should it be implemented

6. Growth inducing impact

7. Mitigation measures to minimize impact

In the application of these guidelines attention must be directed to the following questions. Will the proposed development

1. Cause controversy

2. Noticeably affect the ambient noise level for a significant number of people

3. Displace a significant number of people

4. Have a significant aesthetic or visual effect

5. Divide or disrupt an established community or divide existing uses

6. Have any effect on areas of unique interest or scenic beauty

7. Destroy or derogate important recreational areas

8. Substantially alter the pattern of behavior for a species

9. Interfere with important wildlife breeding, nesting, or feeding grounds

10. Significantly increase air or water pollution

11. Adversely affect the water table of an area

12. Cause excessive congestion on existing ground transportation facilities

13. Adversely affect the land-use plan for the region

The preparation of an environmental impact statement based upon an environmental assessment study is an extremely important part of the airport planning process. The statement should clearly identify the problems that will affect environmental quality and the proposed actions to alleviate them. Unless the statement is sufficiently comprehensive, the entire airport development may be in jeopardy.

Economic and Financial Feasibility

The economic and financial feasibility of alternative plans for a new airport or expansion of an existing site must be clearly demonstrated by the planner. Even if the selected alternative is shown to be economically feasible, then also it is necessary to show that the plan will generate sufficient revenues to cover annual costs of capital investment, administration, operations, and maintenance. This must be determined for each stage or phase of development detailed in the airport master plan.

An evaluation of economic feasibility requires an analysis of benefits and costs. A comparison of benefits and costs of potential capital investment programs indicates the desirability of a project from an economic point of view. The economic criterion used in evaluating an aviation investment is the total cost of facilities, including quantifiable social costs, compared with the value of the increased effectiveness measured in terms of total benefits. The costs include capital investment, administration, operation, maintenance, and any other costs that can be quantified. The benefits include a reduction in aircraft and passenger delays, improved operating efficiency, and other benefits. The costs and benefits are usually determined on an annual basis.

There are a number of techniques for comparing benefits with costs. Most of them consider the time value of money based on an appropriate discount rate which reflects the opportunity cost of capital. The discount rate is a value by which a unit of money received in the future is multiplied to obtain its present value or present worth. In other words a cost incurred in 2010 has a different economic value from that of the same item incurred in 2015.

If the time value of money is not considered, the ratio of benefits to costs is made for each year by merely dividing the benefits in a particular year by the cost of the project in that year. A project is considered economically feasible when the ratio of the benefits to costs is greater than unity, that is, the benefits exceed the costs. The larger the ratio, the more attractive is the project from an economic standpoint. A ratio can also be obtained by comparing the present value of benefits with the present value of costs. This approach recognizes the time value of money. Another approach is to plot the *net present value* (NPV) for each year against time. The net present value is defined as the present value of benefits minus the present value of costs.

The financing of capital improvements for airports is discussed in Chap. 13. In the early years of airport development, substantial capital improvement programs were financed at the local level by sale of general obligation bonds backed by the taxing power of the community. As air transportation became mature and the requirements of the community for capital spending programs increased, airports began to utilize revenue bonds as a source of financing. A financial feasibility study is therefore an analysis to determine if bonds are marketable at reasonable interest rates. It also includes the feasibility of other forms of financing. The analysis requires a thorough evaluation of the revenues to be developed by a proposed improvement and the corresponding costs. Usually this is done in a traffic and earnings study performed over the planning horizon. In such a study, the forecast of demand is utilized and rates and charges established for the various revenue categories. This results in annual revenue projections. To make revenue bonds attractive to buyers a typical airport revenue bond should show an expected coverage by net revenues (gross revenues minus costs) of at least 1.25 times the debt service requirements. If the analysis indicates that the revenues will be insufficient, revisions in the scheduling or scope of the proposed development may have to be made or the rates and charges to the users of the airport may require adjustment.

Continuing Planning Process

A continuous airport planning process is necessary in order to respond to the needs of air transportation in a changing environment [8]. Changes in aviation demand, community policies, new technology, financial constraints, and other factors can alter the need for and the timing of facility improvements. Current data must be continually collected and assessed relative to airport needs, operations and utilization, environmental impact, and financial capabilities. The staging of airport improvements assists in the reevaluation of continuing needs at the points in time when implementation decisions are required.

In the airport planning process, the overall objective of establishing and maintaining a continuous process is to ensure that the airport system plan and airport master plan remain responsive to public needs. As a result, the airport system plan and airport master plan should be formally reviewed and updated at least every 5 years. Specific objectives associated with the continuous airport planning process include [8]

1. Surveillance, maintenance, inventory, and update of the basic data such as aviation activity and socioeconomic and environmental factors relating to the existing airport system and master plan

2. Review and validation of data affecting the airport system and master plan

3. Reappraisal of the airport system and master plan in view of changing conditions

4. Modification of the airport system and master plan to retain its viability

5. Development of a continuous mechanism for ensuring the interchange of information between the system planning and master planning processes

6. Provision of a means for receiving and considering public comment in order to maintain and ensure the public awareness of the role airports play in the transportation system of an area

7. Redefinition of air transportation goals and policies

8. Integration of airport system planning into a multimodal planning process

9. Analysis of special issues

10. Publication of interim reports and formal plan updates

References

1. *Airport Master Plans,* Advisory Circular, AC 150/5070-6B, Federal Aviation Administration, Washington, D.C., July 2005.
2. *Airport Planning and Development Handbook—A Global Survey,* Paul Stephen Dempsey, McGraw-Hill, New York, N.Y., 2000.
3. *Airport Planning and Management,* 5th ed., Alexander T. Wells and Seth B. Young, McGraw-Hill, New York, N.Y., 2004.

4. *Airport Planning Manual, Part 1, Master Planning,* 2d ed., International Civil Aviation Organization, Montreal, Canada, 1987.
5. *Airport Systems—Planning, Design and Management,* Richard de Neufville and Amedeo Odoni, McGraw-Hill, New York, N.Y., 2003.
6. Airport System Capacity, Strategic Choices, Special Report, No. 226, Transportation Research Board, Washington, D.C., 1990.
7. *Airport Systems Planning,* R. DeNeufville, The MIT Press, Cambridge, Mass., 1976.
8. *The Airport System Planning Process,* Advisory Circular, AC 150/5070-7, Federal Aviation Administration, Washington, D.C., November 2004.
9. *Guide for the Planning of Small Airports,* Roads and Transportation Association of Canada, Ottawa, Canada, 1980.
10. *Michigan Airport System Plan,* MASP 2000, Michigan Department of Transportation, Lansing, Mich., January 2000.
11. *National Plan of Integrated Airport Systems (NPIAS)2007-2011,* Federal Aviation Administration, U.S. Department of Transportation, Washington, D.C., September 2006.
12. *Policies and Procedures for Considering Environmental Impacts,* Federal Aviation Administration, Washington, D.C., December 1986.
13. *Southeast Michigan Regional Aviation System Plan,* Technical Report, Southeast Michigan Council of Governments, Detroit, Mich., April 1992.
14. *Strategic Airport Planning,* Robert E. Caves and Geoffrey D. Gosling, Elsevier Science Limited, Oxford, U.K., 1999.

CHAPTER 5

Forecasting for Airport Planning

Introduction

Plans for the development of the various components of the airport system depend to a large extent on the activity levels which are forecast for the future. Since the purpose of an airport is to process aircraft, passengers, freight, and ground transport vehicles in an efficient and safe manner, airport performance is judged on the basis of how well the demand placed upon the facilities within the system is handled. To adequately assess the causes of performance breakdowns in existing airport systems and to plan facilities to meet future needs, it is essential to predict the level and distribution of demand on the various components of the airport system. Without a reliable knowledge of the nature and expected variation in the loads placed upon a component, it is impossible to realistically assess the physical and operational requirements of such a component. For example, a forecast to project the mix of aircraft and the types of aviation activity at an airport site is necessary to identify the critical aircraft which dictates the elements of geometric and structural design, the type and extent of physical facilities, the navigational aid requirements, and any special or unique facility needs at the airport [15, 17].

An understanding of future demand patterns allows the planner to assess future airport performance in light of existing and improved facilities, to evaluate the impact of various quality of service options on the airlines, travelers, shippers, and community, to recommend development programs consistent with the overall objectives and policies of the airport operator, to estimate the costs associated with these facility plans, and to project the sources and level of revenues to support the capital improvement program.

It is essential in the planning and design of an airport to have realistic estimates of the future demand to which airports are likely to be subjected. This is a basic requirement in developing either an airport master plan or an airport system plan. These estimates determine the future needs for which the physical facilities are designed. A financial plan to achieve the recommended staged development along with required land-use zoning usually accompanies the plan. It should be apparent that an airport is designed for a projected level and pattern of demand and changes in the magnitude or characteristics of this demand may require facility modifications or operational measures to meet changing needs. Facility planning is necessary to provide adequate levels of service to airport users.

The development of accurate forecasts requires a considerable expense of time and other resources because of the complex methodologies which must be used and the extensive data acquisition that is often required. The usual justification for a demand forecast in an aviation plan is that the expected level of uncertainty associated with the estimation of essential variables will be reduced, thereby reducing the probability of errors in the planning process and enhancing the decision-making process. The implication, of course, is that the benefits gained due to a better knowledge of the magnitude and fluctuation in demand variables will outweigh the costs incurred in performing the forecast.

To assess the characteristics of future demand, the development of reliable predictions of airport activity is necessary. There are numerous factors that will affect the demand and planners who are preparing forecasts of demand or updating existing forecasts should consider local and regional socioeconomic data and characteristics, demographics, geographic attributes, and external factors such as fuel costs and quality of service parameters. Political developments, including rising international tensions, changes in security, airline delays and congestion, and travel attitudes, will impact demand. Actions taken by local airport authorities, such as changes in user charges, can also stimulate or hinder the demand and investment decisions made as the result of the planning process itself can also produce change by removing physical constraints to growth [2].

Over the years, certain techniques have evolved which enable airport planners and designers to forecast future demand. The principal items for which estimates are usually needed include

- The volume and peaking characteristics of passengers, aircraft, vehicles, and cargo
- The number and types of aircraft needed to serve the above traffic
- The number of based general aviation aircraft and the number of movements generated
- The performance and operating characteristics of ground access systems

Using forecasting techniques, estimates of these parameters and a determination of the peak period volumes of passengers and aircraft movements can be made. From these estimates concepts for the layout and sizing of terminal buildings, runways, taxiways, apron areas, and ground access facilities may be examined.

Forecasting demand in an industry as dynamic as aviation is an extremely difficult matter, and if it could be avoided it undoubtedly would. Nonetheless estimates of traffic must be made as a prelude to the planning and design of facilities. It is very important to remember that forecasting is not a precise science and that considerable subjective judgment must be applied to any analysis no matter how sophisticated the mathematical techniques involved. By anticipating and planning for variations in predicted demand, the airport designer can correct projected service deficiencies before serious deficiencies in the system occur.

Levels of Forecasting

Demand estimates are prepared for a variety of reasons. Broad large-scale aggregated forecasts are made by aircraft and equipment manufacturers, aviation trade organizations, governmental agencies, and others to determine estimates of the market requirements for aviation equipment, trends in travel, personnel needs, air traffic control requirements, and other factors. Similarly, forecasts are made on a smaller scale to examine these needs in particular regions of an area and at specific airports.

In economics, forecasting is done on two levels, aggregate forecasting and disaggregate forecasting, and the same holds true in aviation. From the inception of the planning process for an airport consideration is given at both levels. In airport planning, the designer must view the entire airport system as well as the airport under immediate consideration. Aggregate forecasts are forecasts of the total aviation activity in a large region such as a country, state, or metropolitan area. Typical aggregate forecasts are made for such variables as the total revenue passenger-miles, total enplaned passengers, and the number of aircraft operations, aircraft in the fleet, and licensed pilots in the country. Disaggregate forecasts deal with the activity at individual airports or on individual routes. Disaggregate forecasts for airport planning determine such variables as the number of originations, passenger origin-destination traffic, the number of enplaned passengers, and the number of aircraft operations by air carrier and general aviation aircraft at an airport. Separate forecasts are usually made, depending on the need in a particular study, for cargo movements, commuter service, and ground access traffic. These forecasts are normally prepared to indicate annual levels of activity and are then disaggregated for airport planning purposes to provide forecasts of the peaking characteristics of traffic during the busy hours of the day, days of the week, and months of the year. As appropriate to the requirements of an airport planning study, forecasts of such quantities as the number of general aviation aircraft based at an individual airport and the number of general aviation and military operations are also prepared.

In aggregate forecasting, the entire system of airports is examined relative to the geographic, economic, industrial, and growth characteristics of a region to determine the location and nature of airport needs in a region. The disaggregate forecast then examines the expected demand at local airports and identifies the necessary development of the airside, landside, and terminal facilities to provide adequate levels of service. Within the two levels of forecasting there are certain techniques which enable the planner to project such parameters as annual, daily, and hourly aircraft operations, passenger enplanements, cargo, and general aviation activity. In disaggregate forecasting, there are many significant variables. The forecast of each variable is quite important because it ultimately determines the size requirements of the facilities which will be necessary to accommodate demand. Often, the forecasts of the different variables are linked by a series of steps, that is, originating passengers are forecast first, this then becomes a component in the forecast of enplaned passengers, which leads to a forecast of annual operations, and so on.

The type of forecast and the level of effort depend on the purpose for which the forecast is being used. Forecasts are typically prepared for short-, medium-, and long-term periods. Short-term forecasts, up to 5 years, are used to justify near-term development and support operational planning and incremental improvements or expansion of facilities. Medium-term forecasts, a 6- to 10-year time frame, and long-term forecasts of 10 to 20 years are used to plan major capital improvements, such as land acquisition, new runways and taxiways, extensions of a runway, a new terminal, and ground access infrastructure. Forecasts beyond 20 years are used to assess the need for additional airports or other regional aviation facilities [1].

Forecasting Methods

There are several forecasting methods or techniques available to airport planners ranging from subjective judgment to sophisticated mathematical modeling. The selection of the particular methodology is a function of the use of the forecast, the availability of a database, the complexity and sophistication of the techniques, the resources available, the time frame in which the forecast is required and is to be used, and the degree of precision desirable. There are four major methods:

- Time series method
- Market share method
- Econometric modeling
- Simulation modeling

Time series analysis essentially involves extrapolating or projecting existing historical activity data into the future. Market share forecasting is a simple top-down approach, where current activity at an airport is calculated as a share of some other more aggregate measure for which a forecast has been made (typically a regional, state, or national forecast of aviation activity). Econometric modeling is a multistep process in which a casual relationship is established between a dependent variable (the item to be forecast) and a set of independent variables that influence the demand for air travel. Once the relationship is established, forecasts of independent variables are input to determine a forecast of the dependent variable. These techniques can also be referred to as a bottom-up forecast. Simulation models are often used when one needs very detailed estimates of aircraft, passengers, or vehicles. These models impose precise rules that govern how passengers, aircraft, or vehicles are routed, and then aggregates the results so that planners can assess the needs of the network or a component of the airport to handle the estimated demand. Typically the outputs from the other forecasting methods are used as inputs to simulation models. Forecasts from simulation models represent snapshots of how a given amount of traffic flows across a network or through an airport, rather than a monthly or annual estimate of total traffic.

An important element which should be utilized in any forecasting technique is the use of professional judgment. A forecast prepared through the use of mathematical relationships must ultimately withstand the test of rationality. Frequently a group of professionals knowledgeable about aviation and the factors influencing aviation trends are assembled to examine forecasts from several different sources, and composite forecasts are prepared in accordance with the information in these sources and the collective judgment of the group. In some cases, judgment becomes the principal approach used with or without an evaluation of economic and other factors that are believed to affect aviation activity. A common approach being utilized more often today for preparing forecasts by judgment is known as the Delphi method. In this method a panel of experts on a particular subject matter is asked to rate or otherwise prioritize a series of questions or projections through a survey technique. The results of the survey are then distributed to the members of the panel and an opportunity is provided for each member to reevaluate the original rating based upon the collective ratings of the group. The reevaluation process is often sent through several iterations in order to arrive at a better result. In the Delphi method, the results of the technique do not have to represent a consensus of the panel and, in fact, it is often quite useful to have a forecast which indicates the spread of the panel in reaching conclusions on a particular issue.

The preparation of judgmental forecasts which reflect the collective wisdom of a broad range of professionals has proven to be very successful in many instances principally due to the large number of factors which may be considered in such a process. Though there is often a lack of mathematical sophistication in the process, the knowledge and consideration of the many diverse factors influencing aviation forecasts usually improves the results. The disadvantages of this forecasting technique include the absence of statistical measures on which to base the results and the inability, except in the most obvious cases, to gain a significant consensus relative to the expected performance of the explanatory factors in the future.

Time Series Method

Time series analysis or extrapolation is based upon an examination of the historical pattern of activity and assumes that those factors which determine the variation of traffic in the past will continue to exhibit similar relationships in the future. This technique utilizes times series type data and seeks to analyze the growth and growth rates associated with a particular aviation activity. In practice, trends appear to develop in situations in which the growth rate of a variable is stable in either absolute or percentage terms, there is a gradual increase or decrease in growth rate, or there is a clear indication of market saturation trend over time [11]. Statistical techniques are used to assist in defining the reliability and the expected range in the extrapolated trend. The analysis of the pattern of demand generally requires that upper and lower bounds be placed upon the forecast and statistics are used to define the confidence levels within which specific projections may be expected to be valid. From the variation in the trends and the upper and lower bounds placed on the forecast a preferred forecast is usually developed. Quite often smoothing techniques are incorporated into the forecast to eliminate short run, or seasonal, fluctuations in a pattern of activity which otherwise demonstrates a trend or cyclical pattern in the long run [16].

An illustration of the application of a trend line analysis to forecast annual enplanements at an airport is shown in Example Problem 5-1.

Example Problem 5-1   The historical data shown in Table 5-1 have been collected for the annual passenger enplanements in a region and one of the commercial service airports in this region. It is necessary to prepare a forecast of the annual passenger enplanements at the study airport in the design years 2010 and 2015 using a trend line analysis.

In applying the trend line analysis to these data, a forecast technique of the annual enplanements at the study airport will be made by forecasting the historical trend to the design years. A plot of the trend in the annual passengers enplaned at the study airport is given in Fig. 5-1.

By extrapolating the trend into the future, an estimate of the annual enplaned passengers at the study airport in 2010 is found to be 2,100,000 passengers and in the year 2015 is found to be 2,900,000 passengers.

Year	Annual Enplanements Regional	Airport	Area Population
1998	13,060,000	468,900	250,000
1999	14,733,000	514,300	260,000
2000	16,937,000	637,600	270,000
2001	21,896,000	758,200	280,000
2002	24,350,000	935,200	290,000
2003	28,004,000	995,500	300,000
2004	31,658,000	1,139,700	310,000
2005	37,226,000	1,360,700	320,000
2006	40,753,000	1,488,900	330,000
2007	44,018,000	1,650,600	340,000

TABLE 5-1   Enplanement Data for Airport Demand Forecast for Use in Example Problems 5-1 through 5-3

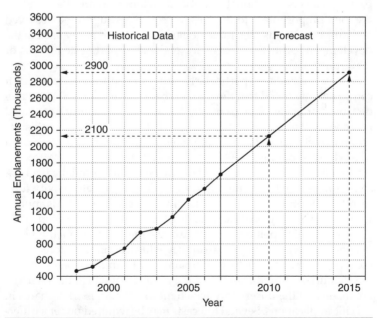

FIGURE 5-1   Trend line forecast of annual enplanements for Example Problem 5-1.

The inability of time series techniques to show a causal relationship between the dependent and independent variables is a serious disadvantage. This is particularly true because, in the absence of such relationships, the degree of uncertainty in such forecasts increases with time. However, the time series method is useful for short-term forecasts in which the response to changes in those factors which stimulate the dependent variables is usually less dynamic. In those cases where cyclic variations may be expected to occur, time series methods may also be quite beneficial.

Market Share Method

Forecasting techniques which are utilized to proportion a large-scale aviation activity down to a local level are called market share, ratio, or top-down models. Inherent to the use of such a method is the demonstration that the proportion of the large-scale activity which can be assigned to the local level is a regular and predictable quantity. This method has been the dominant technique for aviation demand forecasting at the local level and its most common use is in the determination of the share of total national traffic activity which will be captured by a particular region, traffic hub, or airport. Historical data are examined to determine the ratio of local airport traffic to total national traffic and the trends are ascertained. From exogenous sources the projected levels of national activity are determined and these values are then proportioned to the local airport based upon the observed and projected trends. The ratio method is most commonly used in the development of microforecasts for regional airport system plans or for airport master plans.

These methods are particularly useful in applications in which it can be demonstrated that the market share is a regular, stable, or predictable parameter. For example, the number of annual enplaned passengers at major air traffic hubs has been shown to be a consistent and relatively stable factor and, therefore, this method is often used to predict this parameter.

Quite often the application of the market share technique is a two-step process in which a ratio is applied to disaggregate activity forecasts from a national to a regional level and then another ratio is applied to apportion the regional share among the airports in the region.

The most compelling advantage of the market share method is its dependence on existing data sources which minimizes forecasting cost. However, its principal disadvantages lie in its dependence on the stability and predictability of the ratios from which the forecasts are made and the uncertainty which may surround market shares in specific applications. Several forecasts may be required under a differing set of assumptions which are deemed appropriate to the determination of market shares. An illustration of the application of a market share analysis is given in Example Problem 5-2.

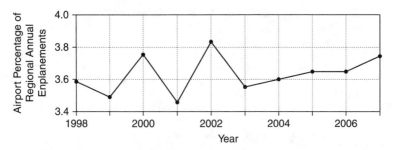

Figure 5-2   Trend line of airport percentage of regional annual enplanements for Example Problem 5-2.

Example Problem 5-2   The historical data shown in Table 5-1 could also be used to prepare a forecast of the annual passenger enplanements at the study airport in the design years 2010 and 2015 using a market share method.

In applying the market share method to these data, a top-down forecast technique will be used. The implicit assumption in such a technique is that the same relationship between regional annual enplanements and the annual enplanements at the study airport will be maintained in the future. To prepare such a forecast, a projection of the percentage of the regional annual enplanements captured at the study airport is performed and then a forecast is made of the regional annual enplanements. The study airport forecast percentage is applied to the regional forecast to arrive at the forecast of the study airport annual enplanements in the design years. A plot of the trend in the percentage of regional annual passengers enplaned at the study airport is given in Fig. 5-2.

Because the variations shown in Fig. 5-2 often make it difficult to determine if trends may exist, a smoothing function is often applied to the data to assist in identifying trends which may not be obvious. In this case, a smoothing of the data was obtained by computing a running 3-year average of the data points. As is shown in Fig. 5-3, this tends to smooth out the variations in the original data and more readily identifies trends in these data.

Though it may not be apparent in the original plot, the smoothing mechanism does indicate a very slight upward trend in the percentage of regional annual passengers captured by the study airport. This trend is shown by the dashed line in Fig. 5-3. This trend line, when projected to the design years, indicates a forecast of

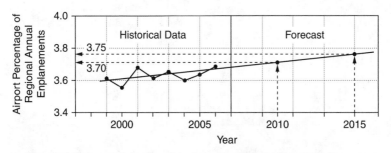

Figure 5-3   Trend line forecast of airport percentage of regional annual enplanements by applying smoothing function to trend data for Example Problem 5-2.

3.70 percent in the year 2010 and 3.75 percent in the year 2015 as the proportion of regional annual enplanements forecast to be captured by the study airport.

To complete the forecast by the market share method, a determination of the regional annual enplanements must be made. This is done by extrapolating the trend in regional annual enplanements as shown in Fig. 5-4. This extrapolation indicates regional annual enplanements in the year 2010 of 52,500,000 and in the year 2015 of 63,500,000.

Therefore, the forecast for the annual enplanements at the study airport becomes 0.0370 × 52,500,000 = 1,942,500 passengers in the year 2010 and 0.0375 × 63,500,000 = 2,381,250 passengers in the year 2015.

Econometric Modeling

The most sophisticated and complex technique in airport demand forecasting is the use of econometric models. Trend extrapolation methods do not explicitly examine the underlying relationships between the projected activity descriptor and the many variables which affect its change. There are a wide range of economic, social, market, and operational factors which affect aviation. Therefore, to properly assess the impact of predicted changes in the other sectors of society upon aviation demand and to investigate the effect of alternative assumptions on aviation, it is often desirable to use mathematical techniques to study the correlations between dependent and independent variables. Econometric models which relate measures of aviation activity to economic and social factors are extremely valuable techniques in forecasting the future.

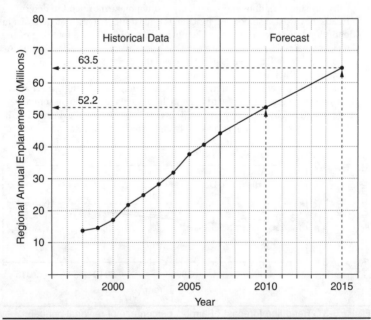

FIGURE 5-4   Trend line forecast of regional annual enplanements for Example Problem 5-2.

There are a great variety of techniques which are used in econometric modeling for airport planning. Classical trip generation and gravity models are quite common in forecasting passenger and aircraft traffic. Simple and multiple regression analysis techniques, both linear and nonlinear, are often applied to a great variety of forecasting problems to ascertain the relationships between the dependent variables and such explanatory variables as economic and population growth, market factors, travel impedance, and intermodal competition. The form of the equations used in multiple linear regression analysis is given in Eq. (5-1).

$$Y_{est} = a_0 + a_1 X_1 + a_2 X_2 + a_3 X_3 + \cdots + a_n X_n \qquad (5\text{-}1)$$

where Y_{est} = dependent variable or variable which is being forecast

$X_1, X_2, X_3, \ldots, X_n$ = dependent variables or variables being used to explain the variation in the dependent variable

$a_0, a_1, a_2, a_3, \ldots, a_n$ = regression coefficients or constants used to calibrate the equation

There are many statistical tests which can be performed to determine the validity of econometric models in accurately portraying historical phenomena and to reliably project demand. Though the constants may be found to define the general equation of the model, it is possible that the range of error associated with the equation may be large or that the explanatory variables chosen do not directly determine the variation in the dependent variable.

There may be a tendency when performing sophisticated mathematical modeling to become disassociated from the significance of the results. It is incumbent upon the analyst to consider the reasonableness as well as the statistical significance of the model. Adequate consideration must be given to the rationality of the functional form and variables chosen for the analysis, and to the logic associated with calibrated constants.

In many cases it is essential to determine the sensitivity of forecasts to changes in the explanatory variables. If a particular design parameter being forecast varies considerably with a change in a dependent variable, and there is a significant degree of unreliability in this independent variable, then a great deal of confidence cannot be placed upon the forecast and, more importantly, the design based upon the forecast. Tests are usually performed to determine the explanatory power of the independent variables and their interrelationships. The analyst should carefully investigate the sensitivity of projections within the expected variation of explanatory variables. It is also possible that certain explanatory variables do not significantly affect the modeling equation and the need for collecting the data associated with these variables required for projections could be eliminated. An illustration of the application of simple linear regression analysis is presented in Example Problem 5-3.

Example Problem 5-3   The historical data shown in Table 5-1 could also be used to prepare a forecast of the annual passenger enplanements at the study airport in the design years 2010 and 2015 using a simple regression analysis.

In applying simple regression analysis to these data, let us assume that a relationship between the study airport annual enplanements (ENP) and the study area population (POP) is to be examined. Therefore, it is assumed that a linear relationship of the form shown in Eq. (5-1) exists between the variables.

$$ENP = a_0 + a_1(POP)$$

Using a standard regression analysis computer program the relationship is found to be

$$ENP = -3,047,032 + 13.8633(POP)$$

where the coefficient of determination R^2 is 0.983815, the coefficient of correlation is 0.991874, and the standard error of the estimate, σ_{yest} is 55,520.9.
The regression line and the data points upon which this regression line is based are shown in Fig. 5-5.

The coefficient of determination indicates that there is an extremely good relationship between the annual enplanements at the study airport and the study area population, that is, 98.4 percent of the variation in the study airport annual enplanements is explained by the variation in the study area population.

The standard error of the estimate, however, indicates that there is a large range of error associated with forecasting with this equation, that is, there is a 68 percent probability that the forecast of annual enplanements at the study airport will have an error range of ± 55,520.9 annual enplanements. This may

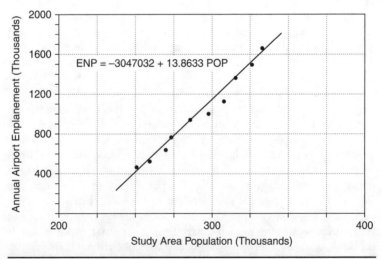

FIGURE 5-5   Trend line forecast of study area population for Example Problem 5-3.

or may not be too high depending on the level of annual operations forecast in the future and the sensitivity of various components of the airport system to such variations.

Using a trend projection, it is forecast that the area population in the year 2010, as shown on Fig. 5-6, is expected to be 363,000. The forecast of the annual enplanements at the airport in the year 2010 can be found by substitution into the regression equation yielding 1,985,300 annual enplanements. Similarly, if the forecast of the area population in the year 2015 is expected to be 410,000, then the forecast of the annual enplanements at the airport in the year 2015 is found to be 2,636,900.

Given the range in the standard error of the estimate, it could be expected that in the year 2010 there is a probability of 68 percent that the forecast could range between 1,985,300 ± 55,500, or from 1,929,800 to 2,040,800 annual enplanements about 68 percent of the time. Similarly, it could be expected that in the year 2015 the forecast could range between 2,636,900 ± 55,500, or from 2,581,900 to 2,692,400. It is likely that this range in the forecasts is acceptable since it represents about a 2 to 3 percent error.

It is interesting to compare the results found by the three different techniques used in Example Problems 5-1 through 5-3. The results compare very well and it gives one some degree of confidence in the results when the three forecasts compare well. This is called redundancy in forecasting.

Based upon the results found in these example problems, a preferred forecast would be developed. If there is no reason to suspect

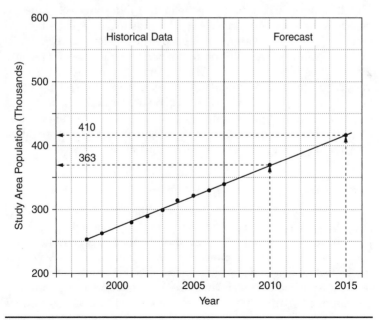

Figure 5-6 Trend line forecast of study area population for Example Problem 5-3.

that one technique is better than another, then a simple average might be used to develop the preferred forecast. If this is done in these examples, then the preferred forecast for the year 2010 is about 2,000,000 annual enplaned passengers and in the year 2015 is 2,600,000 annual enplaned passengers.

The Federal Aviation Administration utilizes econometric models to determine national forecasts of U.S. aviation demand. The FAA Aerospace Forecast [8] provides a 12-year outlook and view of the immediate future for aviation. It is updated in March each year and includes aggregate level forecasts of the following:

- Passenger enplanements, revenue passenger miles, fleet, and hours flown for large carriers and regional commuters
- Cargo revenue ton miles and cargo fleet for large air carriers
- Fleet, hours, and pilots for general aviation
- Activity forecasts for FAA and contract towers by major user category

The FAA Long Range Aerospace Forecasts [9] is a long range forecast that extends the 12-year forecast to a longer time horizon, typically for a period of 25 years. This forecast contains projections of aircraft, fleet and hours, air carrier and regional/commuter passenger enplanements, air cargo freight revenue ton-miles, pilots, and FAA workload measures.

The success in applying mathematical modeling techniques to ascertain the level of future activity depends to a large extent on the certainty associated with the independent variables and the relative influence of these variables on the dependent variable. Simple and multiple regression analysis methods are often applied to a great variety of forecasting problems to determine the relationships between transport related variables and such explanatory factors as economic and population growth, market factors, travel impedance, and competitive forces. Table 5-2 lists many of the variables required for various purposes in aviation planning studies.

Forecasting Requirements and Applications

The specific forecasting needs depend on the nature and scope of the study being undertaken. The requirements for a state aviation system plan are very different than those required for an airport master plan. Facility planning requires projections of the parameters which determine physical design whereas financial planning requires projections of the cost elements and revenue sources associated with physical development. This section outlines the general forecasting requirements for various types of airport studies and discusses the more common methodologies used to arrive at these requirements.

Application in Planning Studies	Forecast Variables Required
Macroforecast	
National airport system needs	Revenue passengers
State or regional airport needs	Revenue passenger-miles
Airlines	Aircraft fleet
Aircraft and equipment manufacturers	Air carrier
Investment planning	General aviation
Research and development needs	Composition
Route planning	Size
Workforce requirements	Capacity
	Enplaned passengers
	Aircraft operations
Microforecast	
Airport facilities	Aircraft operations
Airside	Air carrier
Runways	Fleet mix
Taxiways	Capacity
Apron areas	Peak hour
Navigational aids	General-aviation-based aircraft
General aviation needs	
Landside	Passenger traffic
Gates	Enplaned/deplaned
Terminal facilities	Originating/terminating
Cargo needs	Connecting/transferring
	Airlines
Curb frontage	Peaking characteristics
Parking	Cargo activity
Internal road network	Vehicle traffic
Ground access	
Regional road network	
Public transit	

TABLE 5-2 Typical Air Transportation Forecast Variables and Their Use in Aviation and Airport Planning

The Airport System Plan

The purpose of an airport system plan is to identify the aviation development required to meet the immediate and future aviation needs of a region, state, or metropolitan area [17]. It recommends the general location and characteristics of new and expanded airports, shows the timing, phasing, and estimated cost of development, and identifies revenue sources and legislation for the implementation of the plan. The aviation system plan also provides a basis for the definitive and detailed development of the individual airports in the system.

The primary forecasting requirement for the airport system plan is a projection of the level of aviation activity during the planning period. The forecasts are usually made on an annual basis for the planning entity as a whole and are then proportioned among the various individual airports within this entity. Specific projections are generally made for total aircraft operations, air carrier and general aviation operations, based aircraft, total air cargo, and passenger enplanements for the short-, medium-, and long-range time frames within the planning period. These demand projections are compared to the inventory of physical facilities to determine development needs. It should be emphasized that these projections are normally made in a very aggregate fashion and tend to examine overall determinates of regional activity rather than the specific factors affecting local activity.

The preparation of a forecast for a system plan is initiated by the collection of historical data indicative of the various components of aviation activity. These data normally include broad measures of socioeconomic activity as well as aviation activity statistics. Due to the fact that data collection is very expensive, most of the data collection in a system plan depends primarily on secondary or existing sources and very little survey work is performed. Many of the overall regional projections are made on the basis of trend projections or simple econometric models and these are then apportioned to the individual airports within the region on the basis of ratio methods. In the analysis of observed trends and the preparation of future forecasts, broad indicators are generally used including an examination of the consistency and realization of past trends and a comparison of growth rates and economic indicators.

The Airport Master Plan

The purpose of the airport master plan is to provide the specific details for the future development of an individual airport to satisfy aviation needs consistent with community objectives. The airport master plan requires detailed projections of the level of demand on the various facilities associated with the airport. Various concepts and alternatives for development are examined and evaluated, and recommendations are made relative to the prioritizing, scheduling, and financing of the plan [3]. Though the basis for projections in a system plan is the aggregate level of annual demand, this is not sufficient for the master plan. Projections must be made for magnitude, nature, and variation of demand on a monthly, daily, and hourly basis on the many facilities located at the airport.

Annual forecasts of airport traffic during the planning period are the basis for the preparation of the detailed forecasts in the master plan. These forecasts are made for each type of major airport user including air carrier, commuter, and general aviation aircraft and are often expressed in terms of upper and lower bounds. The master plan forecasts are usually made under both constrained and unconstrained conditions. An unconstrained forecast is one which is made relative to the potential aviation market in which the basic factors which tend to create aviation demand are utilized without regard to any constraining factors that could affect aviation growth at the location. A constrained forecast is one which is made in the context of alternative factors which could limit growth at the specific airport. Constraining factors addressed in master plans include limitations on airport capacity due to land availability and noise restrictions, the development of alternative reliever airports to attract general aviation demand, policies which alter access to airports by general aviation and commuter aircraft operations, and the availability and cost of aviation fuel. The determination of the level of general aviation activity at an airport can be significantly changed when land availability is restricted, thereby placing limits on airside capacity. The available capacity may be utilized in the context of policies which favor commercial over general aviation growth.

Specific forecasts made for a master plan include the annual, daily, and peak hour operations by air carrier, commuter, general aviation, cargo, and military aircraft, passenger enplanements, and annual cargo tonnage, as well as daily and hourly ground access system and parking demand. Projections are also made for the mix or types of aircraft in each of the categories which will utilize the airport during the planning period. In the preparation of these forecasts some variables are projected directly and others are derived from these projections. For example, annual passenger enplanements might be forecasted from an econometric model and then, based upon exogenous estimates of average air carrier fleet passenger capacity and boarding seat load factors, annual air carrier operations could be derived from these data. Guidance is provided by FAA [3, 10] for forecasting the various elements of the master plan and FAA recommends a tabular format for presenting forecasts for review. An example is presented in Table 5-3. Although the activity elements shown in the table refer to annual estimates, master plan forecasts will also require peak period activity levels for the planning of many airport facilities, and depending on the situation, seasonal, monthly, daily, and/or time-of-day demands must be forecast.

A. Forecast Levels and Growth Rates (Sample Data Shown)
Specify Base Year: 2007

	Base Yr Level	Base Yr + 1 yr	Base Yr + 5 yr	Base Yr + 10 yr	Base Yr + 15 yr	Average Annual Compound Growth Rates			
						Base Yr to +1	Base Yr to +5	Base Yr to +10	Base Yr to +15
Passenger Enplanements									
Air Carrier	868,981	904,400	1,021,000	1,273,000	1,587,000	4.1%	3.3%	3.9%	4.1%
Commuter	136,184	143,000	179,000	234,000	306,000	5.0%	5.6%	5.6%	5.5%
Total	1,005,165	1,047,400	1,200,000	1,507,000	1,893,000	4.2%	3.6%	4.1%	4.3%
Operations									
Itinerant									
Air carrier	25,155	25,700	28,000	33,600	40,000	2.2%	2.2%	2.9%	3.1%
Commuter/air taxi	18,100	18,800	22,000	24,700	28,000	3.9%	4.0%	3.2%	3.0%
Total Commercial Operations	43,225	44,500	50,000	58,300	68,000	2.9%	2.9%	3.0%	3.1%
General aviation	40,124	41,600	47,000	52,000	57,500	3.7%	3.2%	2.6%	2.4%
Military	3,124	3,124	3,124	3,124	3,124	0.0%	0.0%	0.0%	0.0%
Local									
General aviation	16,167	16,700	17,500	18,500	19,500	3.3%	1.6%	1.4%	1.3%
Military	2,436	2,436	2,436	2,436	2,436	0.0%	0.0%	0.0%	0.0%
Total Operations	105,106	108,360	120,060	134,360	150,560	3.1%	2.7%	2.5%	2.4%
Instrument Operations	206,391	209,000	220,000	230,000	241,000	1.3%	1.3%	1.1%	1.0%
Peak Hour Operations	40	42	44	47	50	5.0%	1.9%	1.6%	1.5%
Cargo/mail (Enplaned + Deplaned Tons)	16,800	18,010	23,100	30,200	39,500	7.2%	6.6%	6.0%	5.9%

Based Aircraft

	90	91	93	94	95				
Single Engine (Nonjet)	90	91	93	94	95	1.1%	0.7%	0.4%	0.4%
Multi Engine (Nonjet)	14	15	20	25	30	7.1%	7.4%	6.0%	5.2%
Jet Engine	10	11	15	19	23	10.0%	8.4%	6.6%	5.7%
Helicopter	2	2	3	3	4	0.0%	8.4%	4.1%	4.7%
Other	0	0	0	0	0	0.0%	0.0%	0.0%	0.0%
Total	116	119	131	141	152	2.6%	2.5%	2.0%	1.8%

B. Operational Factors

	Base Yr Level	Base Yr + 1 yr	Base Yr + 5 yr	Base Yr + 10 yr	Base Yr + 15 yr
Average Aircraft Size (Seats)					
Air carrier	105.0	106.0	108.0	111.0	115.0
Commuter	36.0	38.0	40.0	46.0	52.0
Average Enplaning Load Factor					
Air carrier	65.8%	66.4%	67.5%	68.2%	69.0%
Commuter	41.8%	40.0%	40.6%	41.2%	42.0%
GA Operations per Based Aircraft	485	490	492	500	507

Note: Show base plus one year if forecast was done. If planning effort did not include all forecast years shown interpolate years as needed, using average annual compound growth rates.

TABLE 5-3 Template for Summarizing and Documenting Airport Planning Forecasts

For the most part, the methods used in the study of new airports are similar to those used for existing airports. However, the principal difference is the inability of the analyst to obtain a local historical database to generate extrapolation trends, market shares, or econometric models. To overcome this deficiency, an attempt is usually made to forecast by drawing an analogy between the subject airport and other existing airports which demonstrate similar traffic experience, and which are located in areas possessing similar socioeconomic, demographic, and geographic characteristics. Forecasts are then made for the airport under consideration by using these airports as surrogates and adjustments are performed to accommodate expected differences between the airports. In the past, the Air Transport Association and the Federal Aviation Administration have collected and tabulated a significant amount of data for many airports. These data have included the number and distribution of commercial air carrier operations, fleet mix, and passengers on a peak and average monthly, daily, and hourly basis. It is apparent that one may expect a rather high degree of uncertainty associated with forecasting through such an analogy.

The Future Aviation Forecasting Environment

Many of the forecasts made by the various aviation-related organizations become biased by the impact of recent events. Forecasts made in the early 1960s showed rather moderate growth, whereas those made in the late 1960s showed fairly ambitious growth. These forecasts were made in the context of expectations which reflected the behavior of aviation at the time when they were made. In the 1970s and 1980s, the overall economic conditions, the availability and cost of petroleum-based fuels, and airline deregulation considerably affected aviation. Forecasts made in this era attempted to analyze the impact of these factors in projecting the demand for aviation in the future.

The Federal Aviation Administration and numerous other transportation organizations are always looking into the future of aviation [2, 12]. As far as future trends in air travel are concerned, it is expected that there will be a greater growth of international air traffic which has been attributed to the globalization of the airline industry and to changing market forces in the United States. The factors which have contributed to the rise of the U.S. air transportation industry are changing. The steady decline in the real cost of air travel has reached a point in which the unit costs in the 1990s will remain steady or slowly rise. The increased quality of service from improvements in the speed, comfort, convenience, and safety of air travel has been largely realized. Past demographic and cultural factors, such as the baby boom, are declining in importance. The rise in family discretionary income has peaked and the use of discretionary income for air travel is meeting competition from mortgages, savings, and luxury goods. Foreign travel by U.S. citizens will be adversely affected by the decline in the exchange value of the dollar. Furthermore, this factor will encourage foreign nationals to travel to the United States [12, 13, 14].

Conferences related to aviation forecasting methodologies [10] concluded that financial concerns and forces concerns are forcing the aviation industry to place more emphasis on short-term forecasting methods. This is not only the case with the airlines but airports are also shortening their planning horizons. Throughout the aviation industry there is a shift to simpler forecasting techniques requiring fewer variables and less detailed data. A wider use is being made of forecasting techniques which depend less on mathematical modeling and more on an analysis of different scenarios, judgment, and market segmentation. Scenarios are used to test basic assumptions and to explore alternatives. Although judgment has always played a significant role in demand forecasting it is becoming more important as a subjective test of the reality associated with forecasting outcomes. There is a growing recognition that airline management strategies are important forces shaping the future of aviation development.

To adequately cope with the uncertainties associated with the traditional air transportation forecasting process and to react in a timely manner to inaccuracies found in estimates, the planning process is emerging into a phase-oriented, continuing process. For example, the FAA prepares annual forecasts of aviation activity on a national and terminal area basis which extend several years into the future [8, 9, 18]. Due to the high costs associated with the traditional planning process and the implementation of physical design changes, and the apparent inability to forecast with any degree of certainty, it is essential that planning techniques be developed which can respond to changes in the demand parameters prior to the investment decision. Perhaps the key to such a process is the recognition of the interaction of demand to supply parameters. The knowledge of the sensitivity of a physical facility component to a variation in demand can lead to more informed decisions and an understanding of the flexibility in facility design. A continued monitoring of the need for physical facilities in light of changing demand requirements provides a sound basis for the investment decision. Recognition of the uncertainties in the demand forecasting process can prevent a wasteful commitment of valuable resources. Explicit treatment of the variability of demand projections and facility modification recommendations though the use of sensitivity and tradeoff analyses is warranted.

References

1. *Airport Aviation Activity Forecasting—A Synthesis of Airport Practice*, ACRP Synthesis 2, Airport Cooperative Research Program, Transportation Research Board, Washington, D.C., 2007.

2. *Airports in the 21st Century*, Conference Proceedings (Washington, April 2000), Transportation Research Circular Number E-C027, Transportation Research Board, Washington, D.C., March 2001.
3. *Airport Master Plans*, Advisory Circular, AC 150/5070-6B, Federal Aviation Administration, Washington, D.C., 2007.
4. *Airport System Capacity, Strategic Choices*, Special Report No. 226, Transportation Research Board, Washington, D.C., 1990.
5. *Assumptions and Issues Influencing the Future Growth of the Aviation Industry*, Circular No. 230, Transportation Research Board, Washington, D.C., August 1981.
6. Aviation Demand Forecasting—a Survey of Methodologies, Transportation Research E-Circular No. E-C040, Transportation Research Board, Washington, D.C., August 2002.
7. *Aviation Forecasting Methodology: A Special Workshop*, Circular No. 348, Transportation Research Board, Washington, D.C., August 1989.
8. *FAA Aerospace Forecast, Fiscal Years 2008–2025*, Federal Aviation Administration, Washington, D.C., March 2008.
9. *FAA Long Range Aerospace Forecasts, Fiscal Years 2020, 2025, and 2030*, Federal Aviation Administration, Washington, D.C., September 2007.
10. *Forecasting Civil Aviation Activities*, Circular No. 372, Transportation Research Board, Washington, D.C., 1991.
11. *Forecasting Methods for Management*, 5th ed., S. G. Makridakis, John Wiley & Sons, Inc., New York, N.Y., 1989.
12. *Future Aviation Activities*, 12th International Workshop, (Washington, September 2002), Circular No. E-C051, Transportation Research Board, Washington, D.C., January 2003.
13. *Future Development of the U.S. Airport Network*, Preliminary Report and Recommended Study Plan, Transportation Research Board, Washington, D.C., 1988.
14. *Future of Aviation*, Circular No. 329, Transportation Research Board, Washington, D.C., 1988.
15. *Guide for the Planning of Small Airports*, Roads and Transportation Association of Canada, Ottawa, Canada, 1980.
16. *Manual on Air Traffic Forecasting*, 3d ed., International Civil Aviation Organization, Montreal, Canada, 2006.
17. *The Airport System Planning Process*, Advisory Circular, AC 150/5070-7, Federal Aviation Administration, Washington, D.C., 2004.
18. *Terminal Area Forecast Summary, Fiscal Years 2007–2025*, Federal Aviation Administration, Washington, D.C., 2007.
19. *Trends and Issues in International Aviation*, Circular No. 393, Transportation Research Board, Washington, D.C., 1992.

PART 2

Airport Design

CHAPTER 6

Geometric Design of the Airfield

Airport Design Standards

In order to provide assistance to airport designers and a reasonable amount of uniformity in the design of airport facilities for aircraft operations, design guidelines have been prepared by the FAA [6] and the ICAO [2, 3, 4]. Any design criteria involving the widths, gradients, separations of runways, taxiways, and other features of the aircraft operations area must necessarily incorporate wide variations in aircraft performance, pilot technique, and weather conditions.

The FAA design criteria provide uniformity at airport facilities in the United States and serve as a guide to aircraft manufacturers and operators with regard to the facilities which may be expected to be available in the future. The FAA design standards are published in Advisory Circulars which are revised periodically as the need arises [1]. The ICAO strives toward uniformity and safety on an international level. Its standards, which are very similar to the FAA standards, apply to all member nations of the Convention on International Civil Aviation and are published as Annex 14 to that convention [2]. Requirements for military services are so specialized that they are not included in this chapter.

The design standards prepared by the FAA and the ICAO are presented in the text which follows under the general headings of airport classification, runways, taxiways, and aprons. The material is organized so that the various criteria may be readily compared. It is incumbent upon airport planners to review the latest specifications for airport design at the time studies are undertaken due to the fact that changes are incorporated as conditions dictate.

The FAA presents guidelines for airfield design in a series of Advisory Circulars. There are more than 200 Advisory Circulars pertaining to different aspects of airport planning and design, a complete list of which may be found on the FAA's website at http://www.faa.gov.

Advisory Circular 150/5300-13 "Airport Design" is the primary source of most airfield design standards. Originally published in 1989, AC 150/5300-13 has been updated 15 times as of 2010. The reader is encouraged to visit the FAA's website for the latest updates to this and any Advisory Circulars when performing airport planning and design work, as they are updated often.

Airport Classification

For the purpose of stipulating geometric design standards for the various types of airports and the functions which they serve, letter and numerical codes and other descriptors have been adopted to classify airports.

For design purposes, airports are classified based on the aircraft they accommodate. While at any airport, a wide variety of aircraft, from small general aviation piston-engine aircraft to heavy air transport aircraft, will use the airfield, airports are designed based on a series of "critical" or "design" aircraft. These aircraft are selected from the fleet using the airport as those most critical to airfield design. The FAA defines the term *critical aircraft* as the aircraft most demanding on airport design that operates at least 500 annual itinerant operations at a given airport. In many cases, more than one critical aircraft will be selected at an airport for design purposes. For example, it is often the smallest aircraft that is critical to the orientation of runways, while the largest aircraft determines most of the other dimensional specifications of an airfield.

As described in Chap. 2, certain dimensional and performance characteristics of the critical aircraft determine the airport's *airport reference code*. The airport reference code is a coding system used to relate the airport design criteria to the operational and physical characteristics of the aircraft intended to operate at the airport. It is based upon the *aircraft approach category* and the *airplane design group* to which the aircraft is assigned. The aircraft approach category, as shown in Table 6-1, is determined by the aircraft approach speed, which is defined as 1.3 times the stall speed in the landing configuration of aircraft at maximum certified landing weight [6].

The airplane design group (ADG) is a grouping of aircraft based upon wingspan or tail height, as shown in Table 6-2. An airplane design group for a particular aircraft is assigned based on the greater (higher Roman numeral) of that associated with the aircraft's wingspan or tail height.

The airport reference code is a two designator code referring to the aircraft approach category and the airplane design group for which the airport has been designed. For example, an airport reference code of B-III is an airport designed to accommodate aircraft

Category	Approach Speed, kn
A	<91
B	91–120
C	121–140
D	141–166
E	>166

1 kn is approximately 1.15 mi/h

TABLE 6-1 Aircraft Approach Categories

with approach speeds from 91 to less than 121 kn (aircraft approach category B) with wingspans from 79 to less than 118 ft or tail heights from 30 to less than 45 ft (airplane design group III). The FAA publishes a list of the airport reference codes for various aircraft in Advisory Circular 150/5300-13 "Airport Design" [6].

As an example, an airport designed to accommodate the Boeing 767-200 which has an approach speed of 130 kn (aircraft approach category C) and a wingspan of 156 ft 1 in (airplane design group IV) would be classified with an airport reference code C-IV.

The ICAO uses a two-element code, *the aerodrome reference code*, to classify the geometric design standards at an airport [2, 3]. The code elements consist of a numeric and alphabetic designator. The aerodrome code numbers 1 through 4 classify the length of the runway available, the *reference field length*, which includes the runway length and, if present, the stopway and clearway. The reference field length is the approximate required runway takeoff length converted to an equivalent length at mean sea level, 15°C, and zero percent gradient. The aerodrome code letters A through E classify the wingspan and outer main gear wheel span for the aircraft for which the airport has been designed.

Group Number	Tail Height, ft	Wingspan, ft
I	<20	<49
II	20–<30	49–<79
III	30–<45	79–<118
IV	45–<60	118–<171
V	60–<66	171–<214
VI	66–<80	214–<262

TABLE 6-2 Aircraft Design Groups

Code Number	Reference Field Length, m	Code Letter	Wingspan, m	Distance between Outside Edges of Main Wheel Gear, m
1	<800	A	<15	<4.5
2	800–<1200	B	15–<24	4.5–<6
3	1200–<1800	C	24–<36	6–<9
4	≥1800	D	36–<52	9–<14
		E	52–<65	9–<14
		F	65–<80	14–<16

TABLE 6-3 ICAO Aerodrome Reference Codes

These aerodrome reference codes are given in Table 6-3. For example, an airport which is designed to accommodate a Boeing 767–200 with an outer main gear wheel span of width of 34 ft 3 in (10.44 m), a wingspan of 156 ft 1 in (48 m), at a maximum takeoff weight of 317,000 lb, requiring a runway length of about 6000 ft (1830 m) at sea level on a standard day, would be classified by ICAO with an aerodrome reference code of 4-D. It will be noted that this classification system does not explicitly include the function of the airport, the service it renders, or the type of aircraft accommodated.

There is an approximate correspondence between the airport reference code of the FAA and the aerodrome reference code of the ICAO [2, 3]. The FAA's aircraft approach category of A, B, C, and D are approximately the same as the ICAO aerodrome code numbers 1, 2, 3, and 4, respectively. Similarly the FAA's airplane design groups of I, II, III, IV, and V approximately correspond to ICAO aerodrome code letters A, B, C, D, and E.

Utility Airports

A *utility airport* is defined as one which has been designed, constructed, and maintained to accommodate approach category A and B aircraft [6]. The specifications for utility airports are grouped for *small aircraft*, those of maximum certified takeoff weights of 12,500 lb or less, and *large aircraft*, those with maximum certified takeoff weight in excess of 12,500 lb.

Design specifications for utility airports are governed by the airplane design group and the types of approaches authorized for the airport runway, that is, visual, nonprecision instrument or precision instrument approaches.

Utility airports for small aircraft are called *basic utility stage I*, *basic utility stage II*, and *general utility stage I*. Utility airports for large aircraft are called *general utility stage II*. Utility airports are further grouped for either visual and nonprecision instrument operations or precision instrument operations. The visual and nonprecision instrument operation utility airports are the basic utility stage I, basic utility stage II, or general utility stage I airports. The precision instrument operation utility airport is the general utility stage II airport.

A basic utility stage I airport has the capability of accommodating about 75 percent of the single engine and small twin engine aircraft used for personal and business purposes. This generally means aircraft weighing on the order of 3000 lb or less is given the airport reference code B-I, which indicates that it accommodates aircraft in aircraft approach categories A and B and aircraft in airplane design group I. A basic utility stage II airport has the capability of accommodating all of the airplanes of a basic utility stage I airport plus some small business and air taxi-type airplanes. This generally means aircraft weighing on the order of 8000 lb or less is also given the airport reference code B-I. A general utility stage I airport accommodates all small aircraft. It is assigned the airport reference code of B-II. A general utility stage II airport serves large airplanes in aircraft approach categories A and B and usually has the capability for precision instrument operations. It is assigned the airport reference code of B-III.

Transport Airports

A *transport airport* is defined as an airport which is designed, constructed, and maintained to accommodate aircraft in approach categories C, D, and E [6]. The design specifications of transport airports are based upon the airplane design group.

Runways

A runway is a rectangular area on the airport surface prepared for the takeoff and landing of aircraft. An airport may have one runway or several runways which are sited, oriented, and configured in a manner to provide for the safe and efficient use of the airport under a variety of conditions. Several of the factors which affect the location, orientation, and number of runways at an airport include local weather conditions, particularly wind distribution and visibility, the topography of the airport and surrounding area, the type and amount of air traffic to be serviced at the airport, aircraft performance requirements, and aircraft noise [2].

Runway Configurations

The term "runway configuration" refers to the number and relative orientations of one or more runways on an airfield. Many runway configurations exist. Most configurations are combinations of several basic configurations. The basic configurations are (1) single runways, (2) parallel runways, (3) intersecting runways, and (4) open-V runways.

Single Runway

This is the simplest of the runway configurations and is shown in Fig. 6-1. It has been estimated that the hourly capacity of a single runway in VFR conditions is somewhere between 50 and 100 operations per hour, while in IFR conditions this capacity is reduced to 50 to 70 operations per hour, depending on the composition of the aircraft mix and navigational aids available [4].

Parallel Runways

The capacities of parallel runway systems depend on the number of runways and on the spacing between the runways. Two, three, and four parallel runways are common. The spacing between parallel runways varies widely. For the purpose of this discussion, the spacing is classified as close, intermediate, and far, depending on the centerline separation between two parallel runways. Close parallel runways are spaced from a minimum of 700 ft (for air carrier airports) to less than 2500 ft [5]. In IFR conditions an operation of one runway is dependent on the operation of other runway. Intermediate parallel runways are spaced between 2500 ft to less than 4300 ft [5]. In IFR conditions an arrival on one runway is independent of a departure on the other runway. Far parallel runways are spaced at least 4300 ft apart [5]. In IFR conditions the two runways can be operated independently for both arrivals and departures. Therefore,

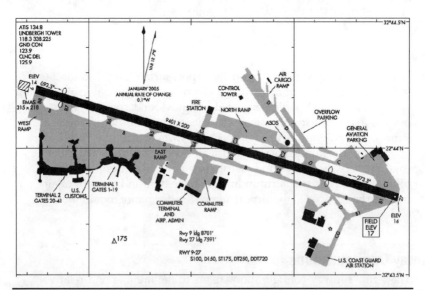

Figure 6-1   Single runway configuration: San Diego International Airport (*NOAA Approach Charts*).

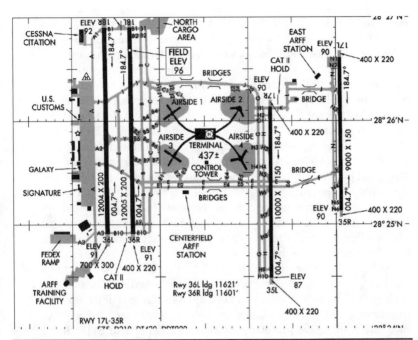

FIGURE 6-2 Example of parallel runway configuration: Orlando International Airport.

as noted earlier, the centerline separation of parallel runways determines the degree of interdependence between operations on each of the parallel runways. It should be recognized that in future the spacing requirements for simultaneous operations on parallel runways may be reduced. If this occurs, new spacing can be applied to the same classifications. Figure 6-2 illustrates an airport with multiple parallel runways with various spacing.

If the terminal buildings are placed between parallel runways, runways are always spaced far enough apart to allow room for the buildings, the adjoining apron, and the appropriate taxiways. When there are four parallel runways, each pair is spaced close, but the pairs are spaced far apart to provide space for terminal buildings.

In VFR conditions, close parallel runways allow simultaneous arrivals and departures, that is, arrivals may occur on one runway while departures are occurring on the other runway. Aircraft operating on the runways must have wingspans less than 171 ft (airplane design groups I through IV, see Table 6-2) for centerline spacing at the minimum of 700 ft [5]. If larger wingspan aircraft are operating on these runways (airplane design groups V and VI), the centerline spacing must be at least 1200 ft for such simultaneous operations [5]. In either case, wake vortex avoidance procedures must be used for simultaneous operations on closely spaced parallel runways. Furthermore, simultaneous arrivals to both runways or simultaneous departures from both runways are not allowed in VFR conditions for closely spaced parallel runways. In IFR conditions, closely spaced parallel runways cannot be used simultaneously but may be operated as dual-lane runways.

Intermediate parallel runways may be operated with simultaneous arrivals in VFR conditions. Intermediate parallel runways may be operated in IFR conditions with simultaneous departures in a nonradar environment if the centerline spacing is at least 3500 ft and in a radar environment if the centerline spacing is at least 2500 ft [5]. Simultaneous arrivals and departures are also permitted if the centerline spacing is at least 2500 ft if the thresholds of the runways are not staggered [5]. There are times when it may be desirable to stagger the thresholds of parallel runways. The staggering may be necessary because of the shape of the acreage available for runway construction, or it may be desirable for reducing the taxiing distance of takeoff and landing aircraft. The reduction in taxiing distance, however, is based on the premise that one runway is to be used exclusively for takeoff and the other for landing. In this case the terminal buildings are located between the runways so that the taxiing distance for each type of operation (takeoff or landing) is minimized. If the runway thresholds are staggered, adjustments to the centerline spacing requirement are allowed for simultaneous arrivals and departures [5]. If the arrivals are on the near threshold then the centerline spacing may be reduced by 100 ft for each 500 ft of threshold stagger down to a minimum centerline separation of 1000 ft for aircraft with wingspans up to 171 ft and a minimum of 1200 ft for larger wingspan aircraft. If the arrivals are on the far threshold the centerline spacing must be increased by 100 ft for each 500 ft of threshold stagger. Simultaneous arrivals in IFR conditions are not permitted on intermediate parallel runways but are permitted on far parallel runways with centerline spacings of at least 4300 ft [5].

The hourly capacity of a pair of parallel runways in VFR conditions varies greatly from 60 to 200 operations per hour depending on the aircraft mix and the manner in which arrivals and departures are processed on these runways [4]. Similarly, in IFR conditions the hourly capacity of a pair of closely spaced parallel runways ranges from 50 to 60 operations per hour, of a pair of intermediate parallel runways from 60 to 75 operations per hour, and for a pair of far parallel runways from 100 to 125 operations per hour [4].

A dual-lane parallel runway consists of two closely spaced parallel runways with appropriate exit taxiways. Although both runways can be used for mixed operations subject to the conditions noted above, the desirable mode of operation is to dedicate the runway farthest from the terminal building (outer) for arrivals and the runway closest to the terminal building (inner) for departures. It is estimated that a dual-lane runway can handle at least 70 percent more traffic than a single runway in VFR conditions and about 60 percent more traffic than a single runway in IFR conditions. It is recommended that the two runways be spaced not less than 1000 ft apart (1200 ft, where particularly larger wingspan aircraft are involved). This spacing also provides sufficient distance for an arrival to stop between the two runways. A parallel taxiway between the runways will provide for a nominal increase in capacity, but is not essential. The major benefit of a dual-lane runway is to provide an increase in IFR capacity with minimal acquisition of land [7, 14].

Intersecting Runways

Many airports have two or more runways in different directions crossing each other. These are referred to as intersecting runways. Intersecting runways are necessary when relatively strong winds occur from more than one direction, resulting in excessive crosswinds when only one runway is provided. When the winds are strong, only one runway of a pair of intersecting runways can be used, reducing the capacity of the airfield substantially. If the winds are relatively light, both runways can be used simultaneously. The capacity of two intersecting runways depends on the location of the intersection (i.e., midway or near the ends), the manner in which the runways are operated for takeoffs and landings, referred to as the runway use strategy, and the aircraft mix. The farther the intersection is from the takeoff end of the runway and the landing threshold, the lower is the capacity. The highest capacity is achieved when the intersection is close to the takeoff and landing threshold. Figure 6-3 provides an example of intersecting runways with the intersection closer to the runway thresholds.

Open-V Runways

Runways in different directions which do not intersect are referred to as open-V runways. This configuration is shown in Fig. 6-4. Like intersecting runways, open-V runways revert to a single runway when winds are strong from one direction. When the winds are light, both runways may be used simultaneously.

The strategy which yields the highest capacity is when operations are away from the V and this is referred to as a diverging pattern. In VFR the hourly capacity for this strategy ranges from 60 to 180 operations per hour, and in IFR the corresponding capacity is from 50 to 80 operations per hour [4]. When operations are toward the V it is referred to as a converging pattern and the capacity is reduced to 50 to 100 operations per hour in VFR and to between 50 and 60 operations per hour in IFR [4].

Combinations of Runway Configurations

From the standpoint of capacity and air traffic control, a single-direction runway configuration is most desirable. All other things being equal,

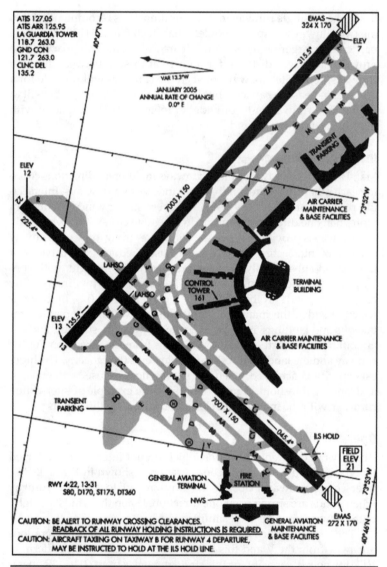

Figure 6-3 Example of intersecting runways: LaGuardia Airport, New York.

this configuration will yield the highest capacity compared with other configurations. For air traffic control the routing of aircraft in a single direction is less complex than routing in multiple directions. Comparing the divergent configurations, the open-V runway pattern is more desirable than an intersecting runway configuration. In the open-V configuration an operating strategy that routes aircraft away from the V will yield higher capacities than if the operations are reversed. If intersecting runways cannot be avoided, every effort

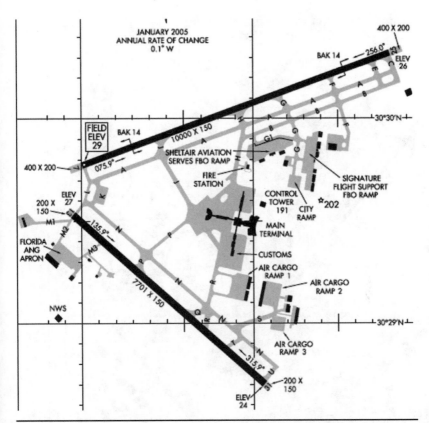

Figure 6-4 Example of open-V runways: Jacksonville International Airport.

should be made to place the intersections of both runways as close as possible to their thresholds and to operate the aircraft away from the intersection rather than toward the intersection.

Figure 6-5 illustrates the complex runway configuration of Chicago's O'Hare Field, with multiple parallel, intersecting, and non-intersecting runways. It should be noted that a large capital improvement program is being undertaken to simplify the runway configuration, by adding additional parallel runways and removing many intersecting runways. This runway redesign is being done with the intention of improving the capacity and efficiency of airport operations at the airport. The runway configuration redesign is illustrated in Fig. 6-6.

Runway Orientation

The orientation of a runway is defined by the direction, relative to magnetic north, of the operations performed by aircraft on the runway. Typically, but not always, runways are oriented in such a manner that they may be used in either direction. It is less preferred to

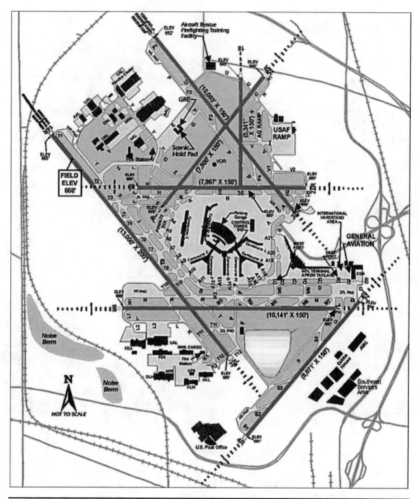

FIGURE 6-5 Example of complex runway system: Chicago O'Hare International Airport.

orient a runway in such a way that operating in one direction is precluded, normally due to nearby obstacles.

In addition to obstacle clearance considerations, which will be discussed later in this chapter, runways are typically oriented based on the area's wind conditions. As such, an analysis of wind is essential for planning runways. As a general rule, the primary runway at an airport should be oriented as closely as practicable in the direction of the prevailing winds. When landing and taking off, aircraft are able to maneuver on a runway as long as the wind component at right angles to the direction of travel, the crosswind component, is not excessive.

The FAA recommends that runways should be oriented so that aircraft may be landed at least 95 percent of the time with

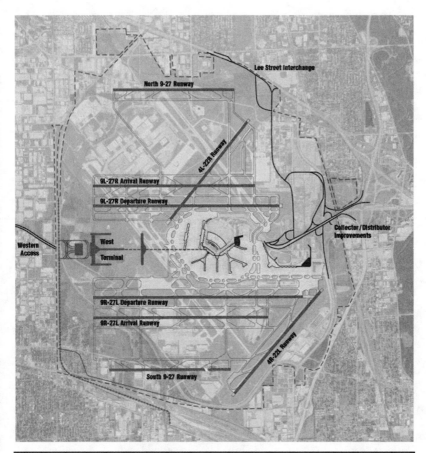

FIGURE 6-6 Planned simplified runway configuration: Chicago O'Hare International Airport (*Courtesy Chicago O'Hare Modernization program*).

allowable crosswind components not exceeding specified limits based upon the airport reference code associated with the critical aircraft that has the shortest wingspan or slowest approach speed. When the wind coverage is less than 95 percent a crosswind runway is recommended.

The allowable crosswind is 10.5 kn (12 mi/h) for Airport Reference Codes A-I and B-I, 13 kn (15 mi/h) for Airport Reference Codes A-II and B-II, 16 kn (18.5 mi/h) for Airport Reference Codes A-III, B-III, C-I, C-II, C-III and C-IV, and 20 knots (23 mph) for Airport Reference Codes A-IV through D-VI [5].

ICAO also specifies that runways should be oriented so that aircraft may be landed at least 95 percent of the time with crosswind components of 20 kn (23 mph) for runway lengths of 1500 m more, 13 kn (15 mi/h) for runway lengths between 1200 and 1500 m, and 10 kn (11.5 mi/h) for runway lengths less than 1200 m [1, 2].

Once the maximum permissible crosswind component is selected, the most desirable direction of runways for wind coverage can be determined by examination of the average wind characteristics at the airport under the following conditions:

1. The entire wind coverage regardless of visibility or cloud ceiling
2. Wind conditions when the ceiling is at least 1000 ft and the visibility is at least 3 mi
3. Wind conditions when ceiling is between 200 and 1000 ft and/or the visibility is between ½ and 3 mi.

The first condition represents the entire range of visibility, from excellent to very poor, and is termed the all weather condition. The next condition represents the range of good visibility conditions not requiring the use of instruments for landing, termed visual meteorological condition (VMC). The last condition represents various degrees of poor visibility requiring the use of instruments for landing, termed instrument meteorological conditions (IMC).

The 95 percent criterion suggested by the FAA and ICAO is applicable to all conditions of weather; nevertheless it is still useful to examine the data in parts whenever this is possible.

In the United States, weather records can be obtained from the Environmental Data and Information Service of the National Climatic Center at the National Oceanic and Atmospheric Administration located in Ashville, N.C., or from various locations found on the Internet.

Weather data are collected from weather stations throughout the United States on an hourly basis and recorded for analysis. The data collected include ceiling, visibility, wind speed, wind direction, storms, barometric pressure, the amount and type of liquid and frozen precipitation, temperature, and relative humidity. A report illustrating the tabulation and representation of some of the data of use in airport studies was prepared for the FAA [15]. The weather records contain the percentage of time certain combinations of ceiling and visibility occur (e.g., ceiling, 500 to 900 ft; visibility, 3 to 6 mi), and the percentage of time winds of specified velocity ranges occur from different directions (e.g., from NNE, 4 to 7 mi/h). The directions are referenced to true north.

The Wind Rose

The appropriate orientation of the runway or runways at an airport can be determined through graphical vector analysis using a wind rose. A standard wind rose consists of a series of concentric circles cut by radial lines using polar coordinate graph paper. The radial lines are drawn to the scale of the wind magnitude such that the area between each pair of successive lines is centered on the wind direction.

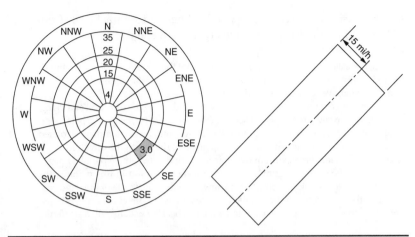

FIGURE 6-7   Wind rose coordinate system and template.

A typical wind rose polar coordinate system is shown on the left side of Fig. 6-7. The shaded area indicates that the wind comes from the southeast (SE) with a magnitude between 20 and 25 mi/h. A template is also drawn to the same radial scale representing the crosswind component limits. A template drawn with crosswind component limits of 15 mi/h is shown on the right side of Fig. 6-7. On this template three equally spaced parallel lines have been plotted. The middle line represents the runway centerline, and the distance between the middle line and each outside line is, to scale, the allowable crosswind component (in this case, 15 mi/h). The template is placed over the wind rose in such a manner that the centerline on the template passes through the center of the wind rose.

By overlaying the template on the wind rose and rotating the centerline of the template through the origin of the wind rose one may determine the percentage of time a runway in the direction of the centerline of the template can be used such that the crosswind component does not exceed 15 mi/h. Optimum runway directions can be determined from this wind rose by the use of the template, typically made on a transparent strip of material. With the center of the wind rose as a pivot point, the template is rotated until the sum of the percentages included between the outer lines is a maximum. If a wind vector from a segment lies outside either outer line on the template for the given direction of the runway, that wind vector must have a crosswind component which exceeds the allowable crosswind component plotted on the template. When one of the outer lines on the template divides a segment of wind direction, the fractional part is estimated visually to the nearest 0.1 percent. This procedure is consistent with the accuracy of the wind data and assumes that the wind percentage within the sector is uniformly distributed within that sector. In practice, it is usually easier to add the percentages contained in the sectors outside of the two outer parallel lines and subtract these from 100 percent to find the percentage of wind coverage.

Example Problem 6-1   As an example, assume that the wind data for all conditions of visibility are those shown in Table 6-4. This wind data is plotted to scale as indicated above to obtain a wind rose, as shown in Fig. 6-8.

The percentage of time the winds correspond to a given direction and velocity range is marked in the proper sector of the wind rose by means of a polar coordinate scale for both wind direction and wind magnitude. The template is rotated about the center of the wind rose, as explained earlier, until the direction of the centerline yields the maximum percentage of wind between the parallel lines.

Once the optimum runway direction has been found in this manner, the next step is to read the bearing of the runway on the outer scale of the wind rose where the centerline on the template crosses the wind direction scale. Because true north is used for published wind data, this bearing usually will be different

Sector	True Azimuth	Wind Speed Range, mi/h				Total
		4–15	15–20	20–25	25–35	
		Percentage of Time				
N	0.0	2.4	0.4	0.1	0.0	2.9
NNE	22.5	3.0	1.2	1.0	0.5	5.7
NE	45.0	5.3	1.6	1.0	0.4	8.3
ENE	67.5	6.8	3.1	1.7	0.1	11.7
E	90.0	7.1	2.3	1.9	0.2	11.5
ESE	112.5	6.4	3.5	1.9	0.1	11.9
SE	135.0	5.8	1.9	1.1	0.0	8.8
SSE	157.5	3.8	1.0	0.1	0.0	4.9
S	180.0	1.8	0.4	0.1	0.0	2.3
SSW	202.5	1.7	0.8	0.4	0.3	3.2
SW	225.0	1.5	0.6	0.2	0.0	2.3
WSW	247.5	2.7	0.4	0.1	0.0	3.2
W	270.0	4.9	0.4	0.1	0.0	5.4
WNW	292.5	3.8	0.6	0.2	0.0	4.6
NW	315.0	1.7	0.6	0.2	0.0	2.5
NNW	337.5	1.7	0.9	0.1	0.0	2.7
Subtotal		60.4	19.7	10.2	1.6	91.9
Calms						ˎ 8.1
Total						100.0

TABLE 6-4   Example Wind Data

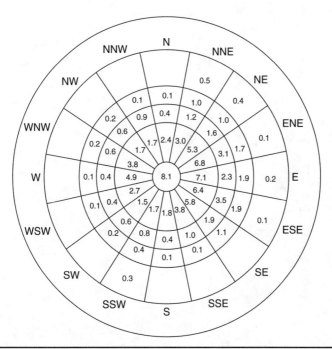

FIGURE 6-8   Wind data in wind rose format.

from that used in numbering runways since runway designations are based on the magnetic bearing. As illustrated in Fig. 6-9, a runway oriented on an azimuth to true north of 90° to 270° (N 90° E to S 90° W true bearing) will permit operations 90.8 percent of the time with the crosswind components not exceeding 15 mi/h.

Should the wind analysis not give the desired wind coverage, the template may then be used to determine the direction of a second runway, a crosswind runway, which would increase the wind coverage to 95 percent. This is done by blocking out the area between the two outer parallel lines for the direction of the primary runway (since this has already been counted in the wind coverage for the primary runway) and rotating the template until the percentages between the outer parallel lines for the remaining area for another direction is maximized. If this is done in this problem it is found that the crosswind runway should be located in an orientation of 12° to 192° (N 12° E to S 12° W true bearing). This will permit an additional wind coverage of 6.2 percent above that provided by the runway oriented 90° to 270° for a total wind coverage for both runways of 97.0 percent.

Let us say that because of noise-sensitive land uses in the direction of the optimal crosswind runway, a crosswind runway will be located at the airport in the orientation of 30° to 210° direction which results in an additional wind coverage of 5.8 percent. This runway orientation, called runway 3–21, is shown in Fig. 6-10. The total wind coverage for both runways is then 96.6 percent. The total wind coverage for a runway in the orientation of 30° to 210° direction is found to be 84.8 percent from Fig. 6-11. The combined wind coverage of 96.6 percent for the use of either runway is shown in Fig. 6-11.

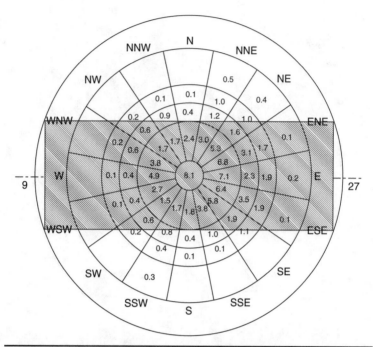

FIGURE 6-9   Wind coverage for runway 9–27, Example Problem 6–1.

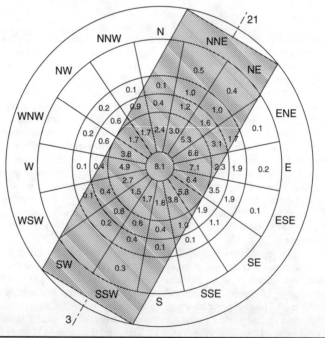

FIGURE 6-10   Wind coverage for runway 3–21, Example Problem 6-1.

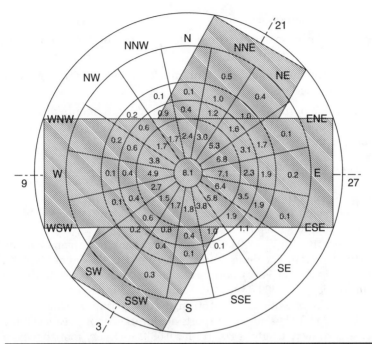

Figure 6-11 Wind coverage for runways 9–27 and 3–21, Example Problem 6-1.

Estimating Runway Length

Other than orientation, planning and designing the length of a runway is critical to whether or not a particular aircraft can safely use the runway for takeoff or landing. Furthermore, designing a runway to accommodate a given aircraft is a difficult task, given the fact that an aircraft's required runway length will vary based on aircraft weight, as well as on several ambient conditions.

As a guide to airport planners, the FAA has published Advisory Circular 150/5325-4b, "Runway Length Requirements for Airport Design" [17]. In this publication, procedures are defined for estimating the design runway length of aircraft, based on their maximum takeoff weights (MTOW), certain aircraft performance specifications, and the airport's field elevation and temperature. The airport design runway length is found for the critical aircraft, defined as the aircraft which flies the greatest nonstop route segment from the airports at least 500 operations per year and requires the longest runway. The FAA's procedure for estimating runway length is based on the following data:

1. Designation of a critical aircraft

2. The maximum takeoff weight of the critical aircraft at the airport

3. The airport elevation

4. The mean daily maximum temperature for the hottest month at the airport

5. The maximum different in elevation along the runway centerline.

For the purposes of estimating runway length requirements, the FAA groups aircraft by MGTOW. Based on the MGTOW of the critical aircraft, the following procedures are defined:

Aircraft Less than 12,500 lb MGTOW

Critical aircraft less than or equal to 12,500 lb MGTOW are considered "small airplanes" for the purposes of estimating runway length requirements. For these small aircraft, design runway length is based on the aircraft's reference approach speed, V_{ref}.

Aircraft with $V_{ref} < 30$ kn are considered short takeoff and landing (STOL) aircraft. The design runway length for STOL aircraft is 300 ft (92 m) at sea level. For airports at elevation above sea level, the design runway length is 300 ft plus 0.03 ft for every foot above sea level.

For aircraft with $30 \leq V_{ref} < 50$ kn, the design runway length at sea level is 800 ft (244 m). For airports at elevation above sea level, the design runway length is 800 ft plus 0.08 ft for every foot above sea level.

For aircraft with $V_{ref} \geq 50$ kn, the design runway length is based on the number of passenger seats in the aircraft. For those aircraft with less than 10 passenger seats, Fig. 6-12 is referenced. This figure has two sets of curves, one representing "95 percent of fleet," to be used at airports serving small communities, and one representing "100 percent of fleet," to be applied at airports near larger metropolitan areas.

Figure 6-12 is illustrated with an example case where the mean daily maximum temperature at the hottest month at the airport is 59°F and elevation is sea level. A vertical line is drawn from the point on the horizontal axis associated with 59°F to the sea level field elevation curve. A horizontal line is then drawn from the associated location on the elevation curve to the right side of the figure. The value at the end of the horizontal line on the right side of the figure is the recommended design runway length. In this case, applying the 95 percent of fleet curve results in a design runway length of 2700 ft, while the 100 percent of fleet curve resulting in a design runway length of 3200 ft.

For those aircraft with 10 or more passenger seats at airports at elevation 3000 ft AMSL or less, Fig. 6-13 is referenced. At airports at elevation greater than 3000 ft AMSL, Fig. 6-12 "100 percent of fleet" is referenced.

Figure 6-13 is illustrated with an example case where the mean daily maximum temperature at the hottest month at the airport is

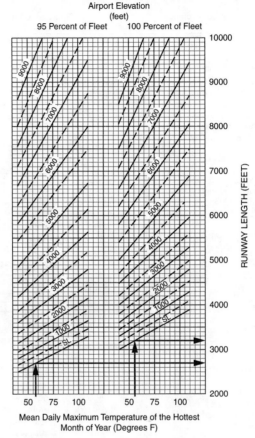

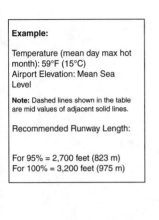

Example:

Temperature (mean day max hot month): 59°F (15°C)
Airport Elevation: Mean Sea Level

Note: Dashed lines shown in the table are mid values of adjacent solid lines.

Recommended Runway Length:

For 95% = 2,700 feet (823 m)
For 100% = 3,200 feet (975 m)

Airport Elevation
(feet)

95 Percent of Fleet   100 Percent of Fleet

RUNWAY LENGTH (FEET)

Mean Daily Maximum Temperature of the Hottest
Month of Year (Degrees F)

FIGURE 6-12   Small airplanes with fewer than 10 passenger seats (*FAA AC 150/5325-4b*).

90°F and elevation is 1000 ft AMSL. A vertical line is drawn from the point on the horizontal axis associated with 90°F to the 1000 ft AMSL field elevation curve. A horizontal line is drawn from the associated point on the field elevation curve to the right side of the figure, where the runway length is estimated. In this example, the design runway length is estimated to be 4400 ft.

Aircraft Greater than 12,500 lb but Less than or Equal to 60,000 lb MGTOW

For aircraft greater than 12,500 lb but less than or equal to 60,000 lb MGTOW, the critical aircraft is located on Table 6-5 "75 percent of fleet," or Table 6-6, "100 percent of fleet." Table 6-5 represents aircraft that generally require less than 5000 ft of runway, while Table 6-6 represents aircraft that generally require 5000 ft or more of runway.

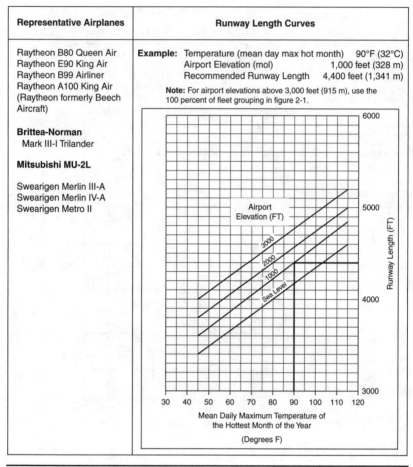

Representative Airplanes	Runway Length Curves
Raytheon B80 Queen Air Raytheon E90 King Air Raytheon B99 Airliner Raytheon A100 King Air (Raytheon formerly Beech Aircraft) **Brittea-Norman** Mark III-I Trilander **Mitsubishi MU-2L** Swearigen Merlin III-A Swearigen Merlin IV-A Swearigen Metro II	**Example:** Temperature (mean day max hot month) 90°F (32°C) Airport Elevation (mol) 1,000 feet (328 m) Recommended Runway Length 4,400 feet (1,341 m) **Note:** For airport elevations above 3,000 feet (915 m), use the 100 percent of fleet grouping in figure 2-1.

FIGURE 6-13 Small airplanes having 10 or more passenger seats (*FAA AC 150/5325-4b*).

As such, if the selected critical aircraft is found on Table 6-5, it is said that the runway length estimated will be able to accommodate 75 percent of the fleet. If the selected critical aircraft is found on Table 6-6, it is said that the runway length estimated will be able to accommodate 100 percent of this size fleet.

For the design aircraft, a "useful load" of either 60 or 90 percent is selected. A 60 percent useful load represents the condition where the critical aircraft typically operates at 60 percent load factors, or performs shorter range operations, requiring less fuel, while a 90 percent useful load represents the condition where the critical aircraft typically operates at 90 percent load factors, or performs longer range operations.

For aircraft falling within the "75 percent of fleet" group as identified in Table 6-5, Fig. 6-14 is then applied, selecting either the 60 or

Manufacturer	Model	Manufacturer	Model
Aerospatiale	Sn-601 Corvette	Dassault	Falcon 10
Bae	125–700	Dassault	Falcon 20
Beech Jet	400A	Dassault	Falcon 50/50 EX
Beech Jet	Premier I	Dassault	Falcon 900/900B
Beech Jet	2000 Starship	Israel Aircraft Industries (LAI)	Jet Commander 1121
Bombardier	Challenger 300	IAI	Westwind 1123/1124
Cessua	500 Citation/ 501 Citation Sp	Learjet	20 Series
Cessna	Citation I/II/III	Learjet	31/31A/31A ER
Cessna	525A Citation II (CJ-2)	Learjet	35/35A/36/36A
Cessna	350 Citation Bravo	Learjet	40/45
Cessna	550 Citation II	Mitsubishi	Mu-300 Diamond
Cessna	551 Citation II/Special	Raytheon	390 Premier
Cessna	552 Citation	Raytheon Hawker	400/400 XP
Cessna	560 Citation Encore	Raytheon Hawker	600
Cessna	560/560 XL Citation Excel	Sabreliner	40/60
Cessna	560 Citation V Ultra	Sabreliner	75A
Cessna	650 Citation VII	Sabreliner	80
Cessna	680 Citation Sovereign	Sabreliner	T-39

Source: FAA AC 150/5235-4b.

TABLE 6-5 Airplanes that Make Up 75 Percent of the Fleet

90 percent useful load sides of the figure, and applied based on the mean daily maximum temperature of the hottest month (in Fahrenheit), and the elevation of the airfield (in feet AMSL).

Figure 6-14 illustrates two examples, one for an airport at sea level with average high temperature during the hottest month at 59°F and a critical aircraft falling within the 75 percent of fleet category at

Manufacturer	Model
Bae	Corporate 800/1000
Bombardier	600 Challenger
Bombardier	601/601-3A/3ER Challenger
Bombardier	604 Challenger
Bombardier	BD-100 Continental
Cessna	S550 Citation S/II
Cessna	650 Citation III/IV
Cessna	750 Citation X
Dassault	Falcon 900C/900EX
Dassualt	Falcon 2000/2000EX
Israel Aircraft Industries (IAI)	Astra 1125
IAI	Galaxy 1126
Learjet	45 XR
Learjet	55/55B/55C
Learjet	60
Raytheon/Hawker	Horizon
Raytheon/Hawker	800/800 XP
Raytheon/Hawker	1000
Sabreliner	65/75

TABLE 6-6   Aircraft that (Including Those in Table 6-1) Make Up 100 Percent of the Fleet

60 percent useful load, and one for an airport at 1000 ft AMSL, average high temperature during the hottest month at 100°F, and a critical aircraft falling within the 75 percent of fleet category at 90 percent useful load. For aircraft falling within the "100 percent of fleet" group as identified in Table 6-6, Fig. 6-15 is similarly applied.

Figure 6-15 is illustrated with two examples, one illustrating an airport at 2000 ft AMSL with average high temperature during the hottest month at 59°F and a critical aircraft falling within the 100 percent of fleet category at 60 percent useful load, and one illustrating an airport at 3000 ft AMSL, average high temperature during the hottest month at 100°F, and a critical aircraft falling within the 100 percent of fleet category at 90 percent useful load.

Based on the runway lengths found in either Fig. 6-14 or Fig. 6-15, an adjustment is made for any nonlevel runway gradient. Specifically, the runway length found in Fig. 6-14 or Fig. 6-15 is increased by

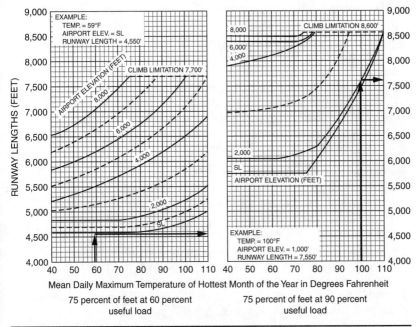

75 percent of feet at 60 percent
useful load

75 percent of feet at 90 percent
useful load

Mean Daily Maximum Temperature of Hottest Month of the Year in Degrees Fahrenheit

FIGURE 6-14   Seventy-five percent of fleet at 60 or 90 percent useful load.

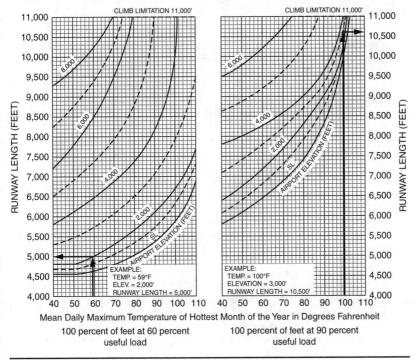

100 percent of feet at 60 percent
useful load

100 percent of feet at 90 percent
useful load

Mean Daily Maximum Temperature of Hottest Month of the Year in Degrees Fahrenheit

FIGURE 6-15   Hundred percent of fleet at 60 or 90 percent useful load.

10 ft for every foot in elevation difference between the lowest point and highest point on the runway.

At higher elevations, it is often the case that the required runway length is greater for aircraft less than 12,500 lb MGTOW than for aircraft greater than 12,500 lb. If this is the case, the design runway length would be that for the lighter aircraft. As such, airport planners estimating runway lengths at high elevation airports should perform runway length estimations for the smallest aircraft, in addition to that for the selected critical aircraft.

Aircraft Greater than 60,000 lb MGTOW

For aircraft greater than 60,000 lb MGTOW, runway lengths are estimated based on the specific performance specifications of the critical aircraft. These performance specifications may be found in the published aircraft "airport planning manuals." These manuals may be found on the Internet sites of the major aircraft manufacturers.

Within the aircraft airport planning manuals are performance charts that are used to determine the aircraft's required runway lengths for both takeoff and landing, based on the aircraft's operating configuration, its estimated weights during takeoff and landing, as well as the airport elevation and average high temperature during the hottest month.

Example Problem 6-2 illustrates the procedure for estimating runway length using these charts.

Example Problem 6-2   Consider the situation where an airport with elevation 1000 ft AMSL and mean daily maximum temperature of the hottest month of 84°F, is planning for a new runway to be designed for the Boeing 737-900 aircraft, equipped with Pratt & Whitney CFM56-7B27 engines. At the airport a runway gradient of 20 ft is projected.

According to the performance specification chart, illustrated in Fig. 6-16, found in the Boeing 737-900 airport planning manual, the maximum design landing weight for the aircraft is 146,300 lb and the maximum design takeoff weight is 174,200 lb.

First, estimation of required runway length for landing is performed using the landing runway length performance chart for the aircraft. As with most landing performance charts, runway length requirements found landing may be found under both dry and wet runway conditions. For airport planning purposes, design runway length for landing is estimated by considering wet runway conditions. If a landing runway length performance chart does not include wet runway conditions, the design runway length is estimated as the runway length found under dry runway conditions, plus 15 percent.

Figure 6-17 illustrates this example. Applying the case example, a vertical line is drawn from the base of the horizontal axis at the location of the maximum design landing weight (146,300 lb), up to an interpolated point between the "sea level" and "2000 ft" (to represent the example airport's 1000 ft elevation) wet-runway curves, and then a horizontal line is drawn to the vertical axis, where the estimated required runway length may be found. In this example, the estimated runway length for landing is approximately 6600 ft.

CHARACTERISTICS	UNITS	737–900	
MAX DESIGN TAXI WEIGHT	POUNDS	164,500	174,700
	KILOGRAMS	74,616	79,243
MAX DESIGN TAKEOFF WEIGHT	POUNDS	164,000	174,200
	KILOGRAMS	Takeoff Weight	79,016
MAX DESIGN LANDING WEIGHT	POUNDS		146,300
	KILOGRAMS		66,361
MAX DESIGN ZERO FUEL WEIGHT	POUNDS	138,300	140,300
	KILOGRAMS		63,639
OPERATING EMPTY WEIGHT (1)	POUNDS	Landing Weight	94,580
	KILOGRAMS		42,901
MAX STRUCTURAL PAYLOAD	POUNDS	43,720	45,720
	KILOGRAMS	19,831	20,738
SEATING CAPACITY (1)	TWO-CLASS	177	177
	ALL-ECONOMY	189	189
MAX CARGO - LOWER DECK	CUBIC FEET	1,835	1,835
	CUBIC METERS	52.0	52.0
USABLE FUEL	US GALLONS	6875	6875
	LITERS	26,022	26,022
	POUNDS	46,063	46,063
	KILOGRAMS	20,894	20,894

NOTE: (1) OPERATING EMPTY WEIGHT FOR BASELINE MIXED CLASS CONFIGURATION. CONSULT WITH AIRLINE FOR SPECIFIC WEIGHTS AND CONFIGURATIONS.

FIGURE 6-16   Boeing 737–900 general airplane characteristics (*Boeing Corp. document #D6-58325-3 and FAA AC 150/5325-4B*).

These charts are designed for level runways. An adjustment for runway gradient must be made by adding 10 ft of runway length for every foot of runway gradient. In this example, an additional 200 ft of runway length is added, resulting in an adjusted runway length for landing of 6800 ft.

Second, estimation of required runway length for takeoff is performed using the takeoff runway length performance chart for the aircraft. Oftentimes, an aircraft will have multiple takeoff performance charts, typically for different average high temperatures. The chart associated with the temperature nearest the airport's average high during the hottest month is used.

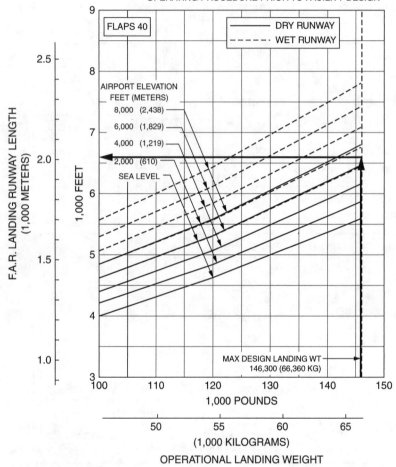

NOTES:
- STANDARD DAY
- AUTO SPOILERS OPERATIVE
- ANTI-SKID OPERATIVE
- ZERO WIND
- CONSULT USING AIRLINE FOR SPECIFIC OPERATING PROCEDURE PRIOR TO FACILITY DESIGN

FIGURE 6-17 Landing runway length for Boeing 737–900 (CFM56-7B27 Engines, 40° Flaps) (Ref: Boeing Doc. D6-58325-3).

Figure 6-18 illustrates this example. Applying the case example, a vertical line is drawn from the base of the horizontal axis at the location of the maximum design takeoff weight (174,200 lb), up to an interpolated point between the "sea level" and "2000 ft" curves, and then a horizontal line is drawn to the vertical axis, where the estimated required runway length for takeoff may be found. In this example, the estimated runway length for takeoff is approximately 8800 ft. Considering the example's runway gradient, an additional 200 ft of runway length is added, resulting in an adjusted runway length for takeoff of 9000 ft.

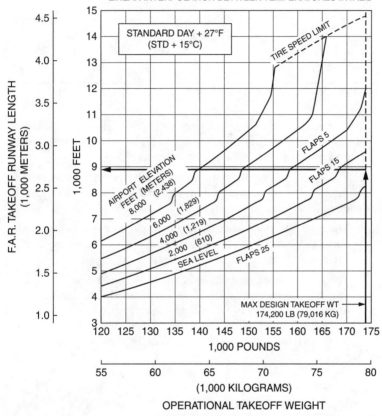

NOTES:
- CFM56–7827 ENGINES RATED AT 27,300 LB SLST
- NO ENGINE AIR BLEED FOR AIR CONDITIONING
- ZERO WIND, ZERO RUNWAY GRADIENT
- DRY RUNWAY SURFACE
- CONSULT WITH USING AIRLINE FOR SPECIFIC OPERATING PROCEDURE PRIOR TO FACILITY DESIGN
- LINEAR INTERPOLATION BETWEEN ALTITUDES INVALID
- LINEAR INTERPOLATION BETWEEN TEMPERATURES INVALID

FIGURE 6-18 Takeoff runway length for Boeing 737–900 (CFM56-7B27 Engines) (Ref: Boeing Doc. D6-58325-3).

For design purposes, the design runway length is the longer of the required runway lengths for landing and for takeoff. In this case, the design runway length for this example is 9000 ft.

Runway System Geometric Specifications

The runway system at an airport consists of the structural pavement, the shoulders, the blast pad, the runway safety area, various obstruction-free surfaces, and the runway protection zone, as shown in Figs. 6-19 and 6-20.

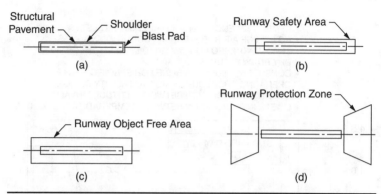

FIGURE 6-19 Runway system dimensions.

1. The runway *structural pavement* supports the aircraft with respect to structural load, maneuverability, control, stability, and other operational and dimensional criteria.

2. The *shoulder* adjacent to the edges of the structural pavement resists jet blast erosion and accommodates maintenance and emergency equipment.

3. The *blast pad* is an area designed to prevent erosion of the surfaces adjacent to the ends of runways due to jet blast or propeller wash.

4. The *runway safety area* (RSA) is an area surrounding the runway prepared or suitable for reducing the risk of damage to aircraft in the event of an undershoot, overshoot, or excursion from the runway. ICAO refers to an area similar to the runway safety area as the *runway strip* and the *runway end safety*

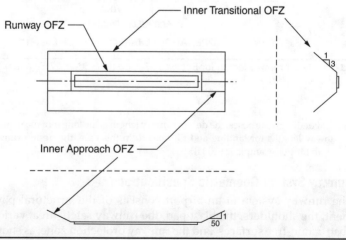

FIGURE 6-20 Object-free zone dimensions.

area. The runway safety area includes the structural pavement, shoulders, blast pad, and stopway, if provided. This area should be capable of supporting emergency and maintenance equipment as well as providing support for aircraft. The runway safety area is cleared, drained, and graded and should have no potentially hazardous ruts, humps, depressions, or other surface variations. It should be free of objects except for objects that are required to be located in the runway safety area because of their function. These objects are required to be constructed on frangible mounted structures at the lowest possible height with the frangible point no higher than 3 in above grade.

5. The runway *object-free area* (OFA) is defined by the FAA as a two-dimensional ground area surrounding the runway which must be clear of parked aircraft and objects other than those whose location is fixed by function.

6. The runway *obstacle-free zone* (OFZ) is a defined volume of airspace centered above the runway which supports the transition between ground and airborne operations. The FAA specifies this as the airspace above a surface whose elevation is the same as that of the nearest point on the runway centerline and extending 200 ft beyond each end of the runway.

7. The *inner approach obstacle-free zone,* which applies only to runways with approach lighting systems, is the airspace above a surface centered on the extended runway centerline beginning 200 ft beyond the runway threshold at the same elevation as the runway threshold and extending 200 ft beyond the last light unit on the approach lighting system. Its width is the same as the runway obstacle-free zone and it slopes upward at the rate of 50 horizontal to 1 vertical.

8. The *inner transitional obstacle-free zone,* which applies only to precision instrument runways, is defined by the FAA as the volume of airspace along the sides of the runway and the inner approach obstacle-free zone. The surface slopes at the rate of 3 horizontal to 1 vertical out from the edge of the runway obstacle-free zone and the inner approach obstacle-free zone until it reaches a height of 150 ft above the established airport elevation.

9. The *runway protection zone* (RPZ) is an area on the ground used to enhance the protection of people and objects near the runway approach.

The FAA runway standards related to the pavement and shoulder width, the safety area, the blast pad, and the obstacle-free surfaces are given in Tables 6-7 and 6-8. Similar data for the ICAO are given in Table 6-9.

	Approach Type									
	Visual and Nonprecision Instrument, Airplane Design Group					Precision Instrument, Airplane Design Group				
	I*	I	II	III	IV	I*	I	II	III	IV
Runway width	60	60	75	100	150	75	100	100	100	150
Shoulder width	10	10	10	20	25	10	10	10	20	25
Blast pad										
Width	80	80	95	140	200	95	120	120	140	200
Length	60	100	150	200	200	60	100	150	200	200
Safety area										
Width	120	120	150	300	500	300	300	300	400	500
Length†	240	240	300	600	1000	600	600	600	800	1000
Object-free area										
Width	250	400	500	800	800	800	800	800	800	800
Length†	300	500	600	1000	1000	1000	1000	1000	1000	1000
Obstacle-free zone										
Width†	120§	250	250	250	250	300	300	300	300	300
Length¶	200	200	200	200	200	200	200	200	200	200

*Facilities for small airplanes only.

†From end of runway; with the declared distance concept, these lengths begin at the stop end of each ASDA and both ends of the LDA, whichever is greater.

‡For runways serving small aircraft only; for large aircraft the greater of 400 ft or 180 ft plus the wingspan of the most demanding aircraft plus 20 ft for each 1000 ft of airport elevation.

§For runways serving small aircraft with approach speeds of less than 50 kn; increase to 250 ft for runways serving aircraft with approach speeds greater than 50 kn.

¶Beyond the end of each runway.

	Airplane Design Group					
	I	**II**	**III**	**IV**	**V**	**VI**
Runway width	100	100	100[a]	150	150	200
Shoulder[b] width	10	10	20[a]	25	35	40
Blast pad						
Width	120	120	140[a]	200	220	280
Length	100	150	200	200	400	400
Safety area						
Width[c]	500	500	500	500	500	500
Length[d]	1000	1000	1000	1000	1000	1000
Object-free area						
Width	800	800	800	800	800	800
Length[d]	1000	1000	1000	1000	1000	1000
Obstacle-free zone						
Width[e]	400	400	400	400	400	400
Length[f]	200	200	200	200	200	200

[a]For airplane design group III serving aircraft with maximum certified takeoff weight greater than 150,000 lb, the standard runway width is 150 ft, the shoulder width is 25 ft, and the blast pad width is 200 ft.
[b]Airplane design groups V and VI normally require stabilized or paved shoulder surfaces.
[c]For Airport Reference Code C-I and C-II, a runway safety area width of 400 ft is permissible. For runways designed after 2/28/83 to serve aircraft approach category D aircraft, the runway safety area width increases 20 ft for each 1000 ft of airport elevation above mean sea level.
[d]From end of runway; with the declared distance concept, these lengths begin at the stop end of each ASDA and both ends of the LDA, whichever is greater.
[e]For large aircraft the greater of 400 ft or 180 ft plus the wingspan of the most demanding aircraft plus 20 ft for each 1000 ft of airport elevation; for small aircraft 300 ft for precision instrument runways, 250 ft for all other runways serving small aircraft with approach speeds of 50 kn or more, and 120 ft for all other runways serving small aircraft with approach speeds less than 50 kn.
[f]Beyond the end of each runway.

TABLE 6-8   Runway Dimensional Standards, ft—Approach Category C, D, and E Aircraft

Parallel Runway System Spacing

The spacing of parallel runways depends on a number of factors such as whether the operations are in VMC or IMC and, if in IMC, whether it is desired to have the capability of accommodating simultaneous arrivals or simultaneous arrivals and departures. At those airports serving both heavy and light aircraft simultaneous use of runways even in VMC conditions may be dictated by separation requirements to safeguard against wake vortices.

	Aerodrome Code Letter				
	A	**B**	**C**	**D**	**E**
Pavement width					
Aerodrome code number					
1*	18	18	23		
2*	23	23	30		
3	30	30	30	45	
4			45	45	45
Pavement and shoulder width†,‡			60	60	60
	Aerodrome Code Number				
	1	**2**	**3**	**4**	
Runway strip width†					
Precision approach	150	150	300	300	
Nonprecision approach	150	150	300	300	
Visual approach	60	80	150	150	
Clear and graded area width†					
Instrument approach	80	80	150§	150§	
Visual approach	60	80	150	150	

*The width of a precision approach runway should not be less than 30 m where the aerodrome code number is 1 or 2.
†Minimum width of pavement and shoulders when pavement width is less than 60 m.
‡Symmetrical about the runway centerline.
§It is recommended that this be provided for the first 150 m from each end of the runway and that it should be increased linearly from this point to a width of 210 m at a point 300 m from each end of the runway and remain at this width for the remainder of the runway.

TABLE 6-9 ICAO Runway and Runway Strip Dimensional Standards, m

Under VMC, the FAA requires parallel runway centerline separations of 700 ft for all aircraft when the operations are in the same direction and wake vortices are not prevalent. It also recommends increasing the separation to 1200 ft for airplane design group V and VI runways. If wake vortices are generated by heavy jets and it is desired to operate on two runways simultaneously in VMC when little or no crosswind is present, the minimum distance specified by the FAA is 2500 ft.

For operations under VMC, the ICAO recommends that the minimum separations between the centerlines of parallel runways for simultaneous use disregarding wake vortices be 120 m (400 ft) for aerodrome code number 1, 150 m (500 ft) for aerodrome code number 2, and 210 m (700 ft) for aerodrome code number 3 or 4 runways.

In IMC conditions, the FAA specifies 4300 ft and ICAO specifies 1525 m (5000 ft) as the minimum separation between centerlines of parallel runways for simultaneous instrument approaches. However, there is evidence that these distances are conservative and steps are being taken to reduce it. The ultimate goal is to reduce this distance by about one-half. For dependent instrument approaches both the FAA and ICAO recommend centerline separations of 3000 ft (915 m). For triple and quadruple simultaneous instrument approaches, the FAA requires 5000-ft separation between runway centerlines, although will allow 4300 ft separations on a case-by-case basis.

Both the FAA and ICAO specify that two parallel runways may be used simultaneously for radar departures in IMC if the centerlines are separated by at least 2500 ft (760 m). The FAA requires a 3500-ft centerline separation for simultaneous nonradar departures. If two parallel runways are to be operated independently of each other in IMC under radar control, one for arrivals and the other for departures, both the FAA and ICAO specify that the minimum separation between the centerlines is 2500 ft (760 m) when the thresholds are even. If the thresholds are staggered, the runways can be brought closer together or must be separated farther depending on the amount of the stagger and which runways are used for arrivals and departures. If approaches are to the nearest runway, then the spacing may be reduced by 100 ft (30 m) for each 500 ft (150 m) of stagger down to a minimum of 1200 ft (360 m) for airplane design groups V and VI and 1000 ft (300 m) for all other aircraft. However, if the approaches are to the farthest runway, then the runway spacing must be increased by 100 ft (30 m) for each 500 ft (150 m) of stagger.

Sight Distance and Longitudinal Profile

The FAA requirement for sight distance on individual runways requires that the runway profile permit any two points 5 ft above the runway centerline to be mutually visible for the entire runway length. If, however, the runway has a full length parallel taxiway, the runway profile may be such that an unobstructed line of sight will exist from any point 5 ft above the runway centerline to any other point 5 ft above the runway centerline for one-half the runway length.

The FAA recommends a clear line of sight between the ends of intersecting runways. The terrain must be graded and permanent objects designed and sited so that there will be an unobstructed line of sight from any point 5 ft above one runway centerline to any point 5 ft above an intersecting runway centerline within the runway visibility zone. The runway visibility zone is the area formed by imaginary lines connecting the visibility points of the two intersecting runways. The runway visibility zone for intersecting runways is shown in Fig. 6-21. The visibility points are defined as follows:

1. If the distance from the intersection of the two runway centerlines is 750 ft or less, the visibility point is on the centerline at the runway end designated by point a in Fig. 6-21.

2. If the distance from the intersection of the two runway centerlines is greater than 750 ft but less than 1500 ft, the visibility point is on the centerline 750 ft from the intersection of the centerlines designated by point b in Fig. 6-21.

3. If the distance from the intersection of the two runway centerlines is equal to or greater than 1500 ft, the visibility point is on the centerline equidistant from the runway end and the intersection of the centerlines designated by points c and d in Fig. 6-21.

The ICAO requirement for sight distance on individual runways requires that the runway profile permit an unobstructed view between any two points at a specified height above the runway centerline to be mutually visible for a distance equal to at least one-half the runway length. ICAO specifies that the height of these two points be 1.5 m (5 ft) above the runway for aerodrome code letter A runways, 2 m (7 ft) above the runway for aerodrome code letter B runways, and 3 m (10 ft) above the runway for aerodrome code letter C, D, or E runways.

It is desirable to minimize longitudinal grade changes as much as possible. However, it is recognized that this may not be possible for reasons of economy. Therefore both the ICAO and FAA allow changes

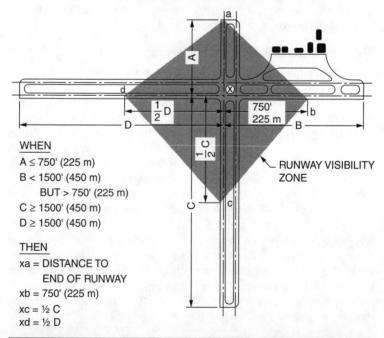

WHEN
A ≤ 750' (225 m)
B < 1500' (450 m)
   BUT > 750' (225 m)
C ≥ 1500' (450 m)
D ≥ 1500' (450 m)

THEN
xa = DISTANCE TO
   END OF RUNWAY
xb = 750' (225 m)
xc = ½ C
xd = ½ D

FIGURE 6-21 Runway visibility zone for intersecting runways (*Federal Aviation Administration*).

	Aircraft Approach Category				
	A	**B**	**C**	**D**	**E**
Gradient (%)					
Pavement longitudinal[a]					
Maximum	2.0	2.0	1.5[b]	1.5[b]	1.5[b]
Maximum change	2.0	2.0	1.5	1.5	1.5
Pavement transverse					
Maximum	2.0	2.0	1.5	1.5	1.5
Shoulder transverse					
Minimum	3.0	3.0	1.5[c]	1.5[c]	1.5[c]
Maximum[d]	5.0	5.0	5.0	5.0	5.0
Runway end safety area					
Maximum longitudinal[e]	3.0	3.0	3.0	3.0	3.0
Maximum longitudinal Grade change	2.0	2.0	2.0	2.0	2.0
Minimum transverse	1.5	1.5	1.5	1.5	1.5
Maximum transverse[d]	5.0	5.0	3.0	3.0	3.0
Vertical curve (ft)					
Minimum length[a,f]	300[g]	300[g]	1000	1000	1000
Minimum distance between points of intersection[a,h]	250	250	1000	1000	1000

[a]Applies also to runway safety area adjacent to sides of the runway.
[b]May not exceed 0.8 percent in the first and last quarter of runway.
[c]A minimum of 3 percent for turf.
[d]A slope of 5 percent is recommended for a 10 ft width adjacent to the pavement areas to promote drainage.
[e]For the first 200 ft from the end of the runway and if it slopes it must be downward. For the remainder of the runway safety area the slope must be such that any upward slope does not penetrate the approach surface or clearway plane and any downward slope does not exceed 5 percent.
[f]For each 1 percent change in grade.
[g]No vertical curve is required if the grade change is less than 0.4 percent.
[h]Distance is multiplied by the sum of the absolute grade grade changes in percent.
Source: Federal Aviation Administration [6].

TABLE 6-10   Runway Surface Gradient Standards

in grade but limit their number and size. The maximum longitudinal grade changes that are permitted by the FAA are listed in Table 6-10 and illustrated in Fig. 6-22. The maximum longitudinal grade changes that are permitted by the ICAO are listed in Table 6-11. Tables 6-10 and 6-11 also list the maximum longitudinal grade. The FAA limits both

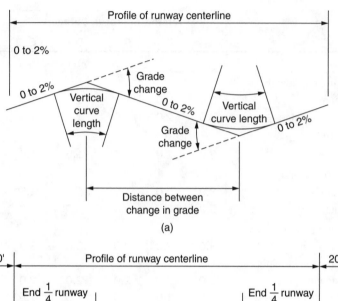

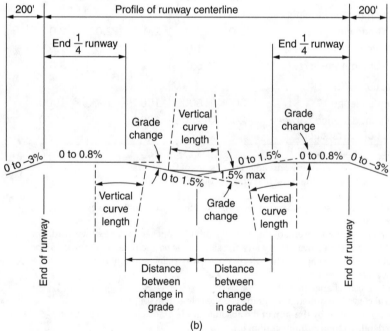

Figure 6-22 Runway longitudinal profile: (*a*) utility airports, (*b*) transport airports.

longitudinal gradient and longitudinal grade changes to 2 percent for runways serving approach category A and B aircraft and 1.5 percent for runways serving approach category C, D, and E aircraft. ICAO limits both longitudinal gradient and longitudinal grade changes to 2 percent for aerodrome code number 1 and 2 runways and 1.5 percent for aerodrome code number 3 runways. For aerodrome code number 4 runways the maximum longitudinal gradient is 1.25 percent and the maximum change in longitudinal gradient is 1.5 percent. In addition, for runways that are equipped to be used in bad weather, the gradient of the first and last quarter of the length of the runway must be very flat for reasons of safety. Both the ICAO and the FAA require that this gradient not exceed 0.8 percent. In all cases it is desirable to keep both longitudinal grades and grade changes to a minimum.

Longitudinal slope changes are accomplished by means of vertical curves. The length of a vertical curve is determined by the magnitude of the changes in slope and the maximum allowable change in the slope of the runway. Both these values are also listed in Tables 6-11 and 6-12.

	Aerodrome Code Number				
	1	**2**	**3**	**4**	
Runway longitudinal Gradient (%)					
Maximum	2.0	2.0	1.5*	1.25*	
Maximum change	2.0	2.0	1.5	1.5	
Maximum effective[†]	2.0	2.0	1.0	1.0	
Vertical curve (m)					
Minimum length of curve[‡]	75	150	300	300	
Minimum distance between points of intersection[§]	50	50	150	300	
Runway strips Gradient (%)					
Maximum longitudinal	2.0	2.0	1.75	1.5	
Maximum transverse	3.0	3.0	2.5	2.5	
	Aerodrome Code Letter				
	A	**B**	**C**	**D**	**E**
Runway transverse gradient (%)					
Maximum	2.0	2.0	1.5	1.5	1.5
Minimum	1.0	1.0	1.0	1.0	1.0
Shoulder transverse gradient (%)					
Maximum	2.5	2.5	2.5	2.5	2.5

*May not exceed 0.8 percent in the first and last quarter of runway for aerodrome code number 4 or for a category II or III precision instrument runway for aerodrome code number 3.
[†]Difference in elevation between high and low point divided by runway length
[‡]For each 1 percent change in grade.
[§]Distance is multiplied by sum of absolute grade changes in percent minimum length is 45 m.
Source: International Civil Aviation Organization [3].

TABLE 6-11   Runway Surface Gradient Standards

	Airplane Design Group				
	I*	I	II	III	IV
Visual or nonprecision runway centerline to					
Taxiway or taxilane centerline†	150	225	240	300	400
Hold line†	125	200	200	200	250
Helicopter touchdown pad	400	400	400	400	400
Aircraft parking area	125	200	250	400	500
Precision instrument runway centerline to					
Taxiway or taxilane centerline†	200	250	300	350	400
Hold line†	175	250	250	250†	250†
Helicopter touchdown pad	400	400	400	400	400
Aircraft parking area	400	400	400	400	500

---

*For facilities for small aircraft only.

†Satisfies the requirement that no part of an aircraft at a holding an increase to these separations may be needed to achieve this result.

‡For sea level up to elevation 6000 ft. Increase by 1 ft for each 100 ft of airport elevation above 6000 ft.

Source: Federal Aviation Administration [6].

TABLE 6-12 Airfield Separation Criteria for Aircraft in Approach Categories A and B, ft

The number of slope changes along the runway is also limited. The FAA requires that the distance between the points of intersection of two successive curves should not be less than the sum of the absolute percentage values of change in slope multiplied by the 250 ft for airports serving aircraft approach category A and B aircraft and 1000 ft for airports serving aircraft approach category C, D, and E aircraft. The ICAO requires that the distance between the points of intersection of two successive curves should not be less than the sum of the absolute percentage values of change in slope multiplied by 50 m (165 ft) for aerodrome code number 1 and 2 runways, 150 m (500 ft) for aerodrome code number 3 runways, and 300 m (1000 ft) for aerodrome code number 4 runways. ICAO also specifies that the minimum distance in all cases is 45 m (150 ft).

For example, for an FAA runway serving transport aircraft, that is, approach category C, D, or E aircraft, if the change in slope was 1.5 percent, the required length of vertical curve would be 1500 ft. Vertical curves are normally not necessary if the change in slope is not more than 0.4 percent. The FAA specifies a minimum length of vertical transition curve of 300 for each 1 percent change in grade for runways

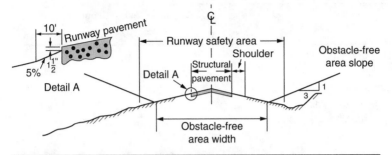

Detail A

Figure 6-23   Runway gradient cross section.

serving approach category A and B aircraft and 1000 ft for each 1 percent change in grade for airport serving approach category C, D, and E aircraft. ICAO specifies a minimum length of vertical transition curve of 75 m for each 1 percent change in grade for aerodrome code number 1 runways, 150 m for each 1 percent change in grade aerodrome code number 2 runways, and 300 m for each 1 percent change in grade for aerodrome code number 4 runways.

Transverse Gradient

A typical cross section of a runway is shown in Fig. 6-23. The FAA and ICAO specifications for transverse slope on the runways are given in Tables 6-10 and 6-11, respectively. It is recommended that a 5 percent transverse slope be provided for the first 10 ft of shoulder adjacent to a pavement edge to ensure proper drainage.

Airfield Separation Requirements Related to Runways

The minimum distance from the runway centerline to parallel taxiways, taxilanes, aircraft holding lines, helicopter touchdown pads, and aircraft parking areas are also specified. These distances are given in Tables 6-12 and 6-13 for the FAA and Tables 6-14 and 6-15 for ICAO.

Obstacle Clearance Requirements

In addition to the geometric standards associated with the design of runways, there are specific requirements concerning the protection of airspace around airfields to provide for the safe navigation of aircraft to and from the airport.

In the United States, the FAA requires that protection zones be provided at the ends of runways. The runway protection zone is the area on the ground beneath the approach surface to a runway from the end of the primary surface to the point where the approach surface is 50 ft above the primary surface, as shown in Fig. 6-24. The dimensions of the runway protection zone are provided in Table 6-16.

	Airplane Design Group					
	I	II	III	IV	V	VI
Visual or nonprecision runway						
Centerline to						
Taxiway or taxilane centerline*	300	300	400	400	400¶	600
Hold line*	250	250	250	250	250	250
Helicopter touchdown pad	400	400	400	400	400	400
Aircraft parking area	400	400	500	500	500	500
Precision instrument runway						
Centerline to						
Taxiway or taxilane centerline*	400	400	400	400	400¶	600
Hold line*,§	250	250	250†	250†	280†	325
Helicopter touchdown pad	400	400	400	400	400	400
Aircraft parking area	500	500	500	500	500	500

*Satisfies the requirement that no part of an aircraft at a holding location or on a taxiway centerline is within the runway safety area or penetrates the obstacle free zone. Accordingly, at higher elevations an increase to these separations may be needed to achieve this result.
†For aircraft in aircraft approach category C and airplane design groups III and IV increase by 1 ft for each 100 ft of airport elevation greater than 3200 ft.
‡For aircraft in aircraft approach category C and airplane design group V increase by 1 ft for each 100 ft of airport elevation above mean sea level.
§For aircraft in aircraft approach category D increase by 1 ft for each 100 ft of airport elevation above mean sea level.
¶For airports at or below an elevation of 1345 ft; increase to 450 ft for airports at elevations between 1345 and 6560 ft and to 500 ft for airports at an elevation above 6560 ft.
Source: Federal Aviation Administration [6].

TABLE 6-13 Airfield Separation Criteria for Aircraft in Approach Categories C and D, ft

When the runway protection zone begins at a location other than 200 ft beyond the end of the runway due to the application of the declared distance concept discussed in Chap. 2, two runway protection zones are usually required, an approach runway protection zone and a departure runway protection zone. The dimensions of the approach runway protection zone are given in Table 6-16 but the departure runway protection zone begins 200 ft beyond the far end of

	Aerodrome Code Letter				
	A	**B**	**C**	**D**	**E**
Runway centerline to parallel taxiway centerline					
Noninstrument runways					
Aerodrome code 1	37.5	42			
Aerodrome code 2	47.5	52			
Aerodrome code 3			93	101	
Aerodrome code 4				101	107.5
Instrument runways					
Aerodrome code 1	82.5	87			
Aerodrome code 2	82.5	87			
Aerodrome code 3			168	176	
Aerodrome code 4				176	182.5

Source: International Civil Aviation Organization [2, 3, 4].

TABLE 6-14   Runway to Taxiway Separation Criteria on the Airfield, m

	Type of Runway				
	Non-instrument	**Non-precision Approach**	**Precision Approach Category**		**Takeoff**
			I	**II & III**	
Aerodrome code 1	30	40	60*	–	30
Aerodrome code 2	40	40	60*	–	40
Aerodrome code 3	75	75	90*,†	90*,†	75
Aerodrome code 4	75	75	90*,†	90*,†	75

*This distance may have to be increased to avoid interference with radio aids; for a precision instrument category III runway this increase may be in the order of 50 m.
†If a holding bay or a taxiway holding position is at a lower elevation compared to the runway threshold the distance may be decreased by 5 m for every meter the holding bay or holding position is lower than the threshold, contingent upon not interfering with the inner transitional surface; if a holding bay or a taxiway holding position is at a higher elevation compared to the runway threshold the distance should be increased by 5 m for every meter the holding bay or holding position is higher than the threshold.
Source: International Civil Aviation Organization [2, 3, 4].

TABLE 6-15   Runway to Holding Line Separation Criteria on the Airfield, m

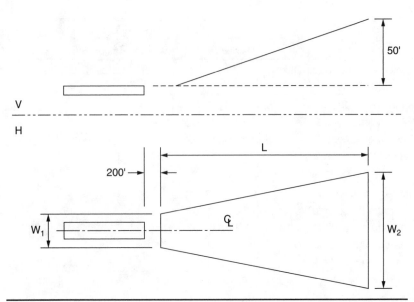

FIGURE 6-24 Runway protection zone.

the takeoff run available and the portion of the runway between the takeoff run available and the end of the runway is declared unavailable and unsuitable for the takeoff run. The dimensions of the departure runway protection zone are

1. For runways serving only small aircraft in aircraft approach categories A and B, the length is 1000 ft, the inner width is 250 ft and the outer width is 450 ft.

2. For runways serving large aircraft in aircraft approach categories A and B, the length is 1000 ft, the inner width is 500 ft and the outer width is 700 ft.

3. For runways serving aircraft in aircraft approach categories C, D, or E, the length is 1700 ft, the inner width is 500 ft and the outer width is 1010 ft.

FAR Part 77

Part 77 of the Federal Aviation Regulations establishes standards for determining what would be considered obstructions to navigable airspace, sets forth the requirements for notice to the FAA due to certain proposed construction or alteration activities, and provides for aeronautical studies of obstructions to air navigation to determine the effect of these obstructions on the safe and efficient use of airspace [8, 9]. The airport operator has the responsibility to ensure that the aerial approaches to the airport will be adequately cleared and protected and that the land adjacent to or in the immediate vicinity of the airport

Aircraft Served	Runway Approach End	Approach* Opposite End	Width			
			Length L, ft	Inner W_1, ft	Outer W_2, ft	Area, acres
Small	V	V	1000	250	450	8.035
		NP	1000	500	650	13.200
		NP+	1000	1000	1050	23.542
		P	1000	1000	1050	23.542
	NP	V	1000	500	800	14.922
		NP	1000	500	800	14.922
		NP+	1000	1000	1200	25.252
		P	1000	1000	1200	25.252
	NP+	V	1700	1000	1510	48.978
		NP	1700	1000	1510	48.978
		NP+	1700	1000	1510	48.978
		P	1700	1000	1510	48.978
	P	V	2500	1000	1750	78.914
		NP	2500	1000	1750	78.914
		NP+	2500	1000	1750	78.914
		P	2500	1000	1750	78.914
Large	V	V	1000	500	700	13.770
		NP	1000	500	700	13.770
		NP+	1000	1000	1100	24.105
		P	1000	1000	1100	24.105
	NP	V	1700	500	1010	29.465
		NP	1700	500	1010	29.465
		NP+	1700	1000	1425	47.320
		P	1700	1000	1425	47.320
	NP+	V	1700	1000	1510	48.978
		NP	1700	1000	1510	48.978
		NP+	1700	1000	1510	48.978
		P	1700	1000	1510	48.978
	P	V	2500	1000	1750	78.914
		NP	2500	1000	1750	78.914
		NP+	2500	1000	1750	78.914
		P	2500	1000	1750	78.914

*V = visual approach; NP = nonprecision instrument approach with visibility minimums more than ¾ statute mile; NP+ = nonprecision instrument approach with visibility minimums as low as ¾ statute mile; P = precision instrument approach.
Source: Federal Aviation Administration [5].

TABLE 6-16 Runway Protection Zone Dimensions

is reasonably restricted to the extent possible through the use of such measures as the adoption of zoning ordinances. A model zoning ordinance to limit height of objects around airports is published by the FAA [6].

Subpart C of FAR Part 77 establishes standards for determining obstructions to air navigation. The standards apply to existing and man-made objects, objects of natural growth, and terrain.

In order to determine whether an object is an obstruction to air navigation, several imaginary surfaces are established with relation to the airport and to each end of a runway. The size of the imaginary surfaces depends on the category of each runway (e.g., utility or transport) and on the type of approach planned for that end of the runway (e.g., visual, nonprecision instrument, or precision instrument).

The principal imaginary surfaces are shown in Fig. 6-25. They are described as follows:

1. *Primary surface.* The primary surface is a surface longitudinally centered on a runway. When the runway is paved, the primary surface extends 200 ft beyond each end of the runway. When the runway is unpaved, the primary surface coincides with each end of the runway. The elevation of the primary surface is the same as the elevation of the nearest point on the runway centerline.

2. *Horizontal surface.* The horizontal surface is a horizontal plane 150 ft above the established airport elevation, the perimeter of which is constructed by swinging arcs of specified radii from the center of each end of the primary surface of each runway and connecting each arcs by lines tangent to those arcs.

3. *Conical surface.* The conical surface is a surface extending outward and upward from the periphery of the horizontal surface at a slope of 20 horizontal to 1 vertical for a horizontal distance of 4000 ft.

4. *Approach surface.* The approach surface is a surface longitudinally centered on the extended runway centerline and extending outward and upward from each end of a runway at a designated slope based upon the type of available or planned approach to the runway.

5. *Transitional surface.* Transitional surfaces extend outward and upward at right angles to the runway centerline plus the runway centerline extended at a slope of 7 to 1 from the sides of the primary surface up to the horizontal surface and from the sides of the approach surfaces. The width of the transitional surface provided from each edge of the approach surface is 5000 ft.

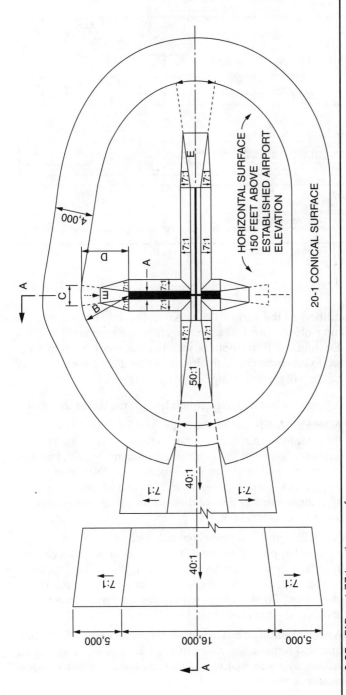

FIGURE 6-25 FAR part 77 imaginary surfaces.

219

OBSTRUCTION IDENTIFICATION SURFACES
FEDERAL AVIATION REGULATIONS PART 77

DIM	ITEM	DIMENSIONAL STANDARDS (FEET)						
		VISUAL RUNWAY		NON - PRECISION INSTRUMENT RUNWAY				PRECISION INSTRUMENT RUNWAY
		A	B	A	B			
					C	D		
A	WIDTH OF PRIMARY SURFACE AND APPROACH SURFACE WIDTH AT INNER END	250	500	500	500	1,000		1,000
B	RADIUS OF HORIZONTAL SURFACE	5,000	5,000	5,000	10,000	10,000		10,000
		VISUAL APPROACH		NON - PRECISION INSTRUMENT APPROACH				PRECISION INSTRUMENT APPROACH
		A	B	A	B			
					C	D		
C	APPROACH SURFACE WIDTH AT END	1,250	1,500	2,000	3,500	4,000		16,000
D	APPROACH SURFACE LENGTH	5,000	5,000	5,000	10,000	10,000		*
E	APPROACH SLOPE	20:1	20:1	20:1	34:1	34:1		*

• A - UTILITY RUNWAYS
• B - RUNWAYS LARGER THAN UTILITY
• C - VISIBILITY MINIMUMS GREATER THAN 3/4 MILE
• D - VISIBILITY MINIMUMS AS LOW AS 3/4 MILE
• * - PRECISION INSTRUMENT APPROACH SLOPE IS 50:1 FOR INNER 10,000 FEET AND 40:1 FOR AN ADDITIONAL 40,000 FEET

FIGURE 6-26   Part 77 Imaginary Surface Dimensions, ft.

Dimensions of the several imaginary surfaces are shown in Fig. 6-26.

In addition to the surfaces defined earlier, other standards for determining obstructions to air navigation are contained in FAR Part 77. Existing and future objects, whether stationary or mobile, are considered to be obstructions to air navigation if they are of greater height than any of the following heights or surfaces:

1. A height of 500 ft above ground level at the site of the object.

2. A height that is 200 ft above ground level or 200 ft above the established airport elevation, whichever is greater, within 3 nautical miles of the established reference point at an airport with its longest runway more than 3200 ft in actual length. This height increases in the ratio of 100 ft for each additional nautical mile of distance from the reference point up to a maximum of 500 ft.

3. A height within a terminal obstacle clearance area, including an initial approach segment, a departure area, and a circling approach area, which would result in the vertical distance between any point on the object and an established minimum instrument flight altitude within that area or segment to be less than the required obstacle clearance.

4. A height within an en route obstacle clearance area, including turn and termination areas, of a federal airway or approved off-airway route, that would increase the minimum obstacle clearance altitude.

5. The surface of a takeoff and landing area of an airport or any of the imaginary surfaces defined earlier.

6. Except for traverse ways on or near an airport with an operative ground traffic control service furnished by the air traffic control tower or by airport management and coordinated with the air traffic control service, the heights of traverse ways must be increased by 17 ft for interstate highways, 15 ft for any other public roadway, 10 ft or the height of the highest mobile object that would normally traverse the road, whichever is greater, for a private road, 23 ft or an amount equal to the height of the highest mobile object that would normally traverse it for railroads, waterways, or any other thoroughfare not previously mentioned.

Subpart B of FAR Part 77 identifies circumstances where notice is required to be given to the FAA when certain construction or alteration activities are proposed. These include the circumstances associated with the standards given above and also any construction or alteration of greater height than an imaginary surface extending outward and upward at one of the following slopes [9]:

1. A slope of 100 horizontal to 1 vertical for a horizontal distance of 20,000 ft from the nearest point of the nearest runway at an airport or seaplane base with at least one runway more than 3200 ft in actual length.

2. A slope of 50 horizontal to 1 vertical for a horizontal distance of 10,000 ft from the nearest point of the nearest runway at an airport or seaplane base with its longest runway no more than 3200 ft in actual length.

3. A slope of 25 horizontal to 1 vertical for a horizontal distance of 5000 ft from the nearest point of the nearest takeoff and landing area for a heliport.

FAR Part 77 imposes strict requirements on both airport sponsors and others associated with construction activities in the vicinity of airports which should be referenced prior to initiating construction activities.

ICAO Annex 14

The ICAO requirements are similar to FAR Part 77 with the following exceptions. ICAO separates arrivals and departures and specifies dimensions for approach surfaces and takeoff climb surfaces for departures. The horizontal surface specified by ICAO is a circle whose center is at the *airport reference point*, whereas in FAR Part 77 it is not a circle nor is the airport reference point used to determine the horizontal surface. The airport reference point is the geometric centroid of the runway system at the airport based upon the lengths of the runways. The height of this surface is 150 ft above the airport elevation, the same as in Part 77. In FAR Part 77 the conical surface extends horizontally 4000 ft at a slope of 20 to 1 irrespective of the type of runway and visibility. In ICAO Annex 14 [1, 2, 3] the slope of the conical surface is the same, but the horizontal distance varies depending upon the aerodrome reference code.

In FAR Part 77 the slope of the transitional surface is a constant 7 to 1, whereas in ICAO Annex 14 this slope is specified for runway reference codes 3 and 4. For other runways the slope is 5 to 1.

TERPS

As defined in FAA Order 8260.3b, TERPS (which stands for terminal instrument approach procedures) is a compilation of criteria used to design published standard procedures for aircraft using instrument-based navigation to depart and approach to airport facilities. These procedures are designed based primarily on the performance characteristics of aircraft, the various types of instrument navigational aids that may be present at or around an airport, and currently existing natural and man-made objects surrounding the airport. As part of these procedures, minimum climb-out gradients for aircraft departures, and minimum descent gradients and safe operating altitudes for aircraft approaches are defined. While TERPS contain standards for creating such procedures, for any given runway at any given airport, one or more approach and departure procedures may be defined, each of which may be entirely unique, based on the airport environment.

With respect to airport design, TERPS defines a "required obstacle clearance" (ROC) value. For aircraft operating within the airport environment, this value is typically as low as 250 ft above the highest object near the runway. The required obstacle clearance values for a published procedure in turn define the TERPS obstacle clearance surface (OCS), as illustrated in Fig. 6-27. The typical slopes for obstacle clearance surfaces for aircraft on approach is on the order of 318 ft/nmi and for departures approximately 200 ft/nmi.

A typical TERPS procedure consists of a series of segments, including climb, en route, initial approach, intermediate approach, final approach, and missed approach segments, that are created based

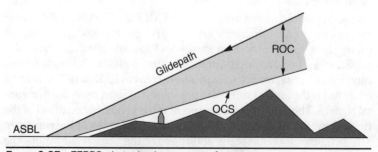

Figure 6-27 TERPS obstacle clearance surface.

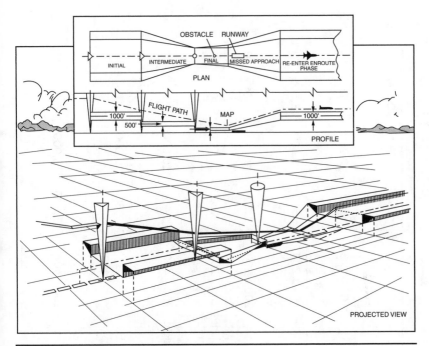

FIGURE 6-28 Typical TERPS procedure segments.

on the above standards and the existing terrain and obstacle environment for any given runway, as illustrated in Fig. 6-28.

It is widely understood that protecting airspace for TERPS is a complex process, often unique to each airport. For planning purposes, however, a slope with 40:1 gradient and 15° from the runway end should be considered as TERPS obstacle clearance surface criteria. Runways with the intention of being supported by published instrument procedures should be designed in such a manner to avoid any natural or man-made obstacles that penetrate this surface. Once a runway exists, airport planners should work to ensure that future development does not conflict with TERPS or FAR Part 77 obstacle clearance requirements.

Runway End Siting Requirements

The specifications for determining obstacles to safe air navigation to existing runways are described in FAR Part 77 and TERPS procedures. However, when locating, or siting a runway, the FAA prescribes a different, yet complimentary set of specifications. These specifications are published in Appendix 2 of Advisory Circular AC 150/5300-13, identified in Table 6-17, and illustrated in Figs. 6-29 through 6-31.

Runway Type	Dimensional Standards* ft					Slope/ OCS
	A	B	C	D	E	
1. Approach end of runways expected to serve small airplanes with approach speeds less than 50 kn (visual runways only, day/night)	0	60	150	500	2,500	15:1
2. Approach end of runways expected to serve small airplanes with approach speeds of 50 kn or more (visual runways only, day/night)	0	125	350	2,250	2,750	20:01
3. Approach end of runways expected to serve large airplanes (visual day/night); or instrument minimums ≥ 1 statute mile (day only)	0	200	500	1,500	8,500	20:1
4. Approach end of runways expected to support instrument night circling[a]	200	200	1,700	10,000	0	20:1
5. Approach end of runways expected to support instrument straight in night operations. Serving approach category A and B aircraft only.[a]	200	200	1,900	10,000[b]	0	20:1
6. Approach end of runways expected to support instrument straight in night operations serving greater than approach category B aircraft.[a]	200	400	1,900	10,000[b]	0	20:1
7[e,f,g,h]. Approach end of runways expected to accommodate approaches with positive vertical guidance (GQS)	0	½ width runway +100	760	10,000[b]	0	30:1
8. Approach end of runways expected to accommodate instrument approaches having visibility minimums ≥ ¾ but < 1 statute mile, day or night	200	400	1,900	10,000[b]	0	20:1

9	Approach end of runways expected to accommodate instrument approaches having visibility minimums < ¾ statute mile or precision approach (ILS, GLS, or MLS) day or night	200	400	1,900	10,000[b]	0	34:1
10	Approach runway ends having category II approach minimums or greater	The criteria are set forth in TERPS. Order 8260.3.					
11	Departure runway ends for all instrument operations	0[d]		See Fig. 6-30			40:1
12	Departure runway ends supporting air carrier operations[c]	0[d]		See Fig. 6-31			625:1

[a] Dimensional standards illustrated in Fig. 6-29

Notes:

[a] Lighting of obstacle penetrations to this surface or the use of a VGSI, as defined by the TERPS order, may avoid displacing the threshold.

[b] 10,000 ft is a nominal value for planning purposes. The actual length of these areas is dependent upon the visual descent point position for 20:1 and 34:1 and decision altitude point for the 30:1.

[c] Any penetration to this surface will limit the runway end to nonprecision approaches. No vertical approaches will be authorized until the penetration(s) is/are removed except obstacles fixed by function and/or allowable grading.

[d] Dimension A is measured relative to departure end of runway (DER) or TODA (to include clearway).

[e] Data collected regarding penetrations to this surface are provided for information and use by the air carriers operating from the airport. These requirements do not take effect until January 1, 2009.

[f] Surface dimensions/obstacle clearance surface (OCS) slope represent a nominal approach with 3° GPA, 50' TCH, < 500' HAT. For specific cases refer to TERPS. The obstacle clearance surface slope (30:1) represents a nominal approach of 3° (also known as the glide path angle). This assumes a threshold crossing height of 50 ft. Three degrees is commonly used for ILS systems and VGSI aiming angles. This approximates a 30:1 approach angle that is between the 34:1 and the 20:1 notice surfaces of Part 77. Surfaces cleared to 34:1 should accommodate a 30:1 approach without any obstacle clearance problems.

[g] For runways with vertically guided approaches the criteria in Row 7 is in addition to the basic criteria established within the table, to ensure the protection of the glide path qualification surface.

[h] For planning purposes, sponsors and consultants determine a tentative decision altitude based on a 3° glide path angle and a 50-ft threshold crossing height.

TABLE 6-17 Runway End Siting Requirements Dimensions Table

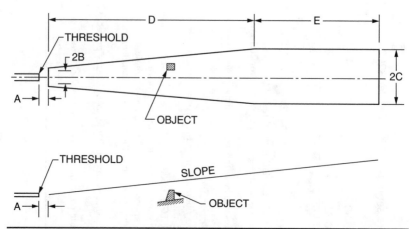

Figure 6-29 Runway end siting requirements.

These specifications are used to site the location of a runway's threshold so that approach and departure procedures associated with that runway are not adversely affected by existing obstacles or terrain. The siting specifications vary depending on a number of runway use conditions, including

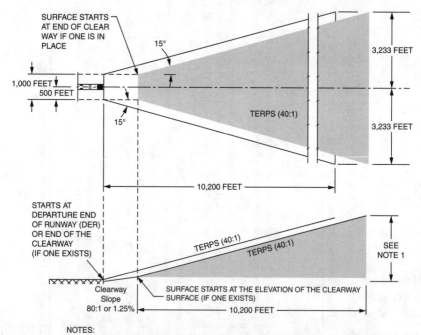

NOTES:
1. THIS IS AN INTERPRETATION OF THE APPLICATION OF THE TERPS SURFACE ASSOCIATED WITH A CLEARWAY.

Figure 6-30 TERPS departure obstacle identification surfaces.

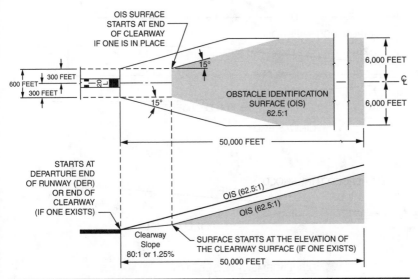

Figure 6-31   One engine inoperative obstacle identification surface (62.5:1).

- The approach speed of arriving aircraft
- The approach category of arriving aircraft
- Day versus night operations
- Types of instrument approaches
- The presence of published instrument departure procedures
- The use of the runway by air carriers

Runway end siting requirements are often the most confusing as well as overlooked element of runway planning. Care should be given to fully understand the purpose of the planned runway, the type of aircraft that will be using the runway, the current and future instrument approach procedures associated with the runway, and of course any terrain or obstacles in the vicinity.

Should an object penetrate any of the surfaces at the site of a runway, the airport planner has the option of displacing the runway threshold, as illustrated in Fig. 6-32. Displacing the threshold allows the airport planner to design runways with sufficient lengths to accommodate aircraft departures, while also allowing arrivals to safely approach the runway by maintaining sufficient clearance from upstream obstacles. Displacing the threshold does carry the penalty of reducing available runway lengths for landing. The FAA recommends avoiding the need for displaced thresholds when possible, but recognizes their benefits in the wake of no other alternatives.

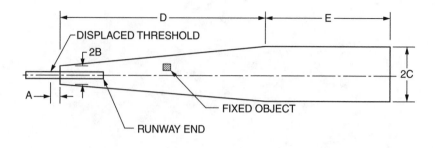

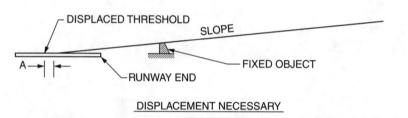

DISPLACEMENT NECESSARY

FIGURE 6-32   Use of displaced threshold, runway siting requirements.

Taxiways and Taxilanes

Taxiways are defined paths on the airfield surface which are established for the taxiing of aircraft and are intended to provide a linkage between one part of the airfield and another. The term "dual parallel taxiways" refers to two taxiways parallel to each other on which airplanes can taxi in opposite directions. An apron taxiway is a taxiway located usually on the periphery of an apron intended to provide a through taxi route across the apron. A taxilane is a portion of the aircraft parking area used for access between the taxiways and the aircraft parking positions. ICAO defines an aircraft stand taxilane as a portion of the apron intended to provide access to the aircraft stands only.

In order to provide a margin of safety in the airport operating areas, the trafficways must be separated sufficiently from each other and from adjacent obstructions. Minimum separations between the centerlines of taxiways, between the centerlines of taxiways and taxilanes, and between taxiways and taxilanes and objects are specified in order that aircraft may safely maneuver on the airfield.

Widths and Slopes

Since the speeds of aircraft on taxiways are considerably less than on runways, criteria governing longitudinal slopes, vertical curves, and sight distance are not as stringent as for runways. Also the lower speeds permit the width of the taxiway to be less than that of the runway. The principal geometric design features of interest are listed in Tables 6-18 and 6-19 for the FAA. ICAO standards are listed in Tables 6-20 and 6-21.

	Airplane Design Group					
	I	**II**	**III**	**IV**	**V**	**VI**
Width	25	35	50[a]	75	75	100
Edge safety margin[b]	5	7.5	10[c]	15	15	20
Shoulder width	10	10	20	25	35[d]	40[d]
Safety area width[e]	49	79	118	171	214	262
Object-free area width						
Taxiway[f]	89	131	186	259	320	386
Taxilane[g]	79	115	162	225	276	334
Separations						
Taxiway centerline to						
taxiway centerline[h]	69	105	152	215	267	324
fixed or movable object[i]	44.5	62.5	93	129.5	160	193
Taxilane centerline to						
taxilane centerline[j]	64	97	140	198	245	298
fixed or movable object[k]	39.5	57.5	81	112.5	138	167

[a]For airplanes in airplane design group III with a wheelbase equal to or greater than 60 ft, the standard taxiway width is 60 ft.
[b]The taxiway edge safety margin is the minimum acceptable between the outside of the airplane wheels and the pavement edge.
[c]For airplanes in airplane design group III with a wheelbase equal or greater than 60 ft, the taxiway edge safety margin is 15 ft.
[d]Airplanes in airplane design groups V and VI normally stabilized or paved taxiway shoulder surfaces.
[e]May use aircraft wingspan in lieu of these values.
[f]May use 1.4 wingspan plus 20 ft in lieu of these values.
[g]May use 1.2 wingspan plus 20 ft in lieu of these values.
[h]May use 1.2 wingspan plus 10 ft in lieu of these values.
[i]May use 0.7 wingspan plus 10 ft in lieu of these values.
[j]May use 1.1 wingspan plus 10 ft in lieu of these values.
[k]May use 0.6 wingspan plus 10 ft in lieu of these values.
Source: Federal Aviation Administration [6].

TABLE 6-18 Taxiway Dimensional Standards, ft

Taxiway and Taxilane Separation Requirements

FAA Separation Criteria

The separation criteria adopted by the FAA are predicated upon the wingtips of the aircraft for which the taxiway and taxilane system have been designed and provide a minimum wingtip clearance on

	Aircraft Approach Category				
	A	B	C	D	E
Gradient (%)					
Taxiway, shoulder and safety area					
Longitudinal					
Maximum	2.0	2.0	1.5	1.5	1.5
Maximum change	3.0	3.0	3.0	3.0	3.0
Taxiway transverse					
Minimum	1.0	1.0	1.0	1.0	1.0
Maximum	2.0	2.0	1.5	1.5	1.5
Shoulder transverse					
Minimum	3.0	3.0	1.5*	1.5*	1.5*
Maximum†	5.0	5.0	5.0	5.0	5.0
Safety area transverse					
Minimum	3.0	3.0	1.5	1.5	1.5
Maximum	5.0	5.0	3.0	3.0	3.0
Vertical curve (ft)					
Minimum length†	100	100	100	100	100
Minimum distance between points of intersection§	100	100	100	100	100

*A minimum of 3 percent for turf.
†A slope of 5 percent is recommended for a 10-ft width adjacent to the pavement areas to promote drainage.
‡For each 1 percent of grade change.
§Distance is multiplied by the sum of the absolute grade changes in percent.
Source: Federal Aviation Administration [6].

TABLE 6-19 Taxiway Gradient Standards

these facilities. The required separation between taxiways, between a taxiway and a taxilane, or between a taxiway and a fixed or movable object requires a minimum wingtip clearance of 0.2 times the wingspan of the most demanding aircraft in the airplane design group plus 10 ft. This clearance provides a minimum taxiway centerline to a parallel taxiway centerline or taxilane centerline separation of 1.2 times the wingspan of the most demanding aircraft plus 10 ft, and between a taxiway centerline and a fixed or movable object of 0.7 times the wingspan of the most demanding aircraft plus 10 ft. This

	Aerodrome Code Letter				
	A	B	C	D	E
Width					
Pavement	7.5	10.5	15*	18†	23
Pavement and shoulder			25	38	44
Edge safety margin, U_1	1.5	2.25	3†	4.5	4.5
Strip	27	39	57	85	93
Graded portion of strip	22	25	25	38	44
Minimum separation					
Taxiway centerline to taxiway centerline	21	31.5	46.5	68.5	81.5
Object	13.5	19.5	28.5	42.5	49
Aircraft stand taxilane to object	12	16.5	24.5	36	42.5

---

*18 m if used by aircraft with a wheelbase equal to or greater than 18 m.
†23 m is used by aircraft with an outer main gear wheel span equal to or greater than 9 m.
‡4.5 m. if intended to be used by airplane with a wheelbase equal to or greater than 18 m.
Source: International Civil Aviation Organization [2, 3, 4].

TABLE 6-20   Taxiway Dimensional Standards, m

separation is also applicable to aircraft traversing through a taxiway on an apron or ramp. This separation may have to be increased to accommodate pavement widening on taxiway curves. It is recommended that a separation of at least 2.6 times the wheelbase of the most demanding aircraft be provided to accommodate a 180° turn when the pavement width is designed for tracking the nose wheel on the centerline.

The taxilane centerline to a parallel taxilane centerline or fixed or movable object separation in the terminal area is predicated on a wingtip clearance of approximately half of that required for an apron taxiway. This reduction in clearance is based on the consideration that taxiing speed is low in this area, taxiing is precise, and special guidance techniques and devices are provided. This requires a wingtip clearance or wingtip-to-object clearance of 0.1 times the wingspan of the most demanding aircraft plus 10 ft. Therefore, this establishes a minimum separation between the taxilane centerlines of 1.1 times the wingspan of the most demanding aircraft plus 10 ft, and between a taxilane centerline and a fixed or movable object of 0.6 times the wingspan of the most demanding aircraft plus 10 ft [6]. Therefore, when dual parallel taxilanes are provided in the terminal apron area, the taxilane object-free area becomes 2.3 the wingspan of the most demanding aircraft plus 30 ft.

| | Aerodrome Code Letter | | | | |
	A	B	C	D	E
Gradient (%)					
Pavement longitudinal					
Maximum	3.0	3.0	1.5	1.5	1.5
Maximum change	4.0	4.0	3.33	3.33	3.33
Pavement transverse					
Maximum	2.0	2.0	1.5	1.5	1.5
Strip					
Maximum transverse					
Graded portion					
Upward	3.0	3.0	2.5	2.5	2.5
Downward	5.0	5.0	5.0	5.0	5.0
Ungraded portion					
Upward	5.0	5.0	5.0	5.0	5.0
Vertical curve (m)					
Minimum length*	25	25	30	30	30

*For each 1 percent of grade change.
Source: International Civil Aviation Organization [2, 3, 4].

TABLE 6-21   Taxiway Gradient Standards

The separation criteria are presented in Table 6-18. The values indicated in this table are based upon the specifications using the largest wingspan in each airplane design group. As noted in this table, the required separations may be reduced to those that would result using the actual wingspan of the design aircraft.

ICAO Separation Criteria

The separation criteria adopted by ICAO are also predicated upon the wingtips of the aircraft for which the taxiway and taxilane system have been designed and providing a minimum wingtip clearance on these facilities, but also consider a minimum clearance between the outer main gear wheel and the taxiway edge. The required separation between taxiways or between a taxiway and a taxilane requires a minimum wingtip clearance, C_1, of 3 m for aerodrome code letter A and B runways, 4.5 m for aerodrome code letter C runways, and

7.5 m for aerodrome code letter D and E runways. The minimum clearance between the edge of each taxiway and the outer main gear wheels, the taxiway edge safety margin U_1, is given in Table 6-20. This clearance provides a minimum taxiway centerline to a parallel taxiway centerline or taxilane centerline separation given by Eq. (6-1).

$$S_{TT} = WS + 2U_1 + C_1 \qquad (6-1)$$

where S_{TT} = minimum taxiway-to-taxiway or taxiway-to-taxilane separation
$\quad$ WS = wingspan of the most demanding aircraft
$\quad$ U_1 = taxiway edge safety margin
$\quad$ C_1 = minimum wingtip clearance

Therefore, for example, an ICAO aerodrome code letter E runway, which accommodates aircraft with wingspans up to 65 m, requires a taxiway centerline to a taxiway centerline or a taxilane centerline separation from Eq. (6-1) of $65 + 2(4.5) + 7.5 = 81.5$ m.

The required separation between a taxiway centerline or an apron taxiway centerline and a fixed or movable object is found from Eq. (6-2).

$$S_{TO} = 0.5\,WS + U_1 + C_2 \qquad (6-2)$$

where S_{TO} is the minimum taxiway or apron taxiway to a fixed or movable object separation and C_2 is the required clearance between a wingtip and an object.

The required clearance between a wingtip and an object C_2 is 4.5 m for aerodrome code letter A runways, 5.25 m for aerodrome code letter B runways, 7.5 m for aerodrome code letter C runways, and 12 m for aerodrome code letter D and E runways.

The required separation between an aircraft stand taxilane centerline and a fixed or movable object is found from Eq. (6-3).

$$S_{ATO} = 0.5\,WS + U_2 + C_1 \qquad (6-3)$$

where S_{ATO} is the minimum aircraft stand taxilane to fixed or movable object separation and U_2 is the aircraft stand safety margin.

Since aircraft moving on the aircraft stand taxilane are moving at low speed and are often under positive ground guidance, the aircraft stand safety margin is less than on the taxiway system. The value for this safety margin U_2 is 1.5 m for aerodrome code letter A and B airports, 2 m for aerodrome code letter C airports, and 2.5 m for aerodrome code letter D or E airports. The taxiway and taxilane separation criteria adopted by ICAO are given in Table 6-20.

Sight Distance and Longitudinal Profile

As in the case of runways, the number of changes in longitudinal profile for taxiways is limited by sight distance and minimum distance between vertical curves.

The FAA does not specify line of sight requirements for taxiways other than those discussed earlier related to runway and taxiway intersections. However, the sight distance along a runway from an intersecting taxiway needs to be sufficient to allow a taxiing aircraft to enter or cross the runway safely. The FAA specifies that from any point on the taxiway centerline the difference in elevation between that point and the corresponding point on a parallel runway, taxiway, or apron edge is 1.5 percent of the shortest distance between the points.

ICAO requires that the surface of the taxiway should be seen for a distance of 150 m from a point 1.5 m above the taxiway for aerodrome code letter A runways, for a distance of 200 m from a point 2 m above the taxiway for aerodrome code letter B runways, and for a distance of 300 m from a point 3 m above the taxiway for aerodrome code letter C, D, or E runways.

In regard to longitudinal profile of taxiways, the ICAO does not specify the minimum distance between the points of intersection of vertical curves. The FAA specifies that the minimum distance for both utility and transport category airports should be not less than the product of 100 ft multiplied by the sum of the absolute percentage values of change in slope.

Exit Taxiway Geometry

The function of exit taxiways, or runway turnoffs as they are sometimes called, is to minimize runway occupancy by landing aircraft. Exit taxiways can be placed at right angles to the runway or some other angle to the runway. When the angle is on the order of 30°, the term high-speed exit is often used to denote that it is designed for higher speeds than other exit taxiway configurations. In this chapter, specific dimensions for high-speed exit, right-angle exit (low-speed) taxiways are presented. The dimensions presented here are the results obtained from research conducted many years ago [13] and subsequent research conducted by the FAA.

The earlier tests [13] were conducted on wet and dry concrete and asphalt pavement with various types of civil and military aircraft in order to determine the proper relationship between exit speed and radii of curvature and the general configuration of the taxiway. A significant finding of the tests was that at high speeds a compound curve was necessary to minimize tire wear on the nose gear and, therefore, the central or main curve radius R_2 should be preceded by a much larger radius curve R_1.

Aircraft paths in the test approximated a spiral. A compound curve is relatively easy to establish in the field and begins to approach the shape of a spiral, thus the reason for suggesting a compound curve. The following pertinent conclusions were reached as a result of the tests [13]:

1. Transport category and military aircraft can safely and comfortably turn off runways at speeds on the order of 60 to 65 mi/h on wet and dry pavements.

2. The most significant factor affecting the turning radius is speed, not the total angle of turn or passenger comfort.

3. Passenger comfort was not critical in any of the turning movements.

4. The computed lateral forces developed in the tests were substantially below the maximum lateral forces for which the landing gear was designed.

5. Insofar as the shape of the taxiway is concerned, a slightly widened entrance gradually tapering to the normal width of taxiway is preferred. The widened entrance gives the pilot more latitude in using the exit taxiway.

6. Total angles of turn of 30° to 45° can be negotiated satisfactorily. The smaller angle seems to be preferable because the length of the curved path is reduced, sight distance is improved, and less concentration is required on the part of the pilots.

7. The relation of turning radius versus speed expressed by the formula below will yield a smooth, comfortable turn on a wet or dry pavement when f is made equal to 0.13.

$$R_2 = \frac{V^2}{15f} \tag{6-4}$$

   where V is the velocity in mi/h and f is the coefficient of friction.

8. The curve expressed by the equation for R_2 should be preceded by a larger radius curve R_1 at exit speeds of 50 to 60 mi/h. The larger radius curve is necessary to provide a gradual transition from a straight tangent direction section to a curved path section. If the transition curve is not provided tire wear on large jet transports can be excessive.

9. The length of the transition curve can be roughly approximated by the relation

$$L_1 = \frac{V^3}{CR_2} \tag{6-5}$$

   where V is in feet per second, R_2 is in feet, and C was found experimentally to be on the order of 1.3.

10. Sufficient distance must be provided to comfortably decelerate an aircraft after it leaves the runway. It is suggested that for the present this distance be based on an average rate of deceleration of 3.3 ft/s². This applies only to transport category aircraft. Until more experience is gained with this type of operation the stopping distance should be measured from the edge of the runway.

A chart showing the relationship of exit speed to radii R_1 and R_2, and length of transition curve L_1 is given in Fig. 6-33.

ICAO has indicated the relationship between aircraft speed and the radius of curvature of taxiway curves as illustrated in Table 6-22. For high-speed exit taxiways ICAO recommends a minimum radius

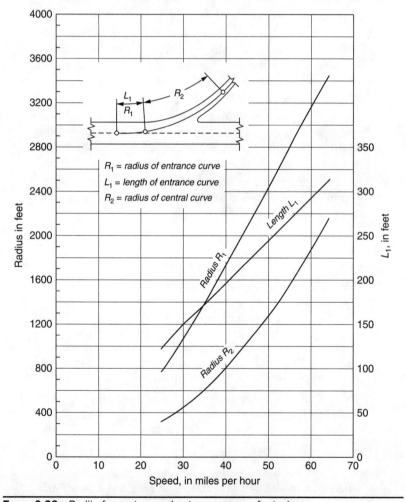

Figure 6-33   Radii of curvature and entrance curves for taxiways.

Taxiing Speed		Radius of Exit curve	
mph	kph	Feet	Meters
10	16	50	15
20	32	200	60
30	48	450	135
40	64	800	240
50	80	1,250	375
60	96	1,800	540

Source: International Civil Aviation Organization [4].

TABLE 6-22 Radii of Curvature for Transport Category Aircraft

of curvature for the taxiway centerline of 275 m (900 ft) for aerodrome code number 1 and 2 runways and 550 m (1800 ft) for aerodrome code number 3 and 4 runways. This will allow exit speeds under wet conditions of 65 km/h (40 mi/h) for aerodrome code number 1 and 2 runways and 93 km/h (60 mi/h) for aerodrome code number 3 and 4 runways. It also recommends a straight tangent section after the turn-off curve to allow exiting aircraft to come to a full stop clear of the intersecting taxiway when the intersection is 30°. This tangent distance should be 35 m (115 ft) for aerodrome code number 1 and 2 runways and 75 m (250 ft) for aerodrome code number 3 and 4 runways [2, 4].

A configuration for an exit speed of 60 mi/h and a turnoff angle of 30° is shown in Fig. 6-34. The FAA recommends that the taxiway centerline circular curve be preceded by a 1400-ft spiral to smooth the transition from the runway centerline to the taxiway exit circular curve. ICAO recommends the same geometry for both of these high-speed exits. Right-angle or 90° exit taxiways, although not desirable from the standpoint of minimizing runway occupancy, are often

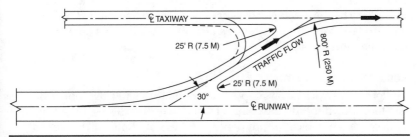

FIGURE 6-34 High-speed exit taxiway.

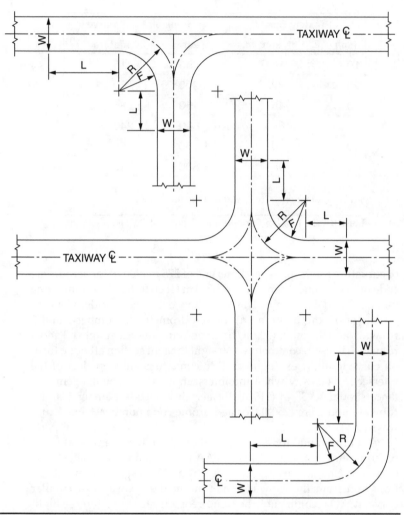

FIGURE 6-35 Common taxiway exit and intersection configurations.

constructed for other reasons. The configurations for a 90° exit and other common taxiway intersection configurations are illustrated in Fig. 6-35. The dimensions labeled in Fig. 6-35 are determined by the aircraft design group of the design aircraft. These dimensional standards are provided in Table 6-23.

Location of Exit Taxiways

The location of exit taxiways depends on the mix of aircraft, the approach and touchdown speeds, the point of touchdown, the exit speed, the rate of deceleration, which in turn depends on the condition of the pavement surface, that is, dry or wet, and the number of exits.

Item	Dim.*	Airplane Design Group					
		I	II	III†	IV	V	VI
Radius of taxiway turn†	R	75	75	100	150	150	170
Length of lead-in to fillet	L	50	50	150	250	250	250
Fillet radius for tracking centerline	F	60	55	55	85	85	85
Fillet radius for judgmental oversteering symmetrical widening§	F	62.5	57.5	68	105	105	110
Fillet radius for judgmental	F	62.5	57.5	60	97	97	100
Oversteering one side widening¶							

*Letters correspond to the dimensions on Fig. 6-35.
†Airplanes in airplane design group III with a wheelbase equal to or greater than 60 ft should use a fillet radius of 50 ft.
‡Dimensions for taxiway fillet designs relate to the radius of taxiway turn specified.
§The center sketch of Fig. 6-35 displays pavement fillets with symmetrical taxiway widening.
¶The lower sketch of Fig. 6-35 displays a pavement fillet with taxiway widening on one side.

TABLE 6-23   FAA Taxiway Curvature Dimensional Standards, ft

While the rules for flying transport aircraft are relatively precise, a certain amount of variability among pilots is bound to occur especially in respect to braking force applied on the runway and the distance from runway threshold to touchdown. The rapidity and the manner in which air traffic control can process arrivals is an extremely important factor in establishing the location of exit taxiways. The location of exit taxiways is also influenced by the location of the runways relative to the terminal area.

Several mathematical analyses or models have been developed for optimizing exit locations. While these analyses have been useful in providing an understanding of the significant parameters affecting location, their usefulness to planners has been limited because of the complexity of the analyses and a lack of knowledge of the inputs required for the application of the models. As a result greater use is made of much more simplified methods.

The landing process can be described as follows. The aircraft crosses the runway threshold and decelerates in the air until the main landing gear touches the surface of the pavement. At this point the nose gear has not made contact with the runway. It may take as long as 3 s to do so. When it does, reverse thrust or wheel brakes or a combination of both are used to reduce the forward speed of the aircraft to exit velocity. Empirical analysis has revealed that the average deceleration of air-carrier aircraft on the runway is about 5 ft/s^2.

In the simplified procedure, an aircraft is assumed to touch down at 1.3 times the stall speed for a landing weight corresponding to 85 percent of the maximum structural landing weight. In lieu of computing the distance from threshold to touchdown, touchdown distances are assumed as fixed values for certain classes of aircraft. Typically these values range from 500 to 1500 ft from the runway threshold. To these distances are added the distances to decelerate to exit speed. These relationships may be approximated by Eqs. (6-6) and (6-7).

$$D = D_{td} + D_e \qquad (6\text{-}6)$$

where D = distance from the runway threshold to the exits
  D_{td} = distance from the runway threshold to the point where the aircraft touches down
  D_e = the distance from the touchdown point to the exit

$$D_e = \frac{V_{td}^2 - V_e^2}{2a} \qquad (6\text{-}7)$$

where V_{td} = aircraft speed at touchdown
  V_e = exit speed of the aircraft
  a = deceleration of the aircraft on the runway

Although approach and touchdown speeds vary, they can be approximated for locating exit taxiways. At predominantly air carrier airports air traffic control authorities request general aviation aircraft to increase their speeds above normal to reduce the wide range in speed between air carriers and general aviation. At these airports, the normal approach speeds for general aviation are probably not applicable.

If it is assumed that the distances to touchdown are 1500 ft for air carrier aircraft and 1000 ft for twin-engine general aviation aircraft, a

Type of Aircraft	Touchdown Speed, kn	Exit Speed, mi/h	
		60	15
Small propeller			
GA single engine	60	2,400	1,800
GA twin engine	95	2,800	3,500
Large jet	130	4,800	5,600
Heavy jet	140	6,400	7,100

TABLE 6-24   Approximate Taxiway Exit Location from Threshold, ft

high-speed exit accommodates aircraft exiting the runway at an exit speed of 60 mi/h and a regular exit accommodates aircraft exiting the runway at 15 mi/h, then using approximate touchdown speeds, the approximate exit locations for various types of aircraft may be found as shown in Table 6-24.

These locations are derived using standard sea-level conditions. Altitude and temperature can affect the location of exit taxiways. Altitude increases distance on the order of 3 percent for each 1000 ft above sea level and temperature increases the distance 1.5 percent for each 10°F above 59°F.

During runway capacity studies conducted for the FAA, data were collected on exit utilization at various large airports in the United States [18]. These data, which are tabulated in Table 6-25, indicate the cumulative percentage of each class of aircraft which have exited the runway at exits located at various distances from the arrival threshold. On the basis of these studies, runway exit ranges from the arrival threshold are used in runway capacity studies [5]. These exit ranges are given in Table 6-26. Comparisons between the approximate relationships given in Table 6-24 and the data given in Tables 6-25 and 6-26 indicate that fairly good correspondence results. Variations which occur are due to pilot technique and preference in choosing exits, the wide range of performance characteristics demonstrated by various aircraft in the aircraft approach categories, altitude and temperature considerations, and the amount of runway available for landing. The latter factor is very important because if pilots recognize that the amount of runway available is near the minimum for a particular aircraft they are more likely to touch down closer to the runway threshold and apply larger than normal deceleration and braking to the aircraft.

It is recommended that the point of intersection of the centerlines of taxiway exits and runways, which are up to 7000 ft in length and accommodate aircraft approach category C, D, and E aircraft, should

Dry Runways								
Distance from Threshold to Exit, ft	Regular Exits Aircraft Class*				High Speed Exits Aircraft Class*			
	A	B	C	D	A	B	C	D
0	0	0	0	0	0	0	0	0
1000	6	0	0	0	13	0	0	0
2000	84	1	0	0	90	1	0	0
3000	100	39	0	0	100	40	0	0
4000		98	8	0		98	26	3
5000		100	49	9		100	76	55
6000			92	71			98	95
7000			100	98			100	100
8000				100				
Wet Runways								
Distance from Threshold to Exit, ft	Aircraft Class*							
	A	B	C	D				
0	0	0	0	0				
1000	4	0	0	0				
2000	60	0	0	0				
3000	96	10	0	0				
4000	100	80	1	0				
5000		100	12	0				
6000			48	10				
7000			88	64				
8000			100	93				
9000				100				

*The aircraft class is the classification of aircraft based upon maximum certified takeoff weight [5].
Source: Federal Aviation Administration [18].

TABLE 6-25 Percentage of Aircraft Exiting at Exits Located at Various Dry Runways

be located about 3000 ft from the arrival threshold and 2000 ft from the stop end of the runway. To accommodate the average mix of aircraft on runways longer than 7000 ft, intermediate exits should be located at intervals of about 1500 ft. At airports where there are extensive

Mix Index*	Exit Range from Arrival Threshold
0–20	2000–4000
21–50	3000–5500
51–80	3500–6500
81–120	5000–7000
121–180	5500–7500

*Mix Index is equal to the percentage of Class C aircraft plus three an aircraft with a maximum certified takeoff weight in excess of class D aircraft, where a class C aircraft is an aircraft with a maximum certified takeoff weight greater than 12,500 lb and up to 300,000 lb and a class D aircraft is an aircraft with a maximum certified takeoff weight in excess of 300,000 lb.
Source: Federal Aviation Administration [5].

TABLE 6-26 Exit Range Appropriate to Runways Serving Aircraft of Different Arrival Mix Indices, ft

operations with aircraft approach category A and B aircraft, an exit located between 1500 and 2000 ft from the landing threshold is recommended.

Planners often find that the runway configuration and the location of the terminal at the airport often preclude placing the exits at locations based on the foregoing analysis. This is nothing to be alarmed about since it is far better to achieve good utilization of the exits than to be too concerned about a few seconds lost in occupancy time.

When locating exits it is important to recognize local conditions such as frequency of wet pavement or gusty winds. It is far better to place the exits several hundred feet farther from the threshold than to have aircraft overshoot the exits a large amount of time. The standard deviation in time required to reach exit speed is on the order of 2 or 3 s. Therefore, if the exits were placed down the runway as much as two standard deviations from the mean, the loss in occupancy time would only be 4 to 6 s. In planning exit locations at specific airports, one needs to consult with the airlines relative to the specific performance characteristics of the aircraft intended for use at the airport.

The total occupancy time of an aircraft can be roughly estimated using the following procedure. The runway is divided into four components, namely, flight from threshold to touchdown of main gear, time required for nose gear to make contact with the pavement after the main gear has made contact, time required to reach exit velocity from the time the nose gear has made contact with the pavement and brakes have been applied, and time required for the aircraft to turn off on to the taxiway and clear the runway. For the first component it can be assumed that the touchdown speed is 5 to 8 kn less than the speed over the threshold. The rate of deceleration in the air is about 2.5 ft/s². The second component is about 3 s and the third component depends upon exit speed. Time to turnoff from the runway will be on the order of 10 s. Thus the total occupancy time in seconds can be approximated by Eq. (6-8).

$$R_i = \frac{V_{ot} - V_{td}}{2a_1} + 3 + \frac{V_{td} - V_e}{2a_2} + t \tag{6-8}$$

where R_i = runway occupancy time, s
   V_{ot} = over the threshold speed, ft/s
   V_{td} = touchdown speed, ft/s
   V_e = exit speed, ft/s
   t = time to turnoff from the runway after exit speed is reached, s
   a_1 = average rate of deceleration in the air, ft/s²
   a_2 = average rate of deceleration on the ground, ft/s²

During the runway capacity studies cited earlier [18] data were also collected on runway occupancy time. These data, which are tabulated in Table 6-27, indicate the total runway occupancy time of each class of aircraft which have exited the runway at exits located at various distances from the arrival threshold. As may be observed in this table, typical runway occupancy times for 60 mi/h high-speed exits are 35 to 45 s. The corresponding time for a 15 mi/h regular exit is 45 to 60 s for air carrier aircraft.

Design of Taxiway Curves and Intersections

The basic design of taxiway curves and intersections for three of the most common types of taxiway intersections have been developed by the FAA [6]. These designs have been taken from this reference and were shown in Fig. 6-35. The dimensions recommended by the FAA for the taxiway width, centerline radius, fillet radius (inner edge radius), and the length of the fillet lead-in are given in Table 6-23. The dimensions given for the fillet radius in this table are related to the taxiway centerline radius.

When an aircraft negotiates a turn with the nose wheel tracking a predetermined curved path, such as a taxiway centerline, the midpoint of the main undercarriage does not follow the same path of the nose gear because of the fairly large distance from the nose gear to the main undercarriage. The relationship between the centerline, which is being tracked by the nose wheel, and position of the main undercarriage are shown in Fig. 6-36. At any point on the curve the distance between the curved path followed by the nose wheel and the midpoint of the undercarriage of main landing gear is referred to as the

Dry Runways								
Distance from Threshold to Exit, ft	Regular Exits				High Speed Exits			
	Aircraft Class*				Aircraft Class*			
	A	B	C	D	A	B	C	D
0	24				19			
1000	24	27			27	24		
2000	34	27			35	24		
3000	44	37	29		43	32	35	35
4000	55	46	38	38		41	35	35
5000	65	56	47	47		49	44	44
6000	76	65	56	56			54	54
7000	76	75	65	65			63	63
8000	76	75	73	73				
9000	76	75	82	82				
10000	76	75	85	85				
11000	76	75	90	90				
Wet Runways								
Distance from Threshold to Exit, ft	Aircraft Class*							
	A	B	C	D				
0	24							
1000	24							
2000	34	27						
3000	44	37	30					
4000	55	47	38					
5000	65	56	47	47				
6000	76	65	56	56				
7000	99	99	65	65				
8000			73	73				
9000			82	82				

*The aircraft class is the classification of aircraft based upon maximum certified takeoff weight [5]

Source: Federal Aviation Administration [8].

TABLE 6-27 Runway Occupancy Time of Aircraft Exiting at Exits Located at Various Distances from the Runway Arrival Threshold, s

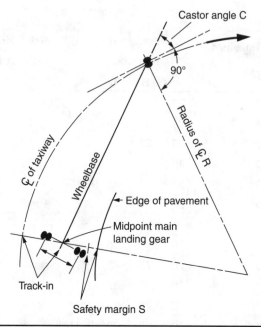

Figure 6-36   Path of main gear on curve.

track-in. The track-in varies, increasing progressively during the turning maneuver. It decreases as the nose gear begins to follow the tangent to the curve. Knowing the path of the main gear, the radius of the fillet can be determined by adding an appropriate taxiway edge safety margin S between the outside edge of the tire on the main landing gear closest to the center of the path followed by the nose wheel and the edge of the pavement.

The nose wheel steering angle, the castor angle C is defined as the angle formed by the longitudinal axis of the aircraft and the direction of movement of the nose wheel, or some other reference point or datum point such as the location of the pilot in the cockpit. For preliminary design it is sufficiently accurate to assume that the datum point is the nose wheel.

The size of the fillet depends not only on the wheelbase of the aircraft, the radius of the curve, the width of the taxiway, and the total change in direction, but also on the path that the aircraft follows on the turn. There are three ways in which an aircraft can be maneuvered on a turn. One is to establish the centerline of the taxiway as the path of the nose gear. This is called nose wheel on centerline tracking. Another is to establish the centerline of the taxiway as the path directly beneath the pilot and assume that this path is followed. This is called maintaining cockpit over the centerline tracking. The last is to assume that the nose gear will follow a path offset outward of the centerline. This is called judgmental oversteering tracking. The latter type of tracking will result in the least amount of taxiway widening but results in greater runway occupancy time and the possibility of pilot error in judging the turn to follow. While there is no agreement on which procedure is desirable, usually maintaining the cockpit over the centerline tracking is preferred.

The principal dimensions of the aircraft related to tracking a curve were given in Fig. 6-36. The geometry of the aircraft tracking the centerline curve from the point of curvature, PC, to the point of tangency, PT, and the various terms used in the equations below to define the movement of the aircraft are given in Fig. 6-37.

The maximum angle formed between the tangent to the centerline and the longitudinal axis of the aircraft will occur at the end of the curve when the nose wheel is at the point of tangency. This angle, called A_{max}, may be approximated by

$$A_{max} = \sin^{-1}(d/R) \qquad (6-9)$$

where d is the distance from the nose wheel or the pilot cockpit position to the center of the main undercarriage; the wheelbase of the aircraft is often used to approximate this distance and R is the radius the nose wheel or the pilot is tracking on the curve.

The maximum nose wheel steering angle, the castor angle, the angle between the longitudinal axis on the nose gear and the longitudinal axis of the aircraft, B_{max}, is given by

$$B_{max} = \tan^{-1}(w/d \tan A_{max}) \qquad (6-10)$$

where w is the wheelbase of the aircraft.

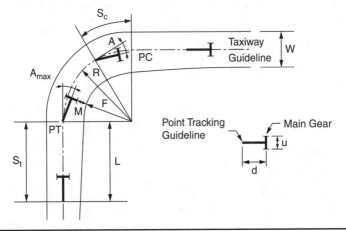

FIGURE 6-37   Taxiway fillet design geometry.

The required fillet radius F is given by

$$F = (R^2 + d^2 - 2Rd \sin A_{max})^{0.5} - 0.5u - M \qquad (6\text{-}11)$$

where u is undercarriage width, that is, the distance between the outside tires on the main gear and M is the minimum distance required between the edge of the outside tire and the edge of the pavement, that is, the edge safety margin.

The length of the lead-in to the fillet is given by

$$L = d \ln \frac{(4d \tan 0.5A_{max})}{W - u - 2M} - d \qquad (6\text{-}12)$$

where W is the taxiway width on the tangent and ln represents the natural logarithm.

These equations may be solved for a given aircraft tracking a curve of radius R to find the necessary lead-in and fillet radius to maintain a minimum edge safety margin between the tire and pavement edge. If the value of the maximum nose wheel steering angle, B_{max}, exceeds 50° it is recommended that the radius of the centerline curve R which the nose wheel is tracking be increased.

The use of these equations in determining the critical taxiway curve design parameters is shown in Example Problem 6-3.

Example Problem 6-3

Determine the minimum lead-in and the radius of the fillet to maintain the cockpit over the centerline for an aircraft with a 156.1-ft wingspan, a wheelbase of 64.6 ft, undercarriage width of 34.25 ft, and the distance between the main undercarriage and cockpit equal to 72.1 ft. The aircraft is moving between two parallel taxiways through a connecting taxiway which has a centerline perpendicular to the parallel taxiways.

With a wingspan of 156.1 ft this aircraft is in airplane design group IV. From Table 6-23, the taxiway width is 75 ft, the minimum safety margin is 15 ft, the recommended centerline radius is 150 ft, the recommended length of the fillet lead-in is 250 ft and the recommended fillet radius is 85 ft.

To verify that these recommended values are acceptable, we can use Eqs. (6-10) through (6-13). From Eq. (6-10), we have

$$A_{max} = \sin^{-1}(d/R)$$

or

$$A_{max} = \sin^{-1}(72.1/150) = 29°$$

From Eq. (6-11),

$$B_{max} = \tan^{-1}(w/d \tan A_{max})$$

or

$$B_{max} = \tan^{-1}[(64.5/72.1) \tan 29°] = 27°$$

Since this is less than 50° the radius is adequate.
From Eq. (6-12), we have

$$F = (R^2 + d^2 - 2Rd \sin A_{max})^{0.5} - 0.5u - M$$

or

$$F = [150^2 + 72.1^2 - (2)(150)(72.1) \sin 29°]^{0.5} - 0.5(34.25) - 15$$

$$F = 100.1 \text{ ft}$$

From Eq. (6-12),

$$L = \frac{d \ln(4d \tan 0.5A_{max})}{W - u - 2M} - d$$

$$L = 72.1 \ln\left[\frac{(4)(72.1)(\tan 14.5°)}{75 - 34.25 - (2)(15)}\right] - 72.1$$

or

$$L = 68 \text{ ft}$$

Therefore, both the required fillet radius and the required length of the lead-in to the fillet for this specific aircraft are both well within those recommended for the airplane design group to which this aircraft is assigned.

End-Around Taxiways

In an effort to reduce the number of times aircraft must cross a runway when traveling around an airfield, the FAA has allowed for the design of taxiways that traverse beyond runway thresholds. These taxiways, known as end-around taxiways, are designed to both reduce the risk of runway incursions and increase the overall efficiency operations on the airfield. For safety considerations, primarily concerned with the transient presence of aircraft immediately off the end of the runway, the FAA has established specific design standards for end-around taxiways. An example of end-around taxiway configuration is shown in Fig. 6-38.

End-around taxiways must remain outside of the runway safety area, and outside of any ILS critical areas. In addition, the tail height of the critical design aircraft at the airport must not exceed any critical Part 77 or TERPS surfaces, when on the end-around taxiway. Furthermore, the location of the end-around taxiway should provide for any aircraft departing on the runway to clear any object on the taxiway by at least 35 ft vertically and 200 ft horizontally from the runway centerline.

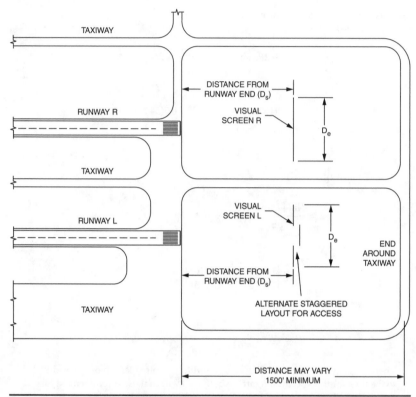

FIGURE 6-38 End-around taxiway.

A diagonal stripe screen is required to be placed between the end of the runway and the end-around taxiway, to provide pilots on the runway visual clarity that any aircraft on the end-around taxiway is not on the runway. An illustration of this screen is found in Fig. 6-39.

Aprons

Holding Aprons

Holding aprons, holding pads, run-up pads, or holding bays as they are sometimes called, are placed adjacent to the ends of runways. The areas are used as storage areas for aircraft prior to takeoff. They are

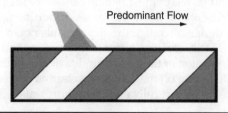

FIGURE 6-39 End-around taxiway screen.

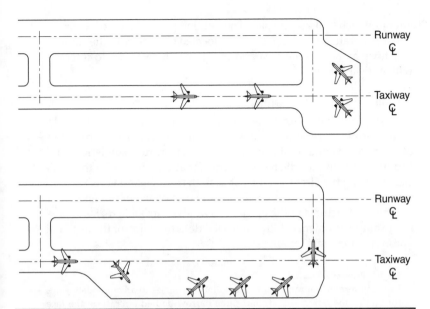

FIGURE 6-40   Typical holding pad configurations.

designed so that one aircraft can bypass another whenever this is necessary. For piston-engine aircraft the holding apron is an area where the aircraft instrument and engine operation can be checked prior to takeoff. The holding apron also provides for a trailing aircraft to bypass a leading aircraft in case the takeoff clearance of the latter must be delayed for one reason or another, or if it experiences some malfunction. There are many configurations of holding aprons, two of which are shown in Fig. 6-40. The important design criteria are to provide adequate space for aircraft to maneuver easily onto the runway irrespective of the position of adjacent aircraft on the holding apron and to provide sufficient room for an aircraft to bypass parked aircraft on the holding apron. The recommendations for the minimum separation between aircraft on holding aprons are the same as those specified for the taxiway object-free area.

The design of a typical flow-through holding pad studied is shown in Fig. 6-41. Holding pads must be designed for the largest

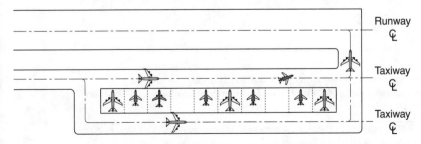

FIGURE 6-41   Flow-through bypass holding pad.

aircraft which will use the pad. The holding pad should be located so that all aircraft using the pad will be located outside both the runway and taxiway object-free area and in a position so as not to interfere with critical ILS signals.

Terminal Aprons and Ramps

Aircraft parking positions, also called aircraft gates or aircraft stands, on the terminal apron or ramp are sized for the geometric properties of a given design aircraft, including wingspan, fuselage length and turning radii, and for the requirements for aircraft access by the vehicles servicing the aircraft at the gates. Both the FAA and ICAO recommend minimum clearances between any part of an aircraft and other aircraft or structures in the apron area as given in Table 6-28.

Example Problem 6-4 illustrates the determination of the terminal apron requirements for aircraft.

Example Problem 6-4

Design a terminal apron with two parallel concourses to accommodate gates for one wide-bodied aircraft and three narrow-bodied aircraft on the face of each of the concourses. The gate design aircraft for the wide-bodied gates is the Boeing 767-200 and the gate design aircraft for the narrow-bodied gates is the McDonnell-Douglas MD-87. Aircraft will park nose-in at each gate and use the gates in a power-in and push-out mode of operation.

The Boeing 767-200 has a fuselage length of 159 ft 2 in and a wingspan of 156 ft 1 in, which places it in airplane design group IV, and the McDonnell-Douglas MD-87 has a fuselage length of 130 ft 5 in and a wingspan of 107 ft 10 in, which places it in airplane design group III.

If the aircraft are arrayed at the concourses as shown in Fig. 6-42, then the size of the terminal apron and the size of each gate position may be determined by referencing the specifications requiring specific separations between aircraft

Airplane Design Group or Aerodrome Code Letter		Minimum Clearance*	
		Feet	Meters
I	A	10	3.0
II	B	10	3.0
III	C	15	4.5
IV	D	25	7.5
V or VI	E	25	7.5

*The FAA recommends the wingtip separation at parking positions to
Source: Federal Aviation Administration [6, 19] and International Civil Aviation Organization [4]

TABLE 6-28   Minimum Clearance between Aircraft and Fixed or Movable Objects at Terminal Apron Parking Positions

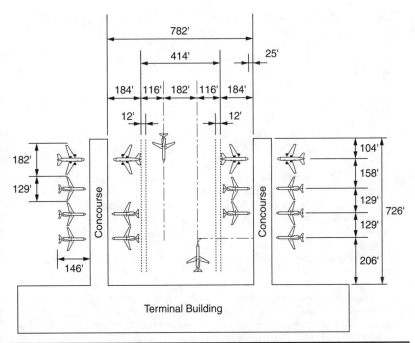

FIGURE 6-42 Terminal apron requirements for Example Problem 6-4.

operating on the taxilanes and between aircraft parked at the concourse gates. In this problem the FAA specifications will be used and the design will be based upon the actual dimensions of the aircraft rather than upon the dimensions of the largest aircraft in the airplane design groups to which these aircraft are assigned. Therefore, the relevant separations are contained in the footnotes in Table 6-18.

The Boeing 767-200 is the greater wingspan aircraft and, therefore, the most demanding aircraft for taxilane dimensions. The separation between taxilane centerlines is equal to 1.1 times the wingspan of the most demanding aircraft plus 10 ft. The recommended separation is then equal to $(1.1)(156.1) + 20 = 182$ ft. A ground access vehicle lane will be provided behind each aircraft for the use of aircraft service vehicles. This lane will be 12 ft wide. The distance from the centerline of each taxilane to a fixed or movable object is equal to 0.6 the wingspan plus 10 ft. Therefore, this distance is $(0.6)(156.1) + 10 = 104$ ft. Considering the ground access vehicle lane, the distance from the centerline of each taxilane to the tail of the aircraft is equal to $104 + 12 = 116$ ft.

Table 6-28 indicates that the recommended clearance between the face of each concourse and the nose of the 767-200 aircraft is 25 ft since this aircraft is in airplane design group IV. The length of the 767-200 is 159 ft 2 in and, therefore, the distance from the face of each concourse to the tail of the aircraft is equal to $25 + 156.2 = 182$ ft.

Considering each of these recommended separations, the width of the terminal apron or ramp between the concourses is found to be 782 ft as shown in Fig. 6-42. The clearance between aircraft wingtips or between the aircraft wing and fixed or movable objects is recommended to be 0.1 times the wingspan plus 10 ft. Therefore, the 767-200 requires a clearance between its wingtips and fixed or movable objects of 0.1 the wingspan plus 10 ft or $(0.1)(156.1) + 10 = 25.6$ ft. This results in a parking position for this aircraft being $156.1 + 25.6 = 182$ ft wide by 182 ft long. The MD-87 requires a clearance of $(0.1)(107.84) + 10 = 20.8$ ft. This results in a parking position for this aircraft being $107.84 + 20.8 = 129$ ft wide. From Table 6-28 this aircraft may be parked as close to 15 ft to the concourse since it is in airplane design group III. This results in the length of a parking position for the MD-87 being $130.42 + 15 = 146$ ft. However, since the parking position for the 767-200 is longer, the actual length provided for the MD-87 is 182 ft.

For aircraft to be moved from their parking positions onto the apron taxilane a separation from the inside edge of the last gate position to the terminal building equal to the fuselage length plus 0.1 times the wingspan plus 10 ft is required. This should allow the aircraft to turn onto the centerline of the taxilane from its gate position and also allow another aircraft to gain access to the gate position before the last aircraft at that gate leaves the apron area. The parking arrangement shown in Table 6-13 shows the MD-87 as the closest aircraft to the terminal building. Therefore, the distance from the centerline of the last gate position to the terminal building is $130.42 + 54 + (0.1)(107.83) + 10 = 206$ ft.

The design must ensure that the clearances provided are adequate based upon the turning capability of the design aircraft. The analyst should verify using the procedures discussed earlier in this chapter that the tracking-in of the aircraft while making a turn from the taxilanes to the parking position will not compromise safe clearances between aircraft. The analyst must also verify for each aircraft that the turning to access gate positions will ensure that no part of the aircraft will compromise these safe clearances.

Using the above dimensions results in a ramp length of 726 ft. The dimensions for all aircraft parking positions or gates are shown in Fig. 6-42.

The terminal apron area for these eight aircraft becomes equal to $(778)(726) = 565,800$ ft^2, which is equal to about 13 acres or a total ramp area per aircraft of about 1.6 acres. A gate for a 767-200 is $(182)(182) = 33,100$ ft^2, which is about 0.75 acres. The required gate for an MD-87 is $(129)(146) = 18,800$ ft^2 which is slightly more than 0.4 acres. Useful rules-of-thumb which allow one to estimate terminal apron gate requirements are a total ramp area of from 1.5 to 2.0 acres per aircraft gate, and an area of from 0.75 to 1.0 acres per gate for a wide-bodied aircraft gate position and an area of about 0.5 acres per gate for a narrow-bodied aircraft gate position.

Terminal Apron Surface Gradients

For fueling, ease of towing and aircraft taxiing, apron slopes or grades should be kept to the minimum consistent with good drainage requirements. Slopes should not in any case exceed 2 percent for utility airports and 1 percent for transport airports. At gates where aircraft are being fueled every effort should be made to keep the apron slope within 0.5 percent.

Control Tower Visibility Requirements

At airports with a permanent air traffic control tower, the runways and taxiways must be located and oriented so that a clear line of sight is maintained to all traffic patterns, the final approaches to all runways, all runway structural pavements, all apron taxiways, and other operational surfaces controlled by the air traffic control tower. A clear line of sight to all taxilane centerlines is desirable. Operational surfaces not having a clear unobstructed line of sight from the tower are designated as uncontrolled or nonmovement areas. At airports without a permanent air traffic control tower, the runways and taxiways should be located and oriented so that a future tower may be sited in accordance with the continuous visibility requirements. This requirement may be satisfied where adequate control of aircraft exists by other means [6].

A typical air traffic control tower site requires between 1 and 4 acres of land. The site must be large enough to accommodate current and future building needs including employee parking. Tower sites must afford maximum visibility to traffic patterns and clear, unobstructed and direct lines of sight to the runway approaches, the landing area, and all runway and taxiway surfaces. Most towers penetrate the FAR Part 77 surfaces and, therefore, are obstructions to aviation and may be a hazard to air navigation unless an FAA study determines otherwise. The tower must not derogate the signal generated by any existing or planned electronic navigational aid or air traffic control facility.

References

1. *Advisory Circular Checklist*, Advisory Circular AC00-2.6, Federal Aviation Administration, Washington, D.C., October 15, 1992 annual.
2. *Aerodromes, Annex 14 to the Convention on International Civil Aviation*, Vol. 1: *Aerodrome Design and Operations*, International Civil Aviation Organization, Montreal, Canada, July 1990.
3. *Aerodrome Design Manual, Part 1: Runways*, 2d ed., Doc 9157-AN/901, International Civil Aviation Organization, Montreal, Canada, 1984.
4. *Aerodrome Design Manual, Part 2: Taxiways, Aprons and Holding Bays*, 2d ed., International Civil Aviation Organization, Montreal, Canada, 1983.
5. *Airport Capacity and Delay*, Advisory Circular AC 150/5060-5, Federal Aviation Administration, Washington, D.C., 1983.
6. *Airport Design*, Advisory Circular AC 150/5300-13, Change 14 Federal Aviation Administration, Washington, D.C., 2008.
7. *A Mathematical Model for Locating Exit Taxiways*, R. Horonjeff, et al., Institute of Transportation and Traffic Engineering, University of California, Berkeley, Cal., 1959.
8. "Calculation of Aircraft Wheel Paths and Taxiway Fillets," J. W. L. van Aswegen, Graduate Student Report, Institute of Transportation and Traffic Engineering, University of California, Berkeley, Cal., July 1973.
9. "Characteristics of High Speed Runway Exit for Airport Design," A. A. Trani, A. G. Hobeika, B. J. Kim, H. Tomita, and D. Middleton, *International Air Transportation*, Proceedings of the 22nd Conference on International Air Transportation, American Society of Civil Engineers, New York, N.Y., 1992.
10. *Criteria for Approving Category I and Category II Landing Minima for FAR Part 121 Operators*, Advisory Circular AC 120-29, Including Changes 1 through 3, Federal Aviation Administration, Washington, D.C., 1974
11. "Determination of the Path Followed by the Undercarriage of a Taxiing Aircraft," Paper prepared by Department of Civil Aviation, Melbourne, Australia.
12. "Determination of Wheel Trajectories," E. Hauer, *Transportation Engineering Journal*, American Society of Civil Engineers, Vol. 96, No. TE4, New York, N.Y., November 1970.

13. *Exit Taxiway Location and Design*, R. Horonjeff, et al., Report prepared for the Airways Modernization Board by the Institute of Transportation and Traffic Engineering, University of California, Berkeley, Cal., 1958.
14. "Movement of Aircraft and Vehicles on the Ground—Taxiway Fillets," Working Paper No. 102, 5th Air Navigation Conference of the International Civil Aviation Organization, Montreal, Canada, October 1967.
15. *Optimization of Runway Exit Locations*, E. S. Joline, R. Dixon Speas, and Associates, Manhasset, N.Y.
16. *Report of the Department of Transportation Air Traffic Control Advisory Committee*, Vols. 1 and 2, Washington, D.C., December 1969.
17. *Runway Length Requirements for Airport Design*, Advisory Circular AC 150/5325-4B, Federal Aviation Administration, Washington, D.C., 2005.
18. *Supporting Documentation for Technical Report on Airport Capacity and Delay Studies*, Report No. FAA-RD-76-162, Federal Aviation Administration, Washington, D.C., 1976.
19. *The Apron and Terminal Building Planning Report*, Report No. FAA-RD-75-191, Federal Aviation Administration, Washington, D.C., 1975.
20. "The Design and Use of Flow-Through Hold Pads," Douglas F. Goldberg, *International Air Transportation*, Proceedings of the 22nd Conference on International Air Transportation, American Society of Civil Engineers, New York, N.Y., 1992.
21. *United States Standard for Terminal Instrument Operations (TERPS)*, 3d ed., FAA Order 8260.3B, Including Changes 1 through 12, Federal Aviation Administration, Washington, D.C., July 1976.

CHAPTER 7

Structural Design of Airport Pavements

Introduction

This chapter briefly describes various methods for designing airfield pavements. The term *structural design of airport pavements* as used in this text refers to the determination of the thickness of the components that make up an airfield pavement structure, rather than the design of pavement materials itself.

Airfield pavement is intended to provide a smooth and safe all-weather riding surface that can support the weights of such heavy objects as aircraft on top of the natural ground base. Airfield pavements are typically designed in layers, with each layer designed to a sufficient thickness to be adequate to ensure that the applied loads will not lead to distress or failure to support its imposed loads. The Federal Aviation Administration provides guidance on the design of airfield pavements within its Advisory Circular AC 150/5320-6E, *Airfield Pavement Design and Evaluation*. Originally published in 1975, this advisory circular was completely revised in 2008 to consider new design methods that are based on recently developed computer software models and appropriate for the heaviest of commercial air carrier aircraft. This chapter provides both an account of the historical pavement design and evaluation methods and details the current method of pavement design and evaluation. As with any element of airport planning and design, appropriate Advisory Circulars and software user guides should be studied and referenced prior to performing any airport pavement analysis. These resources may be downloaded from the FAA at http://www.faa.gov.

Pavement or *pavement structure* is a structure consisting of one or more layers of processed materials. A pavement consisting of a mixture of bituminous material and aggregate placed on high-quality granular materials is referred to as *flexible pavement*. When the pavement consists of a slab of portland cement concrete (PCC), it is referred to as *rigid pavement*. Both structures of pavement are typically found at airports, although often there are preferences to a given type of pavement depending on such factors as the type and frequency of aircraft usage, climatic conditions, and costs of construction and maintenance.

Figure 7-1 illustrates a cross section of a typically layered airfield pavement. As illustrated in Fig. 7-1, airfield pavement, whether flexible or rigid, typically consists of series of layers consisting of a surface course, base course, and one or more subbase courses, resting on the ground, or prepared "subgrade" layer.

The surface course consists of a mixture of bituminous material (generally asphalt) and aggregate ranging in thickness from 2 to 12 in for flexible pavements, and a slab of PCC 8 to 24 in thick for rigid pavements. The principal function of the surface course is to provide for smooth and safe traffic operations, to withstand the effects of applied loads and environmental influences for some prescribed period of operation, and to distribute the applied load to the underlying layers.

The base course may consist of treated or untreated granular material. Like the surface course, it must be adequate to withstand the effects of load and environment and to distribute the applied loads to the underlying layers. Untreated bases consist of crushed or uncrushed aggregates. Treated bases consist of crushed or uncrushed aggregate that has been mixed with a stabilizing material such as cement or bitumen.

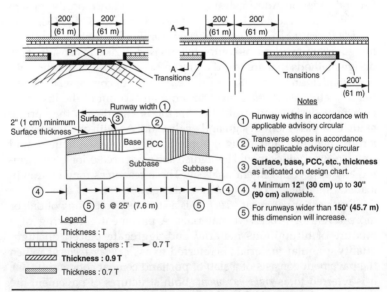

FIGURE 7-1   Typical plan and cross section for airfield pavement.

The subbase course is also composed of treated or untreated material, typically unprocessed pit-run material or material selected form a suitable excavation on the site. The function of the subbase is the same as that of the base. Whether or not a subbase is required, or how many subbase layers are required, is a function of the type of loads on the pavement, as well as the type and quality of soil, or subgrade, on which the pavement will be resting. For most rigid pavements, the surface course rests directly on the subbase.

The design of the thickness of each of the above layers is of primary concern to airport pavement engineers. The two primary factors that contribute to the design thickness of airfield pavement layers are the soil base and the volume and weight of the traffic using the pavement. As such, the first steps in pavement analysis are an investigation of the soil on which the pavement will be placed, and an estimation of the annual traffic volume on the pavement.

Soil Investigation and Evaluation

Accurate identification and evaluation of pavement foundations are essential to the proper design of the pavement structure. The subgrade supports the pavement and the loads placed upon the pavement surface. The function of the pavement is to distribute the loads to the subgrade, and the greater the capability of the subgrade to support the loads, the less the required thickness of the pavement.

Soil investigation consists of a soil survey to determine the arrangement of the different layers of soil in relation to the subgrade elevation, a sampling and testing of the various layers of soil to determine the physical properties of the soil, and a survey to determine the availability and suitability of local materials for use in the construction of the subgrade and pavement. Surveys and sampling are usually accomplished through soil borings to determine the soil of rock profile and its lateral extent. The sampled materials are then tested to determine soil types, gradation or particle sizes, liquid and plastic limits, moisture-density relationships, shrinkage factors, permeability, organic content, and strength properties. In the United States, soil surveys are often conducted using a variety of methods, including referring to U.S. Geological Survey (USGS) geodetic maps, aerial photography, and soil borings. The FAA recommends borings of given spacing and depths for soil surveys as illustrated in Table 7-1.

In the United States, evaluation of sampled soils for the purpose of airfield pavement design is performed according to the U.S. Army Corps of Engineers Unified Soil Classification (USC or "unified") System, as illustrated in Table 7-2. Under the unified system, soils are initially classified as either coarse-grained, fine-grained, or highly organic soils. Coarse-grained soils are those that do not filter through a No. 200 grade sieve. Coarse-grained soils are further divided into gravels and sands, as a function of the percentage of soil that filters

Area	Spacing	Depth
Runways and taxiways	Random across pavement at 200 ft (68 m) intervals	Cut areas—10 ft (3.5 m) below finished grade Fill areas—10 ft (3.5 m) below existing ground
Other areas of pavement	One boring per 10,000 ft² (930 m²) of area	Cut areas—10 ft (3.5 m) below finished grade Fill areas—10 ft (3.5 m) below existing ground
Borrow areas	Sufficient tests to clearly define the borrow material	To depth of borrow excavation

Note: For deep fills, boring depths shall be sufficient to determine the extent of consolidation and/or slippage the fill may cause.

TABLE 7-1　Recommended Soil Boring Spacings and Depths

	Major Divisions		Group Symbols
Coarse-grained soils more than 50% retained on No. 200 sieve	Gravels 50% or more of coarse fraction retained on No. 4 sieve	Clean gravels	GW GP
		Gravels with fines	GM GC
	Sands less than 50% of coarse fraction retained on No. 4 sieve	Clean sands	SW SP
		Sands with fines	SM SC
Fine-grained soils 50% or less retained on No. 200 sieve	Silts and clays liquid limit 50% or less		ML CL OL
	Silts and clays liquid limit greater than 50%		MH CH OH
Highly organic soils			PT

Note: Based on the material passing the 3-in (75-mm) sieve.

TABLE 7-2　Classification of Soils for Airport Pavement Applications

through a No. 4 sieve. Fine grained soils, known also as silts and clays, are subdivided into two groups on the basis of their liquid limits.

These soils are finally grouped into one of 15 different groupings. These groupings are

GW: Well-graded gravels and gravel-sand mixtures, little or no fines

GP: Poorly graded gravels and gravel-sand mixtures, little or no fines

GM: Silty gravels, gravel-sand-silt mixtures

GC: Clayey gravels, gravel-sand-clay mixtures

SW: Well-graded sands and gravelly sands, little or no fines

SP: Poorly graded sands and gravelly sands, little or no fines

SM: Silty sands, sand-silt mixtures

SC: Clayey sands, sand-clay mixtures

ML: Inorganic silts, very fine sands, rock flour, silty or clayey fine sands

CL: Inorganic clays of low to medium plasticity, gravelly clays, silty clays, lean clays

OL: Organic silts and organic silty clays of low plasticity

MH: Inorganic silts, micaceous or diatomaceous fine sands or silts, plastic silts

CH: Inorganic clays or high plasticity, fat clays

OH: Organic clays of medium to high plasticity

PT: Peat, muck, and other highly organic soils

A flowchart illustrating the procedure for the classification of soils by the unified system is given in Fig. 7-2. The uses of the various soil materials for pavement foundations are described in Table 7-3.

It should be noted that column 11 in Table 7-3 refers to the soil's "field CBR" value, or "California Bearing Ratio," a value of the strength of material used in flexible pavement bases, and column 12 in Table 7-3 refers to the soil's "subgrade modulus" or "k value," a value of the bearing capacity of the soil, estimated using a plate bearing test.

The soil's field CBR value is determined by the CBR method of pavement design, which is applied primarily to flexible pavements. The CBR method of design was developed by the California Division in 1928. The method subsequently was adopted for military airport use by the Corps of Engineers, U.S. Army, shortly after the outbreak of World War II. The outbreak of the war required that a decision be made with little delay concerning a design method. At the time, there were no methods available specifically developed for airport pavements. It was apparent that the time required to develop a completed new method of design would preclude its use in a war emergency

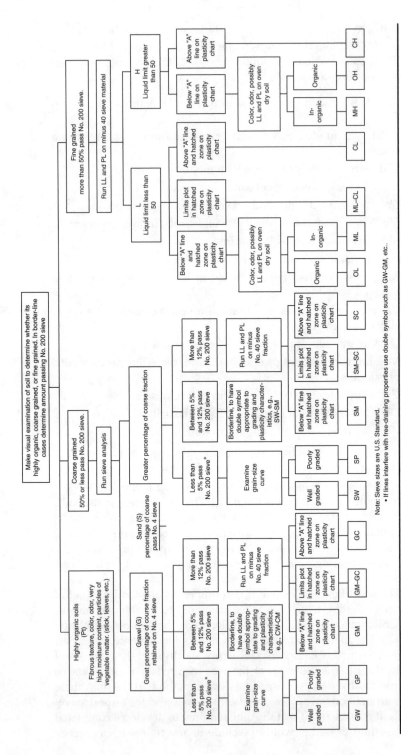

Figure 7-2 Flowchart for the unified soil classification system.

Note: Sieve sizes are U.S. Standard.
* If lines interfere with free-draining properties use double symbol such as GW-GM, etc..

program. Consequently, it was decided to review all available methods for the design of highway pavements and to select one which could be adopted for airfield use. The criteria for selecting a method were many. Among the more important were (1) simplicity in procedures for testing the subgrade and the pavement components, (2) a record of satisfactory experience, and (3) adaptation to the airport problem in a reasonable time. After several months of investigating of suggested methods, the CBR method was tentatively adopted. Application of the CBR method enables the designer to determine the required thickness of the subbase, base, and surface course by entering a set of design curves using the results of a relatively simple soil test.

The CBR Test

The CBR test expresses an index of the shearing strength of soil. Essentially the test consists of compacting about 10 lb of soil into a 6-in-diameter mold, placing a weight, known as a surcharge, on the surface of the sample, immersing the sample in water for 4 days, and penetrating the soaked sample with the steel piston approximately 2 in in diameter at a specified rate of loading. The resistance of the soil to penetration, expressed as a percentage of the resistance for a standard crushed limestone, is the CBR value for the soil. Thus, a CBR of 50 means that the stress necessary for the surcharge to penetrate the soil sample a specified distance is one-half that required for the surcharge to penetrate the same distance in the standard crushed limestone. The relationship is usually based on a penetration of the piston of 0.1 in with 1000 lb/in^2 used as the stress required to penetrate the crushed limestone at 0.1 in penetration. As illustrated in Table 7-3, soils range in CBR values from relatively weak fine-grained and highly organic soils with CBR values as low as 3, to wide-grained coarse soils with CBR values as high as 80 (although CBR testing has been found to be somewhat inaccurate for very gravelly soils, and for application a CBR value of no higher of 50 should be applied).

The Plate Bearing Test

The modulus subgrade of reaction, or k value of the subgrade is determined by what is known as a field plate bearing test. This test consists of applying loads by means of a hydraulic jack through a jacking frame on to a steel plate 30 in in diameter on the soil. By loading the plate, a load-versus-deformation curve is obtained. The k value is determined to be the pressure required to produce a unit deflection of the pavement foundation, measured in pounds per cubic inch. k values range from less than 150 (considered "very poor") to more than 300 (considered "very good"). In general, the greater the coarseness of the soil, the higher k value and the less deflection for a given loading can be expected.

Major Divisions (1)	(2)	Letter (3)	Name (4)	Value as Base Directly under Wearing Surface (5)	Value as Base Directly under Wearing Surface (6)	Potential Frost Action (7)	Compressibility and Expansion (8)	Drainage Characteristic (9)	Unit Dry Weight (pcf) (10)	CBR (11)	Subgrade Modulus k (pci) (12)
Coarse-gravelly soils	Gravel and gravelly soils	GW	Gravel or sandy gravel, well graded	Excellent	Good	None to very slight	Almost none	Excellent	125–140	60–80	300 or more
		GP	Gravel or sandy gravel, poorly graded	Good	Poor to fair	None to very slight	Almost none	Excellent	120–130	35–60	300 or more
		GU	Gravel or sandy gravel, uniformly graded	Good to excellent	Poor	None to very slight	Almost none	Excellent	115–125	25–50	300 or more
		GM	Silty gravel or silty sandy gravel	Good	Fair to good	Slight to medium	Very slight	Fair to poor	130–145	40–80	300 or more
		GC	Clayey gravel or clayey sandy gravel	Good to excellent	Poor	Slight to medium	Slight	Poor to practically impervious	120–140	20–40	200–300
	Sand and sandy soils	SW	Sand or gravelly sand, well graded	Good	Poor to not suitable	None to very slight	Almost none	Excellent	110–130	20–40	200–300
		SP	Sand or gravelly sand, poorly graded	Fair to good	Not suitable	None to very slight	Almost none	Excellent	105–120	15–25	200–300
		SU	Sand or gravelly sand, Poor uniformly Not suitable graded	Fair to good	Poor	None to very slight	Almost none	Excellent	100–115	10–20	200–300
		SM	Silty sand or silty gravelly sand	Good	Not suitable	Slight to high	Very slight	Fair to poor	120–135	20–40	200–300
		SC	Clayey sand or clayey gravelly sand	Fair to good	Not suitable	Slight to high	Slight to medium	Poor to practically impervious	105–130	10–20	200–300

Fine grained soils	Low compress-ibility LL <50	ML	Silts, sandy silts, gravelly silts, or diatomaceous soils	Fair to good	Medium to very high	Not suitable	Slight to medium	Fair to poor	100–125	5–15	100–200
		CL	Lean clays, sandy clays, or gravelly clays	Fair to good	Medium to very high	Not suitable	Medium	Practically impervious	100–125	5–15	100–200
		OL	Organic silts or lean organic clays	Poor	Medium to very high	Not suitable	Medium to high	Poor	90–105	4–8	100–200
	High compress ibility LL<50	MH	Micaccous clays or diatomaceous soils	Poor	Medium to very high	Not suitable	High	Fair to poor	80–105	4–8	100–200
		CH	Fat clays	Poor to very poor	Medium	Not suitable	High	Practically impervious	90–110	3–5	50–100
		OH	Fat organic clays	Poor to very poor	Medium	Not suitable	High	Practically impervious	80–105	3–5	50–100
Peat and other fibrous organic soils		Pt	Peat, humus and other	Not suitable	Slight	Not suitable	Very high	Fair to poor			

TABLE 7-3 Characteristics of Soil Related to Airport Pavement Foundations

Young's Modulus (*E* Value)

The most recent accepted method for determining the strength of subgrade by the FAA is based upon the elastic modulus of the subgrade, or *E*, also known as Young's modulus. In general, a structure's *E* value is its measure of "stiffness" or "elasticity." The greater the *E* value, the more stiff the material, and the less the material is susceptible to deformation under a given stress load. A pavement's *E* value may be empirically estimated by evaluating its stress to strain ratio, according to Eq. (7-1)

$$E \equiv \frac{\text{Tensile stress}}{\text{Tensile strain}} = \frac{\sigma}{\varepsilon} = \frac{F/A_0}{\Delta L/L_0} = \frac{FL_0}{A_0 \Delta L} \tag{7-1}$$

where E = Young's modulus (modulus of elasticity)
F = force applied to the pavement
ΔA_0 = original cross-sectional area through which the force is applied
L = amount by which the shape of the pavement changes
L_0 = original shape of the object

Alternatively, the subgrade modulus E value, in pounds per square inch, may be found using the following conversion formulas: To find E based on CBR value the equation is

$$E = 1500 * \text{CBR} \tag{7-2}$$

To find E based on the modulus of subgrade of reaction, k, the equation is

$$E = 26k^{1.284} \tag{7-3}$$

This conversation is provided by the FAA in its advisory circular primarily to facilitate the transition from the more traditional pavement design methods to the most current software-based method of pavement design and evaluation.

Effect of Frost on Soil Strength

While there are a variety of soil types, the behavioral properties of any given type are relatively similar regardless of other climatic characteristics, such as the average ambient temperature and amount of precipitation. One factor that does significantly impact the strength of soil, however, is the presence of frost on the surface of or within the soil, either on a seasonal or a permanent basis.

Frost action, if severe, results in nonuniform heave of pavements during the winter because of the formation of ice lenses within the subgrade, known as ice segregation, and in loss of supporting capacity of the subgrade during periods of thaw. Figure 7-3 illustrates the process of ice segregation.

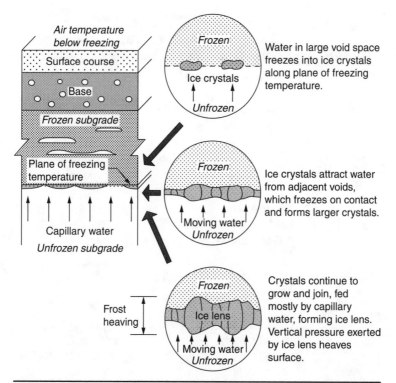

Air temperature below freezing
Surface course
Base
Frozen subgrade

Plane of freezing temperature

Capillary water
Unfrozen subgrade

Frozen
Ice crystals
Unfrozen

Water in large void space freezes into ice crystals along plane of freezing temperature.

Frozen
Moving water
Unfrozen

Ice crystals attract water from adjacent voids, which freezes on contact and forms larger crystals.

Frost heaving

Frozen
Ice lens
Moving water
Unfrozen

Crystals continue to grow and join, fed mostly by capillary water, forming ice lens. Vertical pressure exerted by ice lens heaves surface.

Figure 7-3   The process of ice segregation (*http://www.pavementinteractive.org*).

During periods of thaw, the ice lenses begin to melt, and the water which is released cannot drain through the still-frozen soil at greater depths. Thus, lack of drainage results in loss of strength in the subgrade. It is also possible that a reduction in stiffness will occur in subgrade soils during the thaw period, even though ice lenses may not have formed.

Originally developed by the U.S. Army Corps of Engineers, the FAA categorizes soils into four "frost groups." Soils in frost group 1 are least susceptible to frost and associated soil weakening, while soils in frost group 4 are most susceptible. As illustrated in Table 7-4, those soils with larger particle sizes, such as the gravelly soils, are found in frost group 1, while very fine soils are found in frost group 4.

The design of pavements, both flexible and rigid, is modified slightly depending on the propensity of the soil to encounter frost and the depth of the frost, known as frost penetration. These considerations are described in further detail later in this chapter.

Subgrade Stabilization

In addition to frost, factors such as poor drainage, adverse surface drainage, or merely variations in soil depths, contribute to reductions

Frost Group	Kinds of Soils	Percentage Finer than 0.02 mm by Weight	Soil Classification
FG-1	Gravelly soils	3–10	GW, GP, GW-GM, GP-GM
FG-2	Gravelly soils sands	10–20 3–5	GM, GW-GM, GP-GM. SW, SP, SM, SW-SM, SP-SM
FG-3	Gravelly soils	Over 20	GM, GC
	Sands, except very fine silty sands	Over 15	SM, SC
	Clays, PI above 12	–	CL, CH
FG-4	Very fine silty sands	Over 15	SM
	All silts	–	ML, MH
	Clays, PI = 12 or less	–	CL, CL-ML
	Varved clays and other fine grained baded sediments	–	CL, CH, ML, SM

TABLE 7-4 Frost Design Soil Classification

in the stability of a soil. The FAA allows for the stabilization and treatment of the soils to improve the performance of the constructed pavement. Two types of soil stabilization exist, mechanical stabilization and chemical stabilization. Mechanical stabilization consists of embedding cobble or shot rock sheets within the soil. In some cases, porous concrete or geosynthetics may be used for very soft fine grained soils. Chemical stabilization is achieved by the addition of proper percentages of cement, lime, fly ash, or combinations of these materials to the soil.

FAA Pavement Design Methods

Between 1958 and 2006, the FAA established mandates for aircraft manufacturers to create aircraft, based on their maximum gross take-off weight and landing gear configuration, that produce loads on pavements no greater than 350,000 lb, based on the aircraft at the time that created the heaviest load on airfield pavements, the Douglas DC-8. As aircraft grew in gross weight, landing gear configurations, with additional wheels and spacings, were created to distribute increased gross weights, resulting in equivalent per wheel loads not exceeding the 350,000 lb maximum.

Equivalent Aircraft Method

Historical airfield pavement design methods recommended by the FAA beginning in 1975 took into account the varying weights and landing gear configurations of the fleet of aircraft that may regularly utilize a given airfield's pavement. This historical method involved determining the number of total annual aircraft departures by each type of aircraft and group them into "equivalent annual departures" of each aircraft in terms of the landing gear configuration of a given design aircraft, that is, the aircraft in the fleet mix that requires the greatest pavement strength. This grouping is based on converting the number of annual departures of all aircraft other than the design aircraft to an equivalent number of annual departures by using the multipliers given in Table 7-5.

The equivalent annual departures of the design aircraft were determined by summing the equivalent annual departures of each aircraft in the group, according to the formula given in Eq. (7-4).

$$\text{Log } R_1 = \log R_2 \times \left(\frac{W_2}{W_1}\right)^{1/2} \tag{7-4}$$

where R_1 = equivalent annual departures by the design aircraft
$\quad R_2$ = annual number of departures by an aircraft in terms of design aircraft landing gear configuration
$\quad W_1$ = wheel load of the design aircraft
$\quad W_2$ = wheel load of the aircraft being converted

To Convert From	To	Multiply Departures By
Single wheel	Dual wheel	0.8
Single wheel	Dual tandem	0.5
Dual wheel	Dual tandem	0.6
Double dual tandem	Dual tandem	1.0
Dual tandem	Single wheel	2.0
Dual tandem	Dual wheel	1.7
Dual wheel	Single wheel	1.3
Double dual tandem	Dual wheel	1.7

TABLE 7-5 Factors for Converting Annual Departures by Aircraft to Equivalent Annual Departures by Design Aircraft

Because many of the latest generation aircraft require more complex landing gear configurations that are provided in Table 7-5, a special consideration for very heavy aircraft were made by assigning a gross takeoff weight of 300,000 lb and a dual-tandem landing gear configuration to any aircraft with maximum gross takeoff weight greater than 300,000 lb. This rough approximation was, in part, motivation, to develop an entirely new assessment of fleet mix with respect to airfield pavement design and evaluation.

Cumulative Damage Failure Method

The current method of airfield pavement design and evaluation now considers each type of aircraft that uses the pavement explicitly. The "design aircraft" concept has been replaced by design for fatigue failure expressed in terms of a "cumulative damage factor" (CDF). The CDF for a given aircraft is a value between 0 and 1 which expresses the contribution to ultimate pavement failure of the projected number of uses for each aircraft type that use the pavement. Based on Miner's rule, a traditional theory which estimates the amount of use until failure of a pavement, the CDF for a given fleet of aircraft is determined by Eq. (7-5).

$$CDF = \sum (n_i / N_i) \qquad (7\text{-}5)$$

where n_i is the expected number of annual departures of aircraft i and N_i is the number of departures of aircraft i that would lead to pavement failure for each aircraft i in the mix.

When CDF meets or exceeds 1, the cumulative predicted number of operations for each of the aircraft in the mix will lead to failure of a given pavement system. Any value less than 1 represents the fraction of pavement life that has been effectively "used up." For example, a CDF of 0.75 would indicate that the pavement has used 75 percent of its useful life, and has 25 percent of its life remaining under the predicted traffic usage before fatigue failure.

For both the design of flexible and rigid pavements, the current FAA pavement design method applies a computer software model to estimate the appropriate thickness of designed pavement layers, given the Young's modulus E value of the subgrade and the expected aircraft fleet mix, such that the CDF of the pavement equals 1 after a 20-year life of the designed pavement.

The FAA approved software, *FAA Rigid and Flexible Iterative Elastic Layered Design* or *FAARFIELD* comes equipped with a library of aircraft, their maximum gross weights, landing gear configuration, and contribution to CDF for the given pavement design. Figure 7-4 illustrates the "aircraft" window of FAARFIELD with user inputs of each given aircraft's estimated departures for the to-be-designed pavement system.

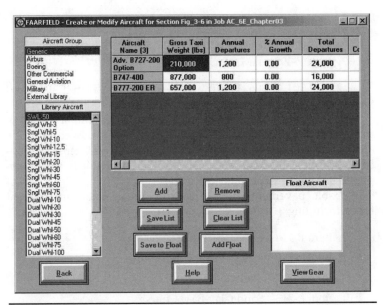

FIGURE 7-4   FAARFIELD aircraft database window.

Design of Flexible Pavements

Flexible pavements consist of a bituminous wearing surface placed upon a base course and, where required by subgrade conditions, a subbase. The entire flexible-pavement structure is ultimately supported by the subgrade. The surface course prevents the penetration of surface water to the base course, provides a smooth, well-bonded surface free of loose particles, resists the shearing stresses caused by aircraft loading, and furnishes a texture of nonskid qualities not causing undue tire wear. The course must also be resistant to fuel spillage and other solvents in areas where maintenance may occur.

The base course is the major structural element of the pavement; it has the function of distributing the wheel loads to the subbase and subgrade. It must be designed to prevent failure in the subgrade, withstand the stresses produced in the base course, resist vertical pressures tending to produce consolidation and deformation of the wearing course, and resist volume changes caused by fluctuations in its moisture content.

Flexible pavement base courses are available in different compositions, or "types," including:

1. Item P-208—Aggregate Base Course

2. Item P-209—Crushed Aggregate Base Course

3. Item P-211—Lime Rock Base Course

4. Item P-304—Cement Treated Base Course

5. Item P-306—Econocrete Subbase Course

6. Item P-401—Plant Mix Bituminous Pavements

7. Item P-403—HMA Base Course

P-211, P-304, P-306, P-401, and P-403 are considered stabilized based courses.

The function of the subbase, when required, is similar to that of the base course, but since the subbase is further removed from the area of load application, it is subjected to lower stress intensities. Subbases are typically required when flexible pavement is to be supported by soils of CBR value less than 20.

Flexible pavement subbase courses are available in different types, including:

1. Item P-154—Subbase Course

2. Item P-210—Caliche Base Course

3. Item P-212—Shell Base Course

4. Item P-213—Sand Clay Base Course

5. Item P-301—Soil Cement Base Course

The subgrade soils are subjected to the lowest loading intensities, and the controlling stresses are usually at the top of the subgrade since stress decreases with depth. However, unusual subgrade conditions, such as layered subgrade materials, can alter the location of controlling stresses.

CBR Method

Prior to 2008, the FAA's standard method for flexible pavement design was known as the CBR method. The CBR method was based on approximation charts that factored in the CBR value of the subgrade and the number and gross weight of equivalent annual departures of the design aircraft. Separate approximation charts were provided by the FAA for different generic aircraft landing gear configurations, and for aircraft greater than 300,000 lb maximum gross weight, specific individual aircraft. Figure 7-5 provides an illustrative example of the CBR method.

The example nomograph found in Fig. 7-5 represents the historical method of estimating the base level thickness of flexible pavement for a Boeing 767. The arrow within the nomograph represents the example for a subgrade with CBR value of 7, a 325,000-lb aircraft gross weight, and 1200 annual equivalent departures, resulting in a required base course of 30 in thickness. The nomograph also provides the necessary thickness for the surface layer, at 4 in thick for critical areas and 3 in thick for noncritical areas, such as pavement shoulders.

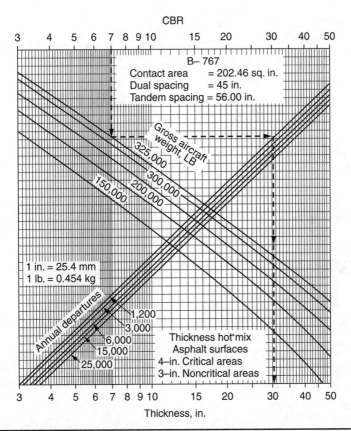

CBR

B– 767
Contact area   = 202.46 sq. in.
Dual spacing   = 45 in.
Tandem spacing = 56.00 in.

Gross aircraft weight, LB

325,000
300,000
200,000
150,000

1 in. = 25.4 mm
1 lb. = 0.454 kg

Annual departures

1,200
3,000
6,000
15,000
25,000

Thickness hot mix
Asphalt surfaces
4–in. Critical areas
3–in. Noncritical areas

Thickness, in.

Figure 7-5 Example approximation chart, CBR method of flexible pavement design.

Layered Elastic Design

Originally applied in 1995 specifically for the heaviest of aircraft, the FAA adopted the layered elastic design (LED) method of flexible pavement design for all pavements designed to accommodate aircraft greater than 30,000 lb in 2008.

Layered elastic design theory considers the fact that the layers of pavement that support loads are impacted by both vertical and horizontal strains and stress, as illustrated in Fig. 7-6. To accommodate the strain, pavement will deflect with the passing of the load. The magnitude of deflection of a given pavement is a function of its elasticity, E, as measured by Young's modulus. In addition, the ratio of transverse to horizontal deflection of a pavement layer, known as Poisson's ratio, μ, is considered.

The layered elastic design and cumulative damage failure methods of pavement design are applied in the FAA's computer pavement design software, FAARFIELD. FAARFIELD uses a Windows-based

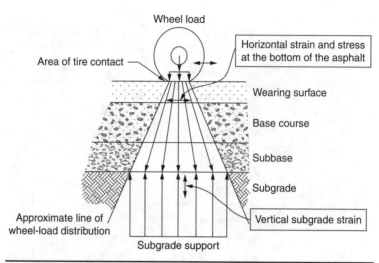

FIGURE 7-6   Visualization of layer elastic design theory.

interface to allow the user to input initial data concerning the sub-grade of the area on which the pavement is to be designed, specifi-cally the Young's modulus of the subgrade, as well as the expected fleet mix that will be using the pavement. As illustrated in Fig. 7-7, FAARFIELD provides the recommended thickness of each layer within the flexible pavement structure, using recommended pavement

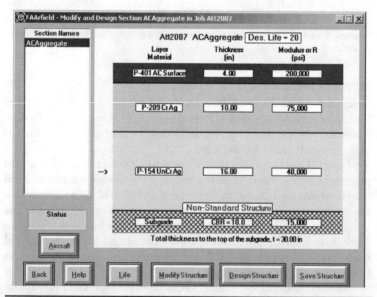

FIGURE 7-7   Example output of FAARFIELD software for flexible pavement using layered elastic design theory.

types (ex. P-401/403 asphalt surfaces), or other user preferred types listed in this section.

Design of Rigid Pavements

Rigid pavements consist of slabs of PCC placed on a subbase that is supported on a compacted subgrade. Like flexible pavements, a properly designed rigid pavement provides a nonskid surface which prevents the infiltration of water into the subgrade, while providing structural support to aircraft which use the pavement.

The subbase under rigid pavements provides uniform stable support for the concrete slabs. As a rule, a minimum thickness of 4 in is required for all subbases under rigid pavements. There are various types of mixtures which are acceptable for rigid pavement subbases including:

> Item P-154—Subbase Course
> Item P-208—Aggregate Base Course
> Item P-209—Crushed Aggregate Base Course
> Item P-211—Lime Rock Base Course
> Item P-301—Soil Cement Base
> Item P-304—Cement Treated Base Course
> Item P-306—Econocrete Subbase Course
> Item P-401—Plant Mix Bituminous Pavements
> Item P-403—HMA Base Course

For rigid pavements accommodating aircraft greater than 100,000 lb maximum gross weight a stabilized subbase is required, which include items P-304, P-306, P-401, and P-403.

Westergaard's Analysis

Similar to the CBR method of design for flexible pavements, prior to 2008, rigid pavement design using nomographs and other approximation charts based on theories developed by H. M. Westergaard was the FAA standard. Westergaard's analysis of pavement design was founded in the mid-1920s and focused on the calculations of stresses and deflections in concrete pavements due to applied loading.

Westergaard assumed the pavement slab to be a thin plate resting on a special subgrade which is considered elastic in the vertical direction only. That is, the reaction is proportional to the deflection of the subgrade $p = kz$, where z is deflection and k is a soil constant, referred to as the *modulus of subgrade reaction*. Other assumptions are that the concrete slab is a homogeneous, isotropic elastic solid and that the wheel load of an aircraft is distributed over an elliptical area. Although

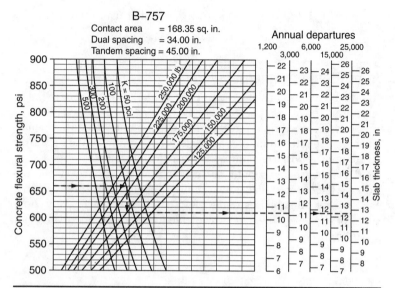

FIGURE 7-8 Example design curve for estimating the slab thickness of rigid pavement using Westergaard's analysis.

these assumptions do not satisfy theory in a strict sense, the results compared reasonably with observations. The Westergaard analysis was used to evaluate stress in a pavement, as well as the deflection of the slab. For airports, Westergaard developed formulas for stresses and deflections in the interior of a slab and at an edge of a slab.

The U.S. Army Corps of Engineers applied Westergaard's formulas toward the creation of approximation charts and design curves. An example is illustrated in Fig. 7-8.

The arrow found in Fig. 7-8 illustrates an example rigid pavement design analysis, considering the use of a PCC mixture of flexural strength of 660 lb/in², a subgrade with k value 100 lb/in³, for a Boeing 757 design aircraft with maximum gross weight of 175,000 lb and 6000 annual equivalent departures, resulting in a design slab thickness of approximately 12 in.

Finite Element Theory

Similar to flexible pavement design, FAARFIELD processes the user defined fleet mix and E value of the subgrade to determine the thickness requirements of the PCC surface. FAARFIELD recommends default subbase layers at 6 in thickness. Multiple subbase layers are recommended for certain subgrades with low E values. Figure 7-9 illustrates an example rigid pavement structure output from FAARFIELD.

FAARFIELD applies finite element theory to estimate the thickness of the PCC surface and any necessary subbase courses. Three-dimensional finite element design theory (3D-FE) is similar to layered

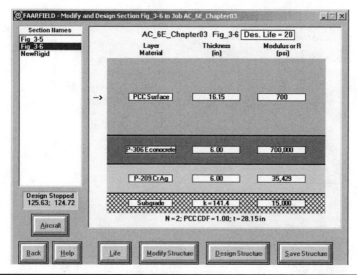

FIGURE 7-9 Example output of FAARFIELD software for rigid pavement design.

elastic design theory in that it takes into account the Young's modulus of the subgrade and the materials used in the slab and subbase courses, and considers a cumulative damage factor in its analysis. 3D-FE design modeling, however, considers the pavement in discrete sections, rather than a continuous material. This perspective allows for more accurate estimation of stresses and strain on the edges of the rigid pavement slabs, which compared to the transverse stress near the center of the slab, is more critical in rigid pavements.

Joints and Joint Spacing

Slabs of PCC rigid pavement are connected by joints to permit expansion and contraction of the pavement, thereby relieving flexural stresses due to curling and friction and to facilitate construction. There are three types of joints: isolation (type A), contraction (type B, C, or E), and construction (type E) joints. The locations of these joints are illustrated in Fig. 7-10, with details as to their specifications found in Fig. 7-11.

The function of type A *isolation joints* is to isolate adjacent pavement slabs and provides space for the expansion of the pavement, thereby preventing the development of very high compressive stresses which can cause the pavement to buckle.

Contraction joints are provided to relieve the tensile stresses due to temperature, moisture, and friction, thereby controlling cracking. If contraction joints were not installed, random cracking would occur on the surface of the pavement. The spacing between contraction joints is dependent on the thickness of the slab, the character of the

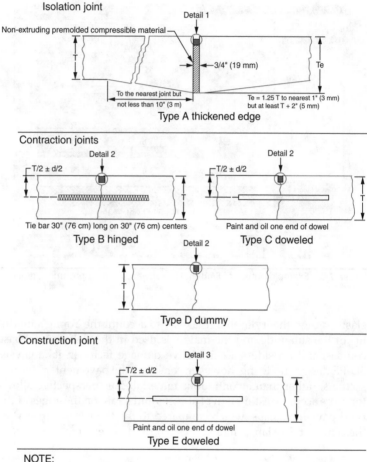

Isolation joint

Detail 1

Non-extruding premolded compressible material

3/4" (19 mm)

To the nearest joint but not less than 10" (3 m)

Te = 1.25 T to nearest 1" (3 mm) but at least T + 2" (5 mm)

Type A thickened edge

Contraction joints

Detail 2

T/2 ± d/2

Tie bar 30" (76 cm) long on 30" (76 cm) centers

Type B hinged

Detail 2

T/2 ± d/2

Paint and oil one end of dowel

Type C doweled

Detail 2

Type D dummy

Construction joint

Detail 3

T/2 ± d/2

Paint and oil one end of dowel

Type E doweled

NOTE:
1. Shaded area is joint sealant.
2. Groove must be formed by sawing.

FIGURE 7-10   Location of rigid pavement joint structures.

aggregate, and whether the slab is plain or reinforced. On the basis of experience, it has been found that for plain slabs 8 to 10 in thick, the spacing should be in the range of 15 to 20 ft. For thicker slabs, the spacing can be increased to 25 ft. Contraction joints may be hinged (type B), doweled (type C), or considered a "dummy" joint (type D). The longitudinal and transverse specifications of these joints are in details in Table 7-6.

Construction joints are required to facilitate construction when two abutting slabs are placed at different times.

All joints in concrete pavements are sealed with sealing compound to prevent infiltration of water or foreign material into the joint spaces. This joint sealant must be capable of withstanding repeated extension and compression as the pavement changes volume.

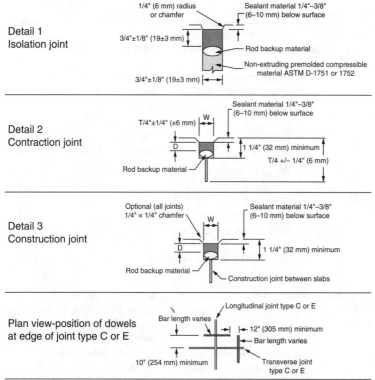

Detail 1
Isolation joint

1/4" (6 mm) radius or chamfer

Sealant material 1/4"–3/8" (6–10 mm) below surface

3/4"±1/8" (19±3 mm)

Rod backup material

Non-extruding premolded compressible material ASTM D-1751 or 1752

3/4"±1/8" (19±3 mm)

Detail 2
Contraction joint

Sealant material 1/4"–3/8" (6–10 mm) below surface

T/4"±1/4" (±6 mm)

W

D

1 1/4" (32 mm) minimum

T/4 +/− 1/4" (6 mm)

Rod backup material

Detail 3
Construction joint

Optional (all joints) 1/4" × 1/4" chamfer

W

Sealant material 1/4"–3/8" (6–10 mm) below surface

D

1 1/4" (32 mm) minimum

Rod backup material

Construction joint between slabs

Plan view-position of dowels at edge of joint type C or E

Longitudinal joint type C or E

Bar length varies

12" (305 mm) minimum

Bar length varies

10" (254 mm) minimum

Transverse joint type C or E

NOTES:
1. Sealant reservoir sized to provide proper shape factor, W/D. Field poured and preformed sealants require different shape factors for optimum performance.
2. Rod backup material must be compatible with the type of sealant used and sized to provide the desired shape factor.
3. Recess sealant 3/8"–1/2" for joints perpendicular to runway grooves.
4. Chamfered edges are recommended when pavements are subject to snow removal equipment of high traffic volumes.

FIGURE 7-11 Rigid pavement joint structure details.

Dowels are load transfer devices which permit joints to open by which prevent differential vertical displacement. Usually dowels are solid, round steel bars, although pipe may also be used. Several different analyses have been proposed for the design of dowels. The spacing of dowels depends on the thickness of the pavement, modulus of subgrade reaction, and the size of the dowel. Table 7-7 contains recommendations for dowel sizes and spacings.

Continuously Reinforced Concrete Pavements

A continuously reinforced concrete pavement (CRCP) is one in which transverse joints have been eliminated (except where the pavement intersects or abuts existing pavements or structures) and the longitudinal reinforcing steel is continuous throughout the length of the pavement. Other than the design of the embedded

Type	Description	Longitudinal	Transverse
A	Thickened edge isolation joint	Use at intersections where dowels are not suitable and where pavements abut structures. Consider at locations along a pavement edge where future expansion is possible.	Use at pavement feature intersections when the respective longitudinal axis intersects at an angle. Use at free edge of pavements where future expansion, using the same pavement thickness is expected.
B	Hinged contraction joint	For all contraction joints in taxiway slabs <9 in (230 mm) thick. For all other contraction joints in slabs <9 in (230 mm) thick, where the joint is placed 20 ft (6 m) or less from the pavement edge.	Not used.
C	Doweled contraction joint	May be considered for general use. Consider for use in contraction joints in slabs >9 in (230 mm) thick, where the joint is placed 20 ft (6 m) or less from the pavement edge.	May be considered for general use. Use on the last three joints from a free edge, and for three joints on either side of isolation joints.
D	Dummy contraction joint	For all other contraction joints in pavement	For all other contraction joints in pavement.
E	Doweled construction joint	Doweled construction joints excluding isolation joints.	Use for construction joints at all locations separating successive paving operations ("headers").

TABLE 7-6 Pavement Joint Types

steel within the pavement, thickness design of CRCP is identical to other rigid pavement types.

The advantages for placing steel in PCC pavements include:

1. Reducing the number of required joints between slabs, resulting in decreased maintenance costs

2. Prolonging service life when pavement is overloaded

3. Reducing pavement deflection

Thickness of Slab	Diameter	Length	Spacing
6–7 in (150–180 mm)	¾ in* (20 mm)	18 in (460 mm)	12 in (305 mm)
8–12 in (210–305 mm)	1 in* (25 mm)	19 in (480 mm)	12 in (305 mm)
13–16 in (330–405 mm)	1¼ in* (30 mm)	20 in (510 mm)	15 in (380 mm)
17–20 in (430–510 mm)	1½ in* (40 mm)	20 in (510 mm)	18 in (460 m)
21–24 in (535–610 mm)	2 in* (50 mm)	24 in (610 mm)	18 in (460 mm)

*Dowels noted may be solid bar or high-strength pipe. High-strength pipe must be plugged on each end with a tight-fitting plastic cap or mortar mix.

TABLE 7-7 Dimensions and Spacing of Steel Dowels in Rigid Pavement

The amount of reinforcing steel required to control volume changes is dependent primarily on the slab thickness, concrete tensile strength, and yield strength of the steel. While several procedures have been proposed for estimating the required amount of steel, experience indicates that it should be approximately 0.6 percent of the gross cross-sectional area and that the yield strength should be at least 60,000 lb/in². The minimum amount may be determined by Eq. (7-6).

$$P_s(\%) = (1.3 - 0.2F)\frac{f_t}{f_s} \qquad (7\text{-}6)$$

where P_s = percentage of embedded steel
  f_t = tensile strength of concrete, lb/in²
  f_s = allowable working stress in steel, lb/in²
  F = coefficient of subgrade friction

The FAA recommends that the cross-sectional area of the reinforcing steel A_s be obtained by Eq. (7-7).

$$A_s = \frac{(3.7)L\sqrt{Lt}}{f_s} \qquad (7\text{-}7)$$

where
  A_s = area of steel per foot of width or length, in²
  L = length or width of slab, ft
  T = thickness of slab, in
  f_s = allowable tensile stress in steel, lb/in²

Note: To determine the area of steel in metric units:
  L should be expressed in meters
  t should be expressed in millimeters
  f_s should be expressed in meganewtons per square meter
  The constant 3.7 should be changed to 0.64.
  f_s will then be in terms of square centimeters per meter.

The longitudinal embedded steel must also be capable of withstanding the forces generated by the expansion and contraction of the pavement due to temperature changes. Equation (7-8) determines the amount of steel required as a function of temperature.

$$P_s = \frac{50 f_t}{f_s - 195T}$$ (7-8)

where

P_s = embedded steel in percent

f_t = tensile strength of concrete, 67 percent of the flexural strength is recommended

f_t = working stress for steel usually taken as 75 percent of specified minimum yield strength

T = maximum seasonal temperature differential for pavement in degrees Fahrenheit

Longitudinal embedded steel is located at mid-depth or slightly above mid-depth of the slab.

Transverse embedded steel is recommended for CPRP airport pavements to control random longitudinal cracking. Equation (7-9) is used to determine the amount of transverse steel, as a percentage of the total slab area.

$$P_s(\%) = \frac{W_s F}{2 f_s} \, 100$$ (7-9)

where

P_s = embedded steel in percent

W_s = width of slab, in ft

F_t = friction factor of subgrade

f_s = allowable working stress in steel, in lb/in². Yield strength of 0.75 is recommended

Transverse steel is designed in the same way as tie bars.

Design of Overlay Pavements

Overlay pavements are required when existing pavements are no longer serviceable due to either deterioration in structural capabilities of a loss in riding quality. They are also required when pavements must be strengthened to carry greater loads or increased repetitions of existing aircraft beyond those anticipated in the original design. Overlays also provide a solution for increased safety. An example would be to provide improved skid resistance and reduced risk of hydroplaning.

There are several types of overlay pavements. A concrete pavement can be overlaid with additional concrete, a bituminous surfacing, or a combination of aggregate base course and a bituminous surfacing. Likewise, a flexible type of pavement can be overlaid with concrete, a bituminous surfacing, or the combination of aggregate base course and a bituminous surfacing. The various types of overlay pavements are defined as follows:

1. *Overlay pavement:* The thickness of a rigid or flexible type of pavement placed on an existing pavement

2. *Portland cement concrete overlay:* An overlay pavement constructed of portland cement concrete

3. *Bituminous overlay:* An overlay consisting entirely of a bituminous surfacing

4. *Flexible overlay:* An overlay consisting of a base course and a bituminous surfacing

Figure 7-12 provides an illustration of typical pavement overlay designs.

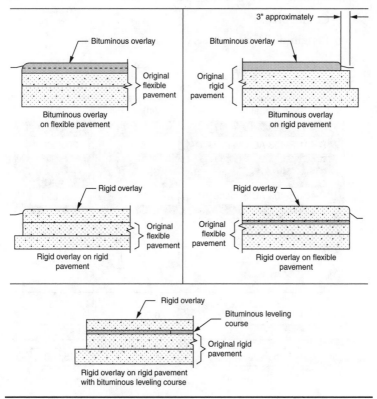

FIGURE 7-12   Typical overlay pavements.

The FAA's FAARFIELD pavement design program includes capabilities for designing airfield pavement overlays. The four types of overlays considered in FAARFIELD are

1. Hot mix asphalt overlay of existing flexible pavement
2. Concrete overlay of existing flexible pavement
3. Hot mix asphalt overlay of existing rigid pavement
4. Concrete overlay of existing rigid pavement

Based on the thickness and condition of the existing pavement layers, FAARFIELD estimates the required thickness of the overlay. Figure 7-13 provides an illustration of FAARFIELD's output for a flexible overlay on an existing flexible pavement.

The design of overlays over an existing rigid pavement is slightly more complex as the condition of the existing rigid pavement plays a significant role in the required thickness of the overlay. The condition of the existing rigid pavement is estimated using a *structural condition index* (SCI), a value which ranges from 0 to 100, in which 100 representing a pavement with no visible structural deficiencies and 0 representing total structural failure. Visible distresses that contribute to a lower SCI include

- Corner breaks
- Longitudinal, transverse, or diagonal cracking

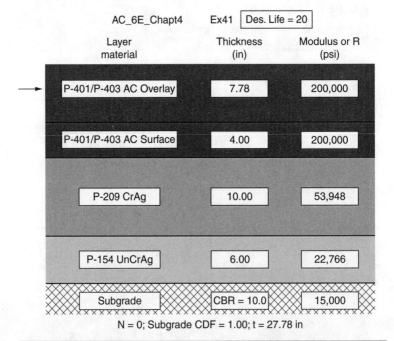

FIGURE 7-13   Design example of flexible overlay on existing flexible pavement.

- Shattered slab
- Shrinkage cracks
- Spalling (cracking, breaking, or chipping of joint/crack edges)

The SCI is estimated using standard pavement structural condition formulas based on empirical analysis that may be found in general concrete structural evaluation references.

In the case when there are no visible or otherwise degradations in structural condition, the FAA calls for the estimation of a *cumulative damage factor used* (CDFU). For aggregate base layers, CDFU may be estimated by Eq. (7-10).

$$CDFU = \frac{L_U}{0.75L_D} \text{ when } L_U < 0.75\,L_D$$

$$= 1 \qquad \text{when } L_U \quad 0.75L_D$$

$$(7-10)$$

where:

L_U = number of years of operation of the existing pavement until overlay

L_D = design life of the existing pavement in years

For rigid pavement bases, CDFU is estimated using the FAARFIELD software, based on the number of years the pavement has been in use to date. Figure 7-14 illustrates the FAARFIELD output for estimating CDFU for rigid pavement bases.

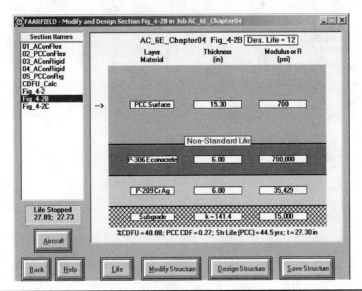

FIGURE 7-14 CDFU estimation for rigid pavement bases using FAARFIELD.

There are several other complexities associated with pavement overlays, particularly with respect to rigid pavements, that are beyond the scope of this text. It is strongly recommended that further study include in-depth review of the FAA Advisory Circular AC 150/5320-6E, "Airfield Pavement Design and Evaluation," as well as familiarization with the FAARFIELD software package.

Pavements for Light Aircraft

Pavements for light aircraft are defined as landing areas intended for personal or other small aircraft engaged in nonscheduled activities, such as recreational, agricultural, or instructional activities, or small aircraft charter operations. Pavements for light aircraft are designed to accommodate aircraft with less than 30,000 lb maximum gross weight. In many cases these aircraft will not exceed 12,500 lb. Figure 7-15 illustrates the composition of light aircraft pavements. Note that, as opposed to pavements for heavier aircraft, light aircraft pavements do not have critical versus noncritical areas and as such the surface thickness of pavement is the same through the paved area.

The FAA FAARFIELD software provides the capability to design pavements for light aircraft, using a similar procedure for typical flexible and rigid pavements. FAARFIELD requires the CBR or modulus E value of the subgrade, the aircraft mix, gross weights, and annual

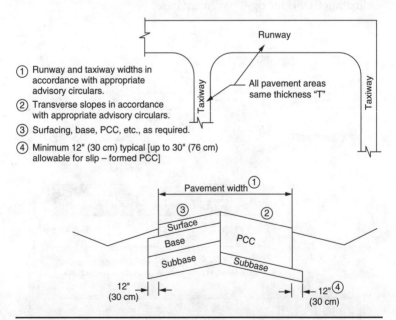

FIGURE 7-15 Typical sections for light aircraft pavements.

departures of all aircraft. For flexible pavements, FAARFIELD will estimate the total thickness of the pavement, including a minimum 2 in surface course. For rigid pavements, FAARFIELD will estimate the slab thickness. In addition FAARFIELD will call for a minimum sub-base thickness of 4 in for aircraft weighing 12,500 lb maximum gross weight or greater.

Other than using flexible or rigid pavement structures, landing facilities for light aircraft may be turf or an aggregate-turf mixture. FAARFIELD also has capabilities for estimating the composition of aggregate-turf mixtures.

Pavement Evaluation and Pavement Management Systems

A pavement management system (PMS) is a mechanism for providing consistent, objective, and systematic procedures for evaluating pavement condition and for determining the priorities and schedules for pavement maintenance and rehabilitation within available resource and budgeting constraints. The pavement management system can also be used to maintain records of pavement condition and to provide specific recommendations for actions which may be required to maintain a pavement network at an acceptable condition while minimizing the cost associated with pavement maintenance and rehabilitation.

A pavement management system evaluates present pavement condition and predicts future condition through the use of a pavement condition indicator. By projecting the rate of deterioration in the pavement condition indicator and adopting some minimum acceptable level for this indicator, a life-cycle cost analysis can be performed for various maintenance and rehabilitation alternatives, and a determination can be made of the optimal time for the application of the most appropriate alternative. The rate of deterioration of a pavement accelerates with time. By implementing a maintenance or rehabilitation strategy to upgrade the pavement condition at the proper time the overall cost of maintenance and rehabilitation can be minimized. As noted by the FAA, the total annual cost to maintain or rehabilitate a pavement in relatively poor condition can be 4 to 5 times that of maintaining or rehabilitating a pavement in relatively good condition.

An effective PMS for use at airports should include the following components:

1. A systematic mechanism for regularly collecting, storing, and retrieving the necessary data associated with pavement use and condition

2. An objective system for evaluating pavement condition at regular intervals

3. Procedures for identifying alternative maintenance and reha- bilitation strategies

4. Mechanisms for predicting and evaluating the impact of pave- ment maintenance and rehabilitation strategies and alternatives on pavement condition, serviceability, and useful service life

5. Procedures for estimating and comparing the costs of various strategies and alternatives

6. Techniques for identifying the optimal strategy or alternative based upon relevant decision criteria

Essential to an effective PMS is the maintenance of a pavement database which should include

1. Information about the pavement structure, including when it was originally constructed, the structural components, the soil conditions, a history of subsequent maintenance and reha- bilitation, and the cost of these actions.

2. A record of the airport pavement traffic including the number of aircraft operations by various types of aircraft using the pavement over its life.

3. The ability to regularly track pavement condition, including measures of pavement distress and the causes of distress. A pavement rating system should be developed based upon the quantity, severity, and type of distress affecting the pavement surface condition. This rating system measures pavement surface performance and has implications for structural per- formance. The periodic collection of condition-rating data is essential to tracking pavement performance

As part of an effective PMS, the evaluation of airport pavements should be a methodical process, which includes a thorough review of construction data and usage history, routine site inspections, and pavement sampling and testing. Types of sampling and testing pro- cedures include: direct sampling, nondestructive testing (NDT), ground penetrating radar, and infrared thermography.

References

1. *Airport Pavement Design and Evaluation*, Advisory Circular AC 150/5320-6E, Federal Aviation Administration, Washington, D.C., 2008.
2. *Airport Pavement Design and Evaluation*, Advisory Circular AC 150/5320-6D includes changes 1 through 4, Federal Aviation Administration, Washington, D.C., 2006.
3. *FAA Finite Element Design Procedure for Rigid Pavements*, FAA AR-07/33, Federal Aviation Administration, Washington, D.C., 2007.
4. "Evolution of Concrete Road Design in the United States," *Pavement Digest*, Vol. 1. Issue 2, 2005.

5. Kawa, Brill, Hayhoe, *FAARFIELD—New FAA Airport Thickness Design Software*, manuscript presenting for the 2007 FAA Worldwide Airport Technology Transfer Conference, Atlantic City, N.J., April 2007.

6. "A Design Procedure for Continuously Reinforced Concrete Pavements for Highways," ACI Subcommittee VII, Title No. 69-32, Vol. 69, pp. 309–319, *Journal of the American Concrete Institute*, Detroit, Mich., 1972.

7. *Aerodrome Design Manual, Part 3, Pavements*, Document No. 9157-AN/901, 2d ed., International Civil Aviation Organization, Montreal, Canada, 1983.

8. *Aerodromes, Annex 14 to the Convention on International Civil Aviation*, Vol. 1, *Aerodrome Design and Operations*, International Civil Aviation Organization, Montreal, Canada, 1990.

9. *Aerodromes, Annex 14 to the Convention on International Civil Aviation*, Vol. 2, *Heliports*, International Civil Aviation Organization, Montreal, Canada, 1990.

10. *Aircraft Loading on Airport Pavements, ACN-PCN, Aircraft Classification Numbers for Commercial Turbojet Aircraft*, Aerospace Industries Association of America, Inc., Washington, D.C., 1983.

11. "Aircraft Pavement Loading: Static and Dynamic," R. C. O'Massey, Research in Airport Pavements, Special Report No. 175, Transportation Research Board, Washington, D.C., 1978.

12. *Airfield Pavement Requirements for Multi-Wheel Heavy Gear Loads*, Report No. FAA-RD-70-77, Federal Aviation Administration, Washington, D.C., 1971.

13. *Airport Design*, Advisory Circular AC 150/5300-13, Federal Aviation Administration, Washington, D.C., 1989.

14. "Analysis of Stresses in Concrete Pavements due to Variations of Temperature," H. M. Westergaard, Proceedings, 6th Annual Meeting, Highway Research Board, Washington, D.C.

15. "Applications of the Results of Research to the Structural Design of Pavements," E. F. Kelley, *Journal of the American Concrete Institute*, Detroit, Mich., 1939.

16. "Characterization of Subgrade Soils in Cold Regions for Pavement Design Purposes," A. T. Bergan and C. L. Monismith, Highway Research Record, No. 431, Highway Research Board, Washington, D.C., 1973.

17. *Computer Aided Design for Flexible Airfield Pavements*, Computer Program FAD (FAA Version F806FAA), U.S. Army Corps of Engineers, Waterways Experiment Station, Vicksburg, Miss., 1992.

18. *Computer Aided Design for Rigid Airfield Pavements*, Computer Program RAD (FAA Version R805FAA), U.S. Army Corps of Engineers, Waterways Experiment Station, Vicksburg, Miss., 1992.

19. *Computer Aided Evaluation for Airfield Pavements*, Computer Program PCN, U.S. Army Corps of Engineers, Waterways Experiment Station, Vicksburg, Miss., 1992.

20. *Computer Program for Airport Pavement Design*, R. G. Packard, Portland Cement Association, Skokie, Ill., 1967.

21. *Computer Program Supplement to Thickness Design Asphalt Pavements for Air Carrier Airports*, Manual Series, No. MS-11A, The Asphalt Institute, Lexington, Ky., 1987.

22. *Design and Construction and Behavior under Traffic of Pavement Test Sections*, Report No. FAA-RD-73-198-I, Federal Aviation Administration, Washington, D.C., 1974.

23. *Design and Construction, Continuously Reinforced Joint and Crack Sealing Materials and Practices*, NCHRP Report, Report No. 38, National Cooperative Highway Research Program, Highway Research Board, Washington, D.C., 1967.

24. *Design and Construction of Airport Pavements on Expansive Soils*, Report No. FAA-RD-76-66, Federal Aviation Administration, Washington, D.C., 1976.

25. *Design and Construction of MESL*, Report No. FAA-RD-73-198-III, Federal Aviation Administration, Washington, D.C., 1974.

26. "Design Considerations for Multi Wheel Aircraft," W. R. Barker and C. R. Gonzalez, International Air Transportation, Proceedings of the 22nd Conference on International Air Transportation, American Society of Civil Engineers, New York, N.Y., 1992.

27. *Design Manual for Continuously Reinforced Concrete Pavements*, Report No. FAA-RD-74-33-III, Federal Aviation Administration, Washington, D.C., 1974.

28. *Design of Civil Airfield Pavement for Seasonal Frost and Permafrost Conditions*, Report No. FAA-RD-74-30, Federal Aviation Administration, Washington, D.C., 1974.

29. *Design of Concrete Airport Pavement*, R. G. Packard, Engineering Bulletin, Portland Cement Association, Skokie, Ill., 1973.

30. *Design of Flexible Airfield Pavements for Multiple-Wheel Landing Gear Assemblies*, Report No. 2, Analysis of Existing Data, Technical Memorandum 3-349, U.S. Army Corps of Engineers, Waterways Experiment Station, Vicksburg, Miss., 1955.

31. "Design of Pavement with High Quality Structural Layers," G. M. Hammitt, Research in Airport Pavements, Special Report, No. 175, Transportation Research Board, Washington, D.C., 1978.

32. *Economic Analysis of Airport Pavement Rehabilitation Alternatives*, Report No. FAA-RD-81-78, Federal Aviation Administration, Washington, D.C., 1981.

33. "Effect of Dynamic Loads on Airport Pavements," R. H. Ledbetter, Research in Airport Pavements, Special Report, No. 175, Transportation Research Board, Washington, D.C., 1978.

34. *ELSYM5—Computer Program for Determining Stresses and Deformation in a Five Layer Elastic System*, G. Ahlborn, University of California, Berkeley, Calif., 1972.

35. "Equivalent Passages of Aircraft with Respect to Fatigue Distress of Flexible Airfield Pavements," J. A. Deacon, Proceedings, Association of Asphalt Paving Technologists, 1971.

36. *Field Survey and Analysis of Aircraft Distribution on Airport Pavements*, Report No. FAA-RD-74-36, Federal Aviation Administration, Washington, D.C., 1975.

37. *Flexible Airfield Pavements*, Technical Manual, TM 5-825-2, U.S. Army Corps of Engineers, A. G. Publication Center, St. Louis, Mo., 1978.

38. *Full-Depth Asphalt Pavements for General Aviation*, Information Series, No. IS-154, The Asphalt Institute, Lexington, Ky., 1973.

39. *Geomechanics Computing Programme, N. 1, Computer Programmes for Circle and Strip Loads on Layered Anisotropic Media*, W. J. Harrison, C. M. Gerrard, and L. J. Wardel, Division of Applied Geomechanics, SCIRO, Australia, 1972.

40. *Guidelines and Procedures for Maintenance of Airport Pavements*, Advisory Circular AC 150/5380-6, Federal Aviation Administration, Washington, D.C., 1982.

41. *Hot Mix Asphalt Paving Handbook*, Advisory Circular AC 150/5370-14, Federal Aviation Administration, Washington, D.C., 1991.

42. "Influence Charts for Rigid Pavements," G. Pickett and G. K. Ray, *Transactions*, American Society of Civil Engineers, Vol. 116, pp. 49–73, New York, N.Y., 1951.

43. "Layered Systems Under Normal Surface Loads," M. G. Peutz, H. P. M. van Kempen, and A. Jones, *Highway Research Record*, No. 228, Highway Research Board, Washington, D.C., 1968.

44. *Measurement, Construction, and Maintenance of Skid Resistant Airport Pavement Surfaces*, Advisory Circular AC 150/5320-12B, Federal Aviation Administration, Washington, D.C., 1991.

CHAPTER 8

Airport Lighting, Marking, and Signage

Introduction

Visual aids assist the pilot on approach to an airport, as well as navigating around an airfield and are essential elements of airport infrastructure. As such, these facilities require proper planning and precise design.

These facilities may be divided into three categories: lighting, marking, and signage. Lighting is further categorized as either approach lighting or surface lighting. Specific lighting systems described in this chapter include

1. Approach lighting
2. Runway threshold lighting
3. Runway edge lighting
4. Runway centerline and touchdown zone lights
5. Runway approach slope indicators
6. Taxiway edge and centerline lighting

The proper placement of these systems is described in this chapter but no attempt has been made to describe in detail the hardware or its installation. Airfield marking and signage includes

1. Runway and taxiway pavement markings
2. Runway and taxiway guidance sign systems

Airfield lighting, marking, and signage facilities provide the following functions:

1. Ground to air visual information required during landing
2. The visual requirements for takeoff and landing
3. The visual guidance for taxiing

In the United States, the Federal Aviation Administration provides guidance for designing standard airfield lighting, marking, and signage, through published Advisory Circulars. These Advisory Circulars are frequently updated. The standards described in this text are current as of 2007. Current advisory circulars may be found at the FAA's website at http://www.faa.gov.

The Requirements for Visual Aids

Since the earliest days of flying, pilots have used ground references for navigation when approaching an airport, just as officers on ships at sea have used landmarks on shore when approaching a harbor. Pilots need visual aids in good weather as well as in bad weather and during the day as well as at night.

In the daytime there is adequate light from the sun, so artificial lighting is not usually required but it is necessary to have adequate contrast in the field of view and to have a suitable pattern of brightness so that the important features of the airport can be identified and oriented with respect to the position of the aircraft in space. These requirements are almost automatically met during the day when the weather is clear.

The runway for conventional aircraft always appears as a long narrow strip with straight sides and is free of obstacles. It can therefore be easily identified from a distance or by flying over the field. Therefore, the perspective view of the runway and other identifying reference landmarks are used by pilots as visual aids for orientation when they are approaching the airport to land. Experience has demonstrated that the horizon, the runway edges, the runway threshold, and the centerline of the runway are the most important elements for pilots to see.

In order to enhance the visual information during the day, the runway is painted with standard marking patterns. The key elements in these patterns are the threshold, the centerline, the edges, plus multiple parallel lines to increase the perspective and to define the plane of the surface.

During the day when visibility is poor and at night, the visual information is reduced by a significant amount over the clear weather daytime scene. It is therefore essential to provide visual aids which will be as meaningful to pilots as possible.

The Airport Beacon

Beacons are lighted to mark an airport. They are designed to produce a narrow horizontal and vertical beam of high-intensity light which is rotated about a vertical axis so as to produce approximately 12 flashes per minute for civil airports and 18 flashes per minute for military airports [28]. The flashes with a clearly visible duration of at least 0.15 s are arranged in a white-green sequence for land airports and a white-yellow sequence for landing areas on water. Military airports use a double white flash followed by a longer green or yellow flash to differentiate them from civil airfields. The beacons are mounted on top of the control tower or similar high structure in the immediate vicinity of the airport.

Obstruction Lighting

Obstructions are identified by fixed, flashing, or rotating red lights or beacons. All structures that constitute a hazard to aircraft in flight or during landing or takeoff are marked by obstruction lights having a horizontally uniform intensity duration and a vertical distribution design to give maximum range at the lower angles ($1.5°$ to $8°$) from which a colliding approach would most likely come. The criteria for determining which structures need to be lighted are published by the FAA [18, 19].

The Aircraft Landing Operation

An aircraft approaching a runway in a landing operation may be visualized as a sequence of operations involving a transient body suspended in a three-dimensional grid that is approaching a fixed two-dimensional grid. While in the air, the aircraft can be considered as a point mass in a three-dimensional orthogonal coordinate system in which it may have translation along three coordinate directions and rotation about three axes. If the three coordinate axes are aligned horizontal, vertical, and parallel to the end of the runway, the directions of motion can be described as lateral, vertical, and forward. The rotations are normally called pitch, yaw, and roll, for the horizontal, vertical, and parallel axes, respectively. During a landing operation, pilots must control and coordinate all six degrees of freedom of the aircraft so as to bring the aircraft into coincidence with the desired approach or reference path to the touchdown point on the runway. In order to do this, pilots need translation information regarding the aircraft's alignment, height, and distance, rotation information regarding pitch, yaw, and roll, and information concerning the rate of descent and the rate of closure with the desired path. The glide path, height, time, and distance relationships during a typical landing are shown in Fig. 8-1.

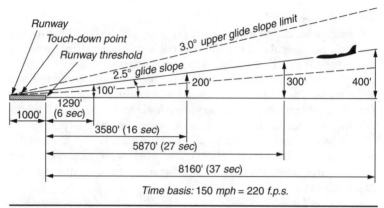

FIGURE 8-1   Glide slope, height, distance, and time relationships.

Alignment Guidance

Pilots must know where their aircraft is with respect to lateral displacement from the centerline of the runway. Most runways are from 75 to 200 ft wide and from 3000 to 12,000 ft long. Thus any runway is a long narrow ribbon when first seen from several thousand feet above. The predominant alignment guidance comes from longitudinal lines that constitute the centerline and edges of the runway. All techniques, such as painting, lighting, or surface treatment that develop contrast and emphasize these linear elements are helpful in providing alignment information.

Height Information

The estimation of the height above ground from visual cues is one of the most difficult judgments for pilots. It is simply not possible to provide good height information from an approach lighting system. Consequently the best source of height information is the instrumentation in the aircraft. However, use of these instruments often requires the availability of precision ground or satellite based navigation technologies. Many airports have no such technologies, and at others only provide lateral approach guidance to certain runways. Consequently two types of ground-based visual aids defining the desired glide path have been developed. These are known as the visual approach slope indicator (VASI) and the precision approach path indicator (PAPI) which are discussed later in this chapter.

Several parameters influence how much a pilot can see on the ground. One of these is the *cockpit cutoff angle*. This is the angle between the longitudinal axis of the fuselage and an inclined plane below which the view of the pilot is blocked by some part of the aircraft, indicated by α in Fig. 8-2. Normally the larger the angle α, the more the pilot can see of the ground. Also important is the *pitch angle*, β,

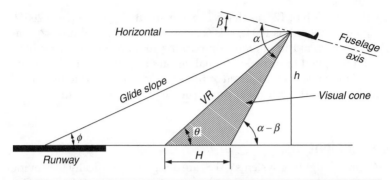

FIGURE 8-2 Visual parameters: φ = glide slope angle, α = cockpit cutoff angle, β = pitch angle, VR = visual range, H = horizontal segment of visual range, h = height of glide slope above the runway, and θ = angle formed by VR with the horizontal.

of the fuselage axis during the approach to the runway. Few aircraft approach a runway with the fuselage angle horizontal; they are either pitched up or down. The larger the angle β (in a pitch-up attitude), the larger must be the angle α to have adequate over-the-nose vision. Approach speed has a profound influence on the angle β. As an example, for some aircraft β can be decreased by about 1° with each 5 kn increase in speed above the reference approach speed.

In Fig. 8-2, VR is the visual range or the maximum distance a pilot can see and some height above the runway h. The horizontal segment of the ground that a pilot can see is H. According to Fig. 8-2,

$$H = VR \cos \Theta - h \cot (\alpha - \beta) \qquad (8\text{-}1)$$

and also

$$\sin \Theta = \frac{h}{VR} \qquad (8\text{-}2)$$

Note from Eq. (8-1) that for a fixed value of VR the ground segment H increases as the height h of the eyes of a pilot above the ground decreases. Typical values of α range from 11° to 16° and typical values of β are ± 0.5°.

It has been found through experience that 3 s is approximately the minimum reaction time for a pilot to cause the aircraft to react after sighting a visual aid [28]. If a minimum of 3 s is necessary for perception, pilot action, aircraft response, and checking the response, and if the approach speed of the aircraft is 150 mi/h (220 ft/s), then the minimum horizontal segment on the ground should not be less than 660 ft. Using Eq. (8-1) with the glide slope angle, φ, of 2.5° and a value of $\alpha - \beta$ of 12° results in H being about 200 ft when h is 200 ft. However, when h is 100 ft, H is 687 ft. Consequently, lighting systems designed to aid in aircraft approaching to land on a runway have been designed to provide optimal visual guidance when aircraft are at relatively low altitudes on approach, and are angled to be consistent with the downward approach angle of arriving aircraft.

Approach Lighting

Approach lighting systems (ALS) are designed specifically to provide guidance for aircraft approaching a particular runway under nighttime or other low-visibility conditions. While under nighttime conditions it may be possible to view approach lighting systems from several miles away, under other low-visibility conditions, such as fog, even the most intense ALS systems may only be visible from as little as 2500 ft from the runway threshold.

Studies of the visibility in fog [3] have shown that for a visual range of 2000 to 2500 ft it would be desirable to have as much as 200,000 candelas (cd) available in the outermost approach lights where the slant range is relatively long. Under these same conditions the optimum intensity of the approach lights near the threshold should be on the order of 100 to 500 cd. A transition in the intensity of the light that is directed toward the pilot is highly desirable in order to provide the best visibility at the greatest possible range and to avoid glare and the loss of contrast sensitivity and visual acuity at short range.

System Configurations

The configurations which have been adopted are the Calvert system [3] shown in Fig. 8-3 which has been widely used in Europe and other parts of the world, the ICAO category II and category III system shown in Fig. 8-4, and the four system configurations which have been adopted by the FAA in the United States shown in Fig. 8-5. The FAA publishes criteria for the establishment of the approach lighting systems [13] and other navigation facilities at airports [6]. Approach lights are normally mounted on frangible pedestals of varying height to improve the perspective of the pilot in approaching a runway.

The first approach lighting system was known as the Calvert system. In this system, developed by E. S. Calvert in Great Britain in 1949, includes a line of single bulb lights spaced on 100-ft centers along the extended runway centerline and six transverse crossbars of lights of variable length spaced on 500-ft centers, for a total length of 3000 ft. The Calvert system was the first approach lighting system to be certified by ICAO, and is also commonly known as the ICAO category I approach lighting system. An illustration of the Calvert system is found in Fig. 8-3. The Calvert system is still used in developing countries.

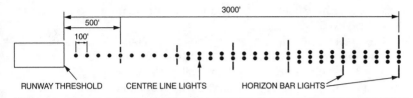

FIGURE 8-3 Calvert approach lighting system.

For operations in very poor visibility, ICAO has certified a modification of the Calvert system, known as the ICAO category II system. The variation calls for a higher lighting intensity to the inner 300 m of the system closest to the runway threshold. The category II and category III system adopted by ICAO shown in Fig. 8-4 consists of two lines of red bars on each side of the runway centerline and a single line of white bars on the runway centerline both at 30 m intervals and both extending out 300 m from the runway threshold. In addition, there are two longer bars of white light at a distance of 150 and 300 m from the runway threshold, and a long threshold bar of green light at the runway threshold. ICAO also recommends that the longer bars of white light also be placed at distances of 450, 500, and 750 m from the runway threshold if the runway centerline lights extend out that distance as shown in Fig. 8-4.

The ALSs currently certified by the FAA for installation in the United States consist of a high-intensity ALS with sequenced flashing lights (ALSF-2), which is required for category II and category III precision approaches, a high-intensity approach lighting system with

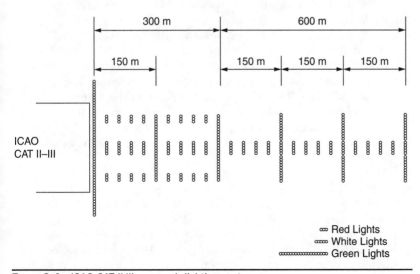

∞ Red Lights
∞ White Lights
∞ Green Lights

FIGURE 8-4 ICAO CAT II-III approach lighting system.

sequenced flashing lights (ALSF-1), and three medium-intensity ALSs (MALSR, MALS, MALSF).

In each of these systems there is a long transverse crossbar located 1000 ft from the runway threshold to indicate the distance from the runway threshold. In these systems roll guidance is provided by crossbars of white light 14 ft in length, placed at either 100- or 200-ft centers on the extended runway centerline. The 14-ft crossbars consist of closely spaced five-bulb white lights to give the effect of a continuous bar of light.

The high-intensity ALS is 2400 ft long (some are 3000 ft long) with various patterns of light located symmetrically about the extended runway centerline and a series of sequenced high-intensity flashing lights located every 100 ft on the extended runway centerline for the outermost 1400 ft. In the high-intensity ALSs the 14-ft crossbars of five-bulb white light are placed at 100-ft intervals and in the medium-intensity ALSs these crossbars of white light are placed at 200-ft intervals both for a distance of 2400 ft from the runway threshold on the extended runway centerline. The high-intensity ALSs have a long crossbar of green lights at the edge of the runway threshold. The ALSF-2 system, shown in Fig. 8-5a, has two additional crossbars consisting of three-bulb white light crossbars which are placed symmetrically about the runway centerline at a distance of 500 ft from the runway threshold and two additional three-bulb red light crossbars are placed symmetrically about the extended runway centerline at 100-ft intervals for the inner 1000 ft to delineate the edges of the runway surface. The ALSF-1 system, shown in Fig. 8-5b, has two additional crossbars consisting of five-bulb red light crossbars which are placed symmetrically about the runway centerline at a distance of 100 ft from the runway threshold to delineate the edge of the runway and two additional three-bulb red light crossbars placed symmetrically about the extended runway centerline at 200 ft from the runway threshold.

The MALSR system, shown in Fig. 8-5c, is a 2400-ft medium-intensity ALS with runway alignment indicator lights (RAILs). The inner 1000 ft of the MALSR is the MALS portion of the system and the outer 1400 ft is the RAIL portion of the system. The system has sequential flashing lights for the outer 1000 ft of the system. It is recommended for category I precision approaches. The simplified short approach lighting system (SSALR) has the same configuration as the MALSR system.

At smaller airports where precision approaches are not required, a medium ALS with sequential flashers (MALSF) or with sequenced flashers (MALS) is adequate. The system is only 1400 ft long compared to a length of 2400 ft for a precision approach system. It is therefore much more economical, an important factor at small airports. The MALSF, similar to the MALSR shown in Fig. 8-5d, is a short approach medium-intensity ALS but the sequenced flashers replace

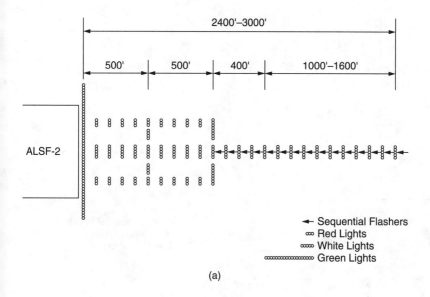

(a)

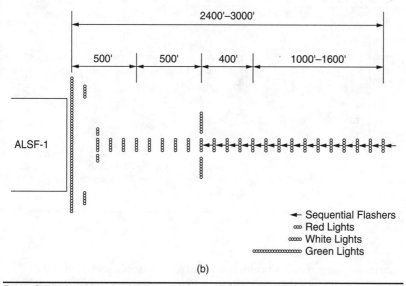

(b)

Figure 8-5a–d   Approach lighting system configurations. (*Continued*)

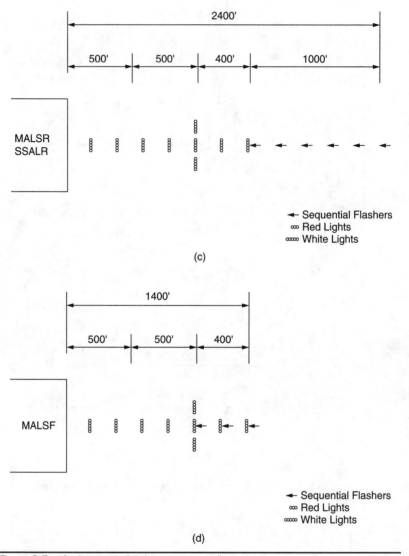

Figure 8-5a–d Approach lighting system configurations.

the runway alignment indicator lights and these are only provided in the outermost 400 ft of the 1400-ft system to improve pilot recognition of the runway approach in areas where there are distracting lights in the vicinity of the airport. The MALS system does not have the runway alignment indicator lights or the sequential flashers.

At international airports in the United States, the 2400-ft ALSs are often extended to a distance of 3000 ft to conform to international specifications.

Sequenced-flashing high-intensity lights are available for airport use and are installed as supplements to the standard approach lighting system at those airports where very low visibilities occur frequently. These lights operate from the stored energy in a capacitor which is discharged through the lamp in approximately 5 ms and may develop as much as 30 million cd of light. They are mounted in the same pedestals as the light bars. The lights are sequence-fired, beginning with the unit farthest from the runway. The complete cycle is repeated every 2 s. This results in a brilliant ball of light continuously moving toward the runway. Since the very bright light can interfere with the eye adaptation of the pilot, condenser discharge lamps are usually omitted in the 1000 ft of the approach lighting system nearest the runway.

Visual Approach Slope Aids

Visual approach slope aids are lighting systems designed to provide a measure of vertical guidance to aircraft approaching a particular runway. The principle of these aids is to provide color-based identification to the pilot indicating their variation from a desired altitude and descent rate while on approach. The two most common visual approach slope aids are the visual approach slope indicator (VASI), and the precision approach path indicator (PAPI).

Visual Approach Slope Indicator

The visual approach slope indicator (VASI) is a system of lights which acts as an aid in defining the desired glide path in relatively good weather conditions. VASI lighting intensities are designed to be visible from 3 to 5 mi during the day and up to 20 mi at night.

There are a number of different VASI configurations depending on the desired visual range, the type of aircraft, and whether large wide bodied aircraft will be using the runway. Each group of lights transverse to the direction of the runway is referred to as a *bar*. The downwind bar is typically located between 125 and 800 ft from the runway threshold, each subsequent bar is located between 500 and 1000 ft from the previous bar. A bar is made up of one, two, or three light units, referred to as *boxes*. The basic VASI-2 system, illustrated in Fig. 8-6, is a two-bar system consisting of four boxes. The bar that is nearest to the runway threshold is referred to as the *downwind bar*, and the bar that is farthest from the runway threshold is referred to as the *upwind bar*. As illustrated in Fig. 8-6, if pilots are on the proper glide path, the downwind bar appears white and the upwind bar appears red; if pilots are too low, both bars appear red; and if they are too high both bars appear white.

In order to accommodate large wide bodied aircraft where the height of the eye of the pilot is much greater than in smaller jets, a third upwind bar is added. For wide bodied aircraft the middle bar becomes

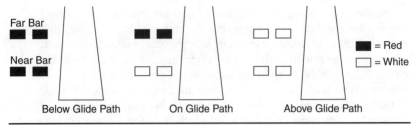

FIGURE 8-6 Two bar VASI system (*FAA/AIM*).

the downwind bar and the third bar is the upwind bar. In other words, pilots of large wide bodied aircraft ignore the bar closest to the runway threshold and use the other two bars for visual reference. The location of the lights for VASI-6 systems is shown in Fig. 8-7.

The more common systems in use in the United States are the VASI-2, VASI-4, VASI-12, and VASI-16. VASI systems are particularly useful on runways that do not have an instrument landing system or for aircraft not equipped to use an instrument landing system.

Precision Approach Path Indicator

The FAA presently prefers the use of another type of visual approach indicator called the *precision approach path indicator* (PAPI) [20]. This system gives more precise indications to the pilot of the approach path of the aircraft and utilizes only one bar as opposed to the minimum of two required by the VASI system. A schematic diagram of the PAPI system is shown in Fig. 8-8.

The system consists of a unit with four lights on either side of the approach runway. By utilizing the color scheme indicated on Fig. 8-8, the pilot is able to ascertain five approach angles relative to the proper glide slope as compared with three with the VASI system. One of the problems with the VASI system has been the lack of an immediate transition from one color indication to another resulting in shades of colors. The PAPI system resolves this problem by providing an instant transition from one color indication to another as a reaction to the

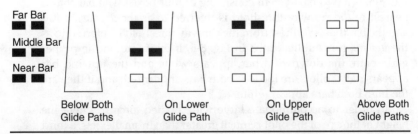

FIGURE 8-7 Three bar VASI-6 system.

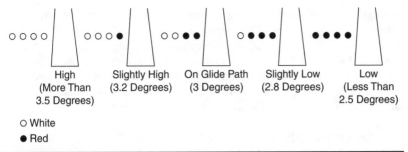

| High (More Than 3.5 Degrees) | Slightly High (3.2 Degrees) | On Glide Path (3 Degrees) | Slightly Low (2.8 Degrees) | Low (Less Than 2.5 Degrees) |

○ White
● Red

FIGURE 8-8 Precision approach path indicator (PAPI) system.

descent path of the aircraft. An advantage of the system is that it is a one-bar system as opposed to the two-bar VASI system. This results in greater operating and maintenance cost economies, and eliminates the need for the pilot to look at two bars to obtain glide slope indications.

Threshold Lighting

During the final approach for landing, pilots must make a decision to complete the landing or "execute a missed approach." The identification of the threshold is a major factor in pilot decisions to land or not to land. For this reason, the region near the threshold is given special lighting consideration. The threshold is identified at large airports by a complete line of green lights extending across the entire width of the runway, as shown earlier in Fig. 8-5, and at small airports by four green lights on each side of the threshold. The lights on either side of the runway threshold may be elevated. Threshold lights in the direction of landing are green but in the opposite direction these lights are red to indicate the end of the runway.

Runway Lighting

After crossing the threshold, pilots must complete a touchdown and roll out on the runway. The runway visual aids for this phase of landing are be designed to give pilots information on alignment, lateral displacement, roll, and distance. The lights are arranged to form a visual pattern that pilots can easily interpret.

At first, night landings were made by floodlighting the general area. Various types of lighting devices were used, including automobile headlights, arc lights, and search lights. Boundary lights were added to outline the field and to mark hazards such as ditches and fences. Gradually, preferred landing directions were developed, and special lights were used to indicate these directions. Floodlighting was then restricted to the preferred landing directions, and runway edge lights were added along the landing strips. As experience was developed, the runway edge lights were adopted as visual aids on a runway. This was followed by the use of runway centerline and touchdown zone lights for operations in very poor visibility. FAA Advisory Circular 150/5340-30C provides guidance for the design and installation of runway and taxiway lighting systems. Those planning and designing such systems should refer to the latest changes to this Advisory Circular, commonly found at http://www.faa.gov.

Runway Edge Lights

Runway edge lighting systems outline the edge of runways during nighttime and reduced visibility conditions. Runway edge lights are classified by intensity, high intensity (HIRL), medium intensity (MIRL), and low intensity (LIRL). LIRLs are typically installed on visual runways and at rural airports. MIRLs are typically installed on visual runways at larger airports and on nonprecision instrument runways, HIRLs are installed on precision-instrument runways.

Recommended standards for the design and installation of runway edge lighting systems are published by the FAA [21] and are contained in ICAO Annex 14 [1]. These light fixtures are usually elevated units but semiflush lights are permitted. Each unit has a specially designed lens which projects two main light beams down the runway. Elevated runway lights are mounted on frangible fittings and project no more than 30 in above the surface on which they are installed. They are located along the edge of the runway not more than 10 ft from the edge of the full-strength pavement surface. The longitudinal spacing is not more than 200 ft. Runway edge lights are white, except that the last 2000 ft of an instrument runway in the direction of aircraft operations these lights are yellow to indicate a caution zone. A typical layout of low-intensity and medium-intensity runway edge lights is shown in Fig. 8-9, and a typical layout of HIRLs are illustrated in Fig. 8-10. If the runway threshold is displaced, but the area that is displaced is usable for takeoffs and taxiing, the runway edge lights in the displaced area in the direction of aircraft operations are red, as shown in Fig. 8-11.

Runway Centerline and Touchdown Zone Lights

As an aircraft traverses over the approach lights, pilots are looking at relatively bright light sources on the extended runway centerline. Over the runway threshold, pilots continue to look along the centerline, but the principal source of guidance, namely, the runway edge lights, has moved far to each side in their peripheral vision. The result is that the central area appears excessively black, and pilots are virtually flying blind, except for the peripheral reference information, and any reflection of the runway pavement from the aircraft's landing lights. Attempts to eliminate this "black hole" by increasing the intensity of runway edge lights have proven ineffective. In order to reduce

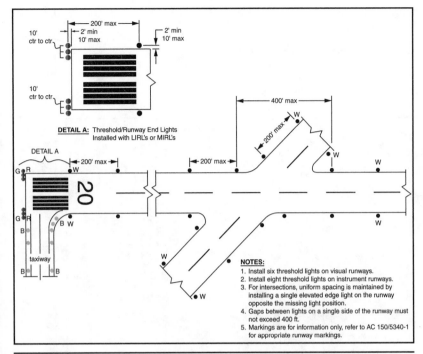

FIGURE 8-9 Low-intensity runway edge lighting specifications (*Federal Aviation Administration*).

the black hole effect and provide adequate guidance during very poor visibility conditions, runway centerline and touchdown zone lights are typically installed in the pavement. An illustration of runway centerline lighting is provided in Fig. 8-12.

These lights are usually installed only at those airports which are equipped for instrument operations. These lights are required for ILS category II and category III runways and for category I runways used for landing operations below 2400 ft runway visual range. Runway centerline lights are required on runways used for takeoff operations below 1600 ft runway visual range. Although not required, runway centerline lights are recommended for category I runways greater than 170 ft in width or when used by aircraft with approach speeds over 140 kn.

When there are displaced thresholds, the centerline lights are extended into the displaced threshold area. If the displaced area is not used for takeoff operations, or if the displaced area is used for takeoff operations and is less than 700 ft in length, the centerline lights are blanked out in the direction of landing. For displaced thresholds greater than 700 ft in length or for displaced areas used for takeoffs, the centerline lights in the displaced area must be capable of being shut off during landing operations.

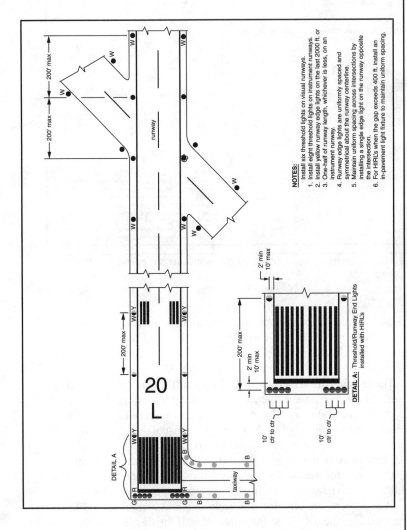

NOTES:

Install six threshold lights on visual runways.
1. Install eight threshold lights on instrument runways.
2. Install yellow runway edge lights on the last 2000 ft. or instrument runway.
3. One-half of runway length, whichever is less, on an instrument runway.
4. Runway edge lights are uniformly spaced and symmetrical about the runway centerline.
5. Maintain uniform spacing across intersections by installing a single edge light on the runway opposite the intersection.
6. For HIRL's when the gap exceeds 400 ft. install an in-pavement light fixture to maintain uniform spacing.

DETAIL A: Threshold/Runway End Lights installed with HIRLs

FIGURE 8-10 High-intensity runway edge lighting specifications *(Federal Aviation Administration)*.

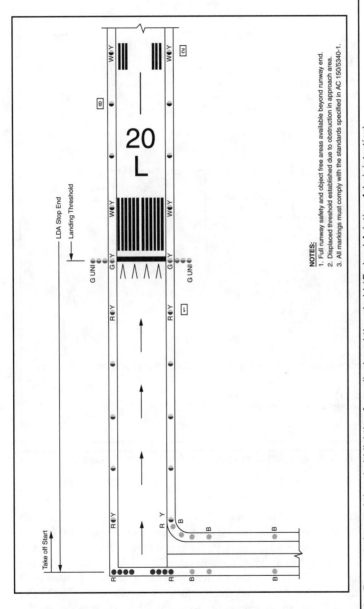

Figure 8-11 Runway edge and threshold lighting for a displaced threshold (*Federal Aviation Administration*).

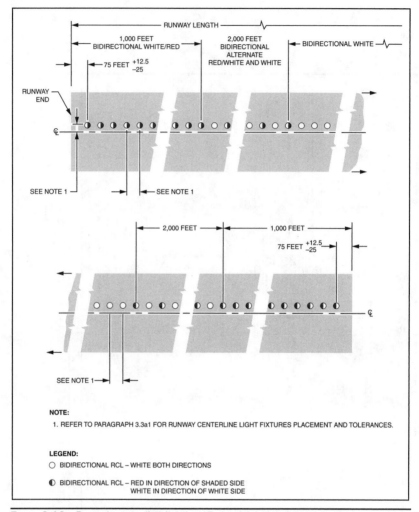

FIGURE 8-12   Runway centerline lighting (*Federal Aviation Administration*).

Runway touchdown zone lights are white, consist of a three-bulb bar on either side of the runway centerline, and extend 3000 ft from the runway threshold or one-half the runway length if the runway is less than 6000 ft long. They are spaced at intervals of 100 ft, with the first light bar 100 ft from the runway threshold, and are located 36 ft on either side of the runway centerline, as shown in Fig. 8-13. The centerline lights are spaced at intervals of 50 ft. They are normally offset a maximum of 2 ft from the centerline to avoid the centerline paint line and the nose gear of the aircraft riding over the light fixtures. These lights are also white, except for the last 3000 ft of runway

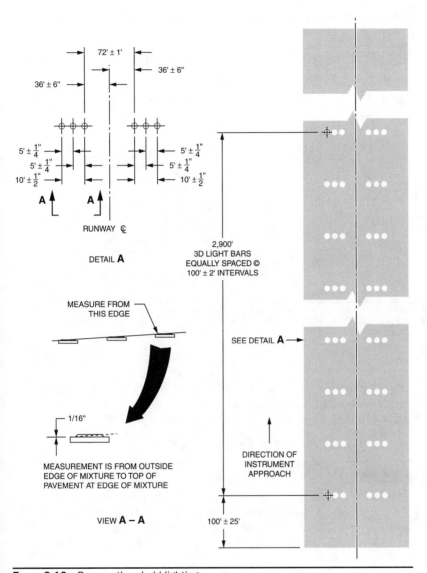

Figure 8-13   Runway threshold lighting.

in the direction of aircraft operations, where they are color coded. The last 1000 ft of centerline lights are red, and the next 2000 ft are alternated red and white.

Runway End Identifier Lights

Runway end identifier lights (REIL) are installed at airports where there are no approach lights to provide pilots with positive visual identification

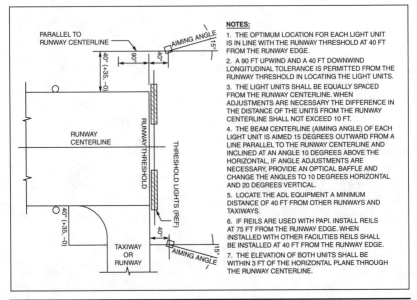

FIGURE 8-14   Typical layout for runway end identifier lights (REILs) (*Federal Aviation Administration*).

of the approach end of the runway. The system consists of a pair of synchronized white flashing lights located on each side of the runway threshold and is intended for use when there is adequate visibility. An illustration and design specifications of REILs may be found in Fig. 8-14.

Taxiway Lighting

Either after a landing or on the way to takeoff, pilots must maneuver the aircraft on the ground on a system of taxiways to and from the terminal and hangar areas. Taxiway lighting systems are provided for taxiing at night and also during the day when visibility is very poor, particularly at commercial service airports.

The following overall guidance should be applied in determining the lighting, marking, and signing visual aid requirements for taxiways:

- In order to avoid confusion with runways, taxiways must be clearly identified.
- Runway exits need to be readily identified. This is particularly true for high-speed runway exits so that pilots can be able to locate these exits 1200 to 1500 ft before the turnoff point.
- Adequate visual guidance along the taxiway must be provided.
- Specific taxiways must be readily identified.

- The intersections between taxiways, the intersections between runways and taxiways, and runway-taxiway crossings need to be clearly marked.

- The complete taxiway route from the runway to the apron and from the apron to the runway should be easily identified.

There are two primary types of lights used for the designation of taxiways. One type delineates the edges of taxiways [21] and the other type delineates the centerline of the taxiway [27]. In addition, there is an increasing use of lighting systems on taxiways, such as runway guard lights (RGLs) and stop bars, to identify intersections with runways, in an effort to reduce accidental incursions on to active runway environments.

Taxiway Edge Lights

Taxiway edge lights are elevated blue colored bidirectional lights usually located at intervals of not more than 200 ft on either side of the taxiway. The exact spacing is influenced by the physical layout of the taxiways. Straight sections of taxiways generally require edge light spacing in 200-ft intervals, or at least three lights equally spaced for taxiway straight line sections less than 200 ft in length.

Closer spacing is required on curves. Light fixtures are located not more than 10 ft from the edge of full strength pavement surfaces. The lights cannot extent more than 30 inches above the pavement surface. The spacing of lights along a curve is shown in Fig. 8-15. Entrance points to runways and exit points from are lighted as shown in Fig. 8-16.

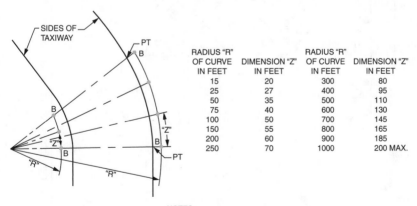

RADIUS "R" OF CURVE IN FEET	DIMENSION "Z" IN FEET	RADIUS "R" OF CURVE IN FEET	DIMENSION "Z" IN FEET
15	20	300	80
25	27	400	95
50	35	500	110
75	40	600	130
100	50	700	145
150	55	800	165
200	60	900	185
250	70	1000	200 MAX.

NOTES:
1. For radii not listed, determine "Z" spacing by linear interpolation.
2. "Z" is the arc length.
3. Uniformly space lights on curved edges. Do not exceed the values determined from the above table.
4. On curved edges in excess of 30 degrees arc, do not install less than three lights including those at the points of tangency (PT).

FIGURE 8-15 Typical taxiway lighting on curved sections.

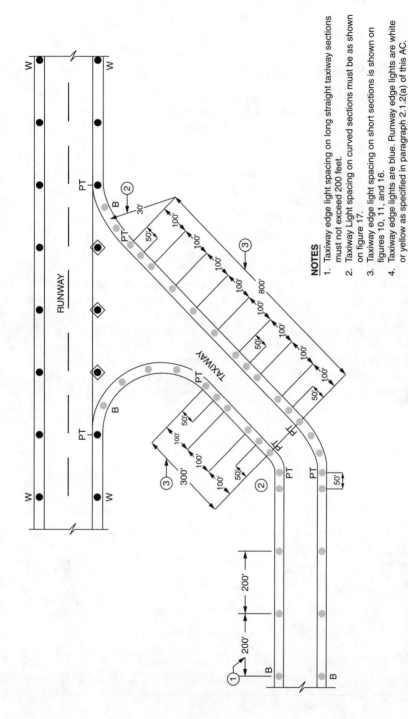

NOTES

1. Taxiway edge light spacing on long straight taxiway sections must not exceed 200 feet.
2. Taxiway Light spacing on curved sections must be as shown on figure 17.
3. Taxiway edge light spacing on short sections is shown on figures 10, 11, and 16.
4. Taxiway edge lights are blue. Runway edge lights are white or yellow as specified in paragraph 2.1.2(a) of this AC.

FIGURE 8-16 Taxiway edge lighting configurations on straight line sections, curves, and runway intersection.

	Maximum Longitudinal Spacing	
	1200 ft (365 m) RVR and Above	Below 1200 ft (365 m) RVR
Radius of curved centerlines		
75 ft (23 m) to 399 ft (121 m)	25 ft (7.6 m)	12.5 ft (4 m) 25 ft (7.6 m)
400 ft (122 m) to 1199 ft (364 m)	50 ft (15 m)	25 ft (7.6 m)
≥1200 ft (365 m)	100 ft (30 m)	50 ft (15 m)
Acute-angled exits	50 ft (15 m)	50 ft (15 m)
Straight segments	100 ft (30 m)	50 ft (15 m)

TABLE 8-1   Taxiway Centerline Lighting Spacing

Taxiway centerline lights are in-pavement bidirectional lights placed in equal intervals over taxiway centerline markings. Taxiway centerline lights are green, except in areas where the taxiway intersects with a runway, where the green and yellow lights are placed alternatively. Research and experience have demonstrated that guidance from centerline lights is superior to that from edge lights, particularly in low visibility conditions. The spacing of the lights on curves and tangents is given in Table 8-1 [21].

For normal exits, the centerline lights are terminated at the edge of the runway. At taxiway intersections the lights continue across the intersection. For long-radius high-speed exit taxiways, the taxiway lights are extended onto the runway from a point 200 ft back from the point of curvature (PC) of the taxiway to the point of tangency of the central curve of the taxiway. Within these limits the spacing of lights is 50 ft. These lights are offset 2 ft from the runway centerline lights and are gradually brought into alignment with the centerline of the taxiway.

Where the taxiways intersect with runways and aircraft are required to hold short of the runway, several yellow lights spaced at 5-ft intervals are placed transversely across the taxiway.

Runway Guard Lights

Runway guard lights (RGLs) are in-pavement lights located on taxiways at intersections of runways to alert pilots and operators of airfield ground vehicles that they are about to enter onto an active runway. RGLs are located across the width of the taxiway, approximately 2 ft from the entrance to a runway, spaced at approximately 10-ft intervals, as illustrated in Fig. 8-17. RGLs are unidirectional, colored yellow for aircraft facing the runway.

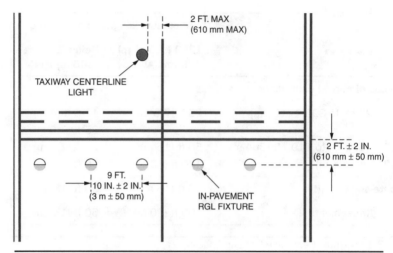

FIGURE 8-17 Runway guard lights.

Runway Stop Bar

Similar to runway guard lights, runway stop bar lights are in-pavement lights on taxiways at intersections with runways. As opposed to RGLs that provide warning to pilots approaching a runway, runway stop bar lights are designed to act as "stop" lights, directing aircraft and vehicles on the taxiway not to enter the runway environment. Runway stop bar lights are activated with red illuminations during periods of runway occupancy or other instances where entrance from the taxiway to the runway is prohibited. In-pavement runway stop bar lighting is typically installed in conjunction with elevated runway guard lights located outside the width of the pavement. An illustration of a typical runway stop bar lighting system is depicted in Fig. 8-18.

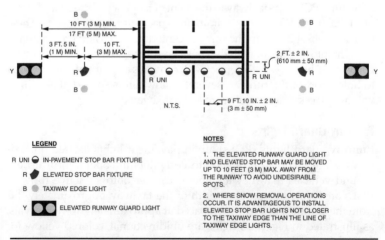

FIGURE 8-18 Runway stop bar lighting.

Runway and Taxiway Marking

In order to aid pilots in guiding the aircraft on runways and taxiways, pavements are marked with lines and numbers. These markings are of benefit primarily during the day and dusk. At night, lights are used to guide pilots in landing and maneuvering at the airport. White is used for all markings on runways and yellow is used on taxiways and aprons. The FAA has developed a comprehensive plan for marking runways and taxiways and they can be found in the FAA Advisory Circular AC 150/5340-1J [17]. Similarly the ICAO recommendations for marking are contained in Annex 14 [2].

Runways

The FAA has grouped runways for marking purposes into three classes: (1) visual, or "basic" runways, (2) nonprecision instrument runways, and (3) precision instrument runways. The visual runway is a runway with no straight-in instrument approach procedure and is intended solely for the operation of aircraft using visual approach procedures. The nonprecision instrument runway is one having an existing instrument approach procedure utilizing air navigation facilities with only horizontal guidance (typically VOR or GPS-based RNAV approaches without vertical guidance) for which a straight-in nonprecision approach procedure has been approved. A precision instrument runway is one having an existing instrument approach procedure utilizing a precision instrument landing system or approved GPS-based RNAV (area navigation) or RNP (required navigation performance) precision approach. Runways that have a published approach based solely on GPS-based technologies are known as GPS runways.

Runway markings include runway designators, centerlines, threshold markings, aiming points, touchdown zone markings, and side stripes. Depending on the length and class of runway and the type of aircraft operations intended for use on the runway, all or some of the above markings are required. Table 8-2 provides the marking requirements for visual, nonprecision, and precision runways.

Figure 8-19 illustrates the required marking for precision runways. Figure 8-20 illustrates the required markings for nonprecision and visual runways (source: FAA AC 150/5340-1J).

Runway Designators

The end of each runway is marked with a number, known as a runway designator, which indicates the approximate magnetic azimuth (clockwise from magnetic north) of the runway in the direction of operations. The marking is given to the nearest 10° with the last digit omitted. Thus a runway in the direction of an azimuth of 163° would be marked as runway 16 and this runway would be in the approximate direction of south-south-east. Therefore, the east end of an east-west runway

Marking Element	Visual Runway	Nonprecision Runway/GPS Nonprecision	Precision Runway/GPS Precision
Designation	X	X	X
Centerline	X	X	X
Threshold marking	X*	X	X
Aiming point	X†	X†	X
Touchdown zone			X
Side stripes	X‡	X‡	X

*Only required on runways used, or intended to be used, by international commercial transport.
†On runways 4000 ft (1200 m) or longer used by jet aircraft.
‡Used when the full pavement width may not be available as a runway.

TABLE 8-2 Required Runway Markings

would be marked 27 (for 270° azimuth) and the west end of an east-west runway would be marked 9 (for a 90° azimuth). If there are two parallel runways in the east-west direction, for example, these runways would be given the designation 9L-27R and 9R-27L to indicate the direction of each runway and their position (L for left and R for right) relative to each other in the direction of aircraft operations. If a third parallel runway existed in this situation it has traditionally been given the designation 9C-27C to indicate its direction and position relative (C for center) to the other runways in the direction of aircraft operations. When there are four parallel runways, one pair is marked with the magnetic azimuth to the nearest 10° while the other pair is marked with the magnetic azimuth to the next nearest 10°. Therefore, if there were four parallel runways in the east-west direction, one pair would be designated as 9L-27R and 9R-27L and the other pair could be designated as either 10L-28R and 10R-28L or 8L-26R and 8R-26L. This type of designation policy is increasingly being applied to three parallel runway configurations, as well. For example, one pair would be designated as 9L-27R and 9R-27L and the third runway may be designated 10-28.

Runway designation markings are white, have a height of 60 ft and a width, depending upon the number or letter used, varying from 5 ft for the numeral 1 to 23 ft for the numeral 7. When more than one number or letter is required to designate the runway the spacing between the designators is normally 15 ft. The sizes of the runway designator markings are proportionally reduced only when necessary due to space limitations on narrow runways and these designation markings should be no closer than 2 ft from the edge of the runway or the runway edge stripes. Specifications for individual runway designators are illustrated in Fig. 8-21.

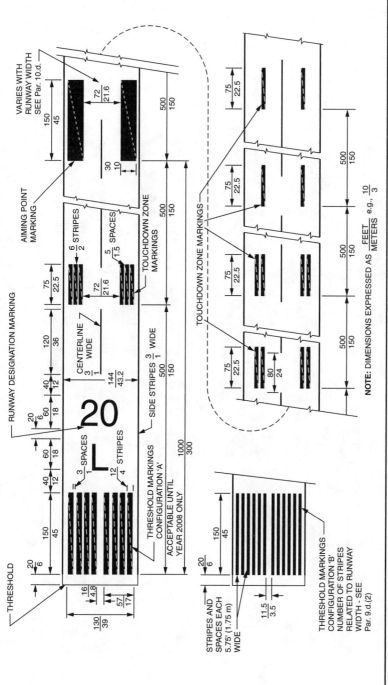

FIGURE 8-19 Precision runway markings.

317

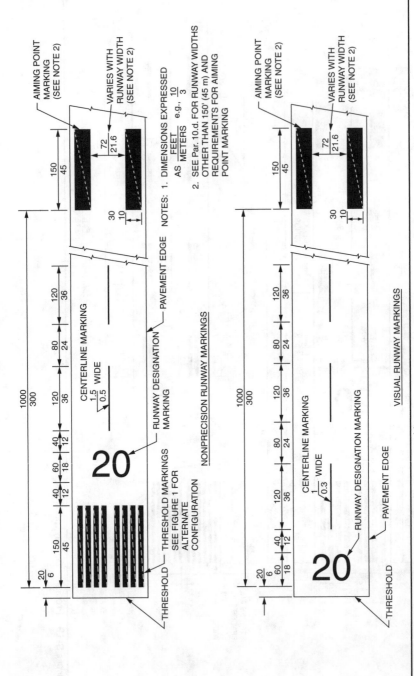

Figure 8-20 Nonprecision and visual runway markings.

318

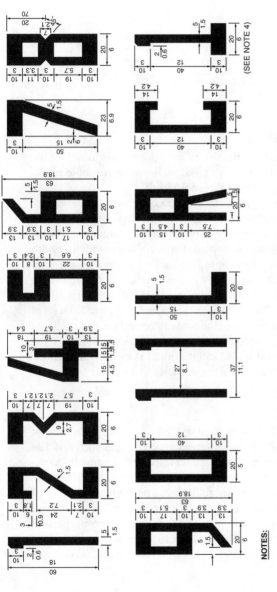

NOTES:

1. ALL NUMERALS EXCEPT THE NUMBER ELEVEN AS SHOWN ARE HORIZONTALLY SPACED 15 FEET (4.5 METERS) APART.

2. SINGLE DIGITS SHALL NOT BE PRECEDED BY A ZERO.

3. DIMENSIONS ARE EXPRESSED THUS: $\dfrac{\text{FEET}}{\text{METERS}}$ e.g., $\dfrac{30}{9}$

4. THE NUMERAL 1, WHEN USED ALONE, CONTAINS A HORIZONTAL BAR TO DIFFERENTIATE IT FROM THE RUNWAY CENTERLINE MARKING.

5. SINGLE DESIGNATIONS ARE CENTERED ON THE RUNWAY PAVEMENT CENTERLINE. FOR DOUBLE DESIGNATIONS, THE CENTER OF THE OUTER EDGES OF THE TWO NUMERALS IS CENTERED ON THE RUNWAY PAVEMENT CENTERLINE.

6. WHERE THE RUNWAY DESIGNATION CONSISTS OF A NUMBER AND A LETTER, THE NUMBER AND LETTER ARE LOCATED ON THE RUNWAY CENTERLINE IN A STACKED ARRANGEMENT AS SHOWN IN FIGURE 1.

FIGURE 8-21 Runway designators.

319

Runway Width	Number of Stripes
60 ft (18 m)	4
75 ft (23 m)	6
100 ft (30 m)	8
150 ft (45 m)	12
200 ft (60 m)	16

TABLE 8-3   Striping Requirements for Runway
Threshold Markings

Runway Threshold Markings

Runway threshold markings identify to the pilot the beginning of the runway that is safe and available for landing. Runway threshold markings begin 20 ft from the runway threshold itself.

Runway threshold markings consist of two series of white stripes, each stripe 150 ft in length and 5.75 ft in width, separated about the centerline of the runway. On each side of the runway centerline, a number of threshold marking stripes are placed, in accordance with the width of the runway, as specified in Table 8-3. Table 8-3 specifies the total number of runway threshold stripes required. For example, for a 100-ft runway, eight stripes are required, in two groups of four are placed about the centerline. Stripes within each set are separated by 5.75 ft. Each set of stripes is separated by 11.5 ft about the runway centerline.

The above specifications for runway threshold markings were adapted by the FAA from ICAO international standards and made mandatory for United States civil use airports in 2008.

Centerline Markings

Runway centerline markings are white, located on the centerline of the runway, and consist of a line of uniformly spaced stripes and gaps. The stripes are 120 ft long and the gaps are 80 ft long. Adjustments to the lengths of stripes and gaps, where necessary to accommodate runway length, are made near the runway midpoint. The minimum width of stripes is 12 in for visual runways, 18 in for nonprecision instrument runways, and 36 in for precision instrument runways. The purpose of the runway centerline markings is to indicate to the pilot the center of the runway and to provide alignment guidance on landing and takeoff.

Aiming Points

Aiming points are placed on runways of at least 4000 ft in length to provide enhanced visual guidance for landing aircraft. Aiming point markings consist of two bold stripes, 150 ft long, 30 ft wide, spaced

Runway Length	Markings on Each End
7990 ft (2436 m) or greater	Full set of markings
6990 ft (2130 m) to 7989 ft (2435 m)	Less one pair of markings
5990 ft (1826 m) to 6989 ft (2129 m)	Less two pairs of markings
4990 ft (1521 m) to 5989 ft (1825 m)	Less three pairs of markings

TABLE 8-4   Touchdown Zone Marking Requirements

72 ft apart symmetrically about the runway centerline, and beginning 1020 ft from the threshold.

Touchdown Zone Markings

Runway touchdown zone markings are white and consist of groups of one, two, and three rectangular bars symmetrically arranged in pairs about the runway centerline. These markings begin 500 ft from the runway threshold. The bars are 75 ft long, 6 ft wide, with 5 ft spaces between the bars, and are longitudinally spaced at distances of 500 ft along the runway. The inner stripes are placed 36 ft on either side of the runway centerline. For runways less than 150 ft in width, the width and spacing of stripes may be proportionally reduced. Where touchdown zone markings are installed on both runway ends on shorter runways, those pairs of markings which would extend to within 900 ft of the runway midpoint are eliminated. In addition, sets of touchdown zone markings are eliminated for shorter runways, as specified in Table 8-4.

Side Stripes

Runway side stripes consist of continuous white lines along each side of the runway to provide contrast with the surrounding terrain or to delineate the edges of the full strength pavement. The maximum distance between the outer edges of these markings is 200 ft and these markings have a minimum width of 3 ft for precision instrument runways and are at least as wide as the width of the centerline stripes on other runways.

Displaced Threshold Markings

At some airports it is desirable or necessary to "displace" the runway threshold on a permanent basis. A displaced threshold is one which has been moved a certain distance from the end of the runway. Most often this is necessary to clear obstructions in the flight path on landing. The displacement reduces the length of the runway available for landings, but takeoffs can use the entire length of the runway. The FAA

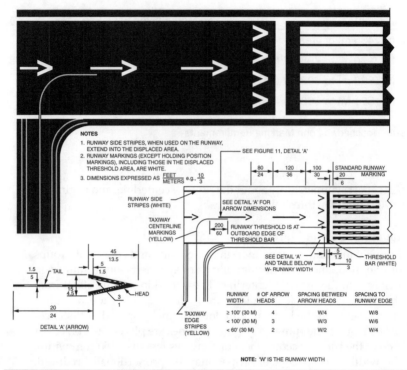

NOTES

1. RUNWAY SIDE STRIPES, WHEN USED ON THE RUNWAY, EXTEND INTO THE DISPLACED AREA.
2. RUNWAY MARKINGS (EXCEPT HOLDING POSITION MARKINGS), INCLUDING THOSE IN THE DISPLACED THRESHOLD AREA, ARE WHITE.
3. DIMENSIONS EXPRESSED AS $\frac{FEET}{METERS}$ e.g., $\frac{10}{3}$

SEE FIGURE 11, DETAIL 'A'

RUNWAY WIDTH	# OF ARROW HEADS	SPACING BETWEEN ARROW HEADS	SPACING TO RUNWAY EDGE
≥ 100' (30 M)	4	W/4	W/8
< 100' (30 M)	3	W/3	W/6
< 60' (30 M)	2	W/2	W/4

NOTE: 'W' IS THE RUNWAY WIDTH

FIGURE 8-22 Displaced threshold markings.

requires that displaced thresholds be marked as shown in Fig. 8-22. These markings consist of arrows and arrow heads to identify the displaced threshold and a threshold bar to identify the beginning of the runway threshold itself. Displaced threshold arrows are 120 ft in length, separated longitudinally by 80 ft for the length of the displaced threshold. Arrow heads are 45 ft in length, placed 5 ft from the threshold bar. The threshold bar is 5 ft in width and extends the width of the runway at the threshold.

Blast Pad Markings

In order to prevent erosion of the soil, many airports provide a paved *blast pad* 150 to 200 ft in length adjacent to the runway end. Similarly, some airport runways have a *stopway* which is only designed to support aircraft during rare aborted takeoffs or landing overruns and is not designed as a full strength pavement. Since these paved areas are not designed to support aircraft and yet may have the appearance of being so designed, markings are required to indicate this. The markings for blast pads and stopways are shown in Fig. 8-23. Likewise the area adjacent to the edge of the runway may have a paved shoulder not capable

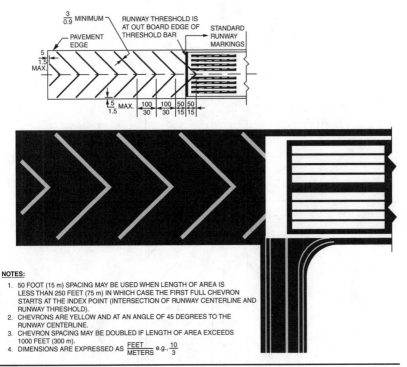

FIGURE 8-23   Blast pad markings.

of supporting aircraft. These areas are marked with a 3-ft-wide stripe, as shown in Fig. 8-24. Yellow color is used for these types of markings.

Taxiway Markings

Taxiway markings consist of centerline markings, holding position markings, and often edge markings. Taxiways are marked as shown in Fig. 8-25.

Centerline and Edge Markings

The centerline of the taxiway is marked with a single continuous 6-in yellow line. On taxiway curves, the taxiway centerline marking continues from the straight portion of the taxiway at a constant distance from the outside edge of the curve. At taxiway intersections which are designed for aircraft to travel straight through the intersection, the centerline markings continue straight through the intersection. At the intersection of a taxiway with a runway end, the centerline stripe of the taxiway terminates at the edge of the runway.

At the intersection between a taxiway and a runway, where the taxiway serves as an exit from the runway, the taxiway marking is usually

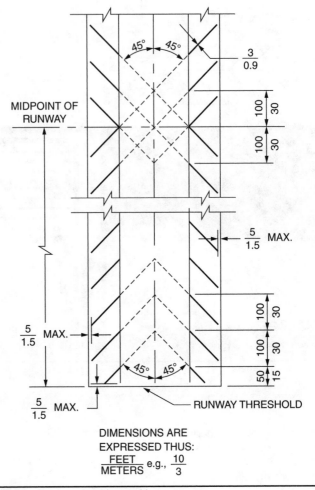

FIGURE 8-24   Runway shoulder markings.

extended on to the runway in the vicinity of the runway centerline marking. The taxiway centerline marking is extended parallel to the runway centerline marking a distance of 200 ft beyond the point of tangency. The taxiway curve radius should be large enough to provide a clearance to the taxiway edge and the runway edge of at least one-half the width of the taxiway. For a taxiway crossing a runway, the taxiway centerline marking may continue across the runway but it must be interrupted for the runway markings.

When the edge of the full strength pavement of the taxiway is not readily apparent, or when a taxiway must be outlined when it is

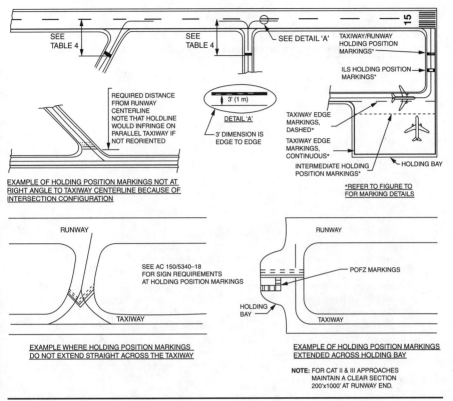

FIGURE 8-25 Taxiway markings.

established on a large paved area such as an apron, the edge of the taxiway is marked with two continuous 6-in wide yellow stripes that are 6 in apart.

Taxiway Hold Markings

For taxiway intersections where there is an operational need to hold aircraft, a dashed yellow holding line is placed perpendicular to and across the centerline of both taxiways.

When a taxiway intersects a runway or a taxiway enters an instrument landing system critical area, a holding line is placed across the taxiway. The holding line for a taxiway intersecting a runway consists of two solid lines of yellow stripes and two broken lines of yellow stripes placed perpendicular to the centerline of the taxiway and across the width of the taxiway. The solid lines are always placed on the side where the aircraft is to hold. The holding line for an instrument landing system critical area consists of two solid lines placed

Aircraft Approach Category and (Airplane Design Group)	Visual and Nonprecision Instrument, ft (m)	Precision Instrument, ft (m)
A and B (I and II) small airplanes only	125 (38)	175 (53)
A and B (I, II, and III)	200 (60)	250 (75)
A and B (IV)	250 (75)	250 (75)
C and D (I through IV)	250 (75)	250 (75)
C and D (V)	250 (75)	280 (85)
C and D (VI)	250 (75)	280 (85)

Source: AC 150/5340-18D

Distances shown above are for planning purposes only. "Hold position markings" must be placed in order to restrict the largest aircraft (tail or body) expected to use the runway from penetrating the obstacle-free zone.

For aircraft approach categories A and B, airplane design group III, this distance is increased 1 ft for each 100 ft above 5100 ft above sea level. For airplane design group IV, precision instrument runways, this distance is increased 1 ft for each 100 ft above sea level.

For aircraft approach category C, airport design group IV, precision instrument runways. This distance is increased 1 ft for each 100 ft above sea level. For airplane design group V, this distance is increased 1 ft for each 100 ft above sea level.

For aircraft approach category D, this distance is increased 1 ft for each 100 ft above sea level.

TABLE 8-5 Location Distances for Holding Position Markings

perpendicular to the taxiway centerline and across the width of the taxiway joined with three sets of two solid lines symmetrical about and parallel to the taxiway centerline. These holding lines are located the minimum distance from the centerline of the runway as indicated in Table 8-5 and illustrated in Fig. 8-26.

Taxiway Shoulders

In some areas on the airfield, the edges of taxiways may not be well-defined due to their adjacency to other paved areas such as aprons and holding bays. In these areas, it is prudent to mark the edges of taxiways with shoulder markings. Taxiway shoulder markings are yellow in color, and are often painted on top of a green background. The shoulder markings consist of 3-ft-long yellow stripes placed perpendicular to the taxiway edge stripes, as illustrated in Fig. 8-27. On straight sections of the taxiway, the marks are placed at a maximum spacing of 100 ft. On curves, the marks are placed on a maximum of 50 ft apart between the curve tangents.

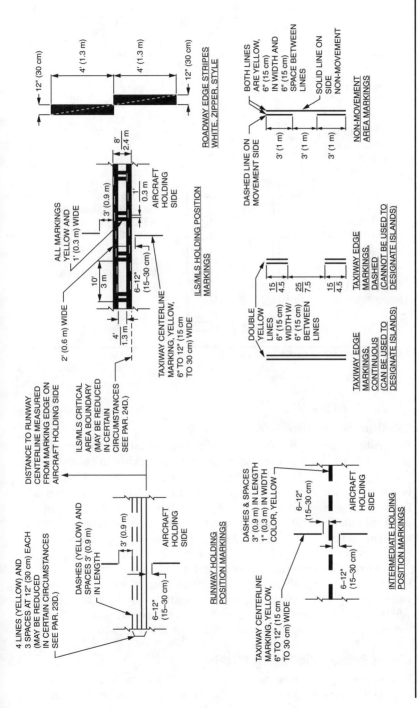

FIGURE 8-26 Taxiway hold short and edge markings.

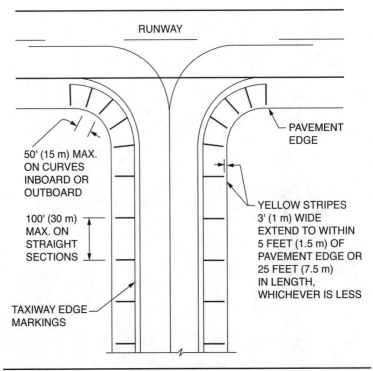

RUNWAY

PAVEMENT EDGE

50' (15 m) MAX.
ON CURVES
INBOARD OR
OUTBOARD

100' (30 m)
MAX. ON
STRAIGHT
SECTIONS

YELLOW STRIPES
3' (1 m) WIDE
EXTEND TO WITHIN
5 FEET (1.5 m) OF
PAVEMENT EDGE OR
25 FEET (7.5 m)
IN LENGTH,
WHICHEVER IS LESS

TAXIWAY EDGE
MARKINGS

Figure 8-27   Taxiway shoulder markings.

Enhanced Taxiway Markings

Beginning in 2008, all airports serving commercial air carriers are required to mark certain critical areas of the airfield with enhanced taxiway markings. These markings are designed to provide additional guidance and warning to pilots of runway intersections. Enhanced markings consist primarily of yellow-painted lines, using paint mixtures with imbedded glass beads to enhance visibility. In addition, yellow markings must be marked on top of a darkened black background.

Taxiway centerlines are enhanced for 150 ft from the runway hold-short markings. The centerline enhancements include dashed yellow lines 9 ft in length, separated longitudinally by 3 ft. These yellow lines are placed 6 in from each end of the existing centerline. An example of enhanced marking is illustrated in Fig. 8-28.

Closed Runway and Taxiway Markings

When runways or taxiways are permanently or temporarily closed to aircraft, yellow crosses are placed on these trafficways. For permanently

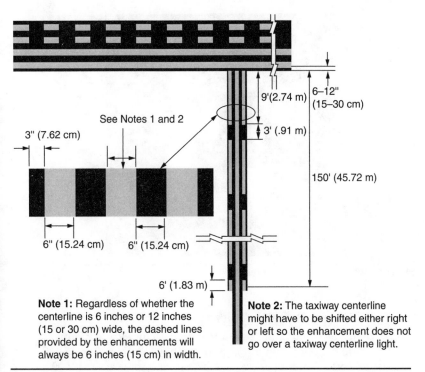

9'(2.74 m)

**6–12"
(15–30 cm)**

See Notes 1 and 2

3" (7.62 cm)

3' (.91 m)

150' (45.72 m)

6" (15.24 cm)   **6" (15.24 cm)**

6' (1.83 m)

Note 1: Regardless of whether the centerline is 6 inches or 12 inches (15 or 30 cm) wide, the dashed lines provided by the enhancements will always be 6 inches (15 cm) in width.

Note 2: The taxiway centerline might have to be shifted either right or left so the enhancement does not go over a taxiway centerline light.

FIGURE 8-28   Example of enhanced taxiway markings.

closed runways, the threshold, runway designation, and touchdown markings are obliterated and crosses are placed at each end and at 1000 ft intervals. For temporarily closed runways, the runway markings are not obliterated, the crosses are usually of a temporary type and are only placed at the runway ends. For permanently closed taxiways, a cross is placed on the closed taxiway at each entrance to the taxiway. For temporarily closed taxiways barricades with orange and white markings are normally erected at the entrances.

Airfield Signage

In addition to markings, signage is placed on the airfield to guide and direct pilots and ground vehicle operators to points on the airport. In addition some signage exists to provide the pilots with information regarding their position on the airfield, the distance remaining on a runway, the location of key facilities at the airport, and often informative signage ranging from voluntary procedures to mitigate noise impacts to warnings about nearby security sensitive areas. FAA Advisory Circular 150/53040-18D describes the U.S. federal standards for airport sign systems.

FIGURE 8-29   Runway distance remaining sign.

Runway Distance Remaining Signs

Runway distance remaining signs are placed on the side of a runway and provide the pilot with information on how much runway is left during takeoff or landing operations. These signs are placed at 1000 ft intervals along the runway is a descending sequential order. Normally, these signs consist of white numerals on a black background, as illustrated in Fig. 8-29.

The FAA recommends that the signs be configured in one of three ways [25]. The preferred method of configuration, and the most economical, is to place double-faced signs on only one side of the runway. In this configuration it is recommended that the signs be placed on the left side of the most frequently used direction of the runway. The signs may be placed on the right side of the runway when necessary due to required runway-taxiway separations or due to conflicts between intersecting runways or taxiways. An alternative method is to provide a set of single-faced signs on either side of the runway to indicate the distance remaining when the runway is used in both directions. The advantage of this configuration is that the distance remaining is more accurately reflected when the runway length is not an even multiple of 1000 ft. Another alternative uses double-faced signs on both sides of the runway. The advantage of this method is that the runway distance is displayed on both sides of the runway in each direction which is an advantage when a sign on one side needs to be omitted because of a clearance conflict. When the runway distance is not an even multiple of 1000 ft, one half of the excess distance is added to the distance on each sign on each runway end. For example, if the runway length available is 8250 ft, the last sign is located at

Sign Size	Legend, in (cm)	Face, in (cm)	Installed (max.), in (cm)	Distance from Defined Pavement Edge
4	40 (100)	48 (120)	60 (152)	50–75 (15–22.5)
5	25 (64)	30 (76)	42 (107)	20–35 (6–10.5)

TABLE 8-6   Runway Distance Remaining Sign Heights and Location Distances

a distance of 1000 plus 125 ft from the end of the runway. A tolerance of ±50 ft is allowed for the placement of runway distance remaining signs. These signs should be illuminated anytime the runway edge lights are illuminated. The recommended sizes and placement of these signs is given in Table 8-6.

Taxiway Guidance Sign System

The primary purpose of a taxiway guidance sign system is to aid pilots in taxiing on an airport. At controlled airports, the signs supplement the instructions of the air traffic controllers and aid the pilot in complying with those instructions. The sign system also aids the air traffic controller by simplifying instructions for taxiing clearances, and the routing and holding of aircraft. At locations not served by air traffic control towers, or for aircraft without radio contact, the sign system provides guidance to the pilot to major destinations areas in the airport.

The efficient and safe movement of aircraft on the surface of an airport requires that a well-designed, properly thought-out, and standardized taxiway guidance sign system is provided at the airport. The system must provide the pilot with the ability to readily determine the designation of any taxiway on which the aircraft is located, readily identify routings to a desired destination on the airport property, indicate mandatory aircraft holding positions, and identify the boundaries for aircraft approach areas, instrument landing system critical areas, runway safety areas and obstacle free zones. It is virtually impossible, except for holding position signs, to completely specify the locations and types of signs that are required on a taxiway system at a particular airport due to the wide variation in the types of functional layouts for airports. The ICAO also publishes recommendations relative to surface movement guidance and control systems [2, 15].

Taxiway Designations

Taxiway guidance sign systems are in a large part based on a system of taxiway designators which identify the individual taxiway components. While runway designators are based on the magnetic heading of the runway, taxiway designators are assigned based on an alphabetic ordering system, independent of the taxiways direction of movement. Taxiways are typically identified in alphabetic order from east to west or north to south (i.e., the northern or easternmost taxiway would be designated "A", the next southern or western taxiway would be designated "B," and so forth). Entrance and exit taxiways perpendicular to main parallel taxiways are designated by the letter of the main parallel taxiway from which they spur, followed by a numeric sequence. For instance, the northernmost entrance taxiway off of taxiway "A" would be designated "A1," and so forth. The letters "I" and "O" are not used as taxiway designators due to their similarity in form to the numbers "1" and "0." In addition the letter "X" is not used as a taxiway designator due to its similarity to a closed runway marking. An example taxiway designation scheme is illustrated in Fig. 8-30.

The taxiway guidance sign system consists of four basic types of signs: *mandatory instruction* signs, which indicate that aircraft should not proceed beyond a point without positive clearance, *location* signs, which indicate the location of an aircraft on the taxiway or runway system and the boundaries of critical airfield surfaces, *direction* signs, which identify the paths available to aircraft at intersections, and *destination* signs, which indicate the direction to a particular destination.

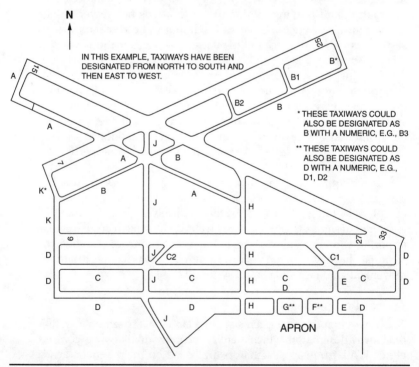

FIGURE 8-30 Example of taxiway designation scheme.

Types of Taxiway Signs

Mandatory Instruction Signs

Mandatory instruction signs denote an entrance to a runway, critical area, or prohibited area. They are used for holding positions signs for runway-taxiway and runway-runway intersections (Fig. 8-31a), instrument landing system critical areas (Fig. 8-31b), runway approach areas (Fig. 8-31c), ILS category II/III critical areas (Fig. 8-31d), and to designate areas for which entry is prohibited by aircraft (Fig. 8-31e). These signs have white inscriptions on a red background and are installed on the left side of the runway or taxiway. In some cases runway-taxiway intersections require a sign on both sides of the taxiway. This includes situations on taxiways which are at least 150 ft wide, where the painted holding line extends across an adjacent holding bay, where the painted holding line markings do not extend straight across the taxiway, and where the painted holding line markings are located a short distance from an intersection with another taxiway. Generally arrows are not permitted on mandatory instruction signs unless they are necessary at the taxiway-runway-runway intersections to indicate directions to these runways. For runway designation these signs normally contain both designations of the runway and the designation on the left is for the runway to the left and the designation on the right is for the runway to the right.

At controlled airports aircraft and ground vehicles are required to hold at these points unless cleared by air traffic control. At uncontrolled airports, these signs are intended to indicate travel beyond these signs is permitted only after appropriate precautions have been taken.

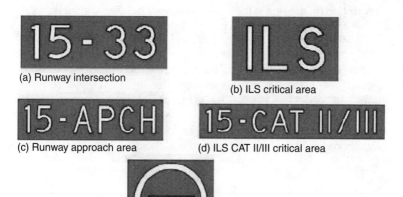

(a) Runway intersection

(b) ILS critical area

(c) Runway approach area

(d) ILS CAT II/III critical area

(e) No entry

FIGURE 8-31 Mandatory instruction signs.

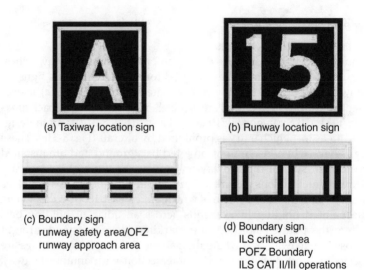

(a) Taxiway location sign

(b) Runway location sign

(c) Boundary sign
runway safety area/OFZ
runway approach area

(d) Boundary sign
ILS critical area
POFZ Boundary
ILS CAT II/III operations

FIGURE 8-32   Location signs.

Location Signs

Location signs are used to identify the taxiway or runway on which an aircraft is located (Fig. 8-32a and 8-32b). These signs consist of a yellow inscription and border on a black background. Location signs are also used to identify the boundary of the runway safety area or obstacle-free zones (Fig. 8-32c), or the instrument landing system critical area (Fig. 8-32d) for a pilot exiting a runway. In the latter cases the signs consist of a black inscription and border on a yellow background and the inscription on the sign is the same as relevant holding line marking.

Direction Signs

Direction signs are used to indicate the direction of other taxiways leading out of an intersection. These signs are used as taxiway direction sign (Fig. 8-33) and runway exit sign. The signs have black inscriptions and borders on a yellow background and always contain arrows. The arrows are oriented in the approximate direction of the turn required. These signs should not be located with holding position signs and should not be located between the holding line and the runway.

FIGURE 8-33   Taxiway direction sign.

FIGURE 8-34   Inbound destination sign (to military facility).

Signs used to indicate the direction of taxiways on the opposite side of the runway should be located on the opposite side of the runway. Runway exit signs should be located prior to the exit on the side of the runway on which the aircraft is expected to exit. If the taxiway crosses the runway and the aircraft could be expected to exit on either side, then a runway exit sign should be installed on either side of the runway.

Destination Signs

Destination signs have black inscriptions on a yellow background and always contain arrows. These signs indicate the general direction to a remote location at the airport, such as an inbound destination (Fig. 8-34), and are generally not required where taxiway direction signs are used. Outbound destination signs are used to identify directions to the takeoff runways. These routes normally begin at the entrance to a taxiway from the apron area. More than one runway number may be used, separated by a dot, if the route is common to more than one runway (Fig. 8-35). Inbound destination signs are often used to indicate the general direction to major airport facilities such as passenger terminal aprons, cargo areas, military aprons, or general aviation facilities. These signs should consist of a minimum of three letters to avoid confusion with taxiway guidance signs.

The typical legends found on taxiway destination signs are:

APRON—general parking, servicing, and loading areas

FUEL—areas where aircraft are fueled or serviced

TERM—gate positions at which aircraft are loaded or unloaded

CIVIL—areas set aside for civil aircraft

MIL—areas set aside for military aircraft

PAX—areas set aside for passenger handling

CARGO—areas set aside for cargo handling

INTL—areas set aside for handling international flights

FBO—fixed-base operator

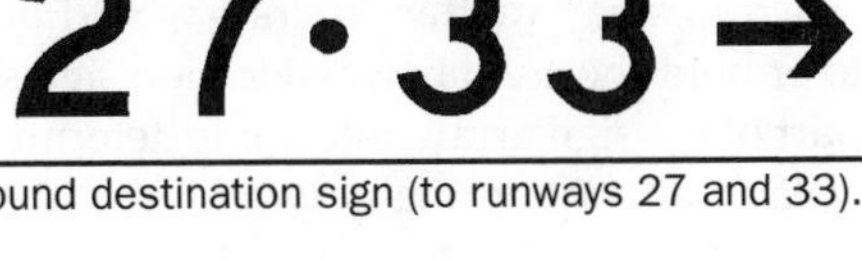

27·33 →

FIGURE 8-35   Outbound destination sign (to runways 27 and 33).

FIGURE 8-36 Taxiway ending marker.

Information Signs

Other types of signs may be necessary on the airfield which are not part of the taxiway guidance systems described before. These signs are called information signs and might be used, for example, to indicate a noise abatement procedure to a pilot ready to takeoff on a specific runway. These signs should have black inscriptions on a yellow background. These signs are not required to be lighted.

Taxiway Ending Sign

The sign system does not provide a sign to indicate that a taxiway does not continue beyond an intersection. A frangible, retroreflective sign should be installed on the far side of the intersection if normal visual cues such as marking and lighting are inadequate. This sign is marked with alternating black and yellow diagonal stripes, as illustrated in Fig. 8-36.

The FAA recommends that the following guidelines be applied when designing a taxiway guidance sign system [25]:

1. A holding position sign and taxiway location sign should be installed at the holding position on any taxiway that provides access to a runway.

2. A holding position sign should be installed on any taxiway at the boundary of the instrument landing system critical area or the runway approach area when it is necessary to protect the navigational signal, airspace, or safety area for a runway. This sign should be placed at the entrance to and the exit from such areas.

3. A holding position sign should be installed on any runway where that runway intersects another runway.

4. A sign array consisting of taxiway direction signs should be installed prior to each intersection between taxiways if an aircraft would normally be expected to turn at or hold short of the intersection. The direction sign in the array should include a sign panel, consisting of a taxiway designation and an arrow, for each taxiway that an aircraft would be expected to turn onto or hold short. A taxiway location sign should be included as part of the sign array unless it is determined to be unnecessary. If an aircraft normally would not be expected to turn at

or hold short of the intersection, the sign array is not needed unless the absence of guidance would cause confusion.

5. A runway exit sign identifying the exit taxiway should be installed along each runway for each normally used runway exit.

6. Destination signs may be substituted for direction signs at the intersection between taxiways or for runway exit signs at uncontrolled airports.

7. Standard highway stop signs should be installed on ground vehicle roadways at the intersection of each roadway with a runway or taxiway. For roadway intersections with taxiways, a standard highway yield sign may be used instead of the stop sign.

8. Additional signs should be installed on the airfield where necessary to eliminate confusion or to provide confirmation relative to location.

Signing Conventions

The FAA recommends the following signing conventions [25]:

1. Signs should be placed on the left side of the taxiway as viewed by the pilot of an approaching aircraft. If signs are installed on both sides of the taxiway at the same location, the sign faces should be identical. Signs may be placed on the right side of the taxiway when necessary to meet clearance requirements or where it is impractical to install them on the left side because of terrain features or conflicts with other objects.

2. Some signs may be installed on the back of other signs even though this may result in the sign being on the right side of the taxiway. Signs which may be installed in this manner include

   a. Runway safety area, obstacle-free zone area, and runway approach area boundary signs may be installed on the back of taxiway-runway intersection and runway approach area holding position signs.

   b. Instrument landing system critical area boundary signs may be installed on the back of instrument landing system critical area holding position signs.

   c. Taxiway location signs, when installed on the far side of the intersection, may be installed on the back of direction signs.

   d. Taxiway location signs may be installed on the back of holding position signs.

   e. Destination signs may be installed on the back of direction signs on the far side of intersections when the destination referred to is straight ahead.

3. Taxiway location signs installed in conjunction with holding position signs for taxiway-runway intersections should always be installed outboard of the holding position sign.

4. Location signs are normally included as part of a direction sign array located prior to the taxiway intersection. Except for the intersection of two taxiways, the location sign is placed in the array so that the designations for all turns to the left would be located to the left side of the location sign and designations for all turns to the right or straight ahead are located to the right of the location sign.

5. All direction signs have arrows. Arrows on signs should be oriented toward the approximate direction of the turn. Each designation appearing in the array of direction signs should only have one arrow. An exception is when the taxiway intersection comprises only two taxiways and then the direction sign for the taxiway may have two arrows.

6. Destination signs should be located in advance of intersections and should not be collocated with other signs. These may also be installed on the far side of the intersection when the taxiway does not continue and direction signs are provided prior to the intersection.

7. Information signs should not be collocated with mandatory, location, direction, or destination signs.

8. Each designation and its associated arrow included in the array of direction signs or destination signs should be delineated from the other designations in the array by a black vertical border.

Sign Size and Location

Taxiway guidance signs are available in three heights as indicated in Table 8-7. The choice of a particular size sign involves several factors including effectiveness, aircraft clearance, jet blast, and snow removal

Sign Size	Legend, in (cm)	Face, in (cm)	Installed (max.), in (cm)	Perpendicular Distance from Defined Taxiway/ Runway Edge to Near Side of Sign, ft (m)
1	12 (30)	18 (46)	30 (76)	10–20 (3–6)
2	15 (38)	24 (61)	36 (91)	20–35 (6–10.5)
3	18 (46)	30 (76)	42 (107)	35–60 (10.5–18)

TABLE 8-7   Taxiway Signage Dimensional Specifications

Aircraft Approach Category and (Airplane Design Group)	Visual and Nonprecision Instrument Runway	Precision Instrument Runway
A and B (I and II) small airplanes only	125 (38)	175 (53)
A and B (I, II, and III)	200 (60)	250 (75)
A and B (IV)	250 (75)	250 (75)
C and D (I through IV)	250 (75)	250 (75)
C and D (V)	250 (75)	280 (85)
C and D (VI)	250 (75)	280 (85)

Perpendicular distance from runway centerline to intersection runway/taxiway centerline is in feet (meters).

TABLE 8-8 Location Distances for Holding Position Markings

operations. Normally, the larger the sign and the closer it is located to the runway or taxiway edge the more effective it is. However, aircraft clearance requirements and jet blast effects require smaller signs when located near the pavement edges, whereas effectiveness requires larger signs when located at further distances. The effects of snow removal operations on the signs should be considered in the choice of sign size and location. The sign used must provide 12 in of clearance between the top of the sign and any part of the most critical aircraft using or expected to use the airport when the wheels of the aircraft are at the defined pavement edge.

The distances shown in Table 8-8 should be used in determining runway holding positions. All signs in an array should be of the same size and at the same height above the ground.

For determining sign locations with respect to intersecting runways, the clearance requirements to other moving aircraft, as given in Table 8-9, should be used. For signs installed at holding positions the signs should be in line with the holding line markings within a tolerance

Airplane Design Group I	Airplane Design Group II	Airplane Design Group III	Airplane Design Group IV	Airplane Design Group V	Airplane Design Group VI
44.5 ft (13.5 m)	65.5 ft (20 m)	93 ft (28.5 m)	129.5 ft (39.5 m)	160 ft (48.5 m)	193 ft (59 m)

TABLE 8-9 Perpendicular Distances for Taxiway Intersection Markings from Centerline of Crossing Taxiway

of 10 ft. Where there is no operational need for taxiway holding line markings the signs may be installed in the area from the taxiway point of tangency to the location where the holding line markings would be installed [25].

Typical locations for taxiway guidance signs are shown in Fig. 8-37. An illustration of the required signs and their placement for a basic airport layout is given in Fig. 8-38 [25].

Sign Operation

Holding positions signs for runways, instrument landing system critical areas, approach areas, and their associated taxiway location signs should be illuminated when the associated runway lights are illuminated. Other taxiway signs should be illuminated when the associated taxiway lights are illuminated.

The installation of retroreflective markings is not mandatory. However, it is quite economical, especially at airports where lights cannot be justified because of the volume or nature of air traffic [12]. The marking is very similar to that used successfully on highways for many years.

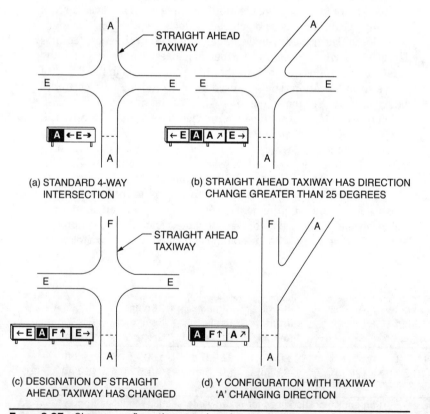

FIGURE 8-37   Signage configuration at taxiway intersections.

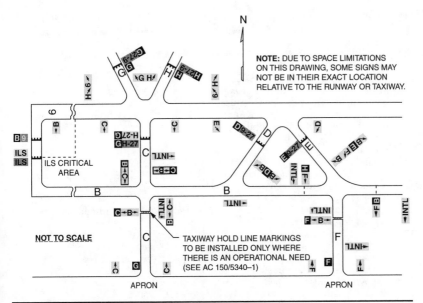

FIGURE 8-38 Typical layout of airfield signage.

References

1. *Aerodromes, Annex 14 to the Convention on International Civil Aviation*, Vol. 1, *Aerodrome Design and Operations*, International Civil Aviation Organization, Montreal, Canada, July 1990.
2. *Aerodrome Design Manual, Part 4, Visual Aids,* 2d ed., International Civil Aviation Organization, Montreal Canada, 1983.
3. "*Airport Approach*, Runway and Taxiway Lighting Systems," E. C. Walter, *Journal of the Air Transport Division*, Vol. 84, No. AT1, American Society of Civil Engineers, New York, N.Y., June 1958.
4. *Airport Design*, Advisory Circular, AC 150/5300-13, Federal Aviation Administration, Washington, D.C., Change 15, 2009.
5. *Airport Miscellaneous Lighting Visual Aids*, Advisory Circular, AC 150/5340-21, Federal Aviation Administration, Washington, D.C., 1971.
6. *Airway Planning Standard Number One—Terminal Air Navigation Facilities and Air Traffic Control Services*, FAA Order 7031.2B, Federal Aviation Administration, Washington, D.C., 1976.
7. "Aviation Ground Lighting for All-Weather Operation," M. Latin, *Airport Forum*, Vol. 7, No. 1, February 1977.
8. *Comparison Between ICAO Annex 14 Standards and Recommended Practices and FAA Advisory Circulars*, Document No. D6-58344, Boeing Commercial Airplane Company, Seattle, Wash., 1979.
9. *Economy Approach Lighting Aids*, Advisory Circular, AC 150/5340-14B, with Changes 1 and 2, Federal Aviation Administration, Washington, D.C., 1970.
10. *Establishment Criteria for Runway End Identification Lights (REIL)*, Report No. FAA-ASP-79-4, Federal Aviation Administration, Washington, D.C., 1979.
11. *Establishment Criteria for Visual Approach Slope Indicator (VASI)*, Report No. FAA-ASP-76-2, Federal Aviation Administration, Washington, D.C., 1977.
12. *FAA Specification L-853, Runway and Taxiway Retroreflective Markers*, Advisory Circular, AC 150/5345-39B, Federal Aviation Administration, Washington, D.C., 1980.

13. *Installation Criteria for the Approach Lighting System Improvement Program (ALSIP)*, Report No. FAA-ASP-78-5, Federal Aviation Administration, Washington, D.C., 1978.

14. *Installation Details for Runway Centerline and Touchdown Zone Lighting Systems*, Advisory Circular, AC 150/5340-4C, with Changes 1 and 2, Federal Aviation Administration, Washington, D.C., 1978.

15. *Manual of Surface Movement Guidance and Control Systems*, Document No. 9476, International Civil Aviation Organization, Montreal, Canada, 1986.

16. *Marking and Lighting of Unpaved Runways*, V. F. Dosch, NAFEC Technical Letter Report, NA-78-34-LR, Federal Aviation Administration, Technical Center, Atlantic City, N.J., 1978.

17. *Standards for Airport Markings*, Advisory Circular, AC 150/5340-1J, Federal Aviation Administration, Washington, D.C., 1995. Change 1, 2008.

18. *Obstruction Marking and Lighting*, Advisory Circular, AC 70/7460-1H, Federal Aviation Administration, Washington, D.C., 1991.

19. *Proposed Construction or Alteration of Objects That May Affect Navigable Airspace*, Advisory Circular, AC 70/7460-2I, Federal Aviation Administration, Washington, D.C., 1988.

20. *Precision Approach Path Indicator (PAPI) Systems*, Advisory Circular, AC 150/5345-28D, Federal Aviation Administration, Washington, D.C., 1985.

21. *Runway and Taxiway Edge Lighting System*, Advisory Circular, AC 150/5340-24, Federal Aviation Administration, Washington, D.C., 1975; and *Design and Installation Details for Airport Visual Aids*, Advisory Circular, AC 150/5340-30C, Federal Aviation Administration, Washington, D.C., 2007.

22. *Runway Visual Range (RVR)*, Advisory Circular, AC 97-1A, Federal Aviation Administration, Washington, D.C., 1977.

23. *Segmented Circle Airport Marker System*, Advisory Circular, AC 150/5340-5B, Federal Aviation Administration, Washington, D.C., 1984.

24. *Specifications for Taxiway and Runway Signs*, Advisory Circular, AC 150/5345-44E, Federal Aviation Administration, Washington, D.C., 1991.

25. *Standards for Airport Sign Systems*, Advisory Circular, AC 150/5340-18D, Federal Aviation Administration, Washington, D.C., 2004.

26. *Taxiway Centerline Lighting System*, Advisory Circular, AC 150/5340-19, Federal Aviation Administration, Washington, D.C., 1968.

27. "The Theory of Visual Judgements in Motion and Its Application to the Design of Landing Aids for Aircraft," E. S. Calvert, *Transactions of the Illuminating Engineering Society*, Vol. 22, No. 10, London, England, 1957.

28. "Working Papers," A. E. Jenks, *International Air Transport Association*, Special Meeting on Visual Aids to Flare and Landing, Amsterdam, Netherlands, November 14–22, 1955.

CHAPTER 9

Airport Drainage

An adequate drainage system for the removal of surface and subsurface water is vital for the safety of aircraft and for the longevity of the pavements. Improper drainage results in the formation of puddles on the pavement surface, which can be hazardous to aircraft taking off and landing. Poor drainage can also result in the early deterioration of pavements. Flat longitudinal and transverse grades and wide pavement surfaces often pose difficulties in making provision for adequate drainage at airports.

The material in this chapter is principally concerned with estimating the amounts of surface and subsurface runoff and not with the hydraulics of pipes or details of installation. These latter items are adequately covered in texts on hydraulics and literature provided by pipe manufacturers.

The FAA and the Corps of Engineers have developed most of the information on airport drainage in the United States and the material presented in this has been drawn from their work. In 2006, several agencies worked together to combine existing surface drainage topics covered in several manuals into one Unified Facilities Criteria (UFC) document. The resulting manual [1] now serves as the design and analysis standard for surface drainage for the FAA.

Purpose of Drainage

The functions of an airport drainage system are as follows:

1. Interception and diversion of surface and groundwater flow originating from lands adjacent to the airport
2. Removal of surface runoff from the airport
3. Removal of subsurface flow from the airport

In very few cases will the natural drainage on a site be sufficient by itself to satisfy these functions; consequently artificial drainage must be installed.

Design Storm for Surface Runoff

The selection of the severity of the storm which the drainage system should accommodate involves economic consideration. An extremely

343

severe storm occurring very infrequently would undoubtedly cause some damage if the system were designed for a storm of lesser severity. However, if serious interruptions in traffic are not anticipated, a system designed for the larger storm may not be economically justified. Taking these factors into account, the FAA recommends that for civil airports the drainage system be designed for a storm whose probability of occurrence is once in 5 years [2]. The design should, however, be checked with a storm of lesser frequency (10 to 15 years) to ascertain if serious damage or interruption of traffic would result from such a storm. Drainage for military airfields is based on a 2-year storm frequency [8].

Ordinarily no ponding is permitted on paved surfaces, but in the intervening areas ponding is permitted, provided it will not result in undesirable saturation of the subgrades underneath the pavements.

Determining the Intensity-Duration Pattern for the Design Storm

The determination of the amount of rainfall which can be expected at the site of the airport is the first step in the design of a drainage system. Rainfall intensity is expressed in inches per hour for various durations of a particular storm. The expected frequency of occurrence is also an important factor to consider. The severity of storms is related to their frequencies; a storm which is expected to occur once in 100 years will be more severe than one having a frequency of occurrence of once in 5 years.

David L. Yarnell of the U.S. Department of Agriculture conducted extensive investigations concerning rainfall intensities, durations, and frequencies throughout the United States [16]. West of the 105th meridian, where the Yarnell information is not as complete, the National Weather Service has compiled rainfall data which appear in Refs. 12 to 14.

Yarnell developed rainfall intensities for 5-, 10-, 15-, 30-, 60-, and 120-min durations for a storm which can be expected to occur once in 5 years and the intensities for a 1-h duration for storms whose expected frequencies of occurrence are once in 2, 5, 10, 25, 50, and 100 years. The intensities for a duration of 1 h for frequencies of 2, 5, 10, 25, 50, and 100 years are shown in Fig. 9-1.

The 1-h intensity does not by itself portray the intensity-duration pattern of a storm. The Corps of Engineers made extensive studies of rainfall patterns in the United States and found that irrespective of frequency, the intensity-duration patterns of storms were largely governed by their 1-h intensities. That is, two storms of different frequency of occurrence whose 1-h intensities are equal will have similar intensity-duration patterns. This is shown in Fig. 9-2. For example, if the 1-h intensities of storms whose frequencies were 5, 10, or 15 years were all exactly 2.0 in/h, the intensity-duration patterns would be expected to follow the pattern indicated by the curve labeled 2.0.

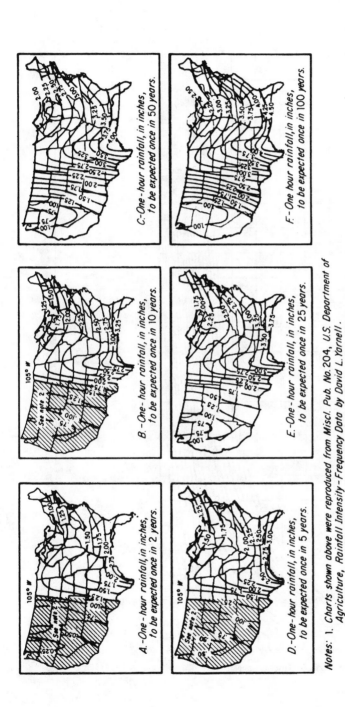

Figure 9-1 One-hour rainfall intensities for the United States (Corps of Engineers).

C.–One-hour rainfall, in inches, to be expected once in 50 years.

F.–One hour rainfall, in inches, to be expected once in 100 years.

B.–One-hour rainfall, in inches, to be expected once in 10 years.

E.–One-hour rainfall, in inches, to be expected once in 25 years.

A.–One-hour rainfall, in inches, to be expected once in 2 years.

D.–One-hour rainfall, in inches, to be expected once in 5 years.

Notes: 1. Charts shown above were reproduced from Miscl. Pub. No. 204, U.S. Department of Agriculture, Rainfall Intensity-Frequency Data by David L. Yarnell.

2. For the Western part of the U.S. West of the 105th Meridian, see One Hour Rainfall Data in Weather Bureau Technical Paper No. 24.

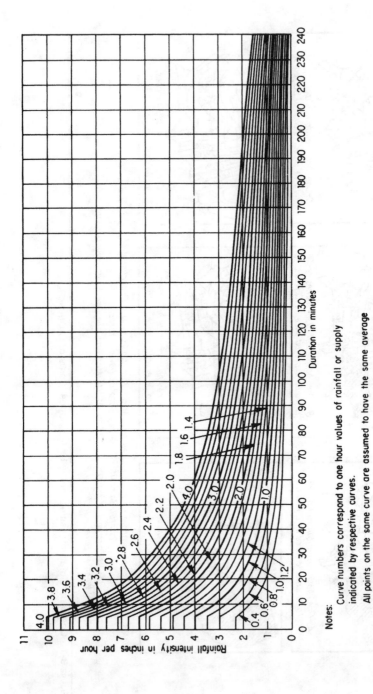

Figure 9-2 Rainfall intensity-duration curves (*Corps of Engineers*).

Notes:

Curve numbers correspond to one hour values of rainfall or supply indicated by respective curves.

All points on the same curve are assumed to have the same average frequency of occurrence.

If the drainage system is to be designed for a storm whose expected frequency of occurrence is once in 5 years and if detailed data concerning the intensity-duration pattern are nonexistent, the pattern can be approximated from Fig. 9-2, provided the 1-h intensity is known. It goes without saying that if sufficient rainfall data are available at an airport site, the intensity-duration frequency data should be developed from this information rather than from other sources. Rarely, however, does a site have such complete rainfall information.

Determining the Amount of Runoff by the FAA Procedure

The FAA analysis of airport surface drainage revolves about the solution of the *rational method* expression

$$Q = CIA \tag{9-1}$$

where Q = runoff from given drainage basin, ft^3/s
C = ratio of runoff to rainfall
I = rainfall intensity for time of concentration of runoff, in/h
A = drainage area, acres

Examples and charts illustrating the FAA procedure for design have been taken largely from the FAA [2].

Time of Concentration

The *time of concentration* is defined as the time taken by water to reach the drain inlet from the most remote point in the tributary area. The *most remote point* refers to the point from which the time of flow is the greatest. The time of concentration is usually divided into two components: inlet time and time of flow. The *inlet time* is the time required for water to flow overland from the most remote point in the drainage area to the inlet. The *time of flow* is the time taken by the water to flow from the drain inlet through the pipes to the point in the system under consideration. Sometimes the inlet time will be the time of concentration; at other times the time of concentration will be the sum of the inlet time and time of flow.

The time of flow can be computed by the use of well-established hydraulic formulas. The inlet time is obtained largely empirically from the relationship

$$D = kT^2 \tag{9-2}$$

Where D = distance, ft
T = time, min
k = dimensional empirical factor which is dependent on slope, roughness of terrain, extent of vegetative cover, and distance to drain inlet

Inlet times can be estimated from Fig. 9-3.

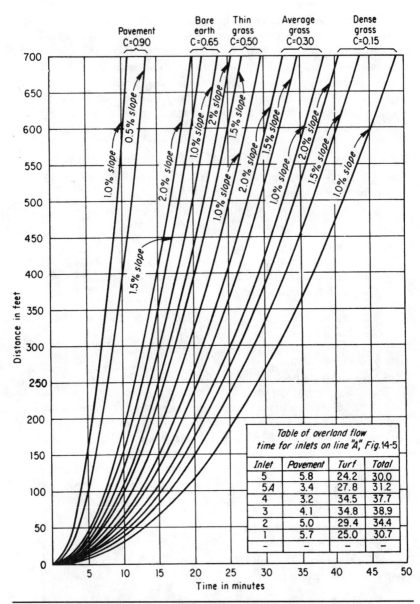

FIGURE 9-3   Inlet time curves (*Federal Aviation Administration [2]*).

Coefficient of Runoff

Application of the rational method requires the exercise of considerable judgment on the part of the engineer. The runoff rate is variable from storm to storm and varies even during a single period of precipitation. The coefficient of runoff depends on antecedent storm conditions, slope and type of surface, and extent of the drainage area. The range of values suggested by the FAA is indicated in Table 9-1.

Types of Surfaces	Factor C
For all watertight roof surfaces	0.75–0.95
For asphalt runway pavements	0.80–0.95
For concrete runway pavements	0.70–0.90
For gravel or macadam pavements	0.35–0.70
For impervious soils (heavy)*	0.40–0.65
For impervious soils with turf*	0.30–0.55
For slightly pervious soils*	0.15–0.40
For slightly pervious soils with turf*	0.10–0.30
For moderately pervious soils*	0.05–0.20
For moderately pervious soils with turf*	0.00–0.10

*For slopes from 1 to 2 percent.
Source: Federal Aviation Administration [2].

TABLE 9-1 Coefficients of Runoff C

For drainage basins consisting of several types of surfaces with different infiltration characteristics, the weighted runoff coefficient should be computed in accordance with

$$C = \frac{A_1 C_1 + A_2 C_2 + A_3 C_3}{A_1 + A_2 + A_3} \tag{9-3}$$

Typical Example—No Ponding

In order that the worst conditions attendant upon the design storm may be used in the design of the pipe system, a separate duration of storm is selected for each subdrainage area tributary to a drain inlet. The duration of the storm is made equal to the sum of the inlet time and time of flow.

Each reach of pipe must be designed to carry the discharge from the inlet at its upstream end plus the contribution from all preceding inlets. For economy of construction, the grade of each reach is determined largely by topography. A minimum mean velocity on the order of 2.5 ft/s should be maintained to provide scouring action so that reduction of the pipe area due to silting will not be a problem.

To clarify the computation of runoff by the FAA method, the following example is presented.

The intensity-duration rainfall pattern for a 5-year-frequency storm at the site of the proposed airport is shown in Fig. 9-4. The layout of the drains on a portion of the airport is shown in Fig. 9-5. Design data for establishing inlet times and coefficients of runoff

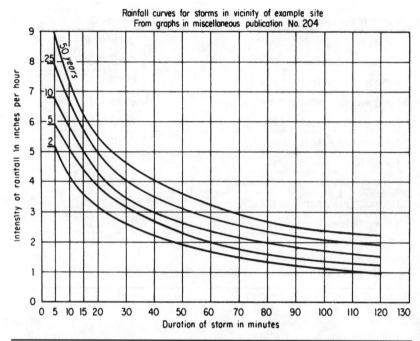

FIGURE 9-4   Intensity-duration rainfall pattern for design storm (*Federal Aviation Administration [2]*).

for drainage areas tributary to drain line *A*, shown in Fig. 9-5, are tabulated in Table 9-2. It is assumed that the coefficients of runoff for pavement and for turf are 0.90 and 0.30, respectively.

From these data, inlet times have been computed by the use of Fig. 9-3 on the basis that the slope of the pavement is 1 percent and the slope of the turfed area is 1.5 percent. The inlet times for this specific problem are shown in Fig. 9-3. The computations for runoff, assuming no ponding, are shown in Table 9-3.

Typical Example—Ponding

In the design of an airfield drainage system, ponding may be used to effect a reduction in the cost of installation. Ponding is simply a means of providing temporary storage of runoff prior to its entry into the underground system. For purposes of design computation, the ponded volume may be assumed to be an inverted pyramid or a truncated pyramid, the height of which is the depth of water above the inlet at any stage. The area of the base of the pyramid is taken as the surface area of the pond. If ponding were permitted, the layout of the drainage system might be as shown in Fig. 9-6. The most remote point to one of the inlets is 950 ft, comprising 100 ft of pavement and 850 ft of turf. The time of concentration is estimated at 4 + 54 = 58 min. The complete

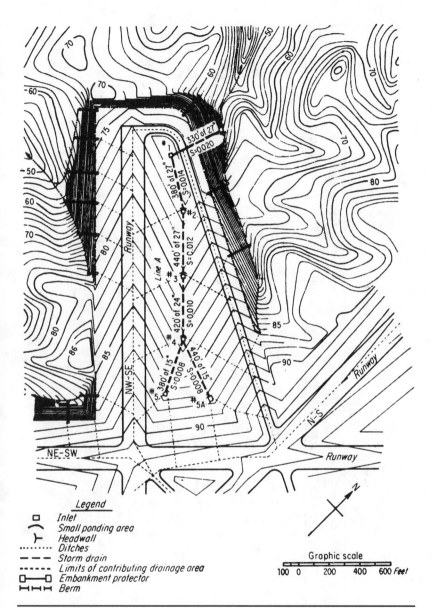

FIGURE 9-5 Portion of airport showing drainage design details (*Federal Aviation Administration [2]*).

drainage area is 31.42 acres, of which 6.44 acres is paved. Assuming that the coefficients of runoff for pavement and turf are 0.90 and 0.30, respectively, the combined C is 0.423. From Fig. 9-4 the rainfall intensities for durations of 5, 10, 15, 20, 30, 60, 90, 120, and 180 min are obtained, and the volumes of runoff are computed as shown in Table 9-4.

Inlets	Tributary Area to Inlets, acres				Distance Remote Point to Inlet, ft				
	Pavement	Turf	Both	Subtotal	Pavement	Turf	Total	Line Segment	Length ft
5	1.27	1.05	2.32	2.32	200	340	540	5-4	380
5A	1.02	1.86	2.88	2.88	70	450	520	5A-4	440
4	1.40	7.35	8.75	13.95	60	690	750	4-3	420
3	0.78	6.46	7.24	21.19	100	700	800	3-2	440
2	0.83	4.56	5.39	26.58	150	500	650	2-1	380
1	1.14	3.70	4.84	31.42	190	360	550	1-outlet	330
Outlet	—	—	—	—	—	—	330		
Total	6.44	24.98	31.42						

Weighted Average for C		
To inlet 5:	To inlet 5A:	To inlet 4:
$\dfrac{1.27}{2.32}(0.90) = 0.49$ $\dfrac{1.05}{2.32}(0.30) = \underline{0.14}$ $C = 0.63$	$\dfrac{1.02}{2.88}(0.90) = 0.32$ $\dfrac{1.86}{2.88}(0.30) = \underline{0.19}$ $C = 0.51$	$\dfrac{1.40}{8.75}(0.90) = 0.14$ $\dfrac{7.35}{8.75}(0.30) = \underline{0.25}$ $C = 0.39$
To inlet 3:	To inlet 2:	To inlet 1:
$\dfrac{0.78}{7.24}(0.90) = 0.10$ $\dfrac{6.46}{7.24}(0.30) = \underline{0.27}$ $C = 0.37$	$\dfrac{0.83}{5.39}(0.90) = 0.14$ $\dfrac{4.56}{5.39}(0.90) = \underline{0.25}$ $C = 0.39$	$\dfrac{1.14}{4.84}(0.90) = 0.21$ $\dfrac{3.70}{4.84}(0.30) = \underline{0.23}$ $C = 0.44$

Source: Federal Aviation Administration [2].

TABLE 9-2 Design Data for Line *A* in Fig. 9-5

	Line Segment	Length of Segment, ft	Inlet Time, min	Flow Time, min	Time of Concentration, min	Runoff Coefficient C	Rainfall Intensity I	Tributary Area A, acres	Remarks
Inlet									
	5A-4	440	31.2	1.8	31.2	0.51	3.10	2.88	n = 0.015
	5-4	380	30.0	1.6	30.0	0.63	3.15	2.32	
	4-3	420	37.7	1.1	37.7	0.39	2.80	8.75	See accumulated runoff computed below.
	3-2	440	38.9	1.0	38.9	0.37	2.70	7.24	Accumulated runoff adjustment negligible.
	2-1	380	34.4	0.8	39.9	0.39	2.65	5.39	
	1-outlet	330	30.7	0.5	40.7	0.44	2.60	4.84	
Outlet									

Calculation of Example

Maximum flow from inlets 5 and 5A will reach inlet 4 in 31.6 and 33.0 min, respectively. All inlet 4 subarea will be contributing to the system only after 37.7 min. Flow from inlets 5 and 5A must be adjusted for 37.7-min time of concentration. Adjusted time of concentration for inlets 5 and 5A (that is, the inlet time for end-of-line structures) equals the flow time through the respective pipe segments.

For inlet 5, adjusted time of concentration = 37.7 − 1.6 = 36.1 min. For inlet 5A, adjusted time of concentration = 37.7 − 1.8 = 35.9 min. By using these adjusted times of concentration, an intensity of rainfall of 2.85 in/h is obtained (slight time difference cannot be read from curves).

Applying these data in the formula $Q = CIA$:

Adjusted flow from inlet 5 = 0.63 × 2.85 × 2.32 = 4.16

Adjusted flow from inlet 5A = 0.51 × 2.85 × 2.88 = 4.18

Flow into inlet 4 from inlets 5 and 5A in 37.7 min = 8.34

Flow from inlet 4 subarea = 9.56

Accumulated flow entering inlet 4 in 37.7 min = 17.90 ft³/s

	Runoff Q ft³/s	Accumulated Runoff, ft³/s	Velocity of Drain, ft/s	Size of Pipe, in	Slope of Pipe, ft/ft	Capacity of Pipe, ft³/s	Invert Elevation	Remarks
Inlet								
5A	4.55	4.55	4.0	15	0.008	5.0	81.52	$n = 0.015$
5	4.60	4.60	4.0	15	0.008	5.0	81.04	
4	9.56	17.90	6.2	24	0.010	20.0	78.00	See accumulated runoff computed below.
3	7.23	25.13	7.4	27	0.012	30.0	73.80	Accumulated runoff adjustment negligible.
2	5.57	30.70	8.0	27	0.014	33.0	68.52	
1	5.54	36.24	9.5	27	0.020	37.5	63.20	
Outlet							56.60	

Source: Federal Aviation Administration [2].

TABLE 9-3 Drainage System Design Data

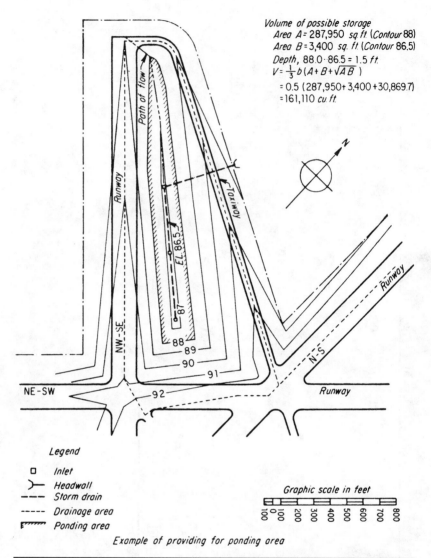

Volume of possible storage
  Area A = 287,950 sq ft (Contour 88)
  Area B = 3,400 sq. ft. (Contour 86.5)
  Depth, 88.0 - 86.5 = 1.5 ft.
  $V = \frac{1}{3}b(A + B + \sqrt{AB})$
      = 0.5 (287,950 + 3,400 + 30,869.7)
      = 161,110 cu ft

Legend

□ Inlet
>— Headwall
- - - Storm drain
- - - - - Drainage area
▟▀▀▀▀ Ponding area

Graphic scale in feet

Example of providing for ponding area

Figure 9-6 Layout of drainage for ponding (*Federal Aviation Administration [2]*).

To visualize the effects of ponding, a comparison is made of the discharge capacity of tentative drainage pipes and the cumulative runoff for the design storm frequency. This comparison is best made as a plot of runoff on the ordinate axis and time on the abscissa. An example of such a plot is shown in Fig. 9-7. The discharge capacity for each of four selected pipe sizes is shown as a straight line. These discharge curves were computed for an assigned slope and roughness

Time, min	Intensity* I	$Q = CIA$, ft³/s	Volume $V = CIAt$, ft³
5	5.80	77.1	23,100
10	4.96	65.9	39,600
15	4.33	57.5	51,800
20	3.95	52.5	63,000
30	3.18	42.3	76,100
60	2.00	26.6	95,700
90	1.62	21.5	116,300
120	1.26	16.7	120,600
180	0.87	11.6	125,000

*Hourly intensities from Fig. 9-4.

TABLE 9-4   Volume of Runoff—Ponding

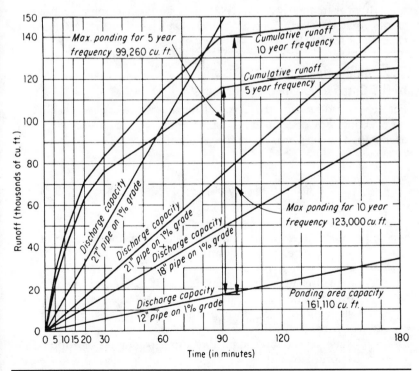

FIGURE 9-7   Cumulative runoff for ponding in Fig. 9-6 (*Federal Aviation Administration [2]*).

coefficient n for each pipe by use of the Manning formula. For this example the pipes were assumed to be concrete with a roughness coefficient n of 0.015 and laid on a 1 percent slope. The discharge in cubic feet per second multiplied by 3600 s is the discharge capacity ordinate in cubic feet at the 60-min abscissa in Fig. 9-7. Each discharge capacity curve must pass through the origin of coordinates, and one point as determined above will define the straight-line relationship.

The significance of the cumulative runoff and discharge capacity curves as plotted in Fig. 9-7 is that the difference in ordinates (cumulative runoff minus discharge capacity) represents the amount of ponding at any instant after the beginning of the storm. The maximum amount of ponding is determined by scaling the largest difference between the cumulative runoff curve and the discharge capacity curve.

It is considered essential that all ponding area edges be kept at least 75 ft from the edges of pavements. In this example, this would mean that the pond should not reach a level above elevation 88.0. The storage capacity below this elevation is 161,000 ft³. If a 12-in-diameter pipe were used, the maximum ponding would amount to 99,260 ft³, considerably less than the available 161,100 ft³. For practical consideration a pipe of lesser diameter is not recommended.

Although not shown in this text, computations were also made for a 10-year-frequency storm. With a 12-in-diameter pipe such a storm would develop a pond of 123,000 ft³, still less than the available capacity of 161,000 ft³.

Determining the Amount of Runoff by the Corps of Engineers Procedure

For determining runoff, the Corps of Engineers uses a relationship for overland flow developed by R. E. Horton [17]. This relationship, as modified by the Corps of Engineers, is as follows:

$$q = (\sigma \tan h^2)\left[0.922t\left(\frac{\sigma}{nL}\right)^{1/2} S^{1/4}\right] \qquad (9\text{-}4)$$

where q = rate of overland flow at lower end of elemental strip of turfed, bare, or paved surface, in/h of ft³/s per acre of drainage area

Q = total discharge from a drainage area, ft³/s; Q equals product of q and drainage area in acres

S = slope of surface or hydraulic gradient, absolute, i.e., 1 percent = 0.01

t = time or duration, min; time from beginning of supply (storm); total time $t = t_c + t_d$

t_c = duration of supply which produces maximum rate of outflow from a drainage area but not in a pipe

t_d = time water flows in pipe
σ = rate of supply, or rainfall in excess of rate of infiltration, in/h
L = effective length of overland or channel flow, ft
n = retardance coefficient

The term t_c is nothing more than the time of concentration for the drainage area under consideration. The term L, the effective length, represents the length of overland sheet flow from the most remote point in the drainage area to the drain inlet, measured in a direction parallel to the maximum slope, before the runoff has reached a defined channel or ponding basin, plus the length of flow in a channel if one is present. If ponding is permitted, L is measured from the most remote point in the drainage area to the mean edge of the pond.

The term n is referred to as the *retardance coefficient*. Typical coefficients are given in Table 9-5.

When a drainage area is composed of two or three types of surfaces, an average retardance coefficient must be computed. For example, if a drainage area consists of 4 acres of average grass cover and 2 acres of pavement, the average retardance coefficient is equal to

$$\frac{4(0.40)+2(0.02)}{6}=0.27$$

Infiltration Rate

Use of the Horton formula requires an estimate of the amount of rainfall which is absorbed in the ground and which therefore does not appear as runoff. This is referred to as *infiltration* and is expressed as a rate in inches per hour. Thus, the intensity of rainfall (in inches per hour) less the infiltration rate is equal to the rate of runoff or the rate of supply σ in the formula for runoff.

Surface	Value of n
Smooth pavements	0.02
Bare packed soil free of stone	0.10
Sparse grass cover, or moderately rough bare surface	0.30
Average grass cover	0.40
Dense grass cover	0.80

Source: Corps of Engineers [8].

TABLE 9-5 Retardance Coefficients

The infiltration rate is dependent largely on the structure of the soil cover, moisture content, and temperature of the air. The infiltration rate is not constant throughout the duration of the storm, but is assumed so in the computations. It is felt that such an assumption is reasonable, especially when the soil is near saturation.

The infiltration rate for paved surfaces is usually assumed to be zero. Infiltration rates for other types of surfaces and soil cover must be estimated from experience. A value of 0.5 in/h has been suggested for turfed areas. Thus, if the rainfall intensity on a turfed area were 2.0 in/h, the rate of supply σ would be 1.5 in/h.

Standard Supply Curves

By use of Eq. (9-4) maximum rates of runoff q for rates of supply σ of 0.8, 1.0, 1.6, and 1.8 in/h are shown in Figs. 9-8 and 9-9. Maximum rates of runoff are also shown for rates of supply of 0.4, 0.6, 1.2, 1.4, 2.0, 2.2, 2.4, 2.6, 2.8, 3.0, 3.2, and 3.4 in/h [8].

Maximum rates of runoff for the curve labeled *supply curve no. 1.0* (Fig. 9-8) were obtained in the following manner. From Fig. 9-2 the intensities of runoff for various durations corresponding to the curve labeled 1.0 are obtained. These intensities are entered as σ in Eq. (9-4), and L is varied to produce the family of curves shown in Fig. 9-8. The curve labeled σ is supply curve no. 1.0, obtained from Fig. 9-2. The dotted line labeled t_c represents the maximum rate of runoff q which would occur from an elemental area with various effective lengths L. For example, the maximum rate of runoff from an area whose effective length L is 60 ft is 2.0 ft^3/s. Multiplying this rate by the drainage area yields the maximum total discharge Q.

Figures 9-8 and 9-9 were prepared for $n = 0.40$ and $S = 1$ percent. If these charts are to be used for other cases, the actual effective L for the area under study must be converted in terms of L for $n = 0.40$ and $S = 1$. A conversion chart is shown in Fig. 9-10. For example, if the actual $n = 0.30$ and $S = 2$ percent and the effective length L is 400 ft, then the equivalent effective L for $n = 0.40$ and $S = 1$ percent is 140 ft.

Typical Example—No Ponding

In the Corps of Engineers procedure, a reach of drain pipe is always designed for a storm whose duration is equal to the time of concentration for the drainage area above the pipe. The time of concentration corresponds to the time necessary to produce maximum flow into a particular inlet (which is the same as the time necessary for water to reach an inlet from the most remote point in the area) plus the flow time in the pipe.

To clarify the computation of runoff by the Corps of Engineers procedure, the following example is presented.

Consider the drainage areas shown in Fig. 9-11. The 1-h intensity of the design storm is assumed to be 2.0 in/h. The infiltration rate for the turfed areas is assumed to be 0.5 in/h. The retardance coefficient

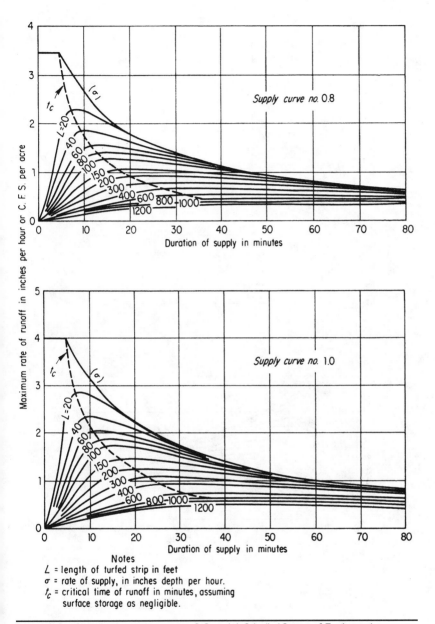

Notes

L = length of turfed strip in feet

σ = rate of supply, in inches depth per hour.

t_c = critical time of runoff in minutes, assuming surface storage as negligible.

FIGURE **9-8**   Standard supply curves, 0.8 and 1.0 in/h (*Corps of Engineers*).

for the pavement is $n = 0.02$, and for the turfed area $n = 0.40$. The drainage areas, retardance coefficients (referred to as *roughness factors*), and actual effective lengths L are shown in Table 9-6. Values of L and S were obtained from a grading plan of the area. The equivalent

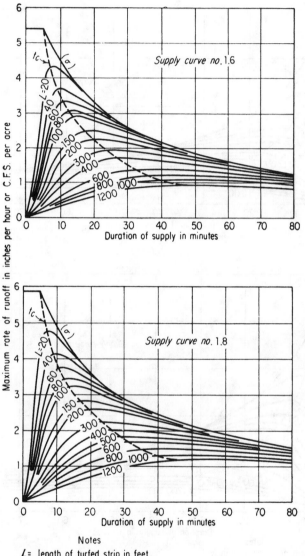

Notes

L = length of turfed strip in feet.

σ = rate of supply, in inches depth per hour.

t_c = critical time of runoff in minutes, assuming surface storage as negligible.

FIGURE 9-9   Standard supply curves, 1.6 and 1.8 in/h (*Corps of Engineers*).

Ls are obtained from Fig. 9-10. Column 14, labeled *adopted for selecting diagrams*, designates the nearest whole number which can be identified on the supply curves (Figs. 9-8 and 9-9). The standard supply curve to be used for the example is obtained by weighting the supply curves for the paved and turfed areas. For example, for inlet 4,

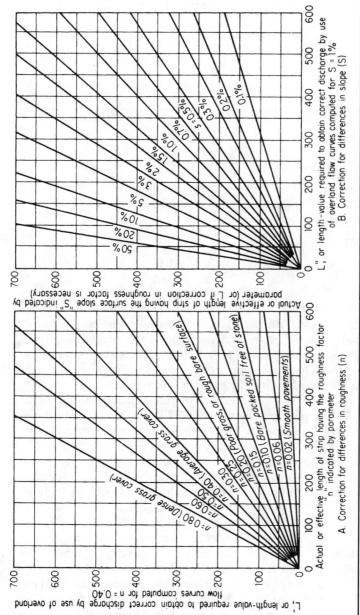

FIGURE 9-10 Modification in L required to compensate for difference in n and S (Corps of Engineers).

363

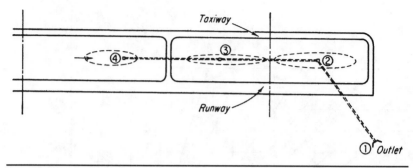

FIGURE 9-11   Portion of airport showing drainage layout (*Corps of Engineers*).

the paved area is 5.97 acres and the supply curve is 2.0 in/h; the turfed area is 26.81 acres and the supply curve is 1.5 in/h. The weighted supply curve is equal to

$$\frac{5.97(2)+26.81(1.5)}{5.97+26.81} = 1.6$$

In columns 20 and 21, the critical inlet time t_c (the time that will produce the maximum discharge) and the corresponding rates of runoff are listed. These values are obtained from Fig. 9-9. In columns 23 and 24, additional rates of runoff for arbitrarily selected times are listed. This is done to facilitate computation for various times of concentration for the several points along a drainage system.

The next step is to compute the volumes of runoff into inlets 4, 3, and 2. The computations are shown in Table 9-7. Obviously the duration of a storm necessary to provide the maximum rate of runoff into inlet 4 is equal to 24 min. The pipe from inlet 4 to inlet 3 is designed for a storm of this duration. At inlet 3 the time of concentration is 24 min plus the flow time in the pipe from inlet 4 to inlet 3 (9.2 min). The pipe from inlet 3 to inlet 2 would be designed for a storm of 33.2-min duration. Enter Fig. 9-9 (supply curves 1.6) with 33 min as the abscissa, and read the rates of runoff for effective lengths L of 280 ft (inlet 3) and 330 ft (inlet 4). Multiply these rates by their respective drainage areas. According to the computations at inlet 3, the area directly tributary to it contributes 62.5 ft³/s, and the area tributary to inlet 4 contributes 59.0 ft³/s. Thus the pipe from inlet 3 to inlet 2 should be designed for a capacity of 59.0 + 62.5 = 121.5 ft³/s. The same process would be repeated for the design of the pipe from inlet 2 to the outlet.

It should be emphasized that the duration of the storm for the analysis of a particular point along the drainage system always corresponds to the time of concentration above this point. Had the inlet time t_c for the area directly tributary to inlet 3 been larger than the sum of the inlet time for the area tributary to inlet 4 plus the flow time to inlet 3, the former would have established the duration of the storm for the design of the pipe from inlet 3 to inlet 2.

Typical Example—Ponding

If ponding is permissible, the first step is to establish the limits of the ponding area. From a grading and drainage plan, the volumes in the various ponds can be computed. These volumes are then expressed in terms of cubic feet per acre of drainage area, as shown in column 9 of Table 9-8. The actual and equivalent L values are determined in the same manner as for the case of no ponding, with one exception. The actual L is measured to the mean edge of the pond rather than to the drain inlet. The actual and equivalent effective lengths are listed in columns 12 and 13.

The Corps of Engineers has developed charts which yield drain inlet capacities to prevent ponds from exceeding certain specified volumes. Typical charts are shown in Figs. 9-12 and 9-13. The volumes are computed for various supply curves (Fig. 9-2), assuming the slope of the basins forming the drainage areas is 1 percent. The supply curves represent the intensity-duration pattern for storms whose 1-h intensities correspond to the supply curve numbers. The volumes of runoff for a specific supply curve are computed in a manner similar to the procedure used by the FAA. The cumulative volumes of runoff are compared with the various capacities of drain inlets to arrive at the volumes of storage shown in Figs. 9-12 and 9-13. Since the volumes of runoff depend on L and S, charts must be prepared for a wide range of L values. Figures 9-12 and 9-13 show drain inlet capacities for L equal to 100, 200, 300, and 400 ft. Additional charts have been prepared for $L = 0$, 40, 600, 800, 1000, and 1200 ft [8].

The physical significance of the charts may be described by reference to the following example. Suppose that L for a large drainage area is 100 ft and that the runoff pattern corresponds to supply curve 2. Assume that the maximum permissible ponding is 300 ft^3/acre of drainage area. From Fig. 9-12 a pipe which has a capacity of 1.0 ft^3/s per acre of drainage area would be adequate to prevent the pond from exceeding a volume of 3000 ft^3 during any part of the storm. The dashed lines labeled 4 are equal to rates of supply corresponding to a duration of 4 h. Although smaller drain inlets are possible, it is felt that the sizes corresponding to a duration of 4 h are about the minimum from a practical standpoint.

The required drain inlet capacities for the drainage layout in Fig. 9-11 were obtained from Figs. 9-12 and 9-13 and are tabulated in Table 9-8. Note that the time of concentration is not a factor in these computations.

TABLE 9-6 Airfield Drainage—Drain Inlet Capacities

	Supply Curve Nos.												
For paved areas	2.0												
For bare areas									Drainage Section Assuming				
For turfed areas	1.5												
	Drainage Area (DA), acres				Permissible Ponding						Length L, ft		
Inlet No.	Paved, n = 0.02	Unpaved Bare	Turf n = 0.40	Total	Depth at Inlet, ft	Pond Area, 1000 ft²	Volume 1000 ft³	Volume, ft³/acre DA	Average Roughness Factor n	Average Slope S, %	Actual or Effective Length, ft	Equivalent L for n = 0.40 and S = 1%	L Adopted for Selecting Diagrams
1	2	3	4	5	6	7	8	9	10	11	12	13	14
4	5.97		26.81	32.78					0.33	2.0	575	330	330
3	5.69		25.54	31.23					0.33	2.8	575	280	280
2	5.69		25.54	31.23					0.33	2.8	575	280	280

Source: Corps of Engineers [8].

TABLE 9-7 Airfield Drainage—Size and Profile of Underground Storm Drains

	Supply Curve Nos.										
For paved areas	2.0										
For bare areas										Drainage Section	
For turfed areas	1.5										
	Point of Design				Critical Runoff Time to Produce Maximum Flow in Underground Drain					Rate of	
	Distance, ft					Drain time, min					
Inlet or Junction	From Main Outlet	From Preceding Inlet	Critical Inlet	t_c, min	Assumed Velocity in Pipe, ft/s	From Preceding Inlet	Accumulation Total	Approximate t_c (col. 5 + col. 8)	Adopted t_c, min	4	3
1	2	3	4	5	6	7	8	9	10	11	12
4	4805	—	4	24	3.0			24	25	59.0	
3	3155	1650	4	24	3.0	9.2	9.2	33	30	59.0	62.5
2	1505	1650	4	24	3.0	9.2	18.4	42	40	55.8	56.2

Source: Corps of Engineers [8].

No Ponding of Runoff

	Standard Supply Curve No. · DA				Drain Inlet Capacity			Critical Contribution to System		
Paved Areas	Unpaved Areas		Total	Weighted Supply Curve (col. 18 ÷ col. 5)	t_c, min	q_d ft.³/s/acre	Q_d ft.³/s (col. 21 · col. 5)	t_c, min	q_d ft.³/s/acre	Q_d ft.³/s (col. 24 · col. 5)
	Bare	Turf								
15	16	17	18	19	20	21	22	23	24	25
11.94		40.22	52.16	1.6	24	1.8	59.0	30	1.8	59.0
								40	1.7	55.8
11.38		38.31	49.69	1.6	23	2.0	62.5	25	2.0	62.5
								30	2.0	62.5
								40	1.8	56.2
11.38		38.31	49.69	1.6	23	2.0	62.5	25	2.0	62.5
								30	2.0	62.5
								40	1.8	56.2

Assuming No Ponding of Runoff

Inflow into Underground Drains, ft³/s, Corresponding to Adopted Value of t_c (col 10)

							Inlet										
2																	Total
13	14	15	16	17	18	19	20	21	22	23	24	25	26	27	28	29	30
																	59.0
																	121.5
56.2																	168.2

	Supply Curve Nos.									
For paved areas	2.0									
For bare areas									**Drainage**	
For turfed areas	1.5									

	Drainage Area (DA), acres				Permissible Ponding						Length L, ft		
Inlet	Paved, $n=$ 0.02	Bare	Unpaved Turf $n=$ 0.40	Total	Depth at Inlet, ft	Pond Area, 1000 ft²	Volume 1000 ft³	Volume, ft³/acre DA	Average Roughness Factor n	Average Slope S, %	Actual or Effective Length, ft	Equivalent L for $n=$ 0.40 and $S=1\%$	L Adopted for Selecting Diagrams
1	2	3	4	5	6	7	8	9	10	11	12	13	14
4	5.97		26.81	32.78	3.0	138	206	6,292	0.33	2.0	525	300	300
3	5.69		25.54	31.23	1.73	145	125	4,016	0.33	2.8	340	200	200
2	5.69		25.54	31.23	2.73	270	368	11,800	0.33	2.8	340	200	200

*Not required when appreciable ponding is permissible.
Source: Corps of Engineers [8].

TABLE 9-8 Airfield Drainage—Drain Inlet Capacities Required to Limit Ponding to Permissible Volumes

Layout of Surface Drainage

A finished grade contour map of the runways, taxiways, and aprons is extremely helpful for the layout of a storm drain system. Several trial drainage layouts may be necessary before the most economical system can be selected. The grades of the storm drain should be such as to maintain a minimum mean velocity on the order of 2.5 ft/s to provide sufficient scouring action to avoid silting. To maintain an adequate cross section for flow at all times, the diameter of the storm drain should not be less than 12 in.

Water from a drainage area is collected into the storm drain by means of inlets. The inlet structure consists of a concrete box, the top of which is covered with a grate made of cast iron, cast steel, or reinforced concrete. The grates must support aircraft wheel loads and

Section East Side of Airfield

Standard Supply Curve No. · DA					Drain Inlet Capacity			Critical Contribution to System		
Paved Areas	Unpaved Areas		Total	Weighted Supply Curve (col. 18 ÷ col. 5)	t_c, min	q_d ft.^{3}/s/acre	Q_d ft.^{3}/s (col. 21 · col. 5)	t_c, min	q_d ft.^{3}/s/acre	Q_d ft.^{3}/s (col. 24 · col. 5)
	Bare	Turf								
15	16	17	18	19	20	21	22	23	24	25
11.94		40.21	51.15	1.6	°	0.52	17.05			
11.38		38.31	49.69	1.6	°	0.52	16.24			
11.38		38.31	49.69	1.6	°	0.52	16.24			

should therefore be designed for contact pressures for the aircraft which will be served by the airport.

On long tangents, drain inlets are usually placed at intervals varying from 200 to 400 ft. The location of the inlets depends on the configuration of the airport and on the grading plan. Normally, if there is a taxiway parallel to the runway, the inlets are placed in a valley between runway and taxiways, as indicated in Fig. 9-11. If there is no parallel taxiway, the drains are placed near the edge of the runway pavement or at the toe of the slope of the graded area. The FAA recommends that the inlets not be closer than 75 ft to the edge of the pavement.

On aprons, inlets are usually placed in the pavement proper. This is the only way a large apron area can be drained. All grates should be securely fastened to the frames so that they will not be jarred loose with the passage of traffic (see Fig. 9-14).

Adequate depths of cover should be provided over the pipes so that the pipes can support traffic. The recommended minimum depths of cover are shown in Table 9-9.

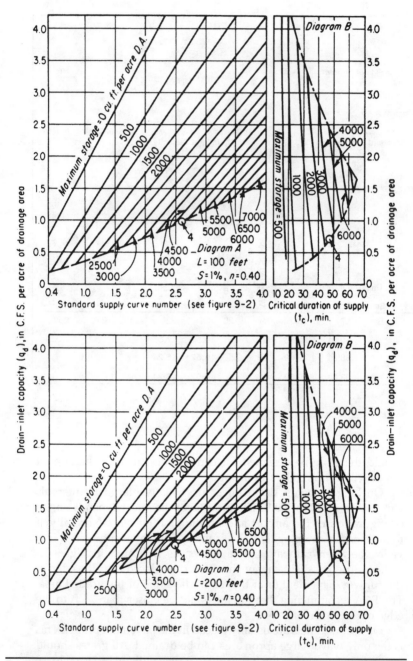

Figure 9-12 Drain inlet capacity versus maximum surface storage, L = 100 ft and L = 200 ft, C.F.S. = cubic feet per second (*Corps of Engineers.*)

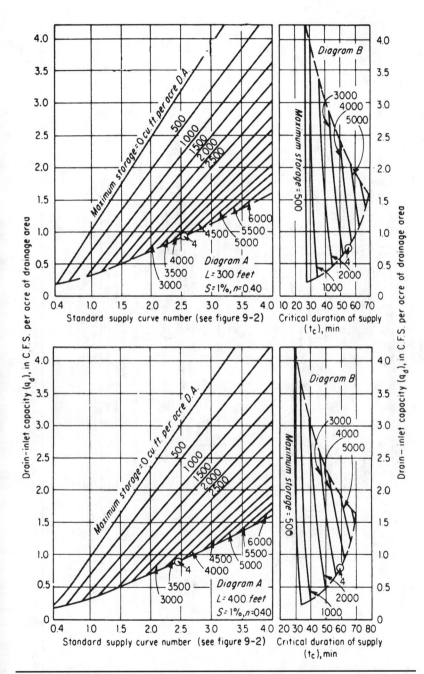

Figure 9-13 Drain inlet capacity versus maximum surface storage, L = 300 ft and L = 400 ft, C.F.S. = cubic feet per second (*Corps of Engineers.*)

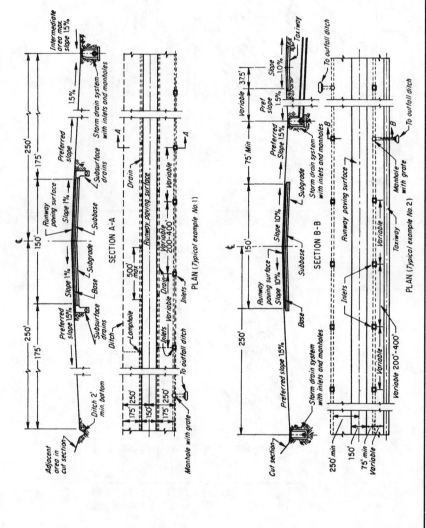

FIGURE 9-14 Recommended pavement drainage sections (*Federal Aviation Administration* [2]).

Flexible Pavement — Pipe Cover — Nominal Diameter of Pipe, in

Kind of Pipe	Wheel Load 15,000 lb								Wheel Load 30,000 lb							
	12	18	24	30	36	42	48	60	12	18	24	30	36	42	48	60
Clay sewer pipe	2.5	3.0	3.0	3.5	3.5				3.0	3.5	4.0	4.5	4.5			
Clay culvert pipe	1.5	1.5	1.5	2.0	2.0				2.5	3.0	3.0	3.0	3.0			
Concrete sewer pipe	2.5	3.0	3.0						3.0	3.5	4.0					
Concrete sewer pipe (extra-strength)	1.5	1.5	2.0						2.5	3.0	3.0					
Reinforced-concrete culvert pipe																
Class I								3.0								4.5
Class II	2.5	2.0	2.0	2.0	2.5	2.5	2.5	2.5	3.0	3.0	3.0	3.0	3.0	3.5	3.5	3.5
Class III	2.0	2.0	2.0	2.0	2.0	2.0	2.0	2.0	2.5	2.5	2.5	2.5	2.5	2.5	3.0	3.0
Class IV	1.5	1.5	1.5	1.5	1.5	1.5	1.5	1.5	2.0	2.0	2.0	2.0	2.0	2.0	2.0	2.0
Class V	1.0	1.0	1.0	1.0	1.0	1.0	1.0	1.0	1.5	1.5	1.5	1.5	1.5	1.5	1.5	1.5
Corrugated metal pipe, gauge no.																
16	1.0	1.5	1.5						1.5	2.0	2.0					
14	1.0	1.0	1.0	1.5	2.0				1.0	1.0	1.5	2.0	2.5			
12	1.0	1.0	1.0	1.0	1.0	1.5	2.0		1.0	1.0	1.5	1.5	2.0	2.5	2.5	2.5
10			1.0	1.0	1.0	1.0	1.5				1.0	1.5	1.5	2.0	2.0	2.5
8					1.0	1.0	1.0	1.5					1.0	1.5	1.5	2.0

TABLE 9-9 Recommended Minimum Depth of Cover for Pipe, ft

Flexible Pavement

Pipe Cover

Nominal Diameter of Pipe, in

Kind of Pipe	Wheel Load 45,000 lb								Wheel Load 60,000 lb							
	12	18	24	30	36	42	48	60	12	18	24	30	36	42	48	60
Clay culvert pipe	3.0	3.5	3.5	4.0	4.0				3.5	4.0	4.5	4.5	5.0			
Concrete sewer pipe (extra-strength)	3.0	3.5	3.5						3.5	4.0	4.5					
Reinforced-concrete culvert pipe																
Class I																
Class II	3.0	3.5	3.5	4.0	4.0	4.5	4.5		3.5	4.0	4.5	4.5	4.5			
Class III	2.5	3.0	3.0	3.5	3.5	3.5	4.0	4.0	3.0	3.5	4.0	4.0	4.5	4.5		
Class IV	2.0	2.5	2.5	2.5	2.5	2.5	3.0	3.0	2.5	3.0	3.0	3.0	3.0	3.5	3.5	4.0
Class V	1.5	1.5	2.0	2.0	2.0	2.0	2.5	2.5	2.0	2.0	2.5	2.5	2.5	3.0	3.0	3.0
Corrugated metal pipe, gauge no.																
16	2.0	2.5	3.0						2.5	3.0	3.5					
14	1.5	2.0	2.5	3.0	3.0				2.0	2.5	3.0	3.5	4.0			
12	1.5	2.0	2.0	2.5	2.5	3.0	3.5		1.5	2.0	2.5	3.0	3.5	4.0	4.0	
10			1.5	2.0	2.0	2.5	3.0	3.5			2.0	2.5	3.0	3.5	3.5	4.0
8					1.5	2.0	2.5	3.0						2.5	3.0	3.5

	Wheel Load 75,000 lb						Wheel Load 100,000 lb					
Reinforced-concrete culvert pipe												
Class I												
Class II												
Class III	3.0	4.0	4.5	5.0			3.5	4.5	5.0	5.5		
Class IV	3.0	3.0	3.5	4.0	4.5	5.0	3.5	3.5	4.0	4.5	5.0	5.5
Class V	2.5	2.5	3.0	3.0	3.5	4.0	3.0	3.0	3.5	3.5	4.0	4.5
Corrugated metal pipe, gauge no.												
16	3.0	3.5	4.0				3.5	4.0	4.5			
14	2.5	3.0	3.5	4.0	4.5		3.0	3.5	4.0	4.5	5.0	
12	2.5	3.0	3.0	3.5	4.0	5.0	3.0	3.5	3.5	4.0	4.5	5.5
10	2.5	3.0	3.5	4.0	4.5	5.0	3.0	3.5	4.0	4.5	5.0	5.5
8	3.0	3.5	4.0	4.5			3.5	4.0	4.5	5.0		

Cover depths measured from top of flexible pavement or unsurfaced areas to top of pipe. Cover for pipe in areas not used by aircraft shall be in accordance with cover requirements for 15,000-lb wheel loads.

Rigid Pavement

Pipe placed under rigid pavements shall have a minimum cover, measured from the bottom of the slab, of 1.0 ft.

Note: The recommended minimum depth of cover for pipe does not provide protection against freezing conditions in seasonal freezing areas.
Source: Federal Aviation Administration [2].

TABLE 9-9 Recommended Minimum Depth of Cover for Pipe, ft (*Continued*)

	n
Pipe	
Clay and concrete	
Good alignment, smooth joints, smooth transitions	0.013
Less favorable flow conditions	0.015
Corrugated metal	
100% of periphery smoothly lined	0.013
Paved invert, 50% of periphery paved	0.018
Paved invert, 25% of periphery paved	0.021
Unpaved, bituminous-coated or noncoated	0.024
Open channels	
Paved	0.015–0.020
Unpaved	
Bare earth, shallow flow	0.020–0.025
Bare earth, depth of flow over 1 ft	0.015–0.020
Turf, shallow flow	0.06–0.08
Turf, depth of flow over 1 ft	0.04–0.06

Source: Federal Aviation Administration [2].

TABLE 9-10   Coefficients of Roughness *n*

As a guide for the design of storm drains, the coefficient of roughness *n* for various types of pipes and open channels is listed in Table 9-10.

Subsurface Drainage

The functions of subsurface drainage are to (1) remove water from a base course, (2) remove water from the subgrade beneath a pavement, and (3) intercept, collect, and remove water flowing from springs or pervious strata.

Base drainage is normally required (1) where frost action occurs in the subgrade beneath a pavement, (2) where the groundwater is expected to rise to the level of the base course, and (3) where the pavement is subject to frequent inundation and the subgrade is highly impervious.

Subgrade drainage is desirable at locations where the water may rise beneath the pavement to less than 1 ft below the base course.

Intercepting drainage is highly desirable where it is known that subsurface waters from adjacent areas are seeping toward the airport pavements.

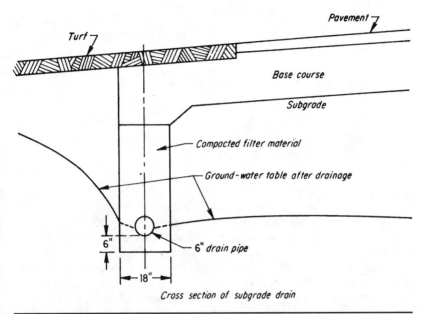

FIGURE **9-15** Subgrade subdrainage details (*Corps of Engineers*).

Methods for Draining Subsurface Water

Base courses are usually drained by installing subsurface drains adjacent to and parallel to the edges of the pavement. The pervious material in the trench should extend to the bottom of the base course, as shown in Fig. 9-15. The center of the drainpipe should be placed a minimum of 1 ft below the bottom of the base course.

Subgrades are drained by pipes installed along the edges of pavement and in some instances, where the groundwater is extremely high, underneath the pavements. The center of the subsurface drain should be placed no less than 1 ft below the level of the groundwater. When subgrade drains are installed along the edges of the pavement, they may also serve for draining the base course.

Intercepting drainage can be accomplished by means of open ditches well beyond the pavement areas. If this is not practical, then subdrains can be used.

Types of Pipe

The following types of pipe have been used for subdrainage:

1. Perforated metal, concrete, or vitrified clay pipe. The joints are sealed. The perforations normally extend over about one-third of the circumference of the pipe. The perforated area is usually placed adjacent to the soil.

2. Bell-and-spigot pipes are laid with the joints open. Vitrified clay, cast iron, and plain concrete are used in the manufacture of bell-and-spigot pipes.

3. Porous concrete pipe collects water by seepage through the concrete wall of the pipe. This type of pipe is laid with the joints sealed.

4. Skip pipe manufactured of both vitrified clay and cast iron is a special type of bell-and-spigot pipe with slots at the bells.

5. Farm tile is made of clay or concrete with the ends separated slightly to permit the entrance of water. This type of pipe is rarely used on airport projects.

Pipe Sizes and Slopes

Experience has shown that a 6-in-diameter drain is adequate, unless extreme groundwater conditions are encountered. If desired, the flow may be estimated by means of the available theories for soil drainage [7]. These theories require knowledge of the effective porosity and coefficient of permeability of the soil which is being drained, as well as the head on the pipe and the distance which the water must flow to reach the drain. Rarely is theory relied on to compute pipe sizes.

The recommended minimum slope for subdrains is 0.15 ft in 100 ft. A minimum thickness of 6 in of filter material should surround the drain. The gradation of the filter material is discussed in succeeding paragraphs.

Utility Holes and Risers

For cleaning and inspection, utility holes and risers are often installed along the drains. The Corps of Engineers recommends that utility holes be placed at intervals of not more than 1000 ft, with one riser approximately midway between the holes [7]. The function of the riser is to be able to insert a hose for flushing the system. The function of a utility hole is to permit inspection of the pipes.

Gradation of Filter Material

The term *filter material* applies to the granular material which is used as backfill in the trenches where subdrains are placed. To permit free water to reach the drain, the filter material must be many times more pervious than the protected soil. Yet if the filter is too pervious, the particles of soil to be drained will move into the filter material and clog it.

On the basis of some general studies conducted by K. Terzaghi, the Corps of Engineers has developed an empirical design for filter material which has been substantiated by tests [10]. The criteria for selecting the gradation of the filter material are as follows:

1. To prevent clogging of a perforated pipe with filter material, the following requirement must be satisfied:

$$\frac{85\% \text{ size of filter material}^*}{\text{Diameter of perforation}} > 1$$

2. To prevent the movement of particles from the protected soil into the filter material, the following conditions must be satisfied:

$$\frac{15\% \text{ size of filter material}}{85\% \text{ size of protected soil}} \leq 5$$

and

$$\frac{50\% \text{ size of filter material}}{50\% \text{ size of protected soil}} \leq 25$$

3. To permit free water to reach the pipe, the following condition must be fulfilled:

$$\frac{15\% \text{ size of filter material}}{15\% \text{ size of protected soil}} \geq 5$$

A typical example of design is shown in Fig. 9-16. Concrete sand has proved to be a satisfactory filter material for the majority of fine soils which are drainable. A single gradation of filter material is preferred for simplicity of construction.

Filter materials tend to segregate as they are placed in trenches. To minimize this tendency, the material should not have a coefficient of uniformity greater than 20. For the same reason, filter materials should not be skip-graded. Filter materials should always be placed in a moist state. The presence of moisture tends to reduce segregation.

Drainability of Soils

Certain types of soils, such as gravelly sands, sand, and sandy loams, are usually self-draining and require very little, if any, subsurface drainage. Subsurface drainage can be effective for draining clay loams, sandy clay loams, and certain silty loams. The amount of sand in these soils largely determines how drainable they are. For soils containing a high percentage of silt and clay, subsurface drainage becomes very problematic.

---

*This means that 85 percent (by weight) is finer than the specified size.

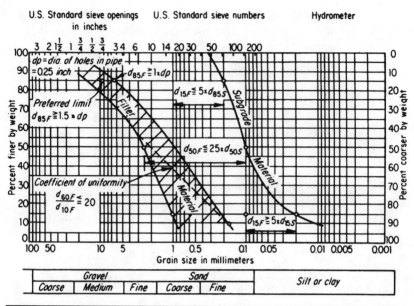

FIGURE 9-16   Design example for filter materials (*Corps of Engineers*).

References

1. *Surface Drainage Design,* Advisory Circular AC 150/5320-5C, Federal Aviation Administration, Washington, 2006.
2. *Airport Drainage,* Advisory Circular AC 150/5320-5B, Federal Aviation Administration, Washington, 1970.
3. *Conduits, Culverts, and Pipes,* Engineering Manual EM 1110-2-2902, Department of the Army, Washington, 1969.
4. "Design of Drainage Facilities for Military Airfields," G. A. Hathaway, *Transactions,* American Society of Civil Engineers, New York, 1949.
5. *Drainage and Erosion Control—Drainage for Areas Other than Airfields,* Tech. Manual TM 5-820-4, Department of the Army, Washington, 1965.
6. *Drainage and Erosion Control—Structures for Airfields and Heliports,* Tech. Manual TM 5-820-3, Department of the Army, Washington, 1965.
7. *Drainage and Erosion Control—Subsurface Drainage Facilities for Airfield Pavements,* Tech. Manual TM 5-820-2, Department of the Army, and Air Force Manual AFM 88-5, Department of the Air Force, Washington, 1979.
8. *Drainage and Erosion Control—Surface Drainage Facilities for Airfields and Heliports,* Tech. Manual TM 5-820-1, Department of the Army, and Air Force Manual AFM 88-5, Washington, 1987.
9. *Drainage of Asphalt Pavement Structures,* Manual Series MS-15, The Asphalt Institute, College Park, Md., 1984.
10. *Filter Experiments and Design Criteria,* Tech. Memo 3-360, U.S. Army Corps of Engineers, Waterways Experiment Station, Vicksburg, Miss., 1953.
11. *On-Site Stormwater Management: Applications for Landscape and Engineering,* B. Ferguson and T. H. Debo, 2d ed., Van Nostrand and Reinhold, New York, 1990.
12. "Pavement Subsurface Drainage Systems," H. H. Ridgeway, *Synthesis of Highway Practice,* No. 96, Transportation Research Board, Washington, 1982.

13. *Precipitation Frequency Atlas of the Western United States,* J. F. Miller, National Weather Service, Washington, 1973.
14. *Rainfall Frequency Atlas of the United States,* Technical Paper 40, U.S. Weather Service, Washington, 1961.
15. *Rainfall Intensities for Local Drainage Design in the United States,* Tech. Paper 24, pts. 1 and 2, U.S. Weather Service, Washington, 1953, 1954.
16. *Rainfall-Intensity Frequency Data,* D. L. Yarnell, Miscellaneous Publication 204, Department of Agriculture, Washington, 1935.
17. "The Interpretation and Application of Runoff Plat Experiments with Reference to Soil Erosion Problems," R. E. Horton, *Proceedings,* vol. 3, Soil Science Society of America, Madison, Wis., 1938.
18. *Urban Hydrology for Small Watersheds,* 2d ed., U.S. Soil Conservation Service, Washington, 1986.

CHAPTER 10

Planning and Design of the Terminal Area

Introduction

The terminal area is the major interface between the airfield and the rest of the airport. It includes the facilities for passenger and baggage processing, cargo handling, and airport maintenance, operations, and administration activities. The passenger processing system is discussed at length in this chapter. Baggage processing, cargo handling, and apron requirements are also discussed relative to the terminal system.

The Passenger Terminal System

The passenger terminal system is the major connection between the ground access system and the aircraft. The purpose of this system is to provide the interface between the passenger airport access mode, to process the passenger for origination, termination, or continuation of an air transportation trip, and convey the passenger and baggage to and from the aircraft.

Components of the System

The passenger terminal system is composed of three major components. These components and the activities that occur within them are as follows:

1. The access interface where the passenger transfers from the access mode of travel to the passenger processing component. Circulation, parking, and curbside loading and unloading of passengers are the activities that take place within this component.

2. The processing component where the passenger is processed in preparation for starting, ending, or continuation of an air transportation trip. The primary activities that take place within

this component are ticketing, baggage check-in, baggage claim, seat assignment, federal inspection services, and security.

3. The flight interface where the passenger transfers from the processing component to the aircraft. The activities that occur here include assembly, conveyance to and from the aircraft, and aircraft loading and unloading.

A number of facilities are provided to perform the functions of the passenger terminal system. These facilities are indicated for each of the components identified above.

The Access Interface

This component consists of the terminal curbs, parking facilities, and connecting roadways that enable originating and terminating passengers, visitors, and baggage to enter and exit the terminal. It includes the following facilities:

1. The enplaning and deplaning curb frontage which provide the public with loading and unloading for vehicular access to and from the terminal building

2. The automobile parking facilities providing short-term and long-term parking spaces for passengers and visitors, and facilities for rental cars, public transit, taxis, and limousine services

3. The vehicular roadways providing access to the terminal curbs, parking spaces, and the public street and highway system

4. The designated pedestrian walkways for crossing roads including tunnels, bridges, and automated devices which provide access between the parking facilities and the terminal building

5. The service roads and fire lanes which provide access to various facilities in the terminal and to other airport facilities, such as air freight, fuel truck stands, and maintenance.

The ground access system at an airport is a complex system of roadways, parking facilities, and terminal access curb fronts. This complexity is illustrated in Fig. 10-1 which shows the various ground access system facilities and directional flows at Greater Pittsburgh International Airport.

The Processing System

The terminal is used to process passengers and baggage for the interface with aircraft and the ground transportation modes. It includes the following facilities:

1. The airline ticket counters and offices used for ticket transactions, baggage check-in, flight information, and administrative personnel and facilities

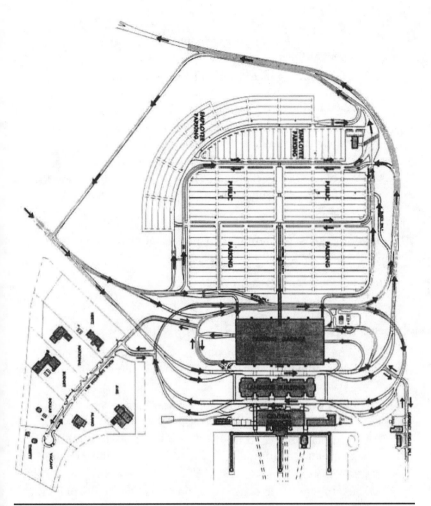

FIGURE 10-1 Ground access system configuration and directional flows for Greater Pittsburgh International Airport (*Tasso Katselas Associates and Michael Baker Jr., Inc. [32]*).

2. The terminal services space which consists of the public and nonpublic areas such as concessions, amenities for passengers and visitors, truck service docks, food preparation areas, and food and miscellaneous storage

3. The lobby for circulation and passenger and visitor waiting

4. Public circulation space for the general circulation of passengers and visitors consisting of such areas as stairways, escalators, elevators, and corridors

5. The outbound baggage space which is a nonpublic area for sorting and processing baggage for departing flights

6. The intraline and interline baggage space used for processing baggage transferred from one flight to another on the same or different airlines

7. The inbound baggage space which is used for receiving baggage from an arriving flight, and for delivering baggage to be claimed by the arriving passenger

8. Airport administration and service areas used for airport management, operations, and maintenance facilities

9. The federal inspection service facilities which are the areas for processing passengers arriving on international flights, as well as performing agricultural inspections, and security functions

The Flight Interface

The connector joins the terminal to parked aircraft and usually includes the following facilities:

1. The concourse which provides for circulation to the departure lounges and other terminal areas

2. The departure lounge or holdroom which is used for assembling passengers for a flight departure

3. The passenger boarding device used to transport enplaning and deplaning passengers between the aircraft door and the departure lounge or concourse

4. Airline operations space used for airline personnel, equipment, and activities related to the arrival and departure of aircraft

5. Security facilities used for the inspection of passengers and baggage and the control of public access to passenger boarding devices

6. The terminal services area providing amenities to the public and those nonpublic areas required for operations such as building maintenance and utilities

The components of the passenger terminal system together with the specific physical facilities corresponding to them are shown in Fig. 10-2. The relative locations of the various physical facilities in the three level landside building of the midfield terminal complex at Greater Pittsburgh International Airport are shown in Figs. 10-3, 10-4, and 10-5. Figure 10-3 shows the enplaning roadway interface with the departure or check-in level. This level also provides access to the commuter aircraft departure lounge. Figure 10-4 shows the transit level which provides airline baggage makeup space, passenger security processing and access to the automated transit system, the interface between the landside building and the airside building. Figure 10-5 shows the deplaning roadway interface with the arrivals or baggage claim level and contains baggage claim facilities and rental car facilities

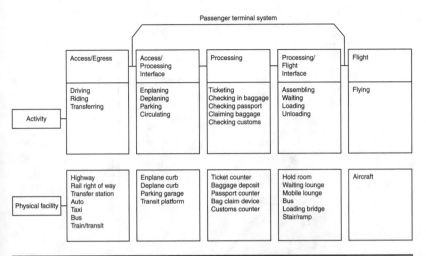

FIGURE 10-2   Components of the passenger terminal system.

as well as a central service building with airport police offices, utility, maintenance, and storage space. The concourse level of the airside building is shown in Fig. 10-6. On this level there is a common core with passenger amenities and four piers providing the departure lounges and boarding devices at the gates providing the interface with aircraft. One of the four piers is for international arrivals and contains the sterile areas for customs and immigration functions required for international passenger processing. The apron level of the airside building is used for airline operations and the lower level provides access to an automated peoplemover transit system.

Design Considerations

In developing criteria for the design of the passenger terminal complex, it is important to realize that there are a number of different factors which enter into a statement of overall design objectives. From these factors general and specific goals are established which set the framework on which design progresses. For example, in designing modifications to the apron and terminal complex at Geneva Intercontinental Airport, the general design objectives included [25]

1. Development and sizing to accomplish the stated mission of the airport within the parameters defined in the master plan

2. Capability to meet the demands for the medium- and long-run time frames

3. Functional, practical, and financial feasibility

4. Maximize the use of existing facilities

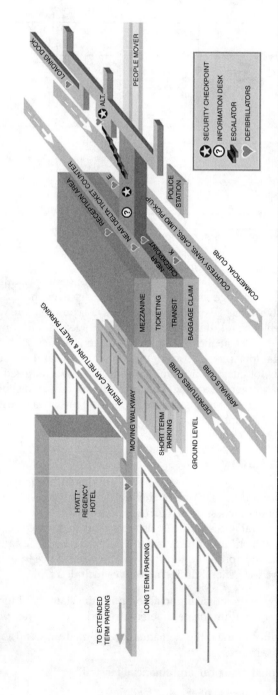

FIGURE 10-3 Landside building enplaning level at Greater Pittsburgh International Airport (*http://www.pitairport.com*).

Ticketing Map

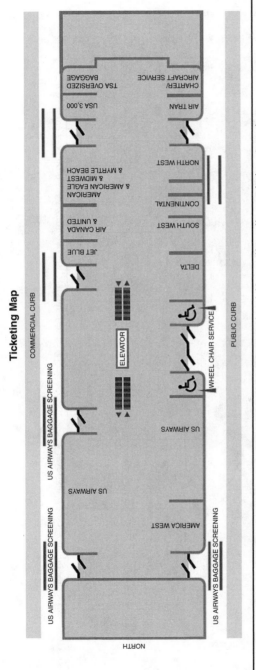

FIGURE 10-4 Landside building transit level at Greater Pittsburgh International Airport (*http://www.pitairport.com*).

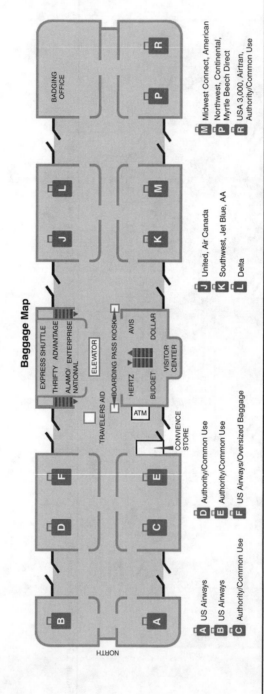

Baggage Map

EXPRESS SHUTTLE
THRIFTY   ADVANTAGE
ALAMO/ ENTERPRISE
NATIONAL

ELEVATOR

TRAVELERS AID

BOARDING PASS KIOSK

ATM

HERTZ   AVIS
BUDGET   DOLLAR
VISITOR CENTER

CONVIENCE STORE

BADGING OFFICE

NORTH

A US Airways
B US Airways
C Authority/Common Use
D Authority/Common Use
E Authority/Common Use
F US Airways/Oversized Baggage

J United, Air Canada
K Southwest, Jet Blue, AA
L Delta

M Midwest Connect, American
P Northwest, Continental, Myrtle Beech Direct
R USA 3,000, Airtran, Authority/Common Use

FIGURE 10-5   Landside building deplaning level at Greater Pittsburgh International Airport (*http://www.pitairport.com*).

390

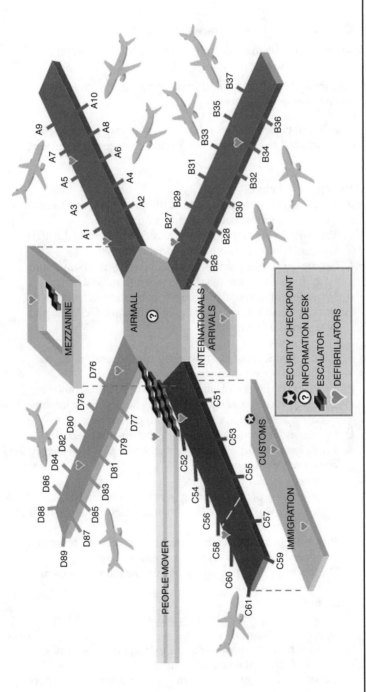

Figure 10-6 Airside building concourse level at Greater Pittsburgh International Airport (*http://www.pitairport.com*).

5. Achievement of a balanced flow between access, terminal, and airfield facilities during the peak hour
6. Consideration of environmental sensitivity
7. Maintenance of the flexibility to meet future requirements beyond the current planning horizon
8. Capability to anticipate and implement significant improvements in aviation technology

Specific design objectives were derived from these general objectives which included the needs of the various categories of airport users. These included

1. Passenger objectives
   a. Responsiveness to the needs of the people relative to convenience, comfort, and personal requirements
   b. Provision of effective passenger access orientation through concise, comprehensive directional graphics
   c. Separation of enplaning and deplaning roadways and curb fronts to ensure maximum operational efficiency
   d. Provision of convenient access to public and employee parking facilities, rental car areas, ancillary facilities, and other on-site facilities
2. Airline objectives
   a. Accommodation of existing and future aircraft fleets with maximum operational efficiency
   b. Provision of direct and efficient means of passenger and baggage flow for all passengers, including domestic and international originating, terminating, and transfer passengers
   c. Provision for economic, efficient, and effective security
   d. Provision of facilities which will embrace the latest energy conservation measures
3. Airport management objectives
   a. Maintenance of the existing terminal operation, access system, runway system, and ancillary facilities during all stages of construction
   b. Provision of facilities which generate maximum revenues from concessionaires and other sources
   c. Provision of facilities which minimize maintenance and operating expenses
4. Community objectives
   a. Render a unique and appropriate expression and impression of the community
   b. Provision of harmony with the existing architectural elements of the total terminal complex
   c. Coordination with the existing and planned off-airport highway system

The designer should consider the combination of these types of objectives in developing specific design criteria for the passenger terminal complex. These criteria should be used as performance measures for the evaluation of design alternatives. In order to generate performance measures, a detailed design should be provided. The analyst can then proceed to calculate the various performance measures using a number of analytical techniques. Some of these techniques are discussed later in this chapter.

Terminal Demand Parameters

The determination of space requirements at passenger terminals is strongly influenced by the quality of service desired by the various airport users and the community. A review of passenger terminals in relation to passenger volumes at existing airports shows a wide range in the configuration and the amount of area provided per passenger. However, some guidelines for the determination of space requirements can be defined. The purpose of these guidelines is to give general orders of magnitude for values that are subject to change depending on the requirements of specific designs.

The following steps should be followed in determining terminal facility space requirements.

Identify Access Modes and Modal Splits

Vehicle volumes are normally derived from projections of passenger and aircraft forecasts. These volumes critically impact the design of highway access facilities, on-airport roadway and circulation systems, curb frontage requirements for private automobiles, buses, limousines, taxis, and rental cars, and parking. Surveys are normally conducted to determine the access modes of passengers and vehicle occupancy rates [26]. In the absence of such surveys, secondary sources may be investigated to ascertain the access characteristics of passengers in similar airport environments [24, 55, 60].

The most important parameters to be obtained include the typical peak hour volumes of vehicles entering and leaving the airport on the design day, the access facilities used and the duration of use, including parking and curb front. Care should be exercised to include employees and visitors as well as passengers in these access studies, and to correlate the peaking characteristics and access modes of each group of airport traveler.

Identify Passenger Volumes and Types

Passenger volumes can be obtained from forecasts normally done in conjunction with airport planning studies. Two measures of volume are used. The first is annual passenger volume, which is used for preliminary sizing of the terminal building. The second is a more detailed hourly volume. It is customary to use typical peak hour passengers as the hourly design volume for passenger terminal design. This parameter is a design index and is usually in the range of 0.03 to 0.05 percent of the annual passenger volume but it is significantly affected by the scheduling practices and fleet mix of the airlines.

The identification of passenger types is necessary because different types of passengers place different demands on the various components. Passenger types are usually broadly classified into domestic and international passengers and then further grouped into originating, terminating, connecting or transfer, through, enplaning and deplaning passengers. These various groupings of passengers are made on the basis of the facilities within the terminal which are normally used by each type of passenger. Airports which are used as airline hubs and have a high proportion of connecting passengers require considerably less ground access and landside facilities than airports with a high proportion of originating and terminating passengers. Historical data and forecasts regarding the proportions of the total volumes that are made up by each of the different types of passengers are useful in obtaining estimates of the parameters needed for the design of the various facilities [4, 58].

Identify Access and Passenger Component Demand

This is done by matching the passenger and vehicle types with facilities in the terminal area. The use of tabulations such as the one shown in Table 10-1 is quite helpful. This table shows which passengers are using which facility. By indicating the volume of each type of passenger in the rows corresponding to the facilities, it is possible to generate the total load on each facility. This is done by taking the row sums of the volumes entered.

Facility Classification

The airport terminal facility may be classified by its principal characteristics relative to its functional role. In general, airports are classified as originating-terminating, transfer, or through airports. The facilities required are considerably different in magnitude and configuration for each.

An originating-terminating airport processes a high level of passengers which are beginning or ending the air transportation trip at the airport. At such airports these passengers may be in the order of 70 to 90 percent of the total passengers. These airports can have a relatively long aircraft ground time for long haul international flights but also may have relatively short ground times for domestic operations and operations by low-cost air carriers. In either case, the main flow of passengers is between the aircraft and the ground transportation system and have relatively high requirements for curb frontage, ticketing and baggage claim facilities, and parking. Typical data indicate that the hourly movements of aircraft per gate at such airports can range from on the order of 1.0 to nearly 3.0 operations per hour per gate.

TABLE 10-1 Determination of Demand for Various Types of Passenger Facilities

Facility *j*	Passenger Type *i*, Arriving			Passenger Type *i*, Departing			Total Volume *V*
	Domestic, No Bags, Auto Driver*	Domestic, with Bags, Auto Passenger†	International, with Bags, Auto Passenger	Domestic, with Bags, Auto Passenger	Domestic, No Bags, Auto Driver	International, with Bags, Auto Driver	
Curb, arrivals	–	$V_{ij}^‡$	V_{ij}	–	–	–	
Curb, departures	–	–	–	V_{ij}	–	V_{ij}	
Domestic lobby	–	V_{ij}	–	V_{ij}	V_{ij}	–	
International lobby	–	–	–	–	–	V_{ij}	
Ticketing counter	–	–	–	V_{ij}	–	V_{ij}	
Assembly	–	–	–	V_{ij}	–	V_{ij}	
Baggage check-in	–	–	–	V_{ij}	V_{ij}	V_{ij}	
Security control	–	–	–	V_{ij}	V_{ij}	V_{ij}	
Customs, health	–	–	V_{ij}	–	–	–	
Immigration	–	–	V_{ij}	–	–	V_{ij}	
Baggage claim	–	V_{ij}	V_{ij}	–	–	–	

*Auto driver = passenger driving a car to and from airport.
†Auto passenger = passenger driven to and from airport.
‡V_{ij} = design volume of passenger type *i* using facility type *j*.

A transfer or connecting airport, on the other hand, has a high percentage of its total passengers connecting between arriving and departing flights. Today many airports in the United States are connecting airports particularly those that are airline hubs. These airports need greater concourse facilities for the processing of connecting passengers and less ground access facility development. Airline ticketing positions and baggage claim facilities are usually less than with originating airports (on a size per passenger basis). However, intraline and interline baggage facilities are usually greater. Care must be exercised in the planning of such airports to locate the gate positions of airlines exchanging passengers in close proximity to each other to minimize central terminal flows and connecting times. Data indicate that such airports demonstrate aircraft activity at the rate of 1.3 to 1.5 aircraft per gate per hour in peak periods.

The through airport combines a high percentage of originating passengers with a low percentage of originating flights. A high percentage of the passengers remain on the aircraft at such points. Aircraft ground times are minimal, averaging between 1.6 and 2.0 hourly movements per gate in peak periods. Departure lounge space, curb frontage, ticketing, security, and baggage facilities are less than at originating airports.

Overall Space Approximations

It is possible to estimate order of magnitude ranges for the overall size of a terminal facility prior to performing more detailed calculations for particular space needs. These estimates allow the planner to broadly define the scope of a project based upon information which summarizes the space provided of other existing facilities.

The FAA has indicated that gross terminal area space requirements of between 0.08 and 0.12 ft² per annual enplaned passenger are reasonable. Another estimate is obtained by applying a ratio of 150 ft² per design hour passenger [43]. Estimates of the level of peak hour passengers, peak hour aircraft operations, and gate positions are also obtained based upon the level of annual enplanements using relationships such as those shown in Fig. 10-7. Others have provided estimating guidelines for total terminal space as shown, for example, in Fig. 10-8 [43].

Approximations of the allocation of space among the various purposes in a terminal building are also useful for preliminary planning. The FAA indicates that approximately 55 percent of terminal space is rentable and 45 percent is non-rentable [49]. An approximate breakdown of these space allocations typically is 35 to 40 percent for airline operations, 15 to 25 percent for concessions and airport administration, 25 to 35 percent for public space, and 10 to 15 percent for utilities, shops, tunnels, and stairways. A final determination of the actual space allocations is obtained through detailed analyses of the performance of the elements of the system as the design process

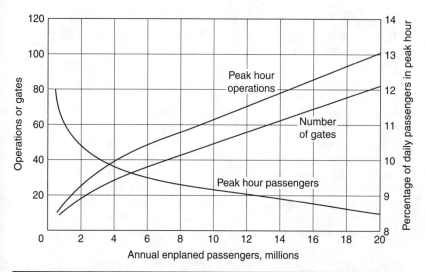

FIGURE 10-7   Estimated peak hour passenger, operations, and gate requirements for intermediate range planning (*Federal Aviation Administration [43]*).

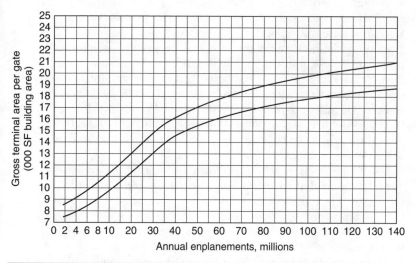

FIGURE 10-8   Gross terminal area estimates for intermediate range planning (*Federal Aviation Administration [49]*).

proceeds from space programming through each of the subsequent phases in the process.

Level of Service Criteria

Considerable research and discussion has taken place in the profession relative to the adoption of level of service standards and associated criteria to evaluate the level of service afforded in the design of landside processing systems. Although it is relatively simple to develop relationships between aircraft delay on the airside and its economic consequences, such relationships are difficult to either define or develop on the airport landside. Much of the difficulty is related to the fact that the various constituent groups associated with airports view quality of service or level of service from different perspectives [37]. Airlines are concerned with such factors as on-time schedules, the allocation of personnel, airport operating costs, and profitability. Passengers are concerned with the completion of an air transportation trip at a reasonable cost, with minimum delay and maximum convenience, without being subjected to excessive levels of congestion. The airport operator is interested in providing a modern airport facility which meets airline and passenger objectives in harmony with the expectations of the community in which the airport is located. Given the number of possible measures of service quality and the differences in airports throughout the country, it is very difficult to adopt level of service criteria on a broad scale.

Many have examined level of service criteria for airports and attempted to define level of service standards [9, 11, 14, 16, 21, 29, 31, 36, 37, 41, 45, 56]. In general, the level of service measures commonly associated with the airport landside system include measures of congestion within the terminal building and the ground access system, passenger delays and waiting line lengths at the various facilities in the terminal building, passenger walking distances, and total passenger processing time. Most of these parameters can be evaluated in a terminal design with the aid of mathematical modeling. However, the various measures of level of service from the perspective of airport users must be balanced in reaching some acceptable solution to the design problem.

As an illustration of the application of a level of service standard, let us say that an airline may desire to limit the percentage of its passengers which must spend more than some increment of time at an airport check-in facility. One could develop a model which computes the percentage of passengers at the check-in facility for various durations of time when the number of check-in counters operated is varied for some peak hour passenger demand [35]. As an example, the results could be shown graphically as is done in Fig. 10-9. From this illustration, if the criteria were to limit the percentage of passengers spending more than 5 min at this facility to 10 percent, then it would have been necessary for the airline to operate nine check-in counters at the airport during the peak hour.

Such formulations can be examined for the various facilities within the airport landside to obtain quantitative measures of component and system performance. These are discussed later in this chapter.

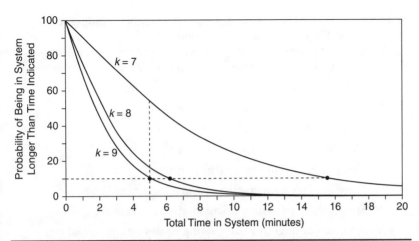

FIGURE **10-9** Impact of the number of check-in counters (k) on passenger time at a check-in facility.

The Terminal Planning Process

The evolution and development of a terminal design is performed in a series of integrated steps. These may be identified as programming, concept development, schematic design, and design development. The terminal facilities are developed in conformity with the planned development of the airside facilities considering the most effective use of the airport site, the potential for physical expansion and operational flexibility, integration with the ground access system, and compatibility with existing and planned land uses near the airport. The planning process explicitly examines physical and operational aspects of the system.

The programming phase defines the objectives and project scope including the rationale for the initiation of the study. It includes a space requirements program, tentative implementation schedules, estimates of the anticipated level of capital investment as well as operating, maintenance, and administrative costs. In concept development, studies are undertaken to identify the overall arrangement of building components, functional relationships, and the characteristics of the terminal building. Schematic design translates the concept and functional relationships into plan drawings which identify the overall size, shape, and location of spaces required for each function. Detailed budget estimates are prepared in schematic design so that comparisons may be made between the space requirements and costs. In design development, the size and character of the entire project is determined and detailed plans of the specific design and allocation of space within the complex are prepared. This phase forms the basis for the preparation of construction documents, bidding, construction, and final project implementation [50].

In the programming and concept development phases of a terminal design project, the following evaluation criteria are typically used to weigh alternatives:

1. Ability to handle expected demand
2. Compatibility with expected aircraft types
3. Flexibility for growth and response to technology changes
4. Compatibility with the total airport master plan
5. Compatibility with on-airport and adjacent land uses
6. Simplicity of passenger orientation and processing
7. Analyses of aircraft maneuvering routes and potential conflicts on the taxiway system and in the apron area
8. Potential for aircraft, passenger, and vehicle delay
9. Financial and economic feasibility

In the schematic and design development phases, more specific design criteria are examined such as:

1. The processing cost per passenger
2. Walking distance for various types of passengers
3. Passenger delays in processing
4. Occupancy levels and degree of congestion
5. Aircraft maneuvering delays and costs
6. Aircraft fuel consumption in maneuvering on the airport between runways and terminals
7. Construction costs
8. Administrative, operating, and maintenance costs
9. Potential revenue sources and the expected level of revenues from each source

Space Programming

The space programming phase of terminal planning seeks to establish gross size requirements for the terminal facilities without establishing specific locations for the individual components. The nature of the processing components is such, however, that approximate locations are indicated for new and existing terminal facilities due to the sequential nature of the processing system. This section provides guidance concerning the spatial requirements to adequately accommodate the several functions carried out within the various areas of the airport terminal.

The Access Interface System

The curb element is the interface between the terminal building and the ground transportation system. A survey of the airport users will establish the number of passengers using each of the available ground transportation modes such as private automobile, taxi, limousine, courtesy car, public bus, rail, or rapid transit. Ratios may be established for both the passenger and vehicle modal choice for airport access.

Terminal Curb

The length of curb required for loading and unloading of passengers and baggage is determined by the type and volume of ground vehicle traffic anticipated in the peak period on the design day. Airports with relatively low passenger levels may be able to accommodate both enplaning and deplaning passengers from one curb front. Airports with higher passenger levels may find it desirable to physically separate the enplaning from the deplaning passengers, horizontally, if space permits, or vertically if space is limited. There is a tendency at large airports to also separate commercial vehicle traffic from private vehicle traffic.

The determination of the amount of curb space which will be required is related to airport policies relative to the assignment of priorities to the use of curb front and the provision of staging areas for taxis, buses, and other public transport vehicles. The parameters required for a preliminary analysis of curb front needs are the number and types of vehicles at the curb, the vehicle length, and the various occupancy times of different types of vehicles at the curb front for arriving and departing passengers.

Normally, a slot for a private automobile is considered to be about 25 ft, whereas for taxis 20 ft, limousines 30 ft, and transit buses 50 ft are used. Reported dwell times for private automobiles range from 1 to 2 min at the enplaning curb and from 2 to 4 min at the deplaning curb. Taxi dwell times lie closer to the lower range of these values, whereas limousines and buses may be at the curb anywhere from 5 to 15 min. These dwell times are highly influenced by the degree of traffic regulation and enforcement in the vicinity of the curb, and should be verified in specific studies. Normally a wide lane, in the order of 18 to 20 ft, is provided to accommodate direct curb access, maneuvering, and standing vehicles. This usually indicates a minimum of one and preferably two additional lanes in the vicinity of terminal entrances and exits to provide adequate capacity for through traffic. Rules of thumb which may be applied to determine curb front needs indicate that the full length of the curb adjacent to the terminal plus about 30 percent of the maneuvering lane may be considered as the available curb front. Therefore, a 100-ft curb may be considered to provide 130 ft of curb front in 1 h or 7800 foot-minutes of vehicle occupancy. If 120 automobiles per hour demand curb space for an average dwell time of 2 min, then 6000 foot-minutes of curb front is required, or the peak hour must provide a curb length of 100 ft. Other methods for approximating curb frontage have been reported in the literature [12, 43, 52, 58, 60].

Roadway Elements

The determination of the vehicular demand for the various on and off airport roadways is essential to ensure that adequate service levels are provided airport users. The main components of the highway system providing for access to airports from population and industrial centers is normally within the jurisdiction of federal, state, and local ground transportation agencies. However, coordination in area-wide planning efforts is essential so that the traffic generation potential of airports may be included within the parameters necessary for the proper planning of regional transportation systems. Guidance on the level and peaking characteristics of airport destined traffic may be found in the literature [55, 60].

The provision of adequate feeder facilities from the regional transport network to the airport is largely within the jurisdiction of the airport operator or owner. Vehicle volumes and peaking characteristics are usually determined by correlating modal preference and occupancy rates with flight schedules. Normally roadway facilities are designed for the peak hour traffic on the design day with adequate provision for the splitting and recirculation of traffic within the various areas of the airport property. The main roadway elements which must be considered are the feeder roads into the terminal area, the enplaning and deplaning roadways, and recirculation roadways.

The *Highway Capacity Manual* [33] provides criteria for level of service design and quantitative methods for determining the volumes which can be accommodated by various types of roadway sections. Unfortunately little guidance is available for the level of service design of airport roadways. For preliminary planning, however, it is reasonable to assume that feeder roads on the airport property provide acceptable service when they are designed to accommodate from 1200 to 1600 vehicles per hour per lane. Roadways providing access to the enplaning and deplaning terminal systems provide adequate service when they are designed to accommodate from 900 to 1000 vehicles per hour per lane. Terminal frontage roads and recirculation roads, however, provide adequate service when they are designed to accommodate from 600 to 900 vehicles per hour per lane. It is recommended that for preliminary planning purposes the above ranges of values be used to establish bounds on the sizes of these facilities for a demand-capacity analysis. In schematic design, analysis of the flow characteristics of individual sections of the roadway elements will yield final design parameters.

Parking

Most large airports provide separate parking facilities for passengers and visitors, employees, and rental car storage. In smaller airports these facilities may be combined in one physical location. Passenger and visitor parking are often segregated into short-term, long-term, and remote parking facilities. Those parking facilities most convenient to the terminal are designated as short term and a premium rate is charged for their use. Long-term parking is usually near the main terminal complex, but not as convenient as short term, and rates are usually discounted for long-term users. Remote parking, on the other hand, is usually quite distant from the terminal complex and provisions are normally made for courtesy vehicle transportation between these areas and the main terminal complex. The rates in these facilities are usually the most economical.

Short-term parkers are normally classified as those which park for 3 h or less and these may account for about 80 percent of the parkers at an airport. However, these short-term parkers account for only 15 to 20 percent of the accumulation of vehicles in the parking facility [43]. Preliminary planning estimates of the number of parking spaces required at an airport may be obtained from Fig. 10-10. The range of public parking spaces provided at existing airports varies from 1000 to 3000 per million originating passengers.

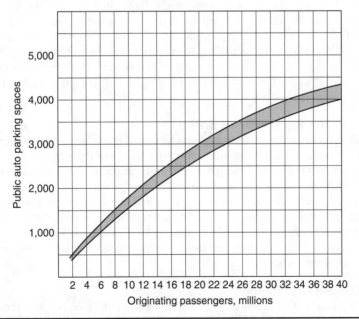

FIGURE 10-10 Public automobile parking space requirements (*Federal Aviation Administration [49]*).

It is recommended, however, that the range for preliminary planning be established between 1000 and 1400 parking spaces per million originating passengers.

More refined estimates of the total amount of parking required and the breakdown of short- and long-term space is obtained from analyses performed in the schematic design phase of the terminal planning process.

Entrances to parking facilities through ticket spitting devices are very common at airports. It has been observed that these devices can process anywhere from 400 to 650 vehicles per hour depending upon the degree of automation used as well as the continuity of the demand flow. It is recommended that the number of entrances be estimated on the basis of 500 vehicles per hour per device in preliminary planning. Parking revenue collection points at parking facility exits process from 150 to 200 vehicles per hour per position.

In parking garages, the capacity of ramps leading from one level to another is important during peak periods when considerable searching for an available space may occur, or vehicles may be directed immediately to a particular level. One-way straight ramps can accommodate about 750 vehicles per hour. However, a reduction in the order of 20 percent should be considered when two-way ramps are utilized. Circular or helical ramps, often used for egress from parking facilities, accommodate about 600 vehicles per hour in one direction.

The precise volume of vehicles which may be accommodated by a particular design will depend to a large extent on the geometric characteristics of the design, continuity of flow, information systems installed, and characteristics of the vehicles and users of the particular facility [22, 40, 60]. Most often, some type of analytical or simulation model is used in the schematic design phase of the project to test a preliminary design.

The Passenger Processing System

The passenger processing system consists of those facilities necessary for the handling of passengers and their baggage prior to and after a flight. It is the element which links the ground access system to the air transportation system. The terminal curbs provide the interface on the ground access side of the system, and the aircraft gates devices provide the interface on the airside of the system. In determining the particular needs of a specific component in this system, knowledge of the types of passengers and the extent of visitors impacting on each component is necessary.

Entryways and Foyers

Entryways and foyers are located along the curb element and serve as weather buffers for passengers entering and leaving the terminal building. The size of an entryway or foyer depends upon its intended usage. As entrances and exits may be relatively small, sheltered public waiting areas should be provided and sized to meet local needs. Designs must accommodate the physically challenged. These facilities are sized to process both passengers and visitors during the peak hour. Although the time of the enplaning and deplaning peaks may be different, it is likely that the deplaning peak will occur over a shorter time duration than the enplaning peak. It is often useful to subject a preliminary design proposal not only to an average peak hour demand but also a peak 20 or 30 min demand, particularly for the deplaning elements of the system. Preliminary design processing rates for automated doors in the vicinity of the enplaning and deplaning curb front can be taken as from 8 to 10 persons per minute per unit. These values may be reduced by 50 percent if the doors are not automated.

Terminal Lobby Area

The functions of significance to an air traveler performed in a terminal lobby are passenger ticketing, passenger and visitor waiting, and baggage check-in and claiming. Airports with less than 100,000 annual enplanements frequently carry out these functions in a single lobby. More active airports usually have separate lobbies for each function. The size of the lobby space depends on whether ticketing and baggage claim lobbies are separate, if passenger and visitor waiting areas are to be provided, and the density of congestion acceptable. In general, the lobby area should provide for passenger queuing, circulation, and waiting. Waiting lobby areas are designed to seat from 15 to 25 percent of the design hour enplaning passengers and visitors if departure lounges are provided for all gates, and from 60 to 70 percent if they are not provided [43, 50]. Usually about 20 ft^2 per person is provided for seating and circulation.

Airline Check-In Counter and Ticket Office

The airline check-in counter and ticket office is the area at the airport where the airline and passenger make final ticket transactions and check-in baggage for a flight. This includes the airline check-in counter, airline ticket agent service area, outbound baggage handling device, and support office area for the airline ticket agents. There are three types of ticketing and baggage check-in facilities, the linear, pass through, and island types. Each of these facilities is shown in a typical arrangement in Fig. 10-11.

The check-in transaction takes place at the check-in counter, which is a stand-up desk. To the left and right of the ticket counter position, a low shelf is provided to deposit, check-in, tag, and weigh baggage, if necessary, for the flight. Subsequently, the baggage is passed back by the agent to an outbound baggage conveyance device located near the counter for security screening, sorting, and loading on the aircraft.

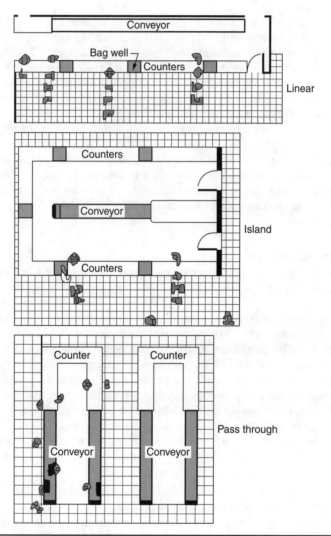

Figure 10-11 Typical check-in counter configurations (*Federal Aviation Administration [49]*).

The total number of check-in counter positions required is a function of the level of peak hour originating passengers, the types of facilities provided (i.e., multipurpose, express baggage check-in, and ticketing only) and the queues and delays acceptable to the airlines. In some markets, a considerable number of passengers may be preticketed and a higher percentage of express check-in positions may be warranted either within the terminal building or at the curb front. This is particularly true at airports serving tourist areas. A reasonable estimate of the number of positions required is gained by assuming the peak loading on the facilities is about 10 percent of the peak hour originating passengers and that a maximum queue length of five passengers per position is a desirable design goal. If this is the case, then 3000 peak hour originating passengers, translates to 300 peak loading passengers, which requires 60 ticketing positions. The sizing of the counter length depends on the mix of position types, but for preliminary estimating purposes a counter length from 10 to 15 ft for two positions is reasonable. Therefore, in this case the total linear footage of counter space would range from 300 to 450 ft. If a queuing depth of 3 ft per passenger is provided, then a minimum queuing depth of 15 ft is required. The counter itself requires a 10-ft depth and a circulation aisle of from 20 to 35 ft is appropriate. Therefore, the area devoted to this function ranges from about 13,500 to 27,000 ft^2. If single row lobby seating is provided in the area, this increases the area to between 16,500 to 30,000 ft^2. Approximations gained from other techniques yield similar results [12, 50].

The above calculations are useful for linear type counters where queues form in lines at each position. For corral type queuing similar results are obtained, the principal advantage of this type of queuing being less restriction to circulation in the ticket lobby area. The pass-through and island type counters result in a similar number of positions and counter length but in different geometric arrangements. The circulation and queuing areas may be modified as shown in Fig. 10-11. Normally, in pass-through ticketing and baggage check-in facilities, queuing is along the length of the counter whereas in island types queuing is in lines at the various ticketing positions.

The final determination of the number and mix of check-in positions is made through consultation with the various airlines to be served and through the use of analytical or simulation models [15, 58]. Figure 10-12 shows the terminal lobby area and airline check-in counters at the Pittsburgh International Airport.

The airline ticket office (ATO) support area may be composed of smaller areas for the operations and functions of accounting and safekeeping of tickets, receipts, and manifests; communications and information display equipment; and personnel areas for rest, personal grooming, and training. On the wall behind the ticket agents are posted information displays for the latest airline information on arriving and departing flights. Typical estimates of ATO space requirements may be obtained by taking the linear footage of counter and multiplying this by a depth of from 20 to 25 ft for this area. This yields a value of from about 6000 to 11,000 ft^2 for 3000 peak hour originating passengers. As the level of peak hour passengers increases various economies of scale may be gained in the use of such facilities. For example, in moving from 3000 to 6000 peak hour

Figure 10-12 Terminal lobby and airline check-in counters at the Pittsburgh International Airport.

originations, the increase in ATO space may be only about 30 percent. Other estimating procedures yield similar results [43, 50]. Again, the final determination of the space requirements is obtained through consultation with the various airlines using the facility.

The proliferation of self-service check-in kiosks at airports, along with the increasing ability for passengers to check-in for flights using the Internet or mobile device has created new challenges for check-in area planning and design. It is clear that there will always be a need for traditional check-in stations staffed by airline personnel. However, the number of staffed stations is becoming smaller compared to the number of installed self-service kiosks.

The implementation of self-service kiosks that have the capability of providing check-in service to more than one air-carrier, known as CUSS—"common use self-service" kiosks are helping to redefine the check-in spacing needs in airport terminals. These CUSS systems may be placed throughout the terminal entry lobby for all passengers to use, regardless of their individual airlines.

Passenger Security Screening

Security screening of passengers is an extremely important function in an airport terminal. The security screening area will include a checkpoint for identification inspection, walk-through metal detectors, and x-ray equipment for carry-on baggage inspection. The location and size of the screening area will be dictated primarily by passenger volume with consideration to issues of queuing, physical search, and passengers requiring additional processing. The equipment, techniques, and procedures may vary with location and are subject to change at any time. A few years ago, greeters, wellwishers, and visitors could be processed at the security area and proceed to the gate areas, but today only ticketed passengers are permitted in the sterile area. In the United States, the Transportation Security Administration (TSA) has prepared extensive guidelines and space planning and analysis tools [57] to assist planners in the design of the security screening area and these are introduced in Chapter 11.

Departure Lounges

The departure lounge serves as an assembly area for passengers waiting to board a particular flight and as the exit passageway for deplaning passengers. It is generally sized to accommodate the number of boarding passengers expected to be in the lounge 15 min prior to scheduled departure time, assuming this is the point in time when aircraft boarding begins. A conservative estimate of the percentage of passengers in the lounge at this time is 90 percent of the boarding passengers. The space should accommodate space for airline processing and information, passenger queues, seating for enplaning passengers, although all need not be seated, and an exitway for deplaning passengers.

Processing queues should not extend into the corridor to the extent that circulation is impaired. Lounge depths of 25 to 30 ft are considered reasonable for holding boarding passengers. Space for a departure lounge is proportioned on the basis of from 10 to 15 ft^2 per boarding passenger. Therefore, if a departure lounge is to accommodate 100 boarding passengers for a flight, its area should range from 1000 to 1500 ft^2. To a greater and greater extent common departure lounge areas are being utilized and these are sized based upon the total peak hour boarding passengers for the gates being served by the common lounge. Since it is likely that boarding for these aircraft will occur at different times in the peak hour, the total space required for separate lounges may be reduced by 20 to 30 percent for common lounges.

The corridor provided for deplaning passengers should be about 10 ft in width. The airline processing area should provide for at least two positions for narrow-bodied aircraft and up to four positions for wide-bodied aircraft to minimize queue lengths extending into the corridor. Processing rates range from one to two passengers per minute at these positions and peak arrival rates range from 10 to 15 percent of the boarding passengers. Therefore, queue depths of about 10 ft are reasonable values for preliminary design for individual departure lounges. For common lounges, the position of the processing area should be such as not to interfere with the circulation of passengers in the vicinity of the entrances and exits from the lounge. Typically these positions are located in the center of a satellite facility or at the end of the corridor

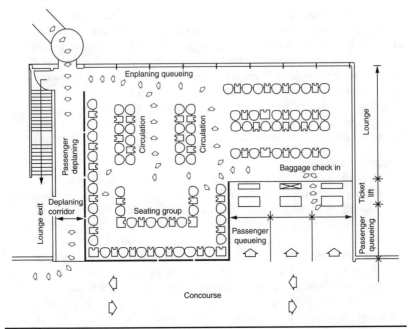

FIGURE 10-13 Departure lounge layout (*Federal Aviation Administration [50]*).

in a pier facility. Figure 10-13 shows the layout for a departure lounge seating about 70 passengers.

Corridors

Corridors provide circulation for passenger and visitor between departure lounges and between departure lounges and the central terminal areas. These should be designed to accommodate physically handicapped persons during the peak periods of high-density flow. Studies have shown that a typical 20-ft-wide corridor will have a capacity ranging from 330 to over 600 persons per minute. For planning purposes, corridor widths should be sized on the basis of about 16.5 passengers per foot of width per minute. The corridor width should be the width required at the most restrictive points, that is, the minimum free-flow width in the vicinity of restaurant entrances, phone booth clusters, or departure lounge check-in points. This standard is based upon a width of 2.5 ft per person and a depth separation of 6 ft between people. The corridor width is adversely impacted by the peaking of deplaning passengers in platoons but the deceased depth separations compensate for the decreased walking rates in these circumstances.

Further guidance on corridor width design is contained in the literature [12, 41, 43, 50].

Baggage Claim Facilities

The baggage claim lobby should be located so checked baggage may be returned to terminating passengers in reasonable proximity to the terminal deplaning curb. At low activity airports, checked baggage may be placed on a shelf for passenger claiming. More active airports have installed mechanical delivery and display equipment similar to that depicted in Fig. 10-14. The number of claim devices required is determined by the number and type of aircraft that will arrive during the peak hour, the time distribution of these arrivals, the number of terminating passengers, the amount of baggage checked on these flights, and the mechanism used to transport baggage from aircraft to

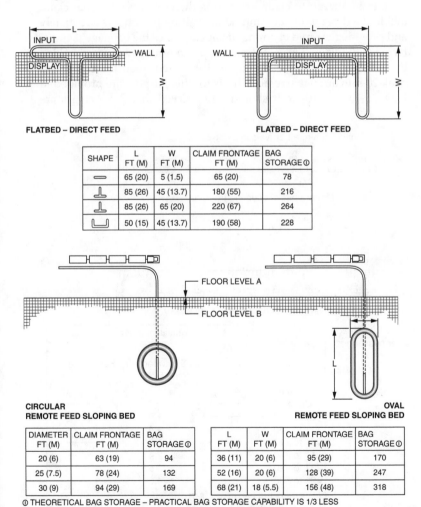

FLATBED – DIRECT FEED

FLATBED – DIRECT FEED

SHAPE	L FT (M)	W FT (M)	CLAIM FRONTAGE FT (M)	BAG STORAGE ①
⌒	65 (20)	5 (1.5)	65 (20)	78
⊥	85 (26)	45 (13.7)	180 (55)	216
⊥	85 (26)	65 (20)	220 (67)	264
⊔	50 (15)	45 (13.7)	190 (58)	228

CIRCULAR REMOTE FEED SLOPING BED

OVAL REMOTE FEED SLOPING BED

DIAMETER FT (M)	CLAIM FRONTAGE FT (M)	BAG STORAGE ①
20 (6)	63 (19)	94
25 (7.5)	78 (24)	132
30 (9)	94 (29)	169

L FT (M)	W FT (M)	CLAIM FRONTAGE FT (M)	BAG STORAGE ①
36 (11)	20 (6)	95 (29)	170
52 (16)	20 (6)	128 (39)	247
68 (21)	18 (5.5)	156 (48)	318

① THEORETICAL BAG STORAGE – PRACTICAL BAG STORAGE CAPABILITY IS 1/3 LESS

FIGURE 10-14 Mechanized baggage claim devices commonly used at airports (*Federal Aviation Administration [43]*).

the claim area. In the ideal situation, a baggage claim device should not be shared between flight arrivals at the same time as this leads to considerable congestion in the vicinity of the device and passenger confusion. Greater utilization of the devices is obtained when airlines time the sharing of claim devices for separate flights. Techniques similar to those used to construct ramp charts are useful in scheduling baggage claim devices and for determining the level of congestion in the baggage claim area.

At the present time, except in the most unusual situations, passenger delays in the baggage claim area can be significant due principally to the fact that passengers can travel from aircraft to claim areas much faster than baggage conveyance systems can transport the baggage from aircraft to claim areas. It is therefore essential that claim lobbies be designed to accommodate waiting passengers adequately and provide for rapid claiming of baggage once the baggage is transferred to the claim device. Estimating procedures have been provided by the FAA for sizing claiming devices based upon the equivalent peak 20-min aircraft arrivals. Charts for this purpose are provided in Figs. 10-15 and 10-16. It should be observed that these charts are

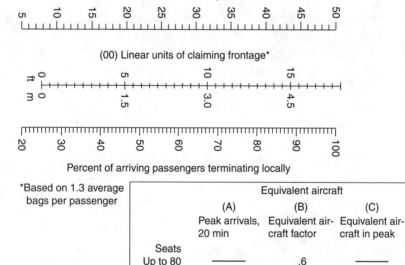

FIGURE 10-15 Estimating nomograph for baggage claim linear claim footage requirements (*Federal Aviation Administration [43]*).

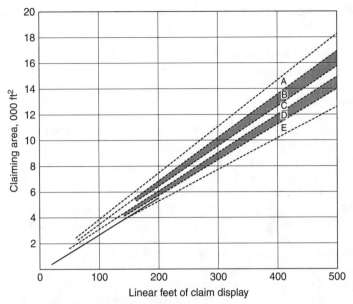

Areas for optimum configurations of

A Round – sloping bed/remote feed
   Tee – flat bed/direct feed
B Tee and u-shape alternating @ 75'
   (flat bed/direct feed)
C Oval – flat bed/direct feed
   Oval – sloping bed/remote feed
D Tee and u-shape alternating @ 60'
   (flat bed/direct feed)
E ⊔ – shape flat bed/direct feed

FIGURE 10-16   Estimating chart for total area of baggage claim facilities (*Federal Aviation Administration [43]*).

based upon checked baggage at the rate of 1.3 bags per person and adjustments may be required as baggage exceeds this rate.

In preliminary planning studies, great care should be exercised in using the above guidelines. Particular attention should be given to the space provided around a claim device and it is specifically recommended that a clear space of from 13 to 15 ft be provided adjacent to the device for active and waiting claim, as well as claim area circulation. In some instances baggage lobbies have been designed with adequate waiting areas and subsequently other facilities have been moved into these areas, considerably diminishing mobility. It is recommended that an additional 15 to 35 ft of circulation space be provided within the deplaning facility to allow for circulation between the claim devices, rental car positions, and deplaning curbs. If ground transportation facilities are located in these areas, they should be physically positioned so as not to restrict passenger flow to and from the claim area.

Intraairport Transportation Systems

The use of automated ground transportation systems within the terminal complex at airports is increasing as airports become larger and both the distance and time for passengers to travel through airports have become excessive. In most cases automated ground transportation also provides a clear delineation between the airside and landside functions of an airport and aids in the location of security processing facilities. Moving walkways and automated people mover (APM) systems have become important features in many large terminals.

An Automated People Mover (APM) is an advanced transportation system in which automated driverless vehicles operate on fixed guideways in exclusive rights-of-way. They differ from other forms of transit in that they operate without drivers or station attendants. These systems have been developed and implemented in various sizes and configurations since the early 1970s and today there are over 40 systems operating at airports worldwide.

Most of the early APM systems were implemented to facilitate passenger movement within a terminal or between the central terminal and a satellite building. Figure 10-17 shows an APM vehicle that operates in an underground tunnel that links the central terminal with a mid-field satellite terminal at the Greater Pittsburgh International Airport. More recently, APM systems have been designed to connect airport terminals

FIGURE 10-17   Automated People Mover (APM) Vehicle at the Greater Pittsburgh International Airport.

and link terminals with landside facilities such as parking, car rental services, regional transportation services, hotels, and other related employment and activity centers. Guidance on the planning of such facilities is contained in recent literature [20, 46, 59].

International Facilities

Airports with international operations require space for the inspection of passengers, crew, baggage, aircraft, and cargo. The area required for customs, immigration, agriculture, and public health services may be in a separate facility or in the terminal building itself. These facilities should be designed so that passenger flow between the aircraft and the initial processing station is unimpeded and as short as possible, there is no possibility of contact with domestic passengers or any unauthorized personnel until processing is complete, there is no possible way for an international arrival to bypass processing, and there is a segregated area for in-transit international passengers.

The size of this facility is based on the projection of hourly passengers requiring processing. It is recommended that the appropriate officials and agencies be contacted during the preliminary deign phase to determine specific design requirements. Some guidance on the processing rates and sizing of these facilities is found in the literature [43, 58].

Other Areas

Most terminals are developed to accommodate several other activities and the space needs should be determined for each airport based upon local requirements. These activities are identified below.

Airline Activities

The following airline activities may exist at all or some terminal facilities and should be discussed with the airlines which plan on utilizing the facility.

1. Outbound baggage makeup and inbound baggage and conveyance system

2. Cabin services and aircraft maintenance

3. Flight operations and crew ready rooms

4. Storage areas for valuable or outsized baggage

5. Air freight pickup and delivery

6. Passenger reservations and VIP waiting areas

7. Administrative offices

8. Ramp vehicle and cart parking and maintenance

Passenger Amenities

The factors which influence the extent of passenger amenities include the passenger volume, community size, the location and extent of off-airport services, interests and abilities of potential concession-aires, and rental rates. These generally include

1. Food and beverage services, and newsstands
2. A variety of stores and services
3. Counters for car rental and flight insurance companies
4. Public lockers and public and courtesy telephones
5. Amusement arcades and vending machines
6. Public restrooms

Airport Operations and Services

These facilities and services are normal to most public buildings and include the following:

1. Offices for airport management and staff functions including police, medical and first aid, and building maintenance
2. Building mechanical systems such as heating, ventilation, and air conditioning
3. Communication facilities
4. Electrical equipment
5. Government offices for air traffic control, weather reporting, public health and immigration, and customs
6. Conference and press facilities

Overall Space Requirements

Guidelines have been presented above for the approximate space requirements for the various components in passenger terminal facilities. Once the facilities have been estimated one might compare the space requirements to the approximations given in Table 10-2. The values in this table present overall space requirements which should provide a reasonable level of service and a tolerable occupancy level for the various facilities indicated.

Concept Development

In this phase of the process, the blocks of spaces determined in space programming are allocated in a general way to the terminal complex. There are a number of ways in which the facilities of the passenger terminal system are physically arranged and in which the various passenger processing activities are performed. Centralized passenger processing means all the facilities of the system are housed in one building and used for processing all passengers using the building. Centralized processing facilities offer economies of scale in that many of the common facilities may be used to service a large number of aircraft gate positions. Decentralized processing, on the other hand, means the passenger facilities are arranged in smaller modular units

Component	Space Required in 1000 ft² or 100 m² per 100 Typical Peak Hour Passengers
Ticket lobby	1.0
Baggage claim	1.0
Departure lounge	2.0
Waiting rooms	1.5
Immigration	1.0
Customs	3.0
Amenities	2.0
Airline operations	5.0
Total gross area	
Domestic	25.0
International	30.0

TABLE 10-2 Typical Terminal Building Space Requirements

and repeated in one or more buildings. Each unit is arranged around one or more aircraft gate positions and serves the passengers using those gate positions. There are four basic horizontal distribution concepts, as well as many variations or hybrids which include combinations of these basic concepts. Each can be used with varying degrees of centralization. These concepts are discussed below.

Horizontal Distribution Concepts

The following terminal concepts should be considered in the development of the terminal area plan. Sketches of the various concepts are shown in Fig. 10-18. Many airports have combined one or more terminal types.

Pier or Finger Concept

The pier concept has an interface with aircraft along piers extending from the main terminal area. Aircraft are usually arranged around the axis of the pier in a parallel or nose-in parking alignment. Each pier has a row of aircraft gate positions on both sides, with a passenger concourse along the axis which serves as the departure lounge and circulation space for both enplaning and deplaning passengers. This concept usually allows for the expansion of the pier to provide additional aircraft parking positions without the expansion of the central passenger and baggage processing facility. Access to the terminal area is at the base of the connector or the pier. If two or more piers are employed, the spacing between the two piers must provide for maneuvering of aircraft on one or two apron taxilanes. When each pier serves a large number of gates, and the probability exists that two or more aircraft may frequently be taxiing between two piers and

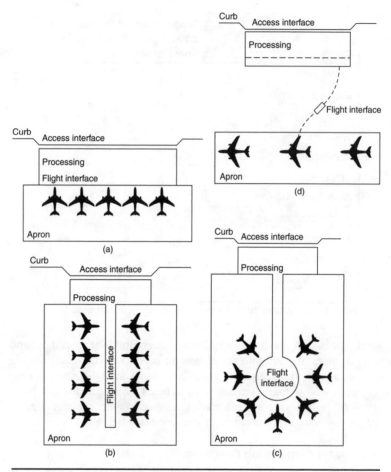

FIGURE 10-18 Horizontal distribution concepts for passenger terminals: (a) linear, (b) pier, (c) satellite, (d) transporter.

will be in conflict with one another, then two taxilanes are advisable. Also, access from this taxiway system by two or more aircraft may require two apron edge taxiways to avoid delays.

The chief advantage of this concept is its ability to be expanded in incremental steps as aircraft or passenger demand warrant. It is also relatively economical in terms of capital and operating cost. Its chief disadvantages are its relatively long walking distance from curb front to aircraft and the lack of a direct curb front relationship to aircraft gate positions.

Satellite Concept

The satellite concept consists of a building, surrounded by aircraft, which is separated from the terminal and is usually reached by means of a surface, underground, or above ground connector. The aircraft are normally parked in radial or parallel positions around the satellite. It often affords the opportunity for simple maneuvering and taxiing patterns for aircraft but requires more apron area than other concepts. It can have common or separate departure lounges. Since enplaning and deplaning from the aircraft is accomplished from a common and often remote area, mechanical systems may be employed to transport passengers and baggage between the terminal and satellite.

The main advantages of this concept lie in its adaptability to common departure lounge and check-in functions and the ease of aircraft maneuverability around the satellite structure. However, construction cost is relatively high due to the need to provide connecting concourses to the satellite. It lacks flexibility for expansion and passenger walking distances are relatively long.

Linear, Frontal or Gate Arrivals Concept

The simple linear terminal consists of a common waiting and ticketing area with exits leading to the aircraft parking apron. It is adaptable to airports with low airline activity which will usually have an apron providing close-in parking for three to six commercial passenger aircraft. The layout of the simple terminal should take into account the possibility of pier, satellite, or linear additions for terminal expansion. In the gate arrivals or frontal concept, aircraft are parked along the face of the terminal building. Concourses connect the various terminal functions with the aircraft gate positions. This concept offers ease of access and relatively short walking distances if passengers are delivered to a point near gate departure by vehicular circulation systems. Expansion may be accomplished by linear extension of an existing structure or by developing two or more terminal units with connectors.

Both of these concepts provide direct access from curb front to aircraft gate positions and afford a high degree of flexibility for expansion. It does not provide convenient opportunities for the use of common facilities and, as this concept is expanded into separate buildings, it may lead to high operating costs.

Transporter, Open Apron or Mobile Conveyance Concept

Aircraft and aircraft servicing functions in the transporter concept are remotely located from the terminal. The connection to the terminal is provided by vehicular transport for enplaning and deplaning passengers. The characteristics of the transporter concept include flexibility in providing additional aircraft parking positions to accommodate increases in schedules or aircraft size, the capability to maneuver an aircraft in and out of a parking position under its own power, the separation of aircraft servicing activities from the terminal, and reduced walking distances for the passenger.

Concept Combinations and Variations

Combinations of concepts and variations are a result of changing conditions experienced from the initial conception of the airport throughout its life span. An airport may have many types of passenger activity, varying from originating and terminating passengers using the full range of terminal services to passengers using limited services on commuter or connecting flights. Each requires a concept that differs considerably from the other. In time, the proportion of traffic handled by flights may change, necessitating modification or expansion of the facilities. Growth of aircraft size or a new combination of aircraft types servicing the same airport will affect the type of concept. In the same way, physical limitations of the site may cause a pure conceptual form to be modified by additions or combinations of other concepts.

Combined concepts acquire certain of the advantages and disadvantages of each basic concept. A combination of concept types can be advantageous where more costly modifications would be necessary to maintain the original concept. For example, an airline might be suitably accommodated within an existing transporter concept terminal while an addition is needed for a commuter operation with rapid turnovers which would be best served by a linear concept extension. In this situation, combined concepts would be desirable. The appearance of concept variations and combinations in a total apron terminal plan may reflect an evolving situation in which altering needs or growth have dictated the use of different concepts. Illustrations of the generation of various horizontal distribution concepts in the concept development phase of Geneva Intercontinental Airport are shown in Fig. 10-19. Applications of several of the concepts to existing airports are shown in Fig. 10-20a through 10-20c.

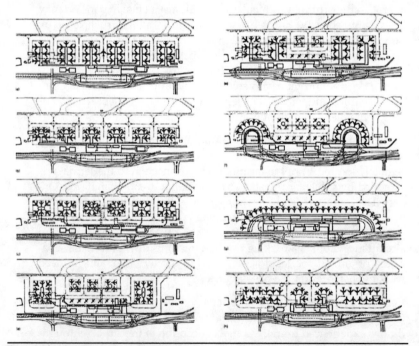

Figure 10-19 Conceptual designs for Geneva Intercontinental Airport (*Reynolds, Smith and Hills [25]*).

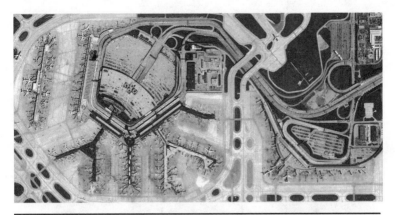

FIGURE 10-20A   Linear, satellite and pier concepts—O'Hare International Airport.

FIGURE 10-20B   Pier and satellite concepts—Cleveland Hopkins International Airport.

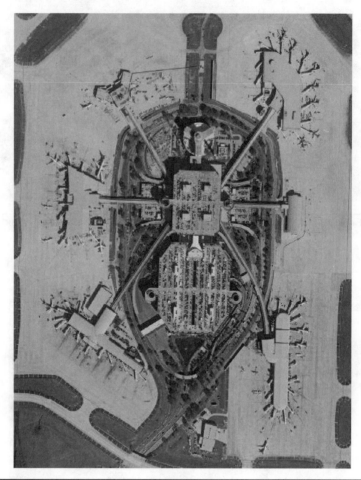

FIGURE 10-20C   Satellite airside/landside concept—Tampa International Airport.

Figure 10-20*a* shows the terminal area layout at O'Hare International Airport which shows the linear, satellite and pier concepts. Concourse B is a linear concept, concourse C is a satellite concept connected to the main terminal building by an underground moving sidewalk, and concourses E, F, G, H, K, and L are pier concepts. Figure 10-20*b* shows the terminal area layout at Cleveland Hopkins International Airport which consists of pier and satellite concepts. Figure 10-20*c* shows the terminal area layout at Tampa International Airport which shows five airside satellites connected to the landside building by automated transit on above ground fixed guideways.

Vertical Distribution Concepts

The basis for distributing the primary processing activities in a passenger terminal among several levels is mainly to separate the flow of arriving and departing passengers. The decision concerning the number of levels a terminal facility should have depends primarily on the volume of passengers and the availability of land for expansion in the immediate vicinity. It may also be influenced by the type of traffic, for example, domestic, international, or commuter passengers being processed, by the terminal area master plan, and by the horizontal processing concept chosen.

With a single level system all processing of passengers and baggage occurs at the level of the apron. Separation between arriving and departing passenger flows is achieved horizontally. Amenities and administrative functions may occur on a second level. With this system, stairs are normally used to load passengers onto aircraft. This system is quite economical and is suitable for relatively low passenger volumes. The single level terminal is shown in Fig. 10-21a.

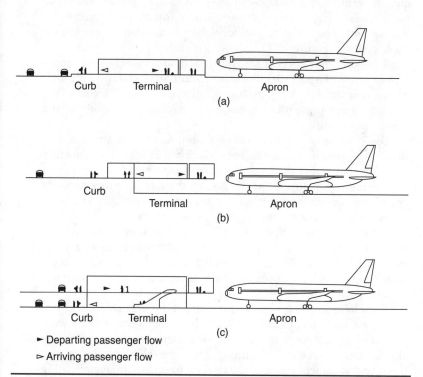

Curb   Terminal   Apron
(a)

Curb
    Terminal   Apron
(b)

Curb   Terminal   Apron
(c)

► Departing passenger flow
▷ Arriving passenger flow

Figure 10-21 Vertical distribution concepts: (a) single level, (b) second level loading, (c) two-level system (*Federal Aviation Administration [50]*).

Two level passenger terminal systems may be designed in a number of different ways. In one type, shown in Fig. 10-21*b*, the two levels are used to separate the passenger processing area and the baggage handling areas. Thus, processing activities including baggage claim occur on the upper level, while airline operations and baggage handling activities occur at the lower apron level. The advantage of raising the passenger handling level is that it becomes compatible with aircraft doorsill heights, allowing convenient interface with the aircraft. Vehicular access occurs on the upper level to facilitate the interface with the processing system.

Another articulation of the two-level system separates the arriving and departing passenger streams. In this case departing passenger processing activities occur on the upper level and arriving passenger processing including baggage claim occurs at the apron level. Airline operations and baggage handling also occur at the lower level. Vehicular access and parking occur at both levels, one for arrivals and one for departures, and parking can be surface or structural. An example of this design is shown in Fig. 10-21*c*.

Variations in these basic designs may occur when traffic volumes or the type of traffic so require. For example, for international airport terminals, a third level may be needed for international passengers. Also, at large airports where intraairport transportation systems operate, a special level may be needed to provide for these systems. Figure 10-22A shows a multilevel system with structural parking, intraairport transportation, and underground mass transit access. Figure 10-22B shows a multilevel system with integrated structural parking. In this variation more direct access to the processing component is attained by providing parking above the processing facility.

Based upon an examination of a large number of airports, it is possible to identify those concepts which are candidates for further consideration. Using the level of annual enplanements and the function nature of the airport, as defined by the relative proportions of originating, terminating, through, or connecting passengers, Fig. 10-23 offers some guidance to the airport planner for the initial identification of appropriate horizontal and vertical distribution concepts. It should be noted, however, that in many instances the constraints of existing terminal facilities, land availability, and the ground access system may restrict the options which are viable alternatives for terminal expansion.

Prior to proceeding with the planning of the airport terminal system, an evaluation of the concepts which have evolved in the conceptual development phase of the project is undertaken, to identify those alternatives which should be brought into the schematic design and design development phase of the project. To do this an overall rating of the various concepts relative to the design criteria is performed.

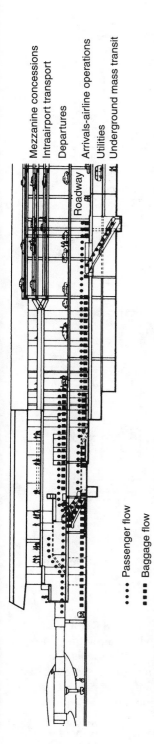

Mezzanine concessions
Intraairport transport
Departures
Arrivals-airline operations
Utilities
Underground mass transit

Roadway

•••• Passenger flow
▪▪▪▪ Baggage flow

Figure 10-22A Multilevel passenger processing system—structural parking adjacent to terminal (*Hamburg Airport Authority*).

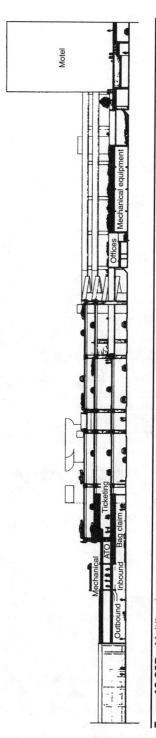

Motel

Offices   Mechanical equipment

Ticketing
ATO   Bag claim
Mechanical
Outbound   Inbound

Figure 10-22B Multilevel passenger processing system—structural parking above processing area (*Reynolds, Smith and Hills*).

Airport size by enplaned pax/year	Concepts applicable				Physical aspects of concepts							
	Linear	Pier	Satellite	Transporter	Single level curb	Multi level curb	Single level terminal	Multi level terminal	Single level connector	Multi level connector	Apron level boarding	Aircraft level boarding
Feeder under 25,000	X				X	X					X	
Secondary 25,000 to 75,000	X				X	X					X	
75,000 to 200,000	X				X	X	X				X	
200,000 to 500,000	X	X			X	X	X				X	
Primary over 75% pax O/D 500,000 to 1,000,000	X	X	X		X		X		X	X	X	X
Over 25% pax transfer 500,000 to 1,000,000	X	X	X		X		X		X	X	X	X
Over 75% pax O/D 1,000,000 to 3,000,000		X	X	X	X	X		X	X	X	X	X
Over 25% pax transfer 1,000,000 to 3,000,000		X	X		X	X		X	X	X	X	X
Over 75% pax O/D over 3,000,000		X	X	X	X	X		X	X	X	X	X
Over 25% pax transfer over 3,000,000		X	X		X	X		X	X	X		X

FIGURE 10-23 Applicable concepts for airport design (*Federal Aviation Administration [50]*).

The concept evaluation rating factors used for Geneva airport are listed in Table 10-3.

Schematic Design

The schematic design process translates the concept development and overall space requirements into drawings which show the general size, location, and shape of the various elements in the terminal plan. Functional relationships between the components are established and evaluated. The adequacy of the overall space program is evaluated by airport users relative to their specific needs. This phase of the process specifically examines passenger and baggage flow routes through the system and seeks to examine the adequacy of the facility from the point of view of flow levels and flow conflicts.

Passenger convenience: 　Walking distance from curb to aircraft 　Walking distance for transfer passengers 　Walking distance from parking to aircraft 　Ease of passenger orientation 　Ease of passenger processing
Operational effectiveness: 　Efficient taxing routes 　Ground flow coordination of vehicles and aircraft 　Apron area maneuverability 　Apron adaptation to future aircraft 　Vehicular access flows 　Direct routes to ancillary facilities
Expansion adaptability: 　Ancillary facilities, flexible land use 　Staging adaptability 　Visual character of increments 　Gross terminal expandability 　Expandability of terminal elements
Economic effectiveness: 　Capital cost 　Maintenance and operating costs 　Ratio of revenue- to non-revenue-producing areas

Source: Reynolds, Smith and Hills [25].

TABLE 10-3   Conceptual Development Rating Factors for Evaluation of Terminal Planning Concepts

Modeling techniques are usually employed in this phase of the process to identify passenger processing, travel, and delay times, and the generation of lines at processing facilities. The main purpose for analyzing passenger and baggage handling systems is the determination of the extent and size of the facilities needed to provide a desired level of convenience to the passenger at reasonable cost. In this analysis alternative layouts can be studied to determine which is the most desirable.

Analysis Methods

A number of systems analysis techniques have proven to be useful for the analysis of facilities for passengers and baggage. These include network models, queuing models, and simulation models.

Network Models

These models are particularly useful for representing and analyzing the interrelationships between the various components of a passenger or baggage processing system. For example, passenger processing can be represented as a network with the nodes representing service facilities and the links representing the travel paths and passenger splits. This type of representation allows estimation of delays to the passenger at various locations within the terminal.

An example of a network that has been applied to the evaluation of arriving passenger delay is the critical path model (CPM) [47]. CPM is used to coordinate the various activities that take place in the system for handling both passengers and baggage. Nodes that represent critical activities, that is, those that take the greatest amount of time, are easily identified and can be analyzed in more detail to determine their effect on the overall performance of the system.

The analysis of the service time and waiting time at each processor in a network model can be obtained through either analytical queuing models or simulation.

Analytical Queuing Models

Queuing theory permits the estimation of delays and queue lengths for service facilities under specified levels of demand. The application of queuing theory yields useful estimates of processing and delay times from which the required sizes of facilities and operating costs may be derived.

Virtually all of the components of the passenger handling system can be modeled as service facilities using queuing models. The diagram in Fig. 10-24 and Example Problem 10-10 showed an example of the application of a deterministic queuing model to the operation of a runway system to determine aircraft delay. A similar type of analysis can be applied to the analysis of passenger processing systems. It is possible to evaluate the impact of adding ticket agents on delays to passengers and on the size of the queues. With this information it is possible to evaluate the feasibility of alternative operating strategies for the ticketing facility.

Diagrams similar to this may be constructed for each of the processors for the passenger and baggage handling system and yield satisfactory results when the demand rate exceeds the service rate.

For the analysis of component delay and queue length when the average demand rate over some period of time is less than the average service rate, queuing theory is used to generate mathematical functions representative of the arrival and service performance of the system. To specify the mathematical formulation of this problem, it is necessary to define the arrival distribution, the service distribution, the number and use of the servers, and the service discipline. Many of the components which service passengers in the airport terminal exhibit a random or Poisson arrival process. The service characteristics are usually exponential, constant, or some general distribution defined by average service times and the variance of average service time. In most cases, there is more than one channel for the performance of passenger service, and the queuing mechanism is based on first come, first served basis. Extensive research has been carried out in recent years to determine mathematical formulations which adequately represent the processing system [2, 10, 14, 19, 21, 53, 56]. Because of the variability associated with passenger behavior at an airport, it is virtually impossible to obtain precise mathematical formulations for delay at processors. However, reasonable estimates of delay and corresponding queue lengths are possible using simple formulations.

One such formulation [17, 35, 52, 56] is that of a multiple station queuing system with a Poisson arrival distribution and a service time distribution which is characterized by the average service time and the variance of the average service time as shown in Eq. (10-1).

$$W_s = \left(\frac{\sigma^2 + t^2}{2t^2}\right) \frac{\lambda^k t^{k+1}}{(k-1)!(k-\lambda t)^2 \sum\limits_{n=0}^{n=k-1} [(\lambda t)^n/n!] + (\lambda t)^k/[(k-1)!(k-\lambda t)]} \tag{10-1}$$

where W_s = average delay per person

$\quad \lambda$ = average demand rate

$\quad t$ = average service time for a processor, the reciprocal of the average service rate of a processor, μ

$\quad \sigma$ = standard deviation of the average service time of a processor

$\quad k$ = number of processors

$\quad n$ = counter in the equation

This equation is valid when the average demand rate on the system of processors λ is less than to total service rate of the processors, $k\mu$; that is, the ratio of the average demand rate to the total service rate ρ is less than 1.

When the number of processors k is equal to 1, this equation reduces to

$$W_s = \left(\frac{\sigma^2 + t_2}{t^2}\right) \frac{\lambda}{2\mu(\mu - \lambda)} \tag{10-2}$$

or since ρ is equal to $\lambda/k\mu$

$$W_s = \left(\frac{\sigma^2 + t^2}{t^2}\right) \frac{\rho}{2\mu(1-\rho)} \tag{10-3}$$

For a single server system which exhibits a Poisson arrival distribution and an exponential service time distribution, Eq. (10-1) reduces to

$$W_s = \frac{\lambda}{\mu(\mu - \lambda)} \tag{10-4}$$

or

$$W_s = \frac{\rho}{\mu(1-\rho)} \tag{10-5}$$

For a single server system which exhibits a Poisson arrival distribution and a constant service time distribution, Eq. (10-1) reduces to

$$W_s = \frac{\lambda}{2\mu(\mu - \lambda)}$$
(10-6)

or

$$W_s = \frac{\rho}{2\mu(1-\rho)}$$
(10-7)

When demand is less than capacity, the following expression gives the average line length N over the period being analyzed and consists of those in service and those waiting for service at a processor.

$$N = \left(W_s + \frac{1}{\mu}\right)\lambda$$
(10-8)

When $\rho > 1$ there is statistical delay plus excess delay which is defined by a deterministic model. For design purposes it is sufficient to estimate delays in such a system in which the statistical delay W_s is computed from the appropriate equation above with $\rho = 0.90$ and the excess delay W_e is added to this from the equation below to compute total processor delay. The rationale for computing the statistical delay with $\rho = 0.9$ is that in reality as demand approached capacity the delay cannot become infinite as airline or airport operating practices will limit delay by utilizing additional servers.

$$W_e = \frac{T(\lambda - k\mu)}{2k\mu}$$
(10-9)

where W_e = delay when the demand exceeds the service rate
   T = time period during which the demand exceeds the service rate
   λ = total demand on the system of processors
   k = number of processors
   μ = service rate of a processor

It is usually assumed that the time required to reduce the demand to capacity T is about one-half of the time period being analyzed. Typically when demand exceeds capacity, the operator of the facility will increase the number of operating service facilities to alleviate the growth in both waiting time and queue lengths. However, the extent to which this is done is a function of airline operating policies and the availability of additional manpower and physical facilities.

The average line length over the period analyzed N including those in service and those waiting for service; when demand exceeds, capacity can be estimated by the following equation.

$$N = \left(W_s + W_e + \frac{1}{\mu}\right)\mu$$
(10-10)

Estimates of the waiting times and queue lengths for multiple service channels may be obtained by proportioning the demand equally among processors with the same service characteristics and utilizing single server models. Better estimates may be obtained through the use of multiple channel queuing models whose mathematical formulation was given above. A representation of the passenger delay at baggage claim facilities is given by the relationship [52]

$$W_t = E(t_2) + \frac{nT}{n+1} - E(t_1) \qquad (10\text{-}11)$$

where $E(t_2)$ = expected value of the time when the first piece of baggage arrives at the claim area

$E(t_1)$ = expected value of the time when passengers arrive at the claim area

n = number of pieces of baggage to be claimed by each passenger

T = length of time from the arrival of the first bag until the arrival of the last bag at the claim device

Others have formulated models which show the buildup and drop-off of passengers and baggage in the claim area [19].

The use of a generalized probability density function called the *Erlang distribution* is recommended as a mechanism for evaluating the service characteristics of the various passenger component processors. By collecting a sample of data relative to a specific component, the constant in the Erlang distribution function can be calculated. This constant determines the particular functional relationships for the server. It is possible that this type of distribution may better describe the queuing characteristics of processors. This distribution has been used successfully in passenger terminal modeling [15].

Great care must be exercised in the application of mathematical models and the interpretation of the results. In most cases, the mathematical representation of the terminal system is best suited for the comparison of alternatives and the identification of those components in the system requiring detailed analyses.

The specification of the service time distribution for use in queuing equations is a function of the distribution of service times demonstrated by the service facility. In general, those processors which exhibit a requirement for small service times, for example, flow through type facilities such as doors, security, and gates are probably best represented by an exponential service time distribution. Those facilities which require a finite service time at a processor such a ticketing, baggage check, seat selection, and rental car checkout, are probably best represented by either a general service time distribution, which is characterized by the average service time and the standard deviation of the average service time, or a constant service time distribution.

In a network model, the average passenger processing time $E(T_p)$ is given by

$$E(T_p) = E(T_w) + E(T_s) + E(T_d) \qquad (10\text{-}12)$$

where $E(T_w)$ = average passenger delay time throughout the system of processors

$E(T_s)$ = average passenger service time throughout the system of processors

$E(T_d)$ = average passenger travel time through the network of processors

Estimates for the observed range of service time for many of the passenger processing components at an airport are given in Table 10-4.

Example Problem 10-1 presents an illustration of the analysis of the enplaning system at an airport using a network model and the above queuing equations.

Example Problem 10-1 The terminal building layout for enplaning passengers at an airport is given in Fig. 10-24.

Two airlines are servicing the airport, North American Airlines (NA) and Western Pacific Airlines (WP). The enplaning passenger processing facilities and their service rates are given in Table 10-5.

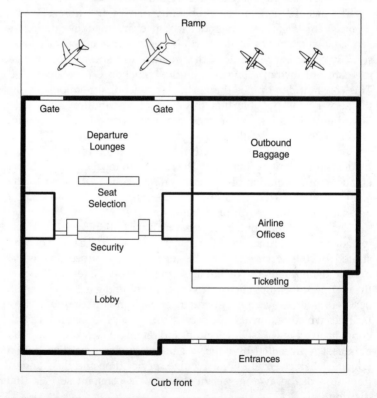

FIGURE 10-24 Layout of enplaning passenger processing system for Example Problem 10-1.

Component Type	Service Rate per Passenger, s	Standard Deviation
Entrance and exit doors		
Automated with baggage	2.0–25	0.5
Automated without baggage	1.0–1.5	0.75
Manual with baggage	3.0–5.0	1.0
Manual without baggage	1.5–3.0	0.75
Stairways	3.0–4.0	1.0
Escalators	1.0–3.0	1.0
Moving sidewalks	1.0–3.0	1.0
Apron doors		
With stairs	4.0–8.0	2.0
Without stairs	3.0–7.0	1.5
Jetway	2.0–6.0	1.0
Ticketing and baggage		
Manual with baggage	180–240	60
Manual without baggage	100–200	30
Baggage only	30–50	10
Information	20–40	10
Automated with baggage	160–220	30
Automated without baggage	90–180	40
Security		
Hand-check baggage	30–60	15
Automated	30–40	10
Seat selection		
Single fights	25–60	20
Multiple fights	35–60	15
Rental car		
Check-in	120–240	60
Checkout	180–300	90
Automated check-in	60–90	20
Baggage claim		
Manual	10–15	8
Automated carousel	5–10	5
Automated racetrack	5–10	5
Automated tee	6–12	5

Sources: Various airport studies.

TABLE 10-4   Observed Service Times for Passenger Processing Facilities at Airports

Type	Airline	Number	Average Service Time per Passenger, s
Entrance doors	All	3	15
Regular ticketing	NC	3	210
	UA	3	180
Express check-in	UA	1	60
Security	All	2	30
Seat selection	NC	1	45
	UA	1	30
Ramp gates	NC	1	20
	UA	1	20

TABLE 10-5 Enplaning Passenger Service Processors for Example Problem 10-1

Demand data collected at the airport indicate that 10 percent of the enplaning passengers proceed directly from curb front to security, 20 percent of Western Pacific Airlines enplaning passengers use the express check-in, there are 0.5 visitors per passenger during the peak hour, and 50 percent of these visitors proceed beyond security.

The peak hour enplaning passenger demand on the design day is expected to be 135 passengers, of which 53 percent use North American Airlines and 47 percent use Western Pacific Airlines. Typically the peak 30 min of the peak hour has 57 percent of the peak hour demand. Assume average walking rates are 1.5 ft/s. Airline operating practices limit the periods when demand exceeds capacity to 15 min.

It is necessary to determine the average enplaning passenger delay and line length at each processor, and the average passenger processing time during the peak 30 min of the peak hour on the design day at this airport.

The link-and-node diagram in Fig. 10-25 shows the relationship between the enplaning passenger processors at the airport, and includes the passenger split between processors, the number of processors, the processing time per passenger, and the distance between processor in feet.

Since the peak 30 min passenger arrival rate is 57 percent of the peak hour arrival rate, the peak 30 min passenger demand into the enplaning passenger system is 0.57(135) = 77 passengers. Since there are 0.5 visitors per passenger, the peak 30 min visitor demand into the enplaning passenger system is 0.5(77) = 39 visitors. Combining these demand parameters with the passenger splits given in Fig. 10-25, recognizing that only 50 percent of the visitors proceed beyond security, the processor flow rates in the peak 30 min are given in Fig. 10-26. In this figure the numbers above the lines connecting processors represent the passengers flowing into the processor and the numbers below these lines represent the visitors flowing into each processor when the processor processes visitors.

Assuming that all processors may be modeled by single processor systems in which the demand is split evenly between processors, the demand rate in persons per minute, the service rate in persons per minute, and the ratio of the

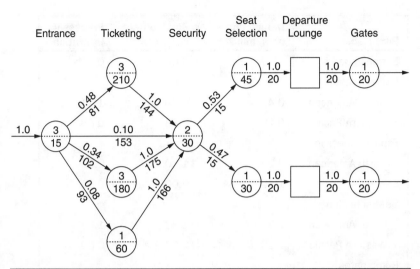

Figure 10-25 Link–and-node diagram representing enplaning passenger processing system for Example Problem 10-1.

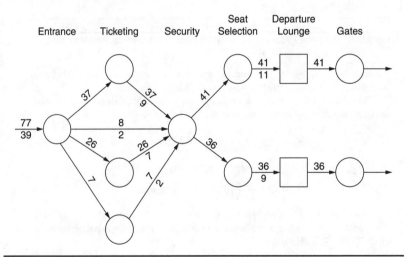

Figure 10-26 Link-and-node diagram representing passenger and visitor flows into passenger processing facilities for Example Problem 10-1.

demand to capacity for each processor, the processor utilization, is computed as shown in Table 10-6.

The departure lounges are accumulating processors and are analyzed differently than the queuing processors. Also note that the North American regular ticketing position and the North American seat selection components have a demand which exceeds capacity since the processor utilization is greater than 1.

Based upon an examination airport processor it was found that those processors which are essentially flow-through processors (such as entrance doors, security and ramp gates) behave as Poisson arrivals and exponential service rate queuing mechanisms, whereas those processors which require a discrete service

Processor	Demand Rate, pers/min	Service Rate pers/min	Processor Utilization
Entrance doors	1.29	4.00	0.32
Regular ticketing			
North American	0.41	0.29	1.41
Western Pacific	0.29	0.33	0.88
Express check-in			
North American	0.23	1.00	0.23
Security	1.62	2.00	0.81
Seat selection			
North American	1.37	1.33	1.03
Western Pacific	1.20	2.00	0.60
Ramp gates			
North American	1.37	3.00	0.46
Western Pacific	1.20	3.00	0.40

TABLE 10-6   Service Processor Characteristics for Example Problem 10-1

function (such as regular ticketing, express check-in, and seat selection) behave as Poisson arrival and constant service rate queuing mechanisms.

Therefore, for the entrance doors, which are Poisson arrivals and exponential services, the average passenger delay is from Eq. (10-4),

$$W_s = \frac{1.29}{4.00(4.00-1.29)} = 0.12 \text{ min}$$

and from Eq. (10-8) the average line length is

$$N = \left(0.12 + \frac{1}{4.00}\right)1.29 = 0.48 \text{ person}$$

For the North American regular ticket counters, which exhibit Poisson arrivals and constant services, since the demand rate is greater than the capacity, we have from Eq. (10-7) with $\rho = 0.9$

$$W_s = \frac{0.9}{2(0.29)(1-0.9)} = 15.52 \text{ min}$$

and from Eq. (10-9)

$$W_e = \frac{15(0.41-0.29)}{2(0.29)} = 3.10 \text{ min}$$

Note that the value of k is equal to 1 because the demand was allocated equally among each of the ticketing positions, and the excess delay period T is equal to $0.5(30) = 15$ min.

Therefore, $W = 15.52 + 3.10 = 18.62$ min of average delay time per passenger. The average line length is, from Eq. (10-10),

$$N = \left(18.62 + \frac{1}{0.29}\right)0.29 = 6.40 \text{ passengers}$$

| Processor | Delay Time | | Line Time, min | Length, pers |
	Statistical, min	Service Excess, min		
Entrance doors	0.12		0.25	0.48
Regular ticketing				
North American	15.52	3.10	3.45	6.40
Western Pacific	10.98		3.00	4.05
Express check-in				
Western Pacific	0.15		1.00	0.26
Security	2.13		0.50	4.26
Seat selection				
North American	3.38	0.23	0.75	5.80
Western Pacific	0.38		0.50	1.06
Ramp gates				
North American	0.28		0.33	0.84
Western Pacific	0.22		0.33	0.66

TABLE 10-7   Average Passenger Delay Time, Service Time, and Processor Line Lengths in Peak 30 Min for Example Problem 10-1

Computing the delays and queue lengths for each processor results in the values shown in Table 10-7.

From this table it can be seen that considerable delay and line formation exist at the both North American and the Western Pacific regular ticketing facilities, the security check-point, and the North American seat selection facility.

From the passenger splits shown in Fig. 10-25 and the distances between components, the expected values of the passenger delay time, the passenger service time, and the travel time for the average airport passenger in the peak 30 min can be computed.

The expected values are then computed for the delay time, service time, and walking time for the average airport passenger to find the average passenger processing time from Eq. (10-11).

$E(T_d) = 1.0(0.12) + 0.48(15.52 + 3.10) + 0.34(10.98) + 0.08(0.15) + 1.0(2.13)$
$\qquad + 0.53(3.38 + 0.28) + 0.47(0.38 + 0.22)$

$E(T_d) = 17.2$ min

$E(T_s) = 1.0(0.25) + 0.48(3.45) + 0.34(3.00) + 0.08(1.00) + 1.0(0.5)$
$\qquad + 0.53(0.75 + 0.33) + 0.47(0.50 + 0.33)$

$E(T_s) = 4.5$ min

$E(D_w) = 0.48(81 + 144) + 0.34(102 + 175) + 0.08(93 + 166) + 0.10(153)$
$\qquad + 0.53(55) + 0.47(55)$

$E(D_w) = 293$ ft

$E(T_w) = \dfrac{293}{1.5(60)} = 3.3$ min

Time before Scheduled Departure, min	Percentage of Passengers at Airport	Time before Scheduled Departure, min	Percentage of Passengers at Airport
95	0	45	31
90	0	40	40
85	1	35	51
80	2	30	61
75	3	25	71
70	5	20	80
65	8	15	88
60	12	10	94
55	17	5	98
50	23	0	100

TABLE 10-8 Typical Arrival Distribution of Originating Passengers at the Airport for Domestic Flights for Example Problem 10-1

The average passenger processing time for the average airport passenger during the peak 30 min is then

$$E(T_p) = 17.2 + 4.5 + 3.3 = 25 \text{ min}$$

Since departure lounges are only waiting areas, there is no queuing system type delay in these facilities. Generally speaking, these facilities are designed with an area requirement per passenger and visitor in the departure lounge. The area required is based upon the number of passengers and visitors which would be in the departure lounge at the moment the aircraft is allowed to be boarded. Typically the square footage requirement is from 15 to 25 ft² per person in the departure lounge at that point.

Given a typical arrival distribution at the airport, such as that shown in Table 10-8, it is assumed that the total number of passengers and visitors arriving at the airport at the moment the flight is called for boarding is also the number in the departure lounge at that point in time.

In this problem, let us assume that the flight is boarded 15 min before scheduled departure and the airline departure lounge requirements are 20 ft² per passenger.

Therefore, at this time 88 percent of the passengers and their visitors would be in the departure lounge. For Western Pacific Airlines, this means that there are 0.88(0.47)(135) passengers and 0.88(0.47)(135)(0.5)(0.5) visitors in the departure lounge at this point in time. This totals 56 passengers and 14 visitors. Therefore, the departure lounge size requirements are 20(56 + 14) = 1400 ft².

Simulation Models

These models become particularly useful when the analysis of the operation of the passenger handling system is to be performed at a relatively detailed level or when it is desired to analyze the operation of the system for extended periods of time. They are useful for the analysis of the whole system, as well as of parts of it [28, 34, 39]. When some important inputs to analysis are unobtainable, such as possible future flight schedules, it is possible with the use of computer simulation to analyze the operation of the system under randomly generated inputs.

Simulation is also particularly useful when analysis is to be repeated for varying operating conditions in order to perform sensitivity studies. Computers allow such repeated analysis which would otherwise be prohibitively expensive and very time consuming. Most computer systems have standard simulation packages available which can be adapted to many physical planning problems including airports.

It is important to note that computer simulation is not a substitute for analysis when information on the system is lacking. In order to construct a simulation model nearly as much detailed information about system operation is necessary as for other analytic techniques. The main feature of simulation is the high speed at which computers can perform lengthy calculations. In the analysis of systems operations computer simulation should be used with caution and several runs made so that the statistical reliability of the results may be determined.

Simulation techniques have been studied by the FAA [10, 21, 24] and have been used in many studies to determine facility needs. An outline of the flow of an enplaning passenger through the airport system which can be modeled by the FAA simulation model is given in Fig. 10-27. In the schematic design phase of the planning project at Fort Lauderdale-Hollywood International Airport, simulation was used to determine the parking requirements at the airport and the number of parking toll collection facilities needed at the exit [30]. The inputs into the simulation was the airline flight schedule, the distribution of the total number of parkers relative to the flight schedule, and the historical distribution of parking duration as obtained from the analysis of parking ticket stubs. Alternative flight schedules were used to generate the arrival distribution at the parking facility, and the parking duration distribution, shown in Fig. 10-28, was used to generate the random service times required by the arrivals. As a result of the simulation, it was possible to determine the peaking characteristics of arrivals and departures at the parking garage, and the accumulation of vehicles within the parking facility.

The following steps are recommended in the design of a simulation model for application to airport terminal projects [50]:

1. Define the scope of the simulation in terms of the questions to be answered, the components to be included, and the level of detail required.

2. Specify the required output so that an interpretation of the results will resolve the questions to be addressed.

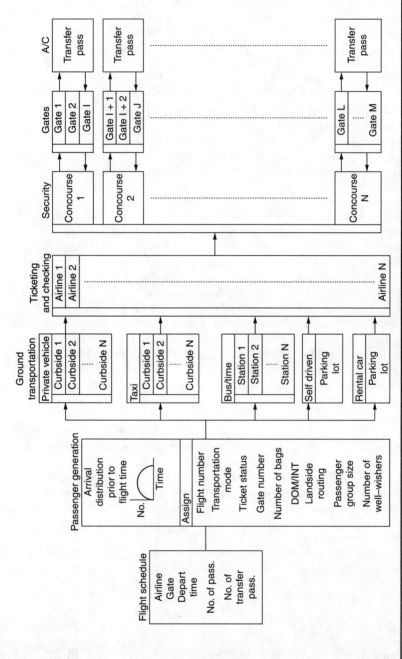

FIGURE 10-27 Flow chart for enplaning passenger simulation model (*Federal Aviation Administration [10]*).

440

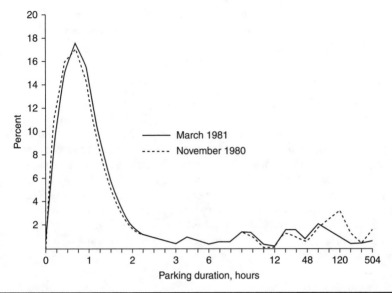

Figure 10-28   Parking duration at Fort Lauderdale-Hollywood International Airport (*Aviation Planning Associates, Inc. [30]*).

3. Structure the model so that the abstract representation of the components in the model and the events and interactions between components are indicative of terminal performance.

4. Define the input data and its variability.

5. Once the model has been developed verify through testing on actual systems.

6. Apply the model and modify the facility design in accordance with the model results.

7. Review the findings and design relative to the degree of variability in the output and through reasonable checks.

Design Development

The final planning phase in terminal projects is called design development. In this phase the size and character of the project is fixed and checked against the findings and recommendations in the prior phases of the project. Acceptance of the project by the airport owner, tenants, and airlines is the final product of this phase. Capital budgeting, operating, maintenance, and administrative costs over the lifetime of the project are determined and a revenue plan is adopted. Agreements are made on rate and charge structures for the airlines, concessionaires, and other tenants. The project moves on toward implementation through the development of construction documents, bid letting and acceptance, and construction following this phase of planning.

The Apron Gate System

The apron provides the connection between the terminal buildings and the airfield. It includes aircraft parking areas, called *ramps*, and aircraft circulation and taxiing areas for access to these ramps. On the ramp, aircraft parking areas are designated as gates. The discussion in this section is limited to the apron gate area for scheduled commercial aircraft operations. The size of the apron gate area depends on four factors, namely, the number of aircraft gates, the size of the gates, the maneuvering area required for aircraft at gates, and the aircraft parking layout in the gate area. The layout of the apron area is discussed in Chap. 6.

Number of Gates

As in the case with other airport facilities, the number of gates is determined in such a way that a predetermined hourly flow of aircraft can be accommodated. Thus, the number of gates required depends on the number of aircraft to be handled during the design hour and on the amount of time each aircraft occupies a gate. The number of aircraft that need to be handled simultaneously is a function of the traffic volume at the airport. As mentioned earlier, it is customary to use the estimated peak hour volume as the input for estimating the number of gates required at the airport. However, in order to achieve a balanced airport design, this volume should not exceed the capacity of the runways.

The amount of time an aircraft occupies a gate is referred to as the *gate occupancy time*. It depends on the size of aircraft and on the type of operation, that is, a through or turnaround flight. Aircraft parked at a gate are there for passenger and baggage processing and for aircraft servicing and preparation for flight. Larger aircraft normally occupy gates a longer time than small aircraft. This is because large aircraft require more time for aircraft servicing, preflight planning, and refueling. The type of operation also affects gate occupancy time by affecting service requirements. Thus an aircraft on a through flight may require little or no servicing and, consequently, the gate occupancy time can be as low as 20 to 30 min. On the other hand, an aircraft on a turnaround flight will require complete servicing, resulting in gate occupancy times ranging from 40 min to more than 1 h. The table shown in Fig. 10-29 lists the activities that normally take place during a turnaround stop, together with a typical time schedule for these activities.

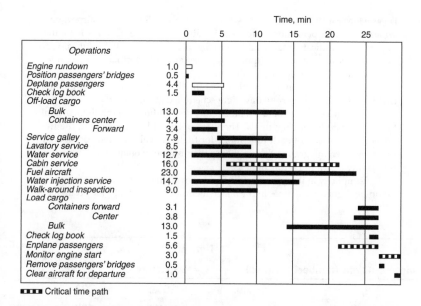

FIGURE 10-29 Typical time schedule of aircraft servicing activities at gate (*Ralph M. Parsons and Federal Aviation Administration [50]*).

If simulation is used a design-day schedule is first forecast and, then based upon the design-day schedule and the practices of the airlines at the airport, a ramp chart based upon the design day is constructed to determine gate requirements.

The average daily gate utilization factor for all gates at an airport usually varies between 0.5 and 0.8. This factor accounts for the fact that it is unlikely that all of the gates available at a terminal building will be used 100 percent of the time. This is caused by the fact that aircraft maneuvering into and out of a gate often blocks other aircraft attempting to move into or out of their gates and by the fact that aircraft schedules often lead to time gaps between the departure of one aircraft and the arrival of another using the same gate. The *gate-use strategy* employed by the airlines at the airport also influences the average gate utilization factor. At airports where gates are used mutually by all airlines, a common gate-use strategy, the gate utilization factor typically varies between 0.6 and 0.8. At airports where groups of gates are used exclusively by different airlines, an exclusive gate-use strategy, the utilization factor drops to about 0.5 or 0.6. The determination of the number of gates needed at an airport should be subjected to the analysis techniques given in Chap. 12 and to the gate-use strategies adopted by the tenant airlines.

An illustration of the use of simulation to generate a design-day schedule from which a ramp chart can be constructed to determine design-day gate requirements in given in Example Problem 10-2.

Example Problem 10-2   The current schedules of airline flights at an airport on a typical day in the peak month are given in Table 10-9. A forecast is made for this airport which indicates that 10 additional flights will be added to the future-year design-day schedule at this airport. It is necessary to perform a simulation for the schedules of these additional flights.

It will be assumed in this problem that the flight schedule of the simulated flights will follow the current day flight schedule distribution.

First it is necessary to obtain a probability distribution for both flight arrival times and gate occupancy times. These are obtained by determining the frequency distribution of each time range for flight arrivals and gate occupancy times and then integrating this frequency distribution to obtain the cumulative probability distribution function.

For simplicity in this problem, the flight arrival time distribution is found by grouping flight arrival times into 1-h increments of time and the gate occupancy

| Airline | Flight Number | Time of | | Aircraft |
		Arrival	Departure	
AE	8/7	7:45 A.M.	9:30 A.M.	727
AE	353	10:30 A.M.	11:15 A.M.	727
AE	319/642	11:30 A.M.	1:00 P.M.	727
AE	421	12:00 P.M.	1:00 P.M.	727
AE	439	1:45 P.M.	2:30 P.M.	727
AE	889	1:45 P.M.	2:30 P.M.	727
AE	852	3:30 P.M.	4:00 P.M.	727
AE	422/660	3:45 P.M.	5:00 P.M.	727
AE	591/544	5:15 P.M.	6:15 P.M.	727
AE	310/390	6:00 P.M.	8:00 P.M.	727
AE	411/428	9:00 P.M.	10:15 P.M.	727
CL	64	7:15 A.M.	7:45 A.M.	737
CL	489	11:15 A.M.	11:45 A.M.	737
CL	41	11:30 A.M.	12:15 P.M.	737
CL	50	1:45 P.M.	2:15 P.M.	737
CL	936	1:45 P.M.	2:15 P.M.	737
CL	81	4:15 P.M.	5:00 P.M.	737
CL	493	8:30 P.M.	9:00 P.M.	737
RX	161	10:15 A.M.	10:45 A.M.	MD8
RX	321/844	4:45 P.M.	5:45 P.M.	MD8

TABLE 10-9 Current Airline Schedule on Typical Day in the Peak Month for Example Problem 10-2

time distribution is found by grouping gate occupancy times into 15 min increments of time.

The flight schedule frequency distribution is then found by counting the number of flights in each 1-h increment of time beginning with 6:00 A.M. and ending with 11:59 P.M. The total number of flights arriving in the periods is shown in Table 10-10.

The cumulative probability distribution function of flight arrival times is then computed by finding the probability that a flight will arrive in a given time period or later beginning with the earliest time period. These values are computed in Table 10-10 and plotted in Fig. 10-30.

By a similar technique, the gate occupancy durations of the flights are grouped as shown in Table 10-11 and the cumulative probability function of these gate occupancy times is plotted in Fig. 10-31.

To simulate a flight arrival and a gate occupancy time for a flight, a table of random digits must be referenced. If this is done, two sets of random numbers, one representing the flight arrival time and one representing the gate occupancy

| Time Period | | Number of Flights | Arriving in Period or Later | |
From	To		Cumulative Number	Cumulative Percentage
6:00 A.M.	6:59 A.M.	0	20	1.00
7:00 A.M.	7:59 A.M.	2	20	1.00
8:00 A.M.	8:59 A.M.	0	18	0.90
9:00 A.M.	9:59 A.M.	0	18	0.90
10:00 A.M.	10:59 A.M.	2	18	0.90
11:00 A.M.	11:59 A.M.	3	16	0.80
12:00 P.M.	12:59 P.M.	1	13	0.65
1:00 P.M.	1:59 P.M.	4	12	0.60
2:00 P.M.	2:59 P.M.	0	8	0.40
3:00 P.M.	3:59 P.M.	2	8	0.40
4:00 P.M.	4:59 P.M.	2	6	0.30
5:00 P.M.	5:59 P.M.	1	4	0.20
6:00 P.M.	6:59 P.M.	1	3	0.15
7:00 P.M.	7:59 P.M.	0	2	0.10
8:00 P.M.	8:59 P.M.	1	2	0.10
9:00 P.M.	9:59 P.M.	1	1	0.05
10:00 P.M.	10:59 P.M.	0	0	0.00
11:00 P.M.	11:59 P.M.	0	0	0.00

TABLE 10-10   Aircraft Arrival Distribution for Example Problem 10-2

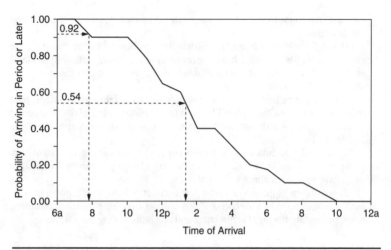

FIGURE 10-30   Flight arrival distribution for Example Problem 10-2.

time, for each of the 10 flights to be simulated are found and these are shown in Table 10-12.

The first simulated flight has an arrival time random number of 0.92 and a gate occupancy time random number of 0.70. Using the flight arrival time

Gate Occupancy Time Period, Min			Duration Range or More	
From	To	Number of Flights	Cumulative Number	Cumulative Percentage
0	14	0	20	1.00
15	29	0	20	1.00
30	44	7	20	1.00
45	59	5	13	0.65
60	74	3	8	0.40
75	89	2	5	0.25
90	104	1	3	0.15
105	119	1	2	0.10
120	134	1	1	0.05
135	149	0	0	0.00
150	164	0	0	0.00
165	179	0	0	0.00
180	194	0	0	0.00

TABLE 10-11   Aircraft Gate Occupancy Time Distribution for Example Problem 10-2

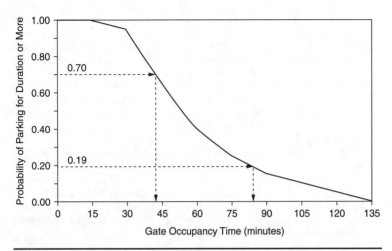

FIGURE 10-31   Flight gate occupancy duration for Example Problem 10-2.

random number of 0.92 with the cumulative probability function of flight arrival times in Fig. 10-30, we find that the simulated arrival time of the first flight is 7:45 A.M. Using the gate occupancy time random number of 0.70 with the cumulative probability function of gate occupancy times in Fig. 10-31, we find that the simulated gate occupancy time of the first flight is 45 min. Both of these numbers are taken to the nearest 15 min time increment for simplicity. Therefore, the first simulated flight has an arrival time of 7:45 A.M. and a flight departure time of 8:30 A.M.

Simulated Flight	Aircraft Arrivals		Gate Occupancy Time		
	Random Number	Arrival Time	Random Number	Duration Time	Departure Time
1	0.92	7.45	0.70	45	8.30
2	0.88	10.15	0.80	45	11.00
3	0.54	1.15	0.19	90	2.45
4	0.18	5.30	0.62	45	4.15
5	0.82	10.45	0.43	60	11.45
6	0.87	10.15	0.86	30	10.45
7	0.75	11.15	0.46	60	12.15
8	0.28	4.15	0.69	45	5.00
9	0.59	1.00	0.87	30	1.30
10	0.60	1.00	0.93	30	1.30

TABLE 10-12   Gate Simulation Results for Example Problem 10-2

This process is continued for each simulated flight. For example, the third simulated flight has an arrival time random number of 0.54 and a gate occupancy time random number of 0.19. Again, using the flight arrival time random number of 0.54 with the cumulative probability function of flight arrival times in Fig. 10-30, we find that the simulated arrival time of the third flight is 1:15 P.M. Using the gate occupancy time random number of 0.19 with the cumulative probability function of gate occupancy times in Fig. 10-31, we find that the simulated gate occupancy time of the third flight is 90 min. Therefore, the third simulated flight has an arrival time of 1:15 P.M. and a flight departure time of 2:45 P.M.

The above simulation process is shown by the dashed lines in Figs. 10-30 and 10-31.

Assuming a simulation was also performed for the airline and the aircraft for each flight, the results of the simulation are tabulated in Table 10-12. The new flight schedule in the design year, with both the current flights and the simulated flights, is shown in Table 10-13. In this table the simulated flights are each given the flight number 9999 for reference purposes.

Ramp Charts

A ramp chart is a graphical representation of the gate occupancy by aircraft throughout the day. Airlines and airports use ramp charts to display the actual gate assignment of aircraft for the flight schedule at the airport.

The ramp chart can also be used to determine the gate requirements at an airport. When it is used for this purpose the ramp charts does not display the actual assignment of aircraft to specific gates for the design-day schedule but only indicates the assignment of aircraft to gates for the determination of the number of gates required at the airport.

There are several factors which influence the gate requirements at an airport. Obviously the flight schedule and the gate occupancy time of aircraft are of paramount importance. However, the scheduling and ramp operating practices of the airlines and the gate-use strategy of the airlines are also important. The scheduling and ramp operating practices give rise to the fact that a gate cannot be used 100 percent of the time as discussed earlier. The gate-use strategy considers whether the gates will be exclusive-use gates, shared-use gates or common gates. Exclusive-use gates are gates which are reserved for the exclusive use of one airline. A shared gate is a gate which is shared by two or three airlines. A common-use gate is a gate which is allocated by the airport based upon the demand for gates and may be used by any airline at the airport.

Gates are sized based upon the geometric properties of the aircraft that will occupy the gates. Therefore, gates may be called *wide-bodied gates* because they are sized to accommodate wide-bodied aircraft. These gates may also be used by narrow-bodied aircraft. Narrow-bodied gates are gates which can only be used by narrow-bodied aircraft. Many airports also have commuter gates which are sized to accommodate commuter aircraft.

Ref No.	Airline	Flight Number	Time of		Aircraft
			Arrival	Departure	
1	AE	8/7	7:45 A.M.	9:30 A.M.	727
2	AE	9999	10:15 A.M.	10:45 A.M.	727
3	AE	9999	10:15 A.M.	11:00 A.M.	727
4	AE	353	10:30 A.M.	11:15 A.M.	727
5	AE	9999	10:45 A.M.	11:45 A.M.	727
6	AE	319/642	11:30 A.M.	1:00 P.M.	727
7	AE	421	12:00 P.M.	1:00 P.M.	727
8	AE	9999	1:00 P.M.	1:30 P.M.	727
9	AE	9999	1:00 P.M.	1:30 P.M.	727
10	AE	439	1:45 P.M.	2:30 P.M.	727
11	AE	889	1:45 P.M.	2:30 P.M.	727
12	AE	852	3:30 P.M.	4:00 P.M.	727
13	AE	422/660	3:45 P.M.	5:00 P.M.	727
14	AE	9999	4:15 P.M.	5:00 P.M.	727
15	AE	591/544	5:15 P.M.	6:15 P.M.	727
16	AE	9999	5:30 P.M.	6:15 P.M.	727
17	AE	310/390	6:00 P.M.	8:00 P.M.	727
18	AE	411/428	9:00 P.M.	10:15 P.M.	727
19	CL	64	7:15 A.M.	7:45 A.M.	737
20	CL	9999	7:45 A.M.	8:30 A.M.	737
21	CL	489	11:15 A.M.	11:45 A.M.	737
22	CL	9999	11:15 A.M.	12:15 P.M.	737
23	CL	41	11:30 A.M.	12:15 P.M.	737
24	CL	9999	1:15 P.M.	2:45 P.M.	737
25	CL	50	1:45 P.M.	2:15 P.M.	737
26	CL	936	1:45 P.M.	2:15 P.M.	737
27	CL	81	4:15 P.M.	5:00 P.M.	737
28	CL	493	8:30 P.M.	9:00 P.M.	737
29	RX	161	10:15 A.M.	10:45 A.M.	MD8
30	RX	321/844	4:45 P.M.	5:45 P.M.	MD8

TABLE 10-13 Simulated Airline Schedule on Typical Day in the Peak Month in the Design Year for Example Problem 10-2

The determination of the number of aircraft gates required at an airport is shown by constructing ramp charts in Example Problem 10-3.

Example Problem 10-3   Let us determine the number of gates required at an airport based upon the design-day schedule shown in Table 10-13. Let us determine the gate requirements under an exclusive gate-use strategy, a shared gate-use strategy, and a common gate-use strategy. Let us assume that the scheduling and operating practices of the airlines require that a minimum time gap of 15 min must be allowed between the departure of a scheduled flight from a gate and the arrival of the next scheduled flight at that gate.

Under an exclusive gate-use strategy each airline will have its own gates which cannot be used by any other airline. To find the number of gates required under this gate-use strategy, a graph is constructed showing time on the X axis and the number of gates on the Y axis. To determine the minimum number of gates required, the flight schedule of each airline must be sorted first by arrival and then departure time. That is, two flights which have the same arrival time are also sorted by the departure time.

In Table 10-13, the flight schedule of each airline has been sorted and a reference number is placed next to each flight for each airline. This reference number will be placed on the ramp chart for illustrative purposes.

To determine the gate required for Alpha Express (AE) Airlines, since the first scheduled flight is scheduled to arrive at 7:45 A.M. and scheduled to depart at 9:30 A.M., a gate is opened on the ramp chart and the block of time from 7:45 A.M. to 9:30 A.M. is filled in on the ramp chart to indicate that the gate is occupied during that period of time. The next scheduled flight by AE Airlines is scheduled to arrive at 10:15 A.M. and depart at 10:45 A.M. Since at the gate just opened up the next scheduled arrival can be placed at this gate 15 min after the departure of the previous aircraft, this flight may be scheduled into that gate and the block of time from 10:15 A.M. to 10:45 A.M. is filled in. The next scheduled flight of AE Airlines is scheduled to arrive at 10:15 A.M. and depart at 11:00 A.M. However, this flight cannot be scheduled into the first gate since that gate is occupied during part of that period of time.

Therefore, a new gate is opened up on the ramp chart to accommodate this aircraft and the block of time from 10:15 A.M. to 11:00 A.M. is filled in at that gate. This process is continued until all of the AE Airline flights have been assigned to gates. As may be seen on the top portion of Fig. 10-32, to accommodate the flights of AE Airlines four gates are required.

The same process is repeated for each airline, and as shown in the middle and bottom portions of Fig. 10-32, this results in three gates being required for Coastal Link (CL) Airlines and one gate being required for Regional Express (RX) Airlines.

Therefore, under an exclusive gate-use strategy, this flight schedule requires eight gates to accommodate the airline schedule at the airport.

For the construction of a shared gate-use strategy ramp chart, let us assume that CL and RX Airlines will share gates at the airport and that AE Airlines will have exclusive-use gates. Therefore, the ramp chart for AE Airlines does not change but the ramp chart for the shared-gate-use airlines must be constructed by first sorting all of the flights of CL and RX Airlines together by arrival time and departure time as shown in Table 10-14. Using the procedure above, Fig. 10-33 is the required ramp chart. It is seen that though AE Airlines requires four gates, RX and CL Airlines now only require at total of three gates. Therefore, the shared-gate-use strategy results in one less gate at the airport to accommodate the design-day flight schedule.

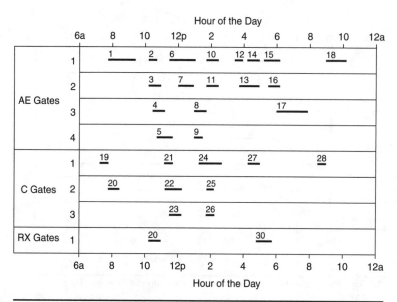

FIGURE 10-32 Ramp chart for exclusive gate use for Example Problem 10-3.

| Ref No. | Flight Airline | Number | Time of | | Aircraft |
			Arrival	Departure	
1	CL	64	7:15 A.M.	7:45 A.M.	737
2	CL	9999	7:45 A.M.	8:30 A.M.	737
3	RX	161	10:15 A.M.	10:45 A.M.	MD8
4	CL	489	11:15 A.M.	11:45 A.M.	737
5	CL	9999	11:15 A.M.	12:15 P.M.	737
6	CL	41	11:30 A.M.	12:15 P.M.	737
7	CL	9999	1:15 P.M.	2:45 P.M.	737
8	CL	50	1:45 P.M.	2:15 P.M.	737
9	CL	936	1:45 P.M.	2:15 P.M.	737
10	CL	81	4:15 P.M.	5:00 P.M.	737
11	RX	321/844	4:45 P.M.	5:45 P.M.	MD8
12	CL	493	8:30 P.M.	9:00 P.M.	737

TABLE 10-14 Simulated Airline Schedule on Typical Day in the Peak Month in the Design Year for CL and RX Airlines for Shared-Gate-Use Strategy for Example Problem 10-3

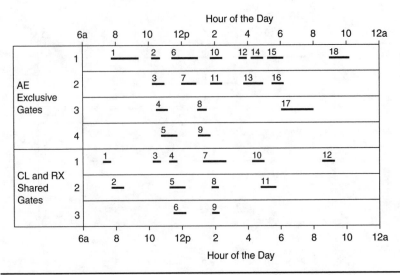

Figure 10-33 Ramp chart for shared gate use for Example Problem 10-3.

For the common use-gate strategy, all of the flights of all of the airlines are sorted together by arrival time and departure time and the ramp chart similarly constructed. If this is done, the ramp chart in Fig. 10-34 results which shows that under a common gate-use strategy only five gates are required at the airport to accommodate the design-day schedule.

The peak hour for gate use occurs around 1:00 P.M. and the peak hour gate utilization is defined as the gate time demanded divided by the gate time supplied in the peak hour. The aircraft demand 210 min of gate time in the peak hour and the gate time supplied is 480 min, 420 min, or 300 min for the exclusive, shared and common gate-use strategies, respectively. Therefore, the peak hour gate utilization is 0.44, 0.50, or 0.70 by each of the gate-use strategies.

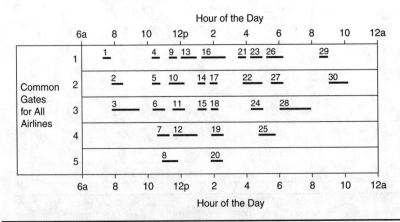

Figure 10-34 Ramp chart for common gate use for Example Problem 10-3.

All of the aircraft on the design-day schedule in this are narrow-bodied aircraft. If some are wide-bodied aircraft, the sorting would be done separately for the wide-bodied and narrow-bodied aircraft. To minimize the number of wide-bodied gates, the wide-bodied aircraft would be assigned to the ramp chart first and then the narrow-bodied aircraft would be assigned, recognizing that a narrow-bodied aircraft may use a wide-bodied gate.

Gates at most airports vary within the range of three to five gates per million annual passengers. The total number of gates may have to be modified if not all gates can handle all types of aircraft. This is particularly important at airports where the aircraft mix includes a considerable amount of large and small aircraft. In such situations, and when data are available, it would be preferable to compute gate requirements separately for the different types of aircraft, keeping in mind that the large gates can be used to handle small aircraft, while the reverse is not true. It is also desirable to calculate the gate requirements separately for different types of traffic. For example, at a large international airport separate calculations may be performed for domestic gates, for international gates, and for charter gates.

Gate Size

The size of a gate depends not only on the size of aircraft which it is to accommodate and but also on the type of parking used, that is, nose-in, parallel, or angled parking. The size of the aircraft determines the space required for parking as well as for maneuvering. Furthermore, the size of aircraft determines the extent and size of the servicing equipment that needs to be provided to service the aircraft. The type of parking used at the gates affects the size since the area required to maneuver into and out of a gate varies depending on the way the aircraft is parked.

In view of the large number of factors that affect the size and exact layout of gates, it is desirable to consult the airlines at an early stage in the design process in order to determine the manner in which they plan to maneuver aircraft and the types of servicing facilities they plan to use.

The design of the gates can be worked out with the aid of procedures and dimensions provided by the FAA [6, 7], ICAO [3], and the International Air Transport Association [12]. Included in these references are diagrams which show various dimensions required for different types of aircraft and various parking and maneuvering conditions. Chapter 6 discusses the layout of ramp areas to accommodate aircraft.

While detailed design of aircraft gates requires the aid of charts such as those found in the airplane characteristics manuals published for aircraft, it is usually sufficient for preliminary planning to adopt uniform dimensions between centers of gates and to use these for sizing the apron gate area. The dimensions depend on the type of aircraft. The typical dimensions for the case where aircraft enter a gate under their own power and are pushed out by a tractor are given in Table 10-15.

	Aircraft Group	Push-Out			Taxi-Out*		
		L	W	Area yd²	L	W	Area yd²
A	FH-227	103 ft 1 in	115 ft 2 in	1319	148 ft 10 in	140 ft 2 in	2318
	YS-11B	106 ft 3 in	124 ft 11 in	1474	171 ft 0 in	149 ft 11 in	2850
	BAC-111	123 ft 6 in	113 ft 6 in	1557	130 ft 0 in	138 ft 6 in	2001
	DC-9-10	134 ft 5 in	109 ft 5 in	1634	149 ft 2 in	134 ft 5 in	2228
B	DC-9-21,30	149 ft 4 in	113 ft 4 in	1880	149 ft 0 in	138 ft 4 in	2290
	727 (all)	173 ft 2 in	128 ft 0 in	2463	194 ft 0 in	153 ft 0 in	3298
	737 (all)	120 ft 0 in	113 ft 0 in	1507	145 ft 4 in	138 ft 0 in	2228
C	B-707 (all)	172 ft 11 in	165 ft 9 in	3188	258 ft 0 in	190 ft 9 in	5468
	B-720	156 ft 9 in	150 ft 10 in	2627	228 ft 0 in	175 ft 10 in	4454
D	DC-8-43, 51	170 ft 9 in	162 ft 5 in	3081	211 ft 10 in	187 ft 5 in	4411
	DC-8-61, 63	207 ft 5 in	168 ft 5 in	3882	252 ft 4 in	193 ft 5 in	5423
E	L-1011	188 ft 8 in	175 ft 4 in	3676	263 ft 6 in	200 ft 4 in	5865
	DC-10	192 ft 3 in	185 ft 4 in	3959	291 ft 0 in	210 ft 4 in	6801
F	B-747	241 ft 10 in	215 ft 8 in	5795	328 ft 0 in	240 ft 8 in	8771

*L = perpendicular to face of building; W = parallel to face of building.

Source: Federal Aviation Administration [50].

TABLE 10-15 Comparison of Apron Parking Envelope Dimensions for Aircraft Push-out and Taxi-out Gate Use for Nose-In Configuration

Aircraft Parking Type

Aircraft parking type refers to the manner in which the aircraft is positioned with respect to the terminal building and to the manner in which aircraft maneuver in and out of parking positions. It is an important factor affecting the size of the parking positions and consequently the apron gate area. Aircraft can be positioned at various angles with respect to the terminal building line and can maneuver into and out of parking positions either under their own power or with the aid of towing equipment. With aircraft towing it is possible to reduce the size of parking positions. It is advisable in choosing among alternative parking types to consult with the airline in question, as different airlines have different preferences for the available systems. It is also advisable in adopting a parking type to take into consideration the objective of protecting passengers from the adverse elements of noise, jet blast, and weather, and the operating and maintenance costs of needed ground equipment.

The aircraft parking types which have been successfully used at a variety of airports and should be evaluated in any airport planning study include nose-in, angled nose-in, angled nose-out, and parallel. These parking types are shown in Fig. 10-35 and are discussed separately below.

Nose-In Parking

In this configuration the aircraft is parked perpendicular to the building line with the nose as close to the building as permissible. The aircraft maneuvers into the parking position under its own power. In order to leave the gate, the aircraft has to be towed out a sufficient distance to allow it to proceed under its own power. The advantages of this configuration are that it requires the smallest gate area for a given aircraft, causes lower noise levels as there is no powered turning movement near the terminal building, sends no jet blast toward the building, and facilitates passenger loading as the nose is near the building. Its disadvantages include the need for towing equipment

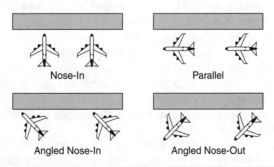

Figure 10-35 Aircraft parking types.

and the nose is too far from the building to effectively use the rear doors for passenger loading.

Angled Nose-In Parking

This configuration is similar to the nose-in configuration except that the aircraft is not parked perpendicular to the building. The configuration has the advantage of allowing the aircraft to maneuver in and out of the gate under its own power. However, it requires a larger gate area than the nose-in configuration and causes a higher noise level.

Angled Nose-Out Parking

In this configuration the aircraft is parked with its nose pointing away from the terminal building. Like the angled nose-in configuration, it has the advantage of allowing aircraft to maneuver in and out of gate positions without towing. It does require a larger gate area than the nose-in position, but less than the angled nose-in. A disadvantage of this configuration is that the breakaway jet blast and noise are pointed toward the building when the aircraft starts its taxiing maneuver.

Parallel Parking

This configuration is the easiest to achieve from the aircraft maneuvering standpoint. In this case noise and jet blast are minimized, as there are no sharp turning maneuvers required. It does require, however, a larger gate position area, particularly along the terminal building frontage.

It is evident that no one parking type can be considered ideal. For any planning situation, all the advantages and disadvantages of the different systems have to be evaluated, taking into consideration the preference of the airline that will be using the gates.

Apron Layout

Another factor that affects apron size and installation requirements is the apron layout. This refers to the manner in which the apron is arranged around the terminal building. The apron layout depends directly on the way the aircraft gate positions are grouped around the buildings and on the circulation and taxiing patterns dictated by the relative locations of the terminal buildings and the airfield system.

Aircraft are grouped adjacent to the terminal building in a variety of ways depending on the horizontal terminal concept used. These groupings are referred to as parking systems and are classified as the frontal or linear system, the finger or pier system, the satellite system, and the open apron or transporter system. Each of these were discussed and illustrated earlier.

The choice of aircraft parking system is, naturally, strongly influenced by the horizontal passenger processing concept adopted. For each there are positive and negative attributes that must be weighed against each other. While the open apron system has the advantage of separating the aircraft and the terminal building from one another, it does require buses or mobile lounges for the conveyance of passengers between them. These vehicles use the apron, and their circulation patterns need careful planning to avoid interference with the flow of aircraft and other service vehicles. While the finger system allows the efficient expansion of gate positions and the efficient use of terminal building space, it may lead to long passenger walking distances if allowed to become excessively long. The frontal system is suitable for the gate arrival processing concept. Other features of these systems were discussed earlier in this chapter.

Apron Circulation

In designing the apron layout it is important to take into account aircraft circulation, particularly the movement of aircraft within the apron gate area and from this area to the taxiways. When the traffic volume is high, it is desirable to provide a taxilane on the periphery of the apron. It is also important to allow sufficient space to permit easy access of aircraft to gates. This is particularly important when pier fingers are used for aircraft parking and the fingers are parallel to each other. Sufficient space must be provided between the fingers to allow aircraft ready access to the gates. The separation between fingers depends on their length and on the size of aircraft to be accommodated. The longer the finger, the more aircraft gates can be accommodated. However, the increase in the number of gates may necessitate the provision of two taxilanes instead of one between the fingers to provide circulation without excessive delay. One taxilane will probably suffice when there are no more than five or six gates on each side of a finger. A large number of gates may require two taxilanes.

Passenger Conveyance to Aircraft

Depending on the passenger processing system used, the type of aircraft parking, and the parking system layout, any of three methods of conveyance can be used between the building and the aircraft. These are walking on the apron, walking through aircraft building connectors such as passenger loading bridges, and by mobile conveyance using any of a variety of apron vehicles.

The first method can be employed with all processing and parking systems. However, as the number of parking positions and the apron size increases, it becomes impractical to use walking for the conveyance of passengers. The economic appeal of this method is overcome by the need to protect the passengers from the elements and from the hazards of walking on the apron.

The second method can be employed for all systems other than where open apron parking is used. A variety of fixed and movable loading systems have been developed for passenger conveyance.

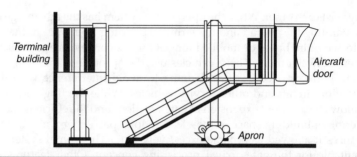

Figure 10-36   Typical aircraft loading bridge.

Most common among these are the nose bridges, which are short connectors suitable for use when the aircraft door comes close to the
building such as with nose-in parking. Another common system is
the telescoping loading bridge. These have the flexibility of extending from the building to reach the aircraft door and of swinging to
accommodate different types of aircraft. A typical boarding bridge is
shown in Fig. 10-36.

Apron Utility Requirements

Aircraft need to be serviced at their respective gates. Thus certain fixed
installations may be required on the apron. Apron congestion is always
a problem and, hence, there is a definite trend at larger airports toward
replacing mobile servicing equipment with fixed facilities.

Aircraft Fueling

Aircraft are fueled at the apron by fuel trucks, fueling pits, and
hydrant systems. At the smaller and even the larger airports the use
of fuel trucks is prevalent, but the pattern is changing in favor of the
hydrant system at airports requiring large amounts of fuel.

The principal advantage of fuel trucks is their flexibility. Aircraft can be fueled anywhere on the apron, the units can be added
or taken away according to need, and the system is relatively economical insofar as airport management and airline operations are
concerned. There are, however, disadvantages associated with the
use of fuel trucks. Large jet transports require a considerable amount
of fuel, from nearly 8000 gal (U.S.) for the McDonnell Douglas MD-88
to almost 50,000 gal (U.S.) for the Boeing 747-100. Two refueler units
are normally required, one under each wing. For the large jets,
standby units are sometimes required if the fuel requirements are in
excess of two units. This means that there are a large number of vehicles on the apron during peak periods, creating a potential hazard
of collision with personnel, other vehicles, and aircraft. Since each
truck carries a considerable quantity of fuel, it also constitutes a
potential fire hazard when moving around on an apron where a

number of other activities are taking place. Trucks are large and awkward and take up valuable space in the operations area. When a truck is empty, it must return to the storage area for refueling before it can be used again. Thus extra trucks must be provided for use during the time when other trucks are being reloaded. When refueling trucks are not in use, parking space must be provided for these vehicles. Modern refuelers are approximately 40 ft in length and weigh as much as 83,000 lb. The capacity of the larger trucks is approximately 8000 gal. For the larger refuelers axle loads are in excess of the legal limits on highways, and, consequently, the airport designer must provide adequate pavement strengths to support these vehicles.

The hydrant system is used at most large airports. In this system, a large fuel storage area, often called a *fuel farm*, is located on the airport property. Fuel is transferred from the fuel storage area to aircraft gate positions through a system of pipes located below the pavement surface. A special valve is mounted in a box in the pavement flush with the pavement surface at each gate position. A special vehicle, a hydrant dispenser, with a hose, meter, filter, and air eliminator is used to connect the fuel supply to the aircraft. One end of the hose has a specially designed valve which is coupled to the valve installed in the pavement. This hose feeds into the meter, filter, and air eliminator, from which another hose, usually on a reel, is led to the fuel intakes on the aircraft.

The principal advantages of the hydrant system are that a continuous supply of fuel is available at the gates, it is safely carried underground, and fuel trucks are eliminated from the apron. The principal disadvantage is that vehicles are not entirely removed from the apron. However, because of their small size, hydrant dispensers reduce possible collision damage to a minimum.

The amounts of fuel required at many airports are so large that, regardless of the type of fueling system used, a central fuel storage area in the vicinity of the landing area is required. If the hydrant system is used, provision must be made for installing pipes from the storage area to the apron.

The location of the hydrant valves at an individual gate will depend upon the location of the fueling connections in the wings of the aircraft occupying the gates. It is desirable that the hose line from the hydrant dispenser to the intakes in the wings not exceed 20 to 30 ft. If a wide variety of aircraft are to be serviced at a gate position, the precise spacing of the hydrant valves should be established in consultation with the airlines. The number of hydrants required per gate position depends not only on the type of aircraft but also on the number of grades of fuel required. Each grade of fuel requires a separate layout.

At a number of airports, hydrant systems are installed by oil companies which contract for fuel with a particular airline or airlines. It is not uncommon to have the hydrant system and fuel trucks operate simultaneously at the same airport. The trend at the large airports is definitely toward the hydrant system.

Electrical Power

Electrical power is required on the apron for the servicing of aircraft prior to engine starting. External electrical power is also often required for starting the engines. Power requirements vary widely for different aircraft. Consequently, it is necessary to consult with the airlines concerning this matter. Power can be supplied by mobile units or by fixed installations in the pavement. The latter is preferable since it removes the need for a vehicle and to some extent reduces noise which emanates from a motor generator set. For a fixed installation, the most satisfactory technique is to bury conduits under the apron, terminating them at supply points some distance from the hydrant valves but convenient to the aircraft.

Recently, there has been a trend toward fixed ground power and air conditioning systems using terminal power sources. The need for these facilities has grown due to the costs of providing power and conditioned air to aircraft during servicing times at the apron gate by using the power generated by the auxiliary power unit on the aircraft. Considerable operating cost economies have been reported in the use of such systems [27].

Aircraft Grounding Facilities

Grounding facilities will be required on the apron to provide protection of parked aircraft and fuel trucks from static discharge, particularly during fueling operations. The location of the grounding facility will be governed by the location of the hydrant valves. With high fueling rates it is essential that grounding facilities be provided.

Apron Lighting and Marking

Adequate lighting and marking are essential on an apron. Wherever possible, each gate should be floodlighted. Floodlighting removes the need for mobile equipment to use headlights, which experience has shown to cause confusion and glare. A system of elevated lights appears to offer the best method of providing apron illumination. Where pier fingers are utilized, the lights can be attached to the fingers. Lighting should be located so as to provide uniform illumination of the apron area yet not cause glare to the pilot.

When personnel are servicing an aircraft, there is a need for lighting its underside and far side, if the floodlights do not provide the necessary illumination. This can be accomplished by installing flush lights in the pavement. When lights of this type are installed, they should be arranged so as not to confuse the pilot insofar as guidance to the gate position is concerned.

Painted guidelines have proved very desirable as aids to maneuvering aircraft accurately on the apron. The best guide appears to be

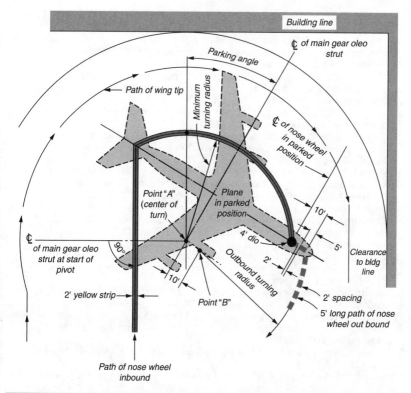

FIGURE 10-37   Typical painted guidelines at gate positions.

a single line, usually the color is yellow, which is followed by the nose gear of the aircraft. A typical layout is shown in Fig. 10-37. It is recognized that a single line will not provide precise guidance for a variety of different aircraft. Usually the guideline is painted for the most critical aircraft using a particular gate position. Smaller aircraft can use the same lines and maneuver without difficulty, especially if personnel on the ground are available to direct the pilot. Because of possible fuel spillage, it is desirable to paint guidelines with special resistant paint in areas where spillage might occur.

References

1. *Access to Commercial Service Airports: The Planning and Design of On-Airport Ground Access System Components*, Final Report, F. X. McKelvey, Federal Aviation Administration, Washington, D.C., 1984.
2. "A Decision Tool for Airport Terminal Building Capacity Analysis," B. F. McCullough and F. L. Roberts, Council for Advanced Transportation Studies, University of Texas at Austin, Presented at 58th Annual Meeting of the Transportation Research Board, Washington, D.C., January 1979.

3. *Aerodrome Design Manual,* Part 2: *Taxiways, Aprons and Holding Bays,* 2d ed., International Civil Aviation Organization, Montreal, Canada, 1983.
4. *Aircraft Movement and Passenger Data for 100 U.S. Airports,* Air Transport Association of America, Washington, D.C., Periodic.
5. *Airline Aircraft Gates and Passenger Terminal Space Approximations,* AD/SC Report No. 4, Air Transport Association of America, Washington, D.C., July 1977.
6. *Airport Capacity and Delay,* Advisory Circular AC 150/5060-5, Federal Aviation Administration, Washington, D.C., 1983.
7. *Airport Design,* Advisory Circular AC 150/5300-13, Federal Aviation Administration, Washington, D.C., 1989.
8. "Airport Gate Position Estimation Under Uncertainty," S. Bandara and S. C. Wirasinghe, *Transportation Research Record,* No. 1199, Transportation Research Board, Washington, D.C., 1988.
9. *Airport Ground Access Planning Guide,* Report No. FAA- EM-80-9, U.S. Department of Transportation, Transportation Systems Center and Federal Aviation Administration, Washington, D.C., 1980.
10. *Airport Landside: The Airport Landside Simulation Model (ALSIM),* U.S. Department of Transportation, Transportation Systems Center and Federal Aviation Administration, Report Nos. FAA-EM-80-8 and DOT-TSC- FAA-84-4, Washington, D.C., 1982.
11. "Airport Landside Level of Service Estimation: Utility Theoretic Approach," K. F. Omer and A. M. Khan, *Transportation Research Record,* No. 1199, Transportation Research Board, Washington, D.C., 1988.
12. *Airport Development Reference Manual,* 9th ed., International Air Transportation Association, Montreal, Canada, January 2004.
13. "Airport Terminal Designs with Automated People Movers," L. David Shen, *Transportation Research Record,* No. 1273, Transportation Research Board, Washington, D.C., 1990.
14. *Airport Terminal Flow Simulation Model,* Transport Canada, Ottawa, Ontario, Canada, 1988.
15. *An Airport Passenger Processing Simulation Model,* S. Hannig-Smith, Aviation Planning Associates, Inc., Cincinnati, Ohio, January 1981.
16. "Analysis of Factors Influencing Quality of Service in Passenger Terminal Buildings," N. Martel and P. N. Seneviratne, *Transportation Research Record,* No. 1273, Transportation Research Board, Washington, D.C., 1990.
17. "Analysis of Passenger and Baggage Flows in Airport Terminal Buildings," R. Horonjeff, *Journal of Aircraft,* American Institute of Aeronautics and Astronautics, Vol. 5, No. 5, 1969.
18. "Analysis of Passenger Delays at Airport Terminals," S. Yager, *Transportation Engineering Journal,* American Society of Civil Engineers, Vol. 99, No. TE4, New York, N.Y., November 1973.
19. "Analytical Models for the Design of Aircraft Terminal Buildings," J. D. Pararas, *Masters Thesis,* Massachusetts Institute of Technology, Cambridge, Mass., January 1977.
20. "Applications for Intra-Airport Transportation Systems," F. X. McKelvey and W. J. Sproule, *Transportation Research Record,* No. 1199, Transportation Research Board, Washington, D.C., 1988.
21. *A Review of Airport Terminal System Simulation Models,* F. X. McKelvey, Final Report, U.S. Department of Transportation, Transportation Systems Center, Cambridge, Mass., November 1989.
22. *Automobile Parking Systems Handbook,* Airports Association Council International-North America, Washington, D.C., 1991.
23. *Chicago Midway Airport Master Plan Study,* Determination of Facility Requirements, Working Paper No. 1, Draft, Landrum and Brown Aviation Consultants, Chicago, Ill., 1990.
24. *Collection of Calibration and Validation Data for an Airport Landside Dynamic Simulation Model,* Wilbur Smith and Associates, Federal Aviation Administration, Washington, D.C., January 1980.
25. *Conceptual Studies for Geneva Intercontinental Airport,* Reynolds, Smith and Hills, Jacksonville, Fla., March 1981.

26. *Demand-Capacity Analysis Ground Access Study*, Draft Report, Terminal Support Working Group, Chicago O'Hare International Airport, Landrum and Brown Aviation Consultants, Chicago, Ill., 1990.
27. *Design Guidebook—400 Hz Fixed Power Systems*, Air Transport Association of America, Washington, D.C., January 1980.
28. "Designing an Improved International Passenger Processing Facility: A Computer Simulation Analysis Approach," V. Gulewicz and J. Browne, *Transportation Research Record*, No. 1273, Transportation Research Board, Washington, D.C., 1990.
29. "Evaluating Performance and Service Measures for the Airport Landside," S. A. Mumayiz, *Transportation Research Record*, No. 1296, Transportation Research Board, Washington, D.C., 1991.
30. *Fort Lauderdale-Hollywood International Airport Parking Analysis*, Working Paper, Aviation Planning Associates, Inc., Cincinnati, Ohio, September 1981.
31. *Ground Transportation Facilities Planning Manual*, Report No. AK-69-13-000, Airport Facilities Branch, Transport Canada, Ottawa, Ontario, Canada, 1982.
32. *Greater Pittsburgh International Airport Expansion Program*, New Terminal Complex Schematic Refinement Phase, Final Report, Tasso Katselas Associates, Inc., and Michael Baker, Jr., Inc., Pittsburgh, Pa., 1986.
33. *Highway Capacity Manual*, HCM2000, Transportation Research Board, Washington, D.C., 2000.
34. "Interactive Airport Landside Simulation: An Object-Oriented Approach", S. A. Mumayiz and R. K. Jain, *Transportation Research Record*, No. 1296, Transportation Research Board, Washington, D.C., 1991.
35. *Introduction to Transportation Engineering and Planning*, E. K. Morlok, McGraw-Hill, Inc., New York, N.Y., 1978.
36. "Level of Service Design Concept for Airport Passenger Terminals: A European View," N. Ashford, *Transportation Research Record*, No. 1199, Transportation Research Board, Washington, D.C., 1988.
37. *Measuring Airport Landside Capacity*, Special Report 215, Transportation Research Board, Washington, D.C., 1987.
38. "Opportunities for Fixed Rail Service to Airports," W. J. Sproule, International Air Transportation, Proceedings of the 22nd Conference on International Air Transportation, American Society of Civil Engineers, New York, N.Y., 1992.
39. "Overview of Airport Terminal Simulation Models," S. A. Mumayiz, *Transportation Research Record*, No. 1273, Transportation Research Board, Washington, D.C., 1990.
40. *Parking Structures: Planning, Design, Construction, Maintenance, and Repair*, A. P. Chrest, M. S. Smith, and S. Bhuyan, Van Nostrand and Reinhold, New York, N.Y., 1989.
41. *Pedestrian Planning and Design*, J. Fruin, Metropolitan Association of Urban Designers and Environmental Planners, New York, N.Y., 1971.
42. "Pier Finger Simulation Model," E. E. Smith and J. T. Murphy, Graduate Report, University of California, Berkeley, Calif, 1972.
43. *Planning and Design Guidelines for Airport Terminal Facilities*, Advisory Circular AC 150/5360-13, Federal Aviation Administration, Washington, D.C., 1988.
44. *Planning and Design of Airport Terminal Facilities at Non-Hub Locations*, Advisory Circular, AC 150/5360-9, Federal Aviation Administration, Washington, D.C., 1980.
45. *Planning Guide for Airport Ground Transportation Facilities*, Report No. AK-69-13, Airport Facilities Branch, Transport Canada, Ottawa, Ontario, Canada, 1982.
46. "Planning of Intra-Airport Transportation Systems," W. J. Sproule, Ph.D. Dissertation, Michigan State University, East Lansing, Mich., 1985.
47. "Simulating the Turnaround Operation of Passenger Aircraft Using the Critical Path Method," J. B. Braaksma, Doctoral Thesis, Waterloo University, Waterloo, Canada, 1970.
48. *Survey of Airport Ground Access*, U.S. Aviation Industry Working Group, Washington, D.C., June 1981.
49. *The Apron and Terminal Building Planning Report*, Report No. FAA-RD-75-191, Federal Aviation Administration, Washington, D.C., July 1975.

50. *The Apron Terminal Complex*, Ralph M. Parsons Company, Federal Aviation Administration, Washington, D.C., September 1973.
51. "The Design of the Airside Concourses (The New Denver International Airport)," J. M. Suehiro, E. K. McCagg, and J. M. Seracuse, International Air Transportation, Proceedings of the 22nd Conference on International Air Transportation, American Society of Civil Engineers, New York, N.Y., 1992.
52. *The FAA's Airport Landside Model, Analytical Approach to Delay Analysis*, Report No. FAA-AVP-78-2, Federal Aviation Administration, Washington, D.C., January 1978.
53. "The Movement of Air Cargo between Cargo Terminals and Passenger Aircraft Gates—Airport Planning Considerations," R. J. Roche, Graduate Report, Institute of Transportation and Traffic Engineering, University of California, Berkeley, Calif., 1972.
54. "The Planning of Passenger Handling Systems," A. Kanafani and H. Kivett, Course Notes, University of California, Berkeley, Calif., 1972.
55. *Trip Generation*, 8th ed., Institute of Transportation Engineers, Washington D.C., 2008.
56. "Use of an Analytical Queueing Model for Airport Terminal Design," F. X. McKelvey, *Transportation Research Record*, No. 1199, Transportation Research Board, Washington, D.C., 1988.
57. Recommended Security Guidelines for Airport Planning, Design and Construction, Transportation Security Administration, Washington, D.C., June 2006.
58. Airport Passenger Terminal Planning and Design, ACRP Report 25, Airport Cooperative Research Program, Transportation Research Board, Washington, D.C., 2010.
59. Guidebook for Planning and Implementing Airport People Mover Systems at Airports, ACRP Project 03-06, Airport Cooperative Research Program, Transportation Research Board, Washington, D.C., 2010.
60. Intermodal Ground Access to Airports—A Planning Guide, Report DOT-T-97-15, US DOT Federal Highway Administration and Federal Aviation Administration, Washington, D.C., December 1996.

PART 3

Special Topics in Airport Planning and Design

CHAPTER 11

Airport Security Planning

Introduction

One of the most significant issues facing airports today is that of airport security. Most users of commercial service airports are subjected to security infrastructure, policies, and procedures within the terminal area; however, airport security concerns all areas and users of the airport.

Safety and security are often considered synonymous; however, the discussion of one invariably invokes reference to the other. Safety is the freedom from the occurrence or risk of injury, danger, or loss to a person and his or her property that is caused unintentionally. A few safety examples in airport design would be actions to prevent a fall on a slippery sidewalk or floor, presence of wildlife on a runway, or loss of an engine due to a bird strike. As aviation grew, government agencies have developed many regulations, standards, and guidelines related to safety. These are covered in numerous documents for airfield, terminal, and ground access planning and design, such as FAA Advisory Circulars, ICAO design manuals, building codes, state highway design manuals, and many others.

Security is the freedom from the occurrence or risk of injury, danger or loss to a person and their property that is caused intentionally through acts of violence. To understand the task of prevention it is necessary to identify the perpetrators of violence and their methods. There three groups most commonly considered are terrorists, criminals, and disruptive passengers.

A *terrorist* is a person or group who uses or advocates the use of violent or threat to intimidate or coerce and these actions are often for political purposes. Terrorist acts are not impulsive acts, but rather are the results of careful planning that evaluates the weak points in the target before taking action. This careful planning aspect of terrorism makes prevention difficult, and agencies have applied the principle of layered security to prevent acts of terrorism. Several layers of security are in place and failure of a single layer does not mean that the entire system will be breached.

A category of terrorist is a suicide bomber and this terrorist presents an even more difficult challenge in security. Suicide terrorism is a simple and low-cost operation in which the terrorist dies. It requires no escape routes or complicated rescue operations. The terrorist can choose the exact time, location, and circumstances of the attack and this has an immense impact on the public due to the overwhelming sense of helplessness. If successful, there will be no terrorists to interrogate because death will be certain.

There may be several definitions of a *criminal*, but for the purpose of airport and aviation protection against acts of crime, criminals are persons who are performing acts that create risk of injury, danger, or loss to persons or property. There are several possible crimes that may occur at an airport. One example is cargo theft. Criminals' intent on cargo theft may carry out an airport invasion, enter the air operations area by force and steal cargo from the aircraft while it is on the ground, or they may board the aircraft, hijack the flight, and force the pilot to land at a predetermined location where the cargo is off-loaded. In both cases, the criminals have foreknowledge of specific cargo, flights, and ways to circumvent security. A subcategory of criminal would be the corrupt insider. This is a person with knowledge about shipments and security procedures who reveals this information to criminals.

A *disruptive passenger* is a person who demonstrates aberrant, abnormal, or abusive behavior at an airport or on a commercial flight. Initially, this person had no intent to cause harm but an event has happened that upsets him or her. It may be caused by alcohol consumption before and during a flight, frustration with airport passenger processing, restrictions, or other reasons.

History of Airport Security

In the early days of civil aviation, the greatest concerns were related to the safety of flight and there was little concern over airport or aviation security. Aviation security first became an issue in 1930 when Peruvian revolutionaries seized a Pan American mail plane with the aim of dropping propaganda leaflets over Lima. Between 1930 and 1958, several hijackings were reported, mostly committed by eastern Europeans seeking political asylum. The world's first fatal aircraft hijacking took place in July 1947 when three Romanians killed an aircrew member.

The first major act of criminal violence against a U.S. air carrier occurred in November 1955, when Jack Graham placed a bomb in baggage belonging to his mother. The bomb exploded in flight, killing all 33 people on board. Graham had hoped to collect on his mother's insurance policy, but instead was found guilty of sabotaging an aircraft and sentenced to death. A second such act occurred in January 1960, when a heavily insured suicide bomber killed all abroad a National Airlines flight. As a result of these two incidents, demand for baggage inspection at airports began.

With the rise of Fidel Castro in Cuba in 1959 came a significant increase in the number of aircraft hijackings, first by those wishing to escape Cuba, then by those hijacking U.S. aircraft to Cuba. Over the next several years, the number of hijacking incidents increased and peaked in the late 1960s. Hijacking became a terrorist act for negotiation with a government body or airline. A program requiring airlines to screen passengers who fit a hijacker profile began in the late 1960s, but hijacking continued so stronger action was taken. The first airport security regulations were implemented in the United States in 1972 and screening of all passengers and their carry-on items began in January 1973.

Under the provisions of Federal Aviation Regulations Part 107—Airport Security, all airports were required to prepare and submit a security program to the FAA that would include the following elements:

- Identification of an air operations area (AOA), that is, those areas used or intended for landing, takeoff, and maneuvering of aircraft

- Identification of those areas with little or no protection against unauthorized access because of lack of adequate fencing, gates, doors, or other controls

- A plan to upgrade the security of air operations with a timetable for each improvement project

Airports were required to implement an airport security plan and were required to have all persons and vehicles that are allowed in the AOA suitably identified. Airport employees allowed in the AOA were subject to background checks prior to receiving proper identification and permission to enter the air operations area.

These measures paid off and the number of hijackings decreased significantly. In June 1985, Lebanese terrorists diverted a TWA flight leaving Athens for Beirut. One passenger was murdered during this two-week ordeal. This hijacking and an upsurge in Middle East terrorism resulted in several U.S. actions including the use of federal air marshals on flights. On December 21, 1988, a bomb destroyed Pan American flight 103 over Lockerbie, Scotland, and all people abroad the London to New York flight were killed. Investigators found that a bomb concealed in a radio-cassette player had been loaded on the plane in Frankfurt, Germany. Security measures were immediately put into effect for U.S. carriers at European and Middle Eastern airports after the Lockerbie bombing and one was the requirements to x-ray or search all checked baggage and reconcile boarded passengers with their checked baggage, in a process known as positive passenger baggage matching. Legislation in the United States also called for increased focus on developing technology and procedures for detecting explosives and weapons. Throughout the 1990s, FAA sponsored research on new equipment to detect bombs and weapons and made several improvements to upgrade security screening procedures at airports.

The most significant event in our generation was the hijacking and crashing of aircraft into the World Trade Center and Pentagon Building on September 11, 2001. In response to this event, the *Aviation and Transportation Security Act* was signed which made several radical changes to airport security in the United States. The Transportation Security Administration (TSA) was formed to develop and enforce new security guidelines for aviation in the United States. In 2003, the TSA along with the Coast Guard, Customs Service, and Immigration and Naturalization Service, was formally moved into the new United States Department of Homeland Security. All regulations regarding the security of airport and other civil aviation operations in the United States are now a TSA responsibility and are published under Title 49 of the Code of Federal Regulations (49 CFR—Transportation). TSA employees were hired and given responsibilities of all passenger and baggage screening at commercial service airports.

Since 2001, there have been a number of additional attempts to perform terrorist acts on the commercial aviation system around the world. As a result, security policies at the world's airports are constantly changing, primarily in reaction to these ever evolving threats.

Airport Security Program

Every airport in the United States that is operating under Federal Aviation Regulations Part 139—Airports Serving Certain Air Carrier Operations, must have an Airport Security Program (ASP). The program defines specific areas of the airport that are subject to various security measures and procedures. These areas include air operations areas, secure areas, sterile areas, SIDA areas, and exclusive areas. Figure 11-1 shows a general depiction of the different areas at a typical commercial airport.

Air operations area (AOA) is the portion of the airport in which security measures are carried out. It includes aircraft movement areas, aircraft parking areas, loading ramps, safety areas for aircraft use, and adjacent areas, such as general aviation.

Secure area is the area where commercial air carriers load and unload passengers and baggage. Specific security measures are specified in 49 CFR Part 1542—Airport Security, 49 CFR 1544—Aircraft Operator Security; Air Carriers and Commercial Operators, and 49 CFR Part 1546—Foreign Air Carrier Security.

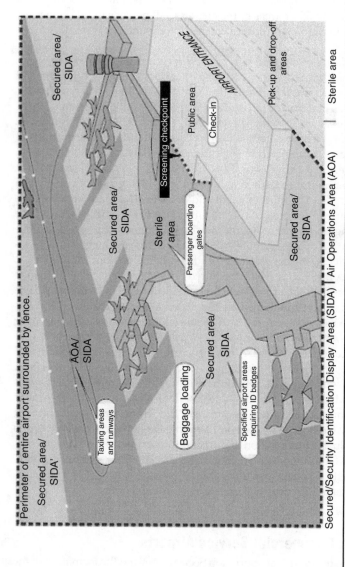

FIGURE 11-1 Airport security areas [9].

Perimeter of entire airport surrounded by fence.

Secured area/ SIDA

Secured area/ SIDA

Secured area/ SIDA

AOA/ SIDA

Taxiing areas and runways

Baggage loading

Secured area/ SIDA

Specified airport areas requiring ID badges

Screening checkpoint

Public area

Check-in

AIRPORT ENTRANCE

Pick-up and drop-off areas

Sterile area

Passenger boarding gates

Secured area/ SIDA

Secured/Security Identification Display Area (SIDA) | Air Operations Area (AOA) | Sterile area

Sterile area is the part of the airport terminal in which passengers have gained access by passing through TSA passenger screening checkpoints or security. In the past, visitors were permitted in this area at some airports, but today only ticketed passengers are permitted in the sterile area.

Security identification display area (SIDA) includes the secure area and possibly other areas of the airports. All persons in this area must display proper identification or be accompanied by an authorized escort.

Exclusive area includes aircraft storage and maintenance hangers, air cargo facilities, and fixed-base operators (FBOs) serving general aviation and charter aircraft.

Areas that do not fall under the above definitions are considered public areas and are not directly subject to TSA security regulations concerning restricted access. These areas would include portions of the airport terminal lobbies, automobile parking areas, and curb frontage.

Planning for security is an integral part of any project undertaken at an airport. The most efficient and cost-effective method of instituting security measures into any facility or operation is through advance planning and continuous monitoring throughout the project. Since the creation of the TSA, the authority to ensure the inclusion of security systems, methods, and procedures is the responsibility of TSA. The TSA must approve the required airport security program which describes how the airport will meet the security requirements of Federal Aviation Regulations Part 139. To assist airport planners, TSA has prepared several documents that present guidance for incorporating security considerations into the planning, design, construction, and modifications for airport infrastructure, facilities, and operational elements [8, 9]. These documents are available on the TSA website at http://www.tsa.gov. The information presented in these documents is expected to be revised and updated periodically as regulations, security requirements, and technology change. The TSA report "Recommended Security Guidelines for Airport Planning, Design, and Construction" is an invaluable introduction or primer on airport security. It contains procedures for examining security issues, "checklists" for security facilities, and methodologies for vulnerability/risk assessment, flow modeling, space planning, and other aspects of security planning.

Security at Commercial Service Airports

The Aviation and Transportation Security Act and the formation of the TSA have contributed to changing rules, regulations, policies, and procedures for airport security. At commercial service airports, many aspects of security will be invisible to air passengers. However, there are a few airport security components that passengers do

experience—passenger screening, baggage screening, employee identification, controlled access, and perimeter security.

Passenger Screening

In the United States, passenger and baggage screening has undergone
a major overhaul following the terrorist attacks of September 11, 2001,
and as of 2003, passenger and baggage security screening is managed
and operated by the TSA. Prior to the TSA, passenger and carry-on
baggage screening fell under the responsibility of the air carriers
whose aircraft provided passenger service at the airport, and the air
carriers would typically subcontract security responsibilities to private security firms. There have been significant impacts on airport terminal planning and operations, and screening policies and procedures
continue to evolve.

Passenger screening facilities (Fig. 11-2) include an automated
screening process, conducted by a magnetometer that attempts to
screen for weapons carried on by a passenger that are metallic in content. As a passenger walks through a magnetometer, the presence of
metal is detected. If a sufficient amount of metal is detected, based on
the sensitivity setting on the magnetometer, an alarm is triggered.
Passengers who trigger the magnetometer are then subject to a manual
search by a screener. A manual search may range from a check with a
handheld wand to a manual pat down. Most recently, advanced
screening technologies have been introduced to screen for nonmetallic
threats, such as powder or liquid explosives.

FIGURE 11-2   Passenger screening checkpoint.

Carry-on baggage screening facilities are located at security screening stations to examine the contents of passengers' carry-on baggage for prohibited items such as firearms, sharp objects that may be used as weapons, or plastic or chemical-based trace explosives. All carry-on baggage is first inspected through the use of an x-ray machine. Bags selected because of suspicions as result of the x-ray examination, or selected on a random basis, are further inspected through the use of explosive trace detection equipment or manual search. In addition, personal electronic items such as laptop computers or cellular phones are frequently inspected by being turned on and operated. In recent years, TSA procedures have mandated more scrutiny, including a wider range of prohibited items, more thorough hand searches, removal of shoes for inspection, and identification checks.

Each airport and airport terminal is unique. The location and size of the passenger screening area depends on several factors including the overall design of the airport and the number of passengers to be processed. As a minimum, a passenger screening area will have one walk-through metal detector and one x-ray device. Figure 11-3 shows the typical layout and elements for a basic station. As the passenger demand grows additional checkpoint lanes and equipment will be required. Guidance and methodologies for planning are provided in TSA reports [8].

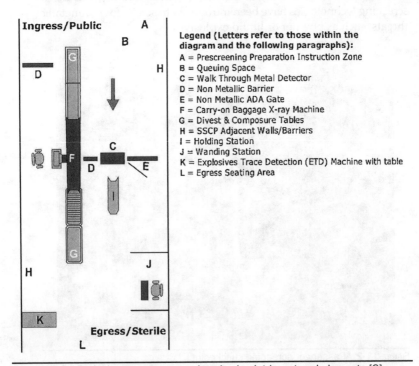

Legend (Letters refer to those within the diagram and the following paragraphs):
A = Prescreening Preparation Instruction Zone
B = Queuing Space
C = Walk Through Metal Detector
D = Non Metallic Barrier
E = Non Metallic ADA Gate
F = Carry-on Baggage X-ray Machine
G = Divest & Composure Tables
H = SSCP Adjacent Walls/Barriers
I = Holding Station
J = Wanding Station
K = Explosives Trace Detection (ETD) Machine with table
L = Egress Seating Area

Figure 11-3 Typical passenger screening checkpoint layout and elements [8].

Baggage Screening

In 2003, TSA mandated that every piece of checked baggage must be screened by certified explosive detection equipment prior to being loaded onto air carrier aircraft. This requirement is known as the 100 percent EDS rule. The primary piece of equipment used to perform checked-baggage screening, the explosive detection system (EDS), uses computed tomography technology similar to the technology found in medical CT scan machines, to detect and identify metal and trace explosives that may be hidden in baggage. EDS equipment has been incorporated into the outgoing baggage processing as shown in Fig. 11-4. The outbound baggage handling system becomes very complex as the number of bags to be processed increases. Due to space constraints and the mandated schedule for implementation,

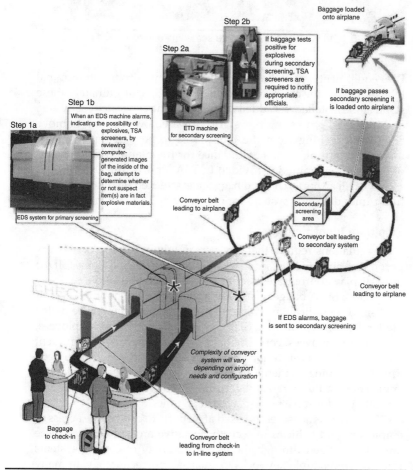

Baggage loaded onto airplane

Step 2b

Step 2a

If baggage tests positive for explosives during secondary screening, TSA screeners are required to notify appropriate officials.

If baggage passes secondary screening it is loaded onto airplane

Step 1b

Step 1a

When an EDS machine alarms, indicating the possibility of explosives, TSA screeners, by reviewing computer-generated images of the inside of the bag, attempt to determine whether or not suspect item(s) are in fact explosive materials.

ETD machine for secondary screening

EDS system for primary screening

Conveyor belt leading to airplane

Secondary screening area

Conveyor belt leading to secondary system

Conveyor belt leading to airplane

If EDS alarms, baggage is sent to secondary screening

Complexity of conveyor system will vary depending on airport needs and configuration

Baggage to check-in

Conveyor belt leading from check-in to in-line system

FIGURE 11-4 Schematic diagram of in-line baggage screening [8].

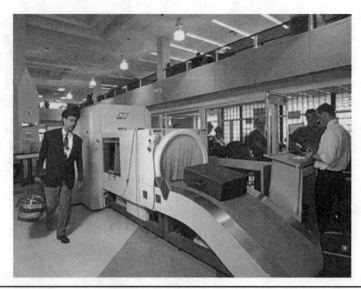

Figure 11-5   Stand alone EDS baggage screening equipment.

EDS equipment was initially installed at many airports in terminal lobbies, next to the check-in counters (Fig. 11-5). Unfortunately, these stand alone installations added confusion to the congested lobby area and increased processing times. Airports are now working to move baggage screening from the lobby area to be part of an in-line baggage handling system. Guidance for the planning and design of baggage screening is provided in TSA reports [8].

At small airports, checked baggage is screened by the use of electronic trace detection systems, or manually by TSA screeners.

Employee Identification

TSA regulations require that any person who wishes to access any portion of an airport's security identification display area (SIDA) must display appropriate identification. This identification, typically known as a SIDA badge, is usually in the form of a laminated credit card-sized identification with a photograph and name of the badge holder. Persons requiring a SIDA badge include airport employees, air carrier employees, concessionaires, contractors, and government employees such as air traffic controllers, airport security, and others. Prior to obtaining an identification badge, persons must complete an application and undergo a fingerprint-based criminal records check. The SIDA badge must be displayed at all times.

A variety of measures are used at airports to control the access of employees and vehicles to security sensitive areas. Access to these areas is provided through the use of a variety of control systems ranging from simple key locks to smart-access technology. In many cases, pass codes are calibrated with a person's SIDA badge and a person must present his or her badge and proper pass code entry to gain access to an area.

Advanced identification verification technologies are being developed to enhance access control at airports. One area of new technologies is biometrics in which human body characteristics, such as fingerprints, eye retinas and irises, voice patterns, facial patterns, and hand measurements are being used for identification authentication purposes. Biometric devices typically consist of a reader or scanning device, software that converts the scanned information into a digital format, and a database that stores the biometric information for comparison.

Perimeter Security

An important part of an airport security plan is its strategy for protecting the airport's perimeter—the area between secured and unsecured areas. The most common methods for securing the airport's perimeter are perimeter fencing, controlled access gates, area lighting, and patrolling of the secured area.

Perimeter fencing is the most common method of creating a barrier around the airport. Fencing can vary in design, height, and type, depending on local security needs. In the United States, standards for perimeter fencing are presented in Advisory Circular 107-1, Aviation Security—Airports.

Controlled access gates provide locations for persons and vehicles to enter the secured area of the airport. The number of access points surrounding an airport's perimeter should be limited to the minimum required for safe and efficient operations of the airport. Controlled access gates typically use some form of controlled access mechanism, ranging from simple key entry or combination locks, to advanced identification authentication machines. Some controlled access gates may be manned by security personnel.

Security lighting is located in and around heavy traffic areas, aircraft service areas, and aircraft operations and maintenance areas at most airports. Security lighting systems will depend on the local situation and the areas to be protected, but typically they help as a deterrent to criminals and terrorists.

Patrolling by airport operations staff and local law enforcement will enhance airport perimeter security. Patrols are usually performed on a routine basis. In addition, most air traffic control towers are situated so that they provide an optimal view of the entire airfield and air traffic controllers can spot potential security threats.

Vulnerability Assessment

An airport vulnerability assessment is an important tool in determining the extent to which an airport facility may require security enhancements and serves to introduce security considerations early in the design process rather than as a more expensive retrofit.

Threats and vulnerabilities cover a wide range of events, none of which can be totally eliminated while still operating the system. Since no system can be totally secure, once threats and vulnerabilities are identified, their impact on the total system must be assessed to determine whether the risk of a particular danger, and the extent to which corrective measures can eliminate or reduce its severity. Security is a process of risk assessment, identifying major threats and considering how vulnerable the system might be. There are several vulnerability assessment tools and methodologies available from government and private organizations.

The threat and vulnerability assessment process is conceptually diagrammed in Fig. 11-6 for a transportation system. These assessments typically use a combination of quantitative and qualitative techniques to identify security requirements, including historical analysis of past events, intelligence assessments, physical surveys,

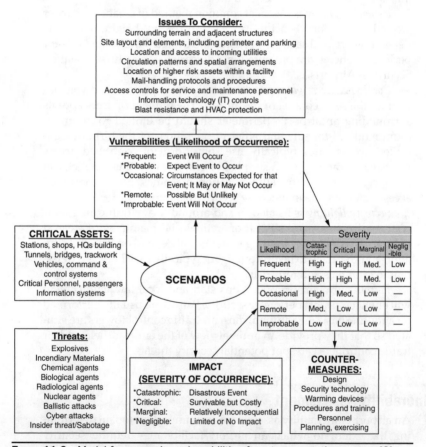

Issues To Consider:
Surrounding terrain and adjacent structures
Site layout and elements, including perimeter and parking
Location and access to incoming utilities
Circulation patterns and spatial arrangements
Location of higher risk assets within a facility
Mail-handling protocols and procedures
Access controls for service and maintenance personnel
Information technology (IT) controls
Blast resistance and HVAC protection

Vulnerabilities (Likelihood of Occurrence):
*Frequent: Event Will Occur
*Probable: Expect Event to Occur
*Occasional: Circumstances Expected for that
               Event; It May or May Not Occur
*Remote: Possible But Unlikely
*Improbable: Event Will Not Occur

CRITICAL ASSETS:
Stations, shops, HQs building
Tunnels, bridges, trackwork
Vehicles, command &
control systems
Critical Personnel, passengers
Information systems

SCENARIOS

Likelihood	Severity			
	Catas-trophic	Critical	Marginal	Neglig-ible
Frequent	High	High	Med.	Low
Probable	High	High	Med.	Low
Occasional	High	Med.	Low	—
Remote	Med.	Low	Low	—
Improbable	Low	Low	Low	—

Threats:
Explosives
Incendiary Materials
Chemical agents
Biological agents
Radiological agents
Nuclear agents
Ballistic attacks
Cyber attacks
Insider threat/Sabotage

IMPACT (SEVERITY OF OCCURRENCE):
*Catastrophic: Disastrous Event
*Critical: Survivable but Costly
*Marginal: Relatively Inconsequential
*Negligible: Limited or No Impact

COUNTER-MEASURES:
Design
Security technology
Warning devices
Procedures and training
Personnel
Planning, exercising

FIGURE 11-6   Model for assessing vulnerabilities for a transportation system [8].

and expert evaluation. When the risk of hostile acts is greater, these analysis methods may draw more heavily on information from intelligence and law enforcement agencies regarding the capabilities and intentions of the aggressors.

Assessments typically include five elements:

1. Asset analysis
2. Target or threat identification
3. Vulnerability assessment
4. Consequence analysis or scenarios
5. Countermeasure recommendations

Assert analysis is an inventory of all airport facilities, operating and maintenance procedures, vehicles, employees, power systems, information systems, and computer network configurations. In reviewing assets, they must be prioritized to determine which assets may require higher or special protection from attack. In making this determination, the airport will consider:

- The value of the asset, including current and replacement value.
- The value of the asset to a potential adversary.
- Where the asset is located and how, when, and by whom an asset is accessed and used.
- If the asset is lost, what is the impact on passengers, employees, public safety organizations, the general public, and airport operations.

A threat is any action with the potential to cause harm. Threat analysis defines the threats against a facility by evaluating the intent, motivation, and possible tactics of those who may carry out the hostile action. The process involves gathering historical data about hostile events and evaluates which information is relevant in assessing the threats against the facility. Some of the questions that are addressed include

- What factors about the system invite hostile action?
- How conspicuous is the transportation facility?
- What political event may generate new hostilities?
- Have similar facilities been targets in the past?

Vulnerability is anything that can be taken advantage of to carry out a threat. This includes vulnerabilities in the design and construction of the facility, operations, administration, and management procedures. Vulnerability analysis identifies specific weaknesses and how hostile actions may occur. Vulnerabilities are usually prioritized through the development of scenarios that pair assets and threats. Using these scenarios, airports can evaluate the effectiveness of their current policies, procedures, and physical protection capabilities to address the consequences.

Scenario analysis requires a methodology that encourages role-playing by airport personnel, emergency responders, contractors, and others to brainstorm ways to attack the airport. By matching threats to critical assets, the airport can identify the capabilities required to support specific types of attacks. For each scenario, consequences are assessed both in terms of severity of impact and probability of loss for a given threat.

Examples of vulnerabilities that may be identified from scenario analysis include the following:

- Accessibility of surrounding terrain and adjacent structures to unauthorized access
- Site layout and elements
- Location and access to utilities
- Building construction with respect to blast resistance
- Sufficiency of lighting, locking controls, alarm systems, venting systems, and facility support control
- Information technology and computer network ease-of-penetration

At the conclusion of the scenario analysis step, the airport will have a list of vulnerabilities for its critical assets. These vulnerabilities will be documented in a confidential report that may be organized as follows:

- Deficiencies in planning
- Deficiencies in the coordination with local emergency responders
- Deficiencies in training
- Deficiencies in physical security—access control, surveillance, blast mitigation, chemical, biological, or radioactive agent protection

Based on the results of the scenario analysis, the airport will identify countermeasures to reduce the vulnerabilities. These actions may be grouped into two general categories:

1. Physical protective measures designed to reduce system asset vulnerability to explosives, ballistics attacks, cyber attacks, and the release of chemical, biological, radiological, or nuclear agents

2. Procedural security measures including procedures to detect, mitigate, and respond to an act of terrorism or extreme violence

Security at General Aviation Airports

Airport security has undergone significant changes over the past 5 years. Regulations, procedures, and the application of new technologies have focused on commercial service airports. However, the TSA mandate is to examine security requirements for all aspects of the transportation system, but to date they have not required general aviation (GA) airports to implement security measures except for three general aviation airports in the Washington, D.C. area. Following the events of September 11, 2001, several aviation groups began work to develop security guidelines for general aviation airports. The Aircraft Owners and Pilots Association (AOPA), state governments, and others prepared guidelines to assist airport managers and the TSA published *Security Guidelines for General Aviation Airports* [11] in 2004. The purpose of the TSA document is to provide owners, operators, sponsors, and other entities charged with oversight of general aviation airports with a set of federally endorsed security best practices and methods for determining when and where these measures may be appropriate. Recognizing the every general aviation airport is unique, TSA has not yet implemented national regulations for GA airport security. The GA industry has developed several security initiatives including awareness programs, reporting methods, and educational courses, and many airports have prepared security plans using principles developed for commercial service airports. Among the security measures taken at general aviation airports include

- Personnel, visitor, aircraft, and vehicle identification procedures
- Perimeter fencing
- Controlled access gates
- Security lighting
- Locks and key control
- Patrolling

Future Security

Protecting airports and aviation against future threats is an imperfect science and, as a result, future airport security will always be an unknown. Concerns for the safe, secure, and efficient travel of passengers and cargo will always be a top priority in civil aviation, and every effort will be taken to make the system as secure as possible for the foreseeable future, and one must anticipate changes in regulations, security requirements, and technologies. Security planning and assessment is a continuing process at every airport.

References

1. *Airport Development Reference Manual*, 9th ed., International Air Transport Association, Montreal, Canada, 2004.
2. *Airport Operations*, 2d ed., Norman Ashford, H. P. Martin Stanton, and Clifton A. Moore, McGraw-Hill, New York, 1997.
3. *Airport Planning and Management*, 5th ed., Alexander T. Wells and Seth B. Young, McGraw-Hill, New York, 2004.
4. *Aviation Security—Airports*, U.S. DOT Federal Aviation Administration, Advisory Circular 107-1, Washington, D.C., 1972.
5. *Airport Services Manual, Part 7—Airport Emergency Planning*, 2d ed., International Civil Aviation Organization, Montreal, Canada, 1991.
6. *General Aviation Safety and Security Practices*, Craig Williams, Airport Cooperative Research Program (ACRP) Synthesis 3, Transportation Research Board, Washington, D.C., 2007.
7. "A New Approach to Airport Security," Sal DePasquale, Proceedings of the 24th International Air Transportation Conference, Louisville, Ky., American Society of Civil Engineers, 1996.
8. *Planning Guidelines and Design Standards for Checked Baggage Inspection Systems*, Transportation Security Administration, Washington, D.C., 2007.
9. *Recommended Security Guidelines for Airport Planning, Design and Construction*, Transportation Security Administration, Washington, D.C., 2006.
10. *Security*, 8th ed., Annex 17, International Civil Aviation Organization, Montreal, Canada, 2006.
11. *Security Guidelines for General Aviation Airports*, Transportation Security Administration, Report A-001, Washington, D.C., 2004.

CHAPTER 12

Airport Airside Capacity and Delay

Introduction

In air transportation, particular concern is focused upon the movement of aircraft, passengers, ground access vehicles, and cargo through both the airport and aviation system. The experienced air traveler has grown accustomed to delayed flights, overbooking, missed connections, ground congestion, parking shortages, and long lines in the terminal building during peak travel periods. For many air transportation trips, the relative advantage of the speed characteristics of aircraft is considerably diminished by ground access, terminal system, and airside delays.

In a more general sense, the unprecedented growth in the demand for air transportation services over the past 30 years has, in many situations, outpaced the ability to provide facilities to adequately accommodate this growth. To a greater extent, elements of the air transport system are being stressed beyond their design capabilities, resulting in significant service deterioration at major airports in this country [5, 6, 9, 10, 18, 21, 27]. It is understandable then that considerable emphasis has been placed upon research to analyze the level and causes of capacity deficiencies. With the maturation of complex computer-based simulation models based on fundamental theories of operations research and queuing theory, it is possible, now more than ever in the history of airport planning and design, to accurately estimate the capability of airport and aviation system components to process demand and to pinpoint the causes of deficiencies in these systems. This knowledge allows one to propose solutions to the problems identified.

Information on airport capacity and delay is important to the airport planner. There is a strong belief within the aviation community that significant gains in air transportation efficiency can be realized through an understanding of the factors causing delays and by the application of technological innovations and operational policies to alleviate delay.

Planners can compare capacity of an airport system, or any of its components, with the existing and forecast demand and ascertain whether improvements to increase capacity will be needed. Comparing the capacity of different configurations at airfields helps determine which are the most efficient. Inadequate capacity leads to increasing delays at airports. Delay is an important factor in a benefit-cost analysis and if an economic value can be placed on delay, the delay reduction savings resulting from an improvement become benefits which can be used to justify the cost of that improvement [19].

Capacity and Delay Defined

The term capacity is used to designate the processing capability of a service facility over some period of time, typically defined as the maximum number of operations that a service facility can accommodate over a defined period of time. For a service facility to realize its maximum or ultimate capacity there must be a continuous demand for service. In the field of aviation, levels of demand that exceed the capacity at a given component of an airport or airspace result in system delays, where delay may be defined as the increase in time required to perform an operation from "normal" nondelayed operations. Additional time required may come in the form of queuing, or waiting, to perform an operation, or a reduction in speed due to congestion. An operation on the airfield is often defined as a takeoff or a landing, while in the terminal an operation may be the processing of a passenger through the terminal. In the airspace, an operation may be considered an aircraft traveling through a certain sector of airspace.

However, the periods of time that demand levels exist to create delays has steadily increased, particularly since the beginning of the twenty-first century. It is this increase in demand to levels that near or exceed capacity over longer periods of time that result in system delays that cause a deterioration in service quality rendering the performance of the aviation system increasingly undesirable. Therefore, airport planners and designers are faced with the problem of providing sufficient capacity to accommodate fluctuating demand with an acceptable level or quality of service. Typically, the design specifications at an airport require that sufficient capacity be provided so that a relatively high percentage of the demand will be subjected to some minimal amount of delay.

To provide sufficient capacity to service a varying demand without delay will normally require facilities which are difficult to economically justify. Therefore, in design, a level of delay acceptable from the perspectives of both the user and the operator is usually established and system components of sufficient capacity are chosen to ensure that these delay criteria are met.

Capacity and Delay in Airfield Planning

In airfield planning, capacity and delay studies are performed to evaluate the ability for an airfield in its current configuration to accommodate current and future levels of demand. As demand is forecast to exceed the current airfield's capacity, airfield planners consider alternative airfield configurations, designed for additional capacity, to measure their effect on mitigating potential future delays.

As the primary objective of capacity and delay studies is to determine effective and efficient means to increase capacity and reduce delay at airports, analyses are conducted to examine the implications of the changes in the nature of the demand, the operating configurations of the airfield and the impact of facility modifications on the quality of service afforded this demand. Some of the typical applications of these analyses might include

1. The effect of alternative runway exit locations and geometry on runway system capacity

2. The impact of airfield restrictions due to noise abatement procedures, limited runway capacity, or inadequate airport navigational aids on aircraft processing rates

3. The consequences of introducing new aircraft into the fleet mix at an airport, and an examination of alternative mechanisms for servicing the mix

4. The investigation of alternative runway-use configurations on the ability to process aircraft

5. The generation of alternatives for new runway or taxiway construction to facilitate aircraft processing

6. The gains which might be realized in system capacity or delay reduction by the diversion of general aviation aircraft to reliever facilities in large air traffic hub areas

According to the United States Government Accountability Office, between 1998 and 2007, delays and cancellations for United States commercial aviation increased by 62 percent, while the number of operations increased by only 38 percent. In 2007 alone, more than 2 million of the nation's 7.5 million annual operations suffered delays or cancellations. In the busiest of regions, such as the New York metropolitan area, delays and cancellations have increased by more than 110 percent, while the number of operations increased by less than 60 percent [31]. Figure 12-1 illustrates the increasing trend of delayed and cancelled operations system wide since 1998.

Delays have also gotten more severe. In 2007, the average length of a flight delay was 56 min, compared to 49 min in 1998. More than 64,000 operations were delayed by more than 3 h in 2007.

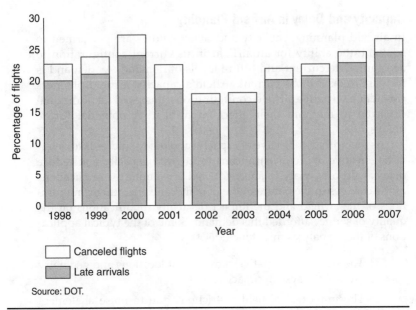

Canceled flights

Late arrivals

Source: DOT.

Figure 12-1   Trends in percentage of late arriving and canceled flights—U.S. system-wide (*U.S. Government Accountability Office*).

The distribution of the causes of delays greater than 15 min are given in Fig. 12-2 for the year 2007, as reported by the U.S. Department of Transportation (DOT).

The operational and economic implications of delay to aircraft increasingly dictate that delay analyses be included in airfield planning studies and that these analyses be conducted well before demand is expected to reach capacity levels.

Approaches to the Analysis of Capacity and Delay

In this chapter, analysis of capacity and delay is confined to the airfield, or aircraft operations area, which is composed of the runways, taxiways, and apron areas. It should be noted that the variations of the principles and tools described in this chapter may also be applied to determining capacity and delay in the airport terminal, as well.

While studies of capacity and delay are most often evaluated by the use of analytical and computer simulation models, the focus of this chapter first is on analytical models, often referred to as mathematical models, which form the basis for more complex computer simulation models.

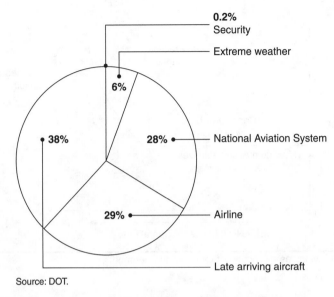

Source: DOT.

Note: Total may not add up to 100 percent due to rounding.

FIGURE 12-2   DOT reported sources of delay, United States, 2007.

Mathematical models of airport operations are tools for understanding the important parameters that influence the operation of systems and investigating specific interactions in systems that are of particular interest. Depending upon the complexity of the system, several conditions may be studied, perhaps more cheaply and quickly than by other methods. To make the mathematics tractable for a complex system, many simplifying assumptions must often be made which may result in unrealistic results. In such a case, one can resort to a computer simulation model or some other technique. Thus it is necessary, when contemplating the formulation and application of a mathematical model, to examine critically the correspondence between the real world being studied and the abstract world of the model and to determine the effect of their differences on the decisions to be made.

For airport planning, airfield capacity has been defined in two ways. One definition which has been used extensively in the United States in the past is that capacity is the number of aircraft operations during a specified interval of time corresponding to a tolerable level of average delay. This is shown in Fig. 12-3 and is referred to as practical capacity. This definition has traditionally been suggested by the Federal Aviation Administration to give rudimentary estimates of delay as a function of ultimate, or throughput capacity, which is defined as the maximum number of operations that a service

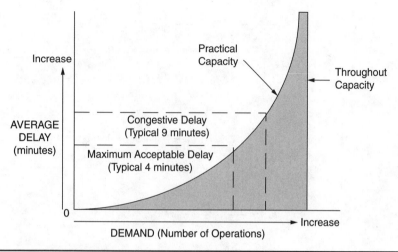

FIGURE 12-3   Delay as a function of capacity and demand.

facility can accommodate over a defined period of time and is also illustrated in Fig. 12-3.

An important difference in these two measures of capacity is that one is defined in terms of delay and the other is not. There are several reasons for considering two definitions of capacity. There has been a general lack of agreement on the specification of acceptable levels of delay applicable to all airports and their airfield components. Because policies, expectations, and constraints differ from airport to airport, the amount of acceptable delay differs from airport to airport. The definition of ultimate capacity does not include delay and reflects the capability of the airfield to accommodate aircraft during peak periods of activity. However, for this definition one does not have an explicit measure of the magnitude of congestion and delay. The magnitude of delay is greatly influenced by the pattern of demand. As an example, when several aircraft wish to use the airfield at the same time, the delay will naturally be larger than if these aircraft were spaced at some interval of time apart. Since the fluctuation of demand within any hour can vary widely, there may be large variations in average delay for the same level of hourly aircraft demand. The shape of the curve in Fig. 12-3 is therefore influenced by the pattern of demand.

Experience has shown that the definition related to ultimate capacity yields values that are slightly larger than the definition which includes delay but the difference is not large. Mathematically, the analysis of ultimate capacity is less complex than that for practical capacity, since the determination of practical capacity implies a definition of the acceptable level of delay.

Factors That Affect Airfield Capacity

There are many factors that influence the capacity of an airfield. In general, capacity depends on the configuration of the airfield, the environment in which aircraft operate, the type and performance characteristics of the aircraft operating on the airfield, the availability and sophistication of aids to navigation, and air traffic control facilities and procedures. A listing of the most important factors includes

1. The configuration, number, spacing, and orientation of the runway system

2. The configuration, number, and location of taxiways and runway exits

3. The arrangement, size, and number of gates in the apron area

4. The runway occupancy time for arriving and departing aircraft

5. The size and mix of aircraft using the facilities

6. Weather, particularly visibility and ceiling, since air traffic rules in good weather are different than in poor weather

7. Wind conditions which may preclude the use of all available runways by all aircraft

8. Noise abatement procedures which may limit the type and timing of operations on the available runways

9. Within the constraints of wind and noise abatement, the strategy which air traffic controllers choose to operate the runway system

10. The number of arrivals relative to the number of departures

11. The number and frequency of touch and go operations by general aviation aircraft

12. The existence and frequency of occurrence of wake vortices which require greater separations when a light aircraft follows a heavy aircraft than when a heavy follows a light aircraft

13. The existence and nature of navigational aids

14. The availability and structure of airspace for establishing arrival and departure routes

15. The nature and extent of the air traffic control facilities

The most significant factor which affects runway capacity is the spacing between successive aircraft. This spacing is dependent on the appropriate air traffic rules, which are, to a large extent, functions of weather conditions and aircraft size and mix.

Formulation of Runway Capacity through Mathematical Theory

In 1960, the FAA contracted with Airborne Instruments Laboratory to develop mathematical models for estimating runway capacity [3]. These models relied on steady-state queuing theory. Essentially there were two models, one for runways serving either arrivals or departures and the other for runways serving mixed operations. For runways used exclusively for arrivals or departures the model was that of a simple Poisson type queue with a first come, first served service discipline. The demand process for arrivals or departures was characterized as a Poisson distribution with a specified mean arrival or departure rate. The runway service process was a general service distribution specified by the mean service time and the standard deviation of the mean service time. For mixed operations, when runways are used for both takeoffs and landings, the process is more complicated, and a preemptive spaced arrivals model was developed. In this model, arrivals have priority over departures for the use of the runways. The takeoff demand process was assumed to follow a Poisson distribution; however, the landing process encountered at the end of the runway is not Poisson but more like the output of an airborne queuing system.

It was recognized that steady-state conditions are rarely achieved at airports; however, it was argued that time-dependent solutions, although possible, were quite complex and were out of the question for the large number of situations required for the preparation of a runway capacity handbook to be used by airport planning and design professionals. Additional support for the use of steady-state solutions came from observations which showed that average delay times yielded by the models were in general agreement with measured delays under a wide variety of operating conditions.

Mathematical Formulation of Delay

The calculation of delay for runways used exclusively by arrivals was computed from Eq. (12-1):

$$W_a = \frac{\lambda_a \left(\sigma_a^2 + 1/\mu_a^2\right)}{2(1 - \lambda_a/\mu_a)} \qquad (12\text{-}1)$$

where W_a = mean delay to arriving aircraft
λ_a = mean arrival rate of aircraft
μ_a = mean service rate for arrivals or the reciprocal of the mean service time
σ_a = standard deviation of the mean service time of the arriving aircraft

The mean service time may be the runway occupancy time or the time separation in the air immediately adjacent to the runway threshold, whichever value is the larger.

The model for departures is identical to arrivals except for a change in subscripts. Equation (12-2) is therefore used for the departure delay:

$$W_d = \frac{\lambda_d \left(\sigma_d^2 + 1/\mu_d^2 \right)}{2 \left(1 - \lambda_d / \mu_d \right)} \tag{12-2}$$

where W_d = mean delay to departing aircraft
λ_d = mean departure rate of aircraft
μ_d = mean service rate for departures, or the reciprocal of the mean service time for departures
σ_d = standard deviation of the mean service time of the departing aircraft

For mixed operations, arriving aircraft are normally given priority and the delay to these aircraft is given by the arrivals of Eq. (12-2). However, the average delay to departures in this situation can be found from Eq. (12-3):

$$W_d = \frac{\lambda_d \left(\sigma_j^2 + j^2 \right)}{2 \left(1 - \lambda_d^j \right)} + \frac{g \left(\sigma_f^2 + f^2 \right)}{2 \left(1 - \lambda_a^f \right)} \tag{12-3}$$

where W_d = mean delay to departing aircraft
λ_a = mean arrival rate of aircraft
λ_d = mean departure rate of aircraft
j = mean interval of time between two successive departures
σ_j = standard deviation of the mean interval of time between successive departures
g = mean rate at which gaps between successive arrivals occur
f = mean value of the interval of time within which no departure can be released
σ_f = standard deviation of the interval of time in which no departure may be released

During busy periods the second term in Eq. (12-3) would be expected to be zero if it is assumed that aircraft are in a queue at the end of the runway and are always ready to go when permission is granted. It must be emphasized that the above equations are only valid when the mean arrival or departure rate is less than the mean service rate which is the condition for which the equations have been derived. The use of the model for the arrivals-only case is illustrated in Example Problem 12-1.

Example Problem 12-1   It is necessary to compute the average delay to arriving aircraft on a runway system which services only arrivals if the mean service time is 60 s per aircraft with a standard deviation in the mean service time of 12 s and the average rate of arrivals is 45 aircraft per hour.

The mean service rate for arrivals μ_a is the reciprocal of the mean service time yielding 1 aircraft per minute of 60 aircraft per hour. Substitution into Eq. (12-1) yields

$$W_a = \frac{45\left[\left(\frac{12}{3600}\right)^2 + 1/60^2\right]}{2\left(1 - \frac{45}{60}\right)} = 0.026\,\text{h} = 1.6\,\text{min}$$

Therefore, the average aircraft delay is about 1.6 min per arrival.

The relationship between delay and capacity can be shown by determining the runway service rate which corresponds to a delay of 4 min using the above equation. Assuming that the standard deviation of the mean service time is the same, we have

$$\frac{4}{60} = \frac{45\left[\left(\frac{12}{3600}\right)^2 + 1/\mu_a^2\right]}{2(1 - 45/\mu_a)}$$

or μ_a is equal to 52 arrivals per hour. If the delay criterion was that arrival delays could not exceed 4 min then the runway capacity related to delay would be 52 arrivals per hour.

It should be observed that an increase in capacity from 52 to 60 arrivals per hour, a 15 percent increase in capacity, results in a delay reduction of 2.4 min, a 60 percent reduction in delay. This is typical at airports nearing saturation. Small increases in capacity can result in significant decreases in delay.

Formulation of Runway Capacity through the Time-Space Concept

The various intervals of time included in the above models are shown on the time-space diagram in Fig. 12-4. The time-space diagram is a very useful device for understanding the sequencing of aircraft operations on a runway system and in the adjacent airspace (Fig. 12-4). Three arrivals and three departures are serviced.

The basic sequencing rules to service these aircraft are

1. Two aircraft may not conduct an operation on the runway at the same time.

2. Arriving aircraft have a priority in the use of the runway over departing aircraft.

3. Departures may be released if the runway is clear and the subsequent arrival is at least a certain distance from the runway threshold.

Examination of the time-space diagram in Fig. 12-4 shows that the mean departure interval j is the average of the interval of time

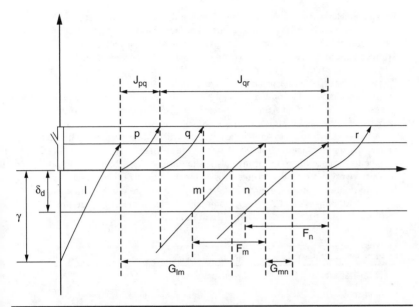

FIGURE 12-4 Time-space diagram concepts for mixed operations on runway system.

between successive departures J_{pq} and J_{qr}. Also, the mean time interval between arrivals, the gap between arrivals I_g during which it may be possible to release g departures—is the average of the quantities G_{lm} and G_{mn}. Finally, the value of the interval of time in which departures cannot be released f is equal to the average of the quantities F_m and F_n.

Several other observations may be made about the sequence of operations shown on this time-space diagram. The initial departure p could have been released, if it was available, before the first arrival l reached the distance δ_d from the runway threshold since the runway was clear. The second departure q was released when the previous departure p cleared the runway, since the next arrival m was more than distance δ_d from the threshold at that point in time. However, the third departure r was not released when that departure cleared the runway because the approaching aircraft m was closer than distance δ_d from the threshold at that point in time. For the same reason, this departure was not released until after the last arrival n cleared the runway. In this figure, the delays which would occur to aircraft are due to the required separations between different types of operational sequences.

The use of the air traffic separation rules is accommodating a series of arrivals and departures may be best understood through a numerical example problem illustrating the time-space concept for processing aircraft on a runway system.

Example Problem 12-2   A runway is to service arrivals and departures. The common approach path is 7 mi long for all aircraft. During a particular interval of time the runway is serving only two types of aircraft, a type A with an approach speed of 120 mi/h and a type B with an approach speed of 90 mi/h. Each arriving aircraft will be on the runway for 40 s before exiting the runway. The air traffic separation rules in effect are given in Table 12-1.

During the period of time to be analyzed five aircraft in an ordered arrival queue of a B, A, A, B, and A aircraft approach the runway. An identical ordered departure queue of aircraft is awaiting clearance to takeoff.

A time-space diagram to service these aircraft will be drawn assuming the first arrival is at the entry gate at time 0 and arrivals are given priority over departures.

The time-space diagram for arrivals is drawn first since these aircraft normally have priority over departures. This is shown on Fig. 12-5. The dashed lines indicate points where the interarrival separation rules are enforced to ensure the minimum interarrival spacing is maintained. The numbers in parentheses indicate the time each aircraft is at the point indicated.

Operational Sequence	Air Traffic Rules
Arrival–departure	Clear runway
Departure–arrival	Arrival at least 2 mi from arrival threshold
Departure–departure	120 s

Arrival–arrival	Miles: Lead
	                    A B
	Trailing A $\begin{bmatrix} 4 & 3 \\ 5 & 3 \end{bmatrix}$
	          B

TABLE 12-1   Air Traffic Separation Rules for Example Problem 12-2

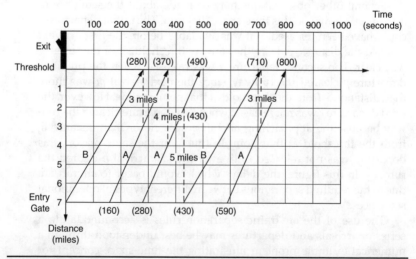

FIGURE 12-5   Time-space diagram for scheduling arrivals in Example Problem 12-2.

Since the first aircraft, a type B aircraft, is at the entry gate at time 0 and it takes 280 s to travel the common approach path from the entry gate to the runway threshold, this aircraft passes over the runway threshold at time 280 s. Immediately behind this aircraft is a type A aircraft which is approaching the runway at a speed of 120 mi/h. In this case, the trailing aircraft A is flying faster than the leading aircraft B and, therefore, it is closing in on the leading aircraft. These two aircraft are closest together when the leading aircraft passes over the arrival threshold or at time 280 s. At this time the trailing aircraft can be scheduled no closer than 3 mi behind the leading aircraft or the trailing aircraft is scheduled to pass over the 3-mi point at time 280 s. Since this aircraft is approaching the runway at a rate of 30 s/mi, it passes over the runway threshold 90 s later or at time 370 s. It passes over the entry gate 210 s earlier or at time 160 s.

This process is continued until all aircraft have been scheduled. It should be observed that when a type B aircraft is trailing a type A aircraft, since the type B is traveling at a speed less than the type A, these two aircraft are closest together when the trailing aircraft passes over the entry gate and the required separation is maintained at that point.

Once all the aircraft are scheduled as shown in Fig. 12-5, it is determined that it will take 800 s to service these five arriving aircraft. The time span at the runway threshold for serving these five arrivals is 800 − 280 = 520 s. In this time span there are four pairs of arrivals. Therefore, the average time between arrivals, the interarrival time, is 520 divided by 4 or 130 s per arrival. The capacity of the runway to service arrivals will be shown later to be

$$C_a = \frac{3600}{130} = 28 \text{ aircraft per hour}$$

The time-space diagram in Fig. 12-6 is then constructed from that in Fig. 12-5 and is used to determine if a departure may be released in the time gaps between arrivals. Each arrival spends 40 s on the runway prior to exiting the

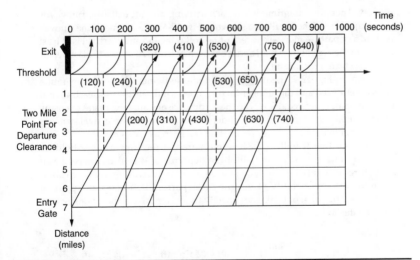

FIGURE 12-6 Time-space diagram for scheduling mixed operations in Example Problem 12-2.

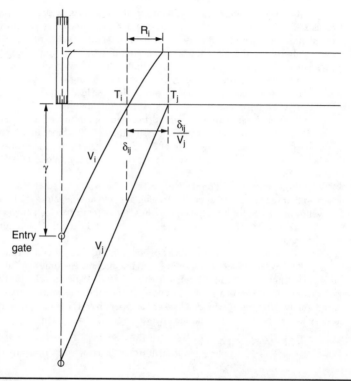

Figure 12-7 Time-space diagram for error-free interarrival spacing for the closing case when $V_i \leq V_j$.

runway. Therefore, the time when each arriving aircraft exits the runway is determined first. The results are shown in Fig. 12-6. At any time, if the runway is clear, a departure may be cleared for takeoff if the incoming arrival is at least 2 mi from the arrival threshold and it has been at least 120 s since the last departure was cleared for takeoff.

Again, in Fig. 12-7, the dashed lines indicate points where the separation rules are enforced and the numbers in parentheses indicate the time each aircraft is at the point indicated. However, these comparisons are now made to ensure that the departure-departure, arrival-departure, and departure-arrival spacings are each maintained.

It is seen that it will take 840 s, measured at the runway threshold, to service all of the arrivals and all of the departures. It is also observed that departures can only be inserted between a pair of arrivals on two occasions. Therefore, the probability of inserting a departure between the 4 pairs of arrivals is 2 out of 4, or 0.50. The capacity to service mixed operations will be shown later to be

$$C_m = \frac{3600}{130}(1.0 + 0.50) = 42 \text{ aircraft per hour}$$

where 1.0 represents the probability of an arrival at the threshold every 130 s and 0.50 represents the probability of inserting a departure in an interarrival time of 130 s.

The capacity of the runway to service departures only will be shown later to be

$$C_d = \frac{3600}{120} = 30 \text{ aircraft per hour}$$

Formulation of Ultimate Capacity

Capacity as defined here expresses the maximum physical capability of a runway system to process aircraft. It is the *ultimate capacity* or maximum aircraft operations rate for a set of specified conditions and it is independent of the level of average aircraft delay. In fact, it has been shown that when traffic volumes reach hourly capacity levels average aircraft delays may range from 2 to 10 min.

Delay is dependent on the capacity as well as the magnitude, nature, and pattern of demand. Delays can occur even when the demand averaged over 1 h is less than the hourly capacity. Such delays occur because demand fluctuates within an hour so that, during some smaller intervals of time, demand is greater than the capacity.

If the magnitude, nature, and pattern of demand are fixed, then delay can be reduced only by increasing capacity. On the other hand, if demand can be manipulated to produce more uniform patterns of demand, then delay can be reduced without increasing capacity. Thus, estimating capacity is an integral step in determining delay to aircraft.

Mathematical Formulation of Ultimate Capacity

These types of models determine the maximum number of aircraft operations that a runway system can accommodate in a specified interval of time when there is continuous demand for service [26]. In these models capacity is equal to the inverse of a weighted average service time of all aircraft being served. For example, if the weighted average service time is 90 s, the capacity of the runway is 1 operation every 90 s or 40 operations per hour. Models treat the common approach path to the runway together with the runway as the runway system. The runway service time is defined as either the separation in the air between arrivals in terms of time, the *interarrival time*, or the *runway occupancy time*, whichever is the largest. The material presented in this section is taken largely from several references [16, 25, 26].

Development of Models for Arrivals Only

The capacity of a runway system used only for arriving aircraft is influenced by the following factors:

1. The aircraft mix which is usually characterized by segregating aircraft into several classes according to their approach speeds

2. The approach speeds of the various classes of aircraft

3. The length of the common approach path from the entry gate to the runway threshold

4. The minimum air traffic separation rules or the practical observed separations if no rules apply

5. The magnitude of errors in arrival time at the entry point to the common approach path, the entry gate, and speed variation of aircraft on the common approach path

6. The specified probability of violation of the minimum air traffic separations considered acceptable or attainable

7. The mean runway occupancy times of the various classes of aircraft in the mix and the magnitude of the variation in these times

The Error-Free Case

With little loss in accuracy and to make the computations simpler, aircraft are grouped into several discrete speed classes V_k. To obtain the weighted service time for arrivals, it is necessary to formulate a matrix of the intervals of time between aircraft arrivals at the runway threshold. Having this matrix and the percentage of the various classes in the aircraft mix, the weighted service time can be computed. The inverse of the weighted service time is the capacity of the runway.

Let the error-free matrix be designated as $[M_{ij}]$, which is made up of the elements m_{ij}, the minimum error-free time interval at the runway threshold for aircraft of speed class i followed by aircraft of class j, the percentage of aircraft of class i in the mix p_i, and the percentage of aircraft of class j in the mix p_j. Then

$$\Delta T_{ij} = T_j - T_i = m_{ij} \qquad (12\text{-}4)$$

where ΔT_{ij} = actual time separation at the runway threshold for two successive arrivals, an aircraft of speed class i followed by an aircraft of speed class j
T_i = time that the leading aircraft i passes over the runway threshold

T_j = time that the trailing aircraft j passes over the runway threshold

m_{ij} = minimum error-free interarrival separation at the runway threshold which is the same as ΔT_{ij} in the error-free case

$$E(\Delta T_{ij}) = \Sigma p_{ij} m_{ij} = \Sigma [p_{ij}][M_{ij}] \qquad (12\text{-}5)$$

where $E(\Delta T_{ij})$ = expected value of the service time, or interarrival time, at the runway threshold for the arrival aircraft mix

p_{ij} = probability that the leading arriving aircraft i will be followed by the trailing arriving aircraft j

$[p_{ij}]$ = matrix of these probabilities

$[M_{ij}]$ = matrix of the minimum interarrival separations m_{ij}

The capacity for arrivals is given by

$$C_a = \frac{1}{E(\Delta T_{ij})} \qquad (12\text{-}6)$$

where C_a is the capacity of the runway to process this mix of arrivals.

To obtain the interarrival time at the runway threshold, it is necessary to know whether the speed of the leading aircraft V_i is greater or less than that of the trailing aircraft V_j, since the separation at the runway threshold will differ in each case. This can be illustrated by drawing time-space diagrams representative of these conditions as shown in Figs. 12-8 and 12-9. In these diagrams the following notation is used:

γ length of the common approach path

δ_{ij} minimum permissible distance separation between two arriving aircraft, a leading aircraft i and a trailing aircraft j, anywhere along the common approach path

V_i approach speed of the leading aircraft i of class k

V_j approach speed of the trailing aircraft j of class k

R_i runway occupancy time of the leading aircraft

The Closing Case ($V_i \le V_j$)

First let us consider the case where the leading aircraft's approach speed is less than that of the trailing aircraft, as shown in Fig. 12-8. The minimum time separation at the threshold may be written in terms of the minimum distance separation δ_{ij} and the speed of the trailing aircraft V_j. However, if the runway occupancy time of the

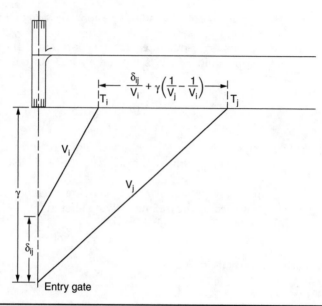

FIGURE 12-8   Time-space diagram for error-free interarrival spacing for the opening case when $V_i > V_j$ for aircraft control from entry gate to arrival threshold.

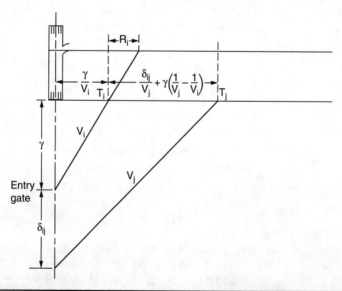

FIGURE 12-9   Time-space diagram for error-free interarrival spacing for the opening case when $V_i > V_j$ for both aircraft separated in vicinity of entry gate.

arrival R_i is greater than the airborne separation, then it would be the minimum separation at the threshold. The equation for this case is

$$\Delta T_{ij} = T_j - T_i = \frac{\delta_{ij}}{V_j} \tag{12-7}$$

The Opening Case ($V_i > V_j$)

Next let us consider the case where the leading aircraft's approach speed V_i is greater than that of the trailing aircraft V_j as shown in Figs. 12-8 and 12-9, the minimum time separation at the threshold is written in terms of the distance δ_{ij} the length of the common approach path γ and the approach speeds of the leading and trailing aircraft V_i and V_j. This corresponds to the minimum distance separation δ_{ij} along the common approach path which now occurs at the entry gate instead of the threshold. The equation for the case shown in Fig. 12-8, when control is exercised only from the entry gate to the arrival threshold is

$$\Delta T_{ij} = T_j - T_i = \frac{\delta_{ij}}{V_i} + \left(\frac{1}{V_j} - \frac{1}{V_i}\right) \tag{12-8}$$

When control is exercised to maintain the separations between both aircraft as the leading aircraft passes over the entry gate, as shown in Fig. 12-10, the equation is

$$\Delta T_{ij} = T_j - T_i = \frac{\delta_{ij}}{V_j} + \left(\frac{1}{V_j} - \frac{1}{V_i}\right) \tag{12-9}$$

It should be carefully noted that the only difference between Eqs. (12-8) and (12-9) is in the first term of the equation, where V_i and V_j are interchanged.

Example Problem 12-3 This problem will solve Example Problem 12-2 using the error-free analytical equations developed above. It is necessary to determine the arrival capacity of the runway in an error-free context where aircraft separations are maintained in the airspace along the common approach path between the entry gate and the arrival threshold.

There are four possible interarrival cases, a leading A and a trailing A, a leading B and a trailing B, a leading A and a trailing B, and a leading B and a trailing A. These cases are governed by Eqs. (12-7) and (12-8). Equation (12-7) gives the minimum time between arrivals at the runway threshold when the leading aircraft is approaching the runway at an approach speed less than or equal to the approach speed of the trailing aircraft.

If we have a leading A and a trailing A, or a leading B and a trailing B, both aircraft are traveling at the same speed. Therefore, Eq. (12-7) applies and we have for a type A following a type A

$$\Delta T_{ij} = \frac{4(3600)}{120} = 120 \text{ s}$$

and for a type B following a type B

$$\Delta T_{ij} = \frac{3(3600)}{90} = 120 \text{ s}$$

When the leading aircraft is type B and the trailing aircraft is type A, Eq. (12-7) also applies and we have

$$\Delta T_{ij} = \frac{3(3600)}{120} = 90 \text{ s}$$

When the leading aircraft is approaching the runway at an approach speed greater than the approach speed of the trailing aircraft, the minimum time between arrivals at the runway threshold is given by Eq. (12-8). This is the case when a type B follows a type A aircraft. Therefore, we have

$$\Delta T_{ij} = \frac{5(3600)}{120} + 7\left(\frac{1}{90} - \frac{1}{120}\right)(3600) = 220 \text{ s}$$

The ordered queue consists of the pairs of arrivals B-A, A-A, A-B, and B-A. Therefore, we have the following interarrival matrix and probability matrix based upon the actual queue of arriving aircraft given:

$[T_{ij}]$:

		Leading	
		A	B
Trailing	A	120	90
	B	220	120

$[p_{ij}]$:

		Leading	
		A	B
Trailing	A	0.25	0.50
	B	0.25	0.00

From Eq. (12-5), since in the error-free case $[Tij]$ is equal to $[Mij]$, we have that the expected value of the interarrival time is

$$E(\Delta T_{ij}) = 0.25(120) + 0.50(90) + 0.25(220) + 0.00(120) = 130 \text{ s}$$

From Eq. (12-6) for the arrival capacity of the runway we then have

$$C_a = \frac{3600}{130} = 28 \text{ operations per hour}$$

This agrees exactly with the results of this problem done by the time-space diagram method in Example Problem 12-2.

Consideration of Position Error

The above models represent the situation of a perfect system with no errors. To take care of position errors, a *buffer time* is added to the minimum separation time to ensure that the minimum interarrival separations are maintained. The size of the buffer depends upon the probability of violation of the minimum separation rules which is acceptable. Figure 12-10 shows the position of the trailing aircraft as it approaches the runway threshold. In the top portion of this illustration, the trailing aircraft is sequenced so as its mean position is exactly determined by the minimum separation between the leading and trailing aircraft. However, if the aircraft position is a random variable there is an equal probability that it can be either ahead or behind schedule.

Naturally if it is ahead of schedule the minimum separation criterion will be violated. If the position error is normally distributed, then the shaded area of the bell-shaped curve would correspond to a probability of violation of the minimum separation rule of 50 percent. Therefore, in order to lower this probability of violation, the aircraft may be scheduled to arrive at this position later by building in a buffer to the minimum separation criterion as shown in the bottom portion of the illustration. In this case, only when the aircraft is so far ahead

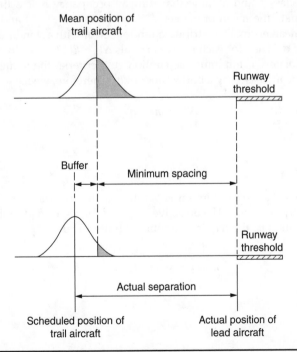

FIGURE 12-10   Illustration of buffer spacing on actual separation between aircraft when position error is considered.

of schedule as to encroach upon the smaller shaded area of the bell-shaped curve would a separation violation occur. There is of course less of a probability of this occurring. In practice, air traffic controllers schedule aircraft with a buffer, as well as to instruct pilots to vary aircraft speeds, so that the probability of violation of the minimum separation rules is at an acceptable level.

As will be shown, in the closing case the buffer is a constant value. However, in the opening case the buffer need not be a constant value and will normally be less than the buffer for the closing case. Having the models for the buffer, a matrix of buffer times $[B_{ij}]$ for aircraft of speed class i followed by aircraft of speed class j is developed. This matrix is added to the error-free matrix to determine the actual inter-arrival time matrix from which the capacity may be found. The relationship is given in Eq. (12-10).

$$E(\Delta T_{ij}) = \Sigma[p_{ij}][M_{ij} + B_{ij}] \tag{12-10}$$

The Closing Case

In this case the leading aircraft's approach speed is less than that of the trailing aircraft and the separations are shown in Fig. 12-7. Let us call ΔT_{ij} the actual minimum interval of time between aircraft of class i and class j, and assume that runway occupancy is less than ΔT_{ij}. Designate the mean or expected value of ΔT_{ij} as $E(\Delta T_{ij})$ and e_0 as a zero-mean normally distributed random error with a standard deviation of σ_0. Then for each pair of arrivals $\Delta T_{ij} = E(\Delta T_{ij}) + e_0$. In order to not violate the minimum separation rule criteria, the value of ΔT_{ij} must be increased by a buffer amount b_{ij}. Therefore, we have

$$\Delta T_{ij} = m_{ij} + b_{ij}$$

and also

$$\Delta T_{ij} = m_{ij} + b_{ij} + e_0$$

For this case the minimum separation at the runway threshold is given by Eq. (12-7). The objective is to find for a specified probability of violation p_v, the required amount of buffer. Thus

$$p_v = \text{Prob}\left(\Delta T_{ij} < \frac{\delta_{ij}}{V_j}\right)$$

or

$$p_v = \text{Prob}\left(\frac{\delta_{ij}}{V_j} + b_{ij} + e_0 < \frac{\delta_{ij}}{V_j}\right)$$

which simplifies to the relationship

$$p_v = \text{Prob}(b_{ij} < -e_0)$$

Using the assumption that errors are normally distributed with standard deviation σ_0, the value of the buffer can be derived as [16]

$$b_{ij} = q_v \sigma_0 \tag{12-11}$$

where q_v is the value for which the cumulative standard normal distribution has the value $(1 - p_v)$. Stated differently, this simply means the number of standard deviations from the mean in which a certain percentage of the area under the normal curve would be found. For example, if $p_v = 0.05$, then q_v is the 95th percentile of the distribution and equals 1.65. Therefore, in the closing case the buffer time is a constant that depends on the magnitude of the dispersion of the position error and the acceptable probability of violation p_v.

The Opening Case

Next consider the case when the leading aircraft is approaching the runway threshold at a speed greater than the trailing aircraft. In this case, the separation between aircraft increases from the entry gate. The model is premised on the supposition that the trailing aircraft should be scheduled not less than a distance δ_{ij} behind the leading aircraft when the latter is at the entry gate, but it is assumed that strict separation is enforced by air traffic control only when the trailing aircraft reaches the entry gate. This assumption was shown in Fig. 12-8.

For this case, the probability of violation is simply the probability that the trailing aircraft will arrive at the entry gate before the leading aircraft is at a specified distance inside the entry gate. This may be expressed mathematically as follows:

$$p_v = \text{Prob}\left(T_j - \frac{\delta_{ij} + \gamma}{V_j} < T_i - \frac{\gamma}{V_i} \right)$$

or

$$p_v = \text{Prob}\left[T_j - T_i < \frac{\delta_{ij}}{V_j} + \left(\frac{\gamma}{V_j} - \frac{\gamma}{V_i} \right) \right]$$

Using Eq. (12-9) with this equation to compute the actual spacing at the arrival threshold, and simplifying

$$b_{ij} = \sigma_0 q_v - \delta_{ij}\left(\frac{1}{V_j} - \frac{1}{V_i} \right) \tag{12-12}$$

Therefore, for the opening case the amount of buffer is reduced from that required in the closing case, as shown in Eq. (12-12). Negative values of buffer are not allowed and, therefore, the buffer is some finite positive number with a minimum of zero. The matrix of the buffer $[B_{ij}]$ for each pair of aircraft with an interarrival buffer of b_{ij} can then be found. The application of position error to the arrivals only runway capacity problem is illustrated in Example Problem 12-4.

Example Problem 12-4   Assume that the aircraft approaching the runway in Example Problem 12-3 have a position error of 20 s which is normally distributed. In this environment the probability of violating the minimum separation rule for arrival spacing is allowed to be 5 percent.

It is necessary to determine the hourly capacity of the runway to service arrivals.

The error-free interarrival matrix $[M_{ij}]$ was found earlier and it is only necessary to now find the buffer matrix $[B_{ij}]$ and solve Eq. (12-10) for the expected value of the interarrival time.

In the closing case where the leading aircraft is slower than the trailing aircraft, Eq. (12-11) gives the buffer. For a 5 percent probability of violation, q_v can be found from statistics tables as 1.65. For each of these cases, the buffer is independent of speed and therefore

$$b_{ij} = 20(1.65) = 33 \text{ s}$$

In the opening case where the leading aircraft is faster than the trailing aircraft, Eq. (12-12) gives the buffer. Therefore, for $V_i = 120$ and $V_j = 90$, we have

$$b_{ij} = 20(1.65) - 5\left(\tfrac{1}{90} - \tfrac{1}{120}\right)3600 = -17 \text{ s}$$

However, the minimum value of the buffer is always 0. Summarizing the values of b_{ij} found for the buffer in a matrix and recalling the $[M_{ij}]$ matrix for the error-free case, we have

$[M_{ij}]$:

		Leading	
		A	B
Trailing	A	120	90
	B	220	120

$[B_{ij}]$:

		Leading	
		A	B
Trailing	A	33	33
	B	0	33

Substitution into Eq. (12-10) then gives the expected value of the interarrival time.

$$E(\Delta T_{ij}) = 0.25(153) + 0.50(123) + 0.25(220) = 155 \text{ s}$$

and from Eq. (12-6), we have

$$C_a = \frac{3600}{155} = 23 \text{ operations per hour}$$

which is a reduction from the arrival capacity found in Example Problem 12-3. Therefore, one may conclude that the existence of position error reduces the arrival capacity of a runway.

Development of a Model for Departures Only

Since departures are normally cleared for takeoff based upon maintaining a minimum time interval between successive departures, the interdeparture time t_d, the departure-only capacity of a runway C_d, is given by

$$C_d = \frac{3600}{E(t_d)} \tag{12-13}$$

and

$$E(t_d) = \Sigma [p_{ij}][t_d] \tag{12-14}$$

where $E(t_d)$ = expected value of the service time, or interdeparture time, at the runway threshold for the departure aircraft mix

$[p_{ij}]$ = matrix of the probabilities that the leading departing aircraft i will be followed by the trailing departing aircraft j

$[t_d]$ = matrix of the interdeparture times

Development of Models for Mixed Operations

This model is based on the same four operating rules as the model developed by Airborne Instruments Laboratory [3]. These may be listed as follows:

Arrivals have priority over departures.

Only one aircraft can occupy the runway at any instant of time.

A departure may not be released if the subsequent arrival is less than a specified distance from the runway threshold, usually 2 nmi in IFR conditions.

Successive departures are spaced at a minimum time separation equal to the departure service time.

A time-space diagram may be drawn to show the sequencing of mixed operations under the rules stated above and this is done in Fig. 12-11. In this figure, T_i and T_j are the times that the leading aircraft i and the trailing aircraft j, respectively, pass over the arrival threshold, δ_{ij} is the minimum separation between arrivals, T_1 is the time when the arriving aircraft clears the runway, T_d is the time when the departing aircraft begins its takeoff roll, δ_d is the minimum distance that an arriving aircraft must be from the threshold to release a departure, T_2 is the time which corresponds to the last instant when a departure can be released, R_i is the runway occupancy time for an

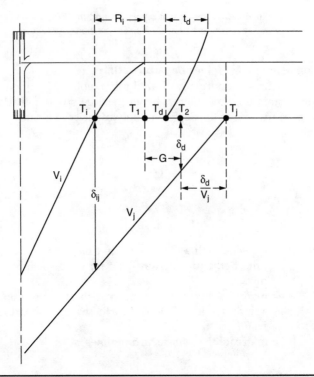

FIGURE 12-11   Time-space diagram for error-free interarrival spacing for mixed operations on a runway system.

arrival, G is the time gap in which a departure may be released, and t_d is the required service time for a departure.

Since arrivals are given priority over departures, the arriving aircraft are sequenced at the minimum interarrival separation and a departure cannot be released unless there is a gap between arrivals G. Therefore, we may write

$$G = T_2 - T_1 \geq 0$$

but we know that

$$T_1 = T_i + R_i$$

and

$$T_2 = T_j - \frac{\delta_d}{V_j}$$

Therefore, we may write

$$T_2 - T_1 \geq \left(T_j - \frac{\delta_d}{V_j} \right) - (T_i + R_i) \geq 0$$

or to release one departure between a pair of arrivals we have

$$T_j - T_i \geq R_i + \frac{\delta_d}{V_j}$$

Through a simple extension of this equation it is apparent that the required mean interarrival time $E(\Delta T_{ij})$ to release n_d departures between a pair of arrivals is given by

$$E(\Delta T_{ij}) \geq E(R_i) + E\left(\frac{\delta_d}{V_j}\right) + (n_d - 1)E(t_d) \qquad (12\text{-}15)$$

It should be noted that the last term of this equation is equal to zero when only one departure is to be inserted between a pair of arrivals. An error term may be added on to the above equation, $\sigma_G q_v$, to account for the violation of the gap spacing. The use of Eq. (12-15) with gap error will be illustrated in Example Problem 12-5.

The capacity for mixed operations is given by the equation

$$C_m = \frac{1}{E(\Delta T_{ij})}\left(1 + \sum n_d p_{nd}\right) \qquad (12\text{-}16)$$

where C_m = capacity of the runway to process mixed operations
$E(\Delta T_{ij})$ = expected value of the interarrival time
n_d = number of departures which can be released each gap between arrivals
p_{nd} = probability of releasing n_d departures in each gap

The application of the equations for mixed operations on a runway is shown in Example Problem 12-6.

Example Problem 12-5   Assume that the aircraft approaching the runway in Example Problem 12-4 have an error in the gap between arrivals of 30 s which is normally distributed. In this environment, the probability of violating the minimum gap in which departures can be released is 10 percent. The runway occupancy time for a type A aircraft is 50 s and for a type B aircraft is 40 s. A departure can be released if the arriving aircraft is at least 2 mi from the arrival threshold. The minimum time between successive departures is 60 s. The arrival mix and the departure mix are identical.

It is necessary to determine the minimum separation between arrivals in order to ensure that one departure can be released between each pair of arrivals.

The required interarrival time to release a departure between every pair of arrivals is given by Eq. (12-15).

There are three type A aircraft and two type B aircraft in both the arrival and departure queue. Therefore, there is a 60 percent chance of a type A aircraft and a 40 percent chance of a type B aircraft. Substitution into Eq. (12-15) yields

$$E(\Delta T_{ij}) \geq 0.6(50) + 0.4(40) + \left(\frac{1}{120} + \frac{1}{90}\right)3600 + (1-1)(60) = 114 \text{ s}$$

Therefore, to release a departure between a pair of arrivals in an error-free context there must be a gap between successive arrivals of 114 s. In a position error context, the error term must be computed and added to this value. For a probability of violation of 10 percent the value of q_v is found to be 1.28. The error in the gap between arrivals is found to be $\sigma_g q_v = 30(1.28) = 38$ s. Therefore, in a position error context there must be a gap between arrivals of $114 + 38 = 152$ s to release a departure between every pair of arrivals.

If the actual interarrival time matrix in Example Problem 12-4 is examined, it can be seen that a departure can be released only when an arrival of a type A aircraft is followed by an arrival of a type B aircraft since this is the only case where the required interarrival time of 152 s is attained. This occurs 25 percent of the time. Therefore, the capacity of the runway to service mixed operations in a position error context is

$$C_m = \frac{3600}{155}[1.0 + 1(0.25)] = 30 \text{ operations per hour}$$

A comprehensive example problem using the analytical equations developed for determining the hourly capacity of a runway is presented to summarize this section.

Example Problem 12-6   A runway is to service arrivals and departures. The common approach path is 6 mi long for all aircraft. During a particular interval of time the runway is serving three types of aircraft with the mix and operating characteristics shown in Table 12-2. The air traffic separation rules in effect are given in Table 12-3.

Assume that the standard deviation of the position of airborne aircraft and the error in the gaps between arrivals are known to be 20 s and that the minimum separation rules may be violated 10 percent of the time.

First let us find the capacity of the runway system to service arrivals only.

The error-free interarrival time equations are given in Eqs. (12-7) and (12-8). Using these error-free interarrival time equations the interarrival matrix can be computed. For example, for a leading B followed by a trailing A, we have

$$\Delta T_{BA} = \frac{3(3600)}{135} = 80 \text{ s}$$

and for a leading A followed by a trailing B, we have

$$\Delta T_{AB} = \frac{5(3600)}{135} + 6\left(\frac{1}{110} - \frac{1}{135}\right)3600 = 170 \text{ s}$$

Aircraft Type	Approach Speed (mph)	Runway Occupancy Time, s	Mix, %	
			Arrival	Departure
A	135	50	20	15
B	110	40	45	55
C	90	30	35	30

TABLE 12-2   Aircraft Mix and Operating Characteristics for Example Problem 12-6

Operational Sequence	Air Traffic Separation Rules		
Arrival–departure	Clear runway		
Departure–arrival	2 mi		
Departure–departure	Seconds	Lead	

Departure–departure:

$$
\begin{array}{c c}
 & \quad\quad \text{Lead} \\
 & \begin{array}{ccc} \text{A} & \text{B} & \text{C} \end{array} \\
\text{Trail} \begin{array}{c} \text{A} \\ \text{B} \\ \text{C} \end{array} &
\begin{bmatrix} 90 & 90 & 60 \\ 90 & 90 & 60 \\ 120 & 90 & 60 \end{bmatrix}
\end{array}
$$

Arrival–arrival (Miles, Lead):

$$
\begin{array}{c c}
 & \quad\quad \text{Lead} \\
 & \begin{array}{ccc} \text{A} & \text{B} & \text{C} \end{array} \\
\text{Trail} \begin{array}{c} \text{A} \\ \text{B} \\ \text{C} \end{array} &
\begin{bmatrix} 4 & 3 & 3 \\ 5 & 4 & 3 \\ 6 & 4 & 3 \end{bmatrix}
\end{array}
$$

TABLE 12-3   Air Traffic Separation Rules for Example Problem 12-6

Continuing these computations for all combinations of leading and trailing aircraft and computing the arrival mix probabilities results in the matrices below.

$[M_{ij}]$:

$$
\begin{array}{c c}
 & \quad\quad \text{Leading} \\
\text{Trailing} \begin{array}{c} \text{A} \\ \text{B} \\ \text{C} \end{array} &
\begin{bmatrix} 107 & 80 & 80 \\ 170 & 131 & 98 \\ 240 & 175 & 120 \end{bmatrix}
\end{array}
$$

$[P_{ij}]$:

$$
\begin{array}{c c}
 & \quad\quad \text{Leading} \\
\text{Trailing} \begin{array}{c} \text{A} \\ \text{B} \\ \text{C} \end{array} &
\begin{bmatrix} 0.04 & 0.09 & 0.07 \\ 0.09 & 0.20 & 0.16 \\ 0.07 & 0.16 & 0.12 \end{bmatrix}
\end{array}
$$

The interarrival buffer time which must be added to the error-free case when position error is present is given by Eqs. (12-11) and (12-12).

If the probability of violation is 10 percent, then $q_v = 1.28$. Using Eqs. (12-11) and (12-12) to solve for the buffer, we have for a leading aircraft B and a trailing aircraft A

$$b_{BA} = 20(1.28) = 26 \text{ s}$$

and for a leading aircraft A followed by a trailing aircraft B

$$b_{AB} = 20(1.28) - 5\left(\tfrac{1}{110} - \tfrac{1}{135}\right)3600 = -5 \text{ s}$$

But the minimum value of the buffer is always 0.

Continuing this for all combinations of leading and trailing aircraft gives the buffer and interarrival time matrices:

$[B_{ij}]$:   Leading

$$
\begin{array}{c}
 & \begin{array}{ccc} A & B & C \end{array} \\
\text{Trailing} \begin{array}{c} A \\ B \\ C \end{array} & \left[\begin{array}{ccc} 26 & 26 & 26 \\ 0 & 26 & 26 \\ 0 & 0 & 26 \end{array}\right]
\end{array}
$$

$[M_{ij} + B_{ij}]$:   Leading

$$
\begin{array}{c}
 & \begin{array}{ccc} A & B & C \end{array} \\
\text{Trailing} \begin{array}{c} A \\ B \\ C \end{array} & \left[\begin{array}{ccc} 133 & 106 & 106 \\ 170 & 157 & 124 \\ 240 & 175 & 146 \end{array}\right]
\end{array}
$$

The expected value of the interarrival time becomes from Eq. (12-10)

$$E(\Delta T_{ij}) = 0.04(133) + 0.09(106) + \cdots + 0.12(146) = 151 \text{ s}$$

The arrival capacity is then from Eq. (12-6)

$$C_a = \frac{3600}{151} = 24 \text{ operations per hour}$$

Next let us find capacity of the runway system to service departures only. The expected value of the departure time is computed from Eq. (12-14) using the departure-departure time matrix given and the departure mix probability matrix below. This matrix is based on actual departure-departure times and always considers error.

$[p_{ij}]$:   Leading

$$
\begin{array}{c}
 & \begin{array}{ccc} A & B & C \end{array} \\
\text{Trailing} \begin{array}{c} A \\ B \\ C \end{array} & \left[\begin{array}{ccc} 0.0225 & 0.0825 & 0.0450 \\ 0.0825 & 0.3025 & 0.1650 \\ 0.0450 & 0.1650 & 0.0900 \end{array}\right]
\end{array}
$$

The expected value of the departure time is then

$$E(t_d) = 0.0225(90) + 0.0825(90) + \cdots + 0.0900(60) = 77 \text{ s}$$

The departure capacity of the runway is given by Eq. (12-13)

$$C_d = \frac{3600}{77} = 47 \text{ operations per hour}$$

Next let us find the probability of releasing a departure after each arrival and the capacity of the runway system to service mixed operations in the case where arrivals are given priority over departures.

To release n_d departures the required interarrival time is given by Eq. (12-15) with a buffer term added. Solving for each term in this equation, we have

$$E(R_i) = 0.20(50) + 0.45(40) + 0.35(30) = 38 \text{ s}$$

$$E\left(\frac{\delta_d}{V_j}\right) = \left[0.20\left(\frac{2}{135}\right) + 0.45\left(\frac{2}{110}\right) + 0.35\left(\frac{2}{90}\right)\right]3600 = 68 \text{ s}$$

$$E(t_d) = 77 \text{ s}$$

$$E(B_{ij}) = 26(0.68) + 0(0.32) = 18 \text{ s}$$

and therefore

$$E(\Delta T_{ij}) \geq 38 + 68 + 18 + 77 \ (n_d - 1)$$

$$E(\Delta T_{ij}) \geq 124 + 77 \ (n_d - 1)$$

For one departure we then have a required interarrival time of 124 s, for two successive departures we have a required interarrival time of 201 s, and for three successive departures we have a required interarrival time of 278 s.

Therefore, anytime the interarrival time is greater than or equal to 124 and less than 201 s, one departure may be released between a pair of arrivals. Anytime the interarrival time is greater than or equal to 201 and less than 278 s, two departures may be released between a pair of arrivals. Anytime the interarrival time is greater than or equal to 278 s, three or more departures may be released between a pair of arrivals.

Examination of the interarrival time matrix, gives the probability of releasing departures between arrivals. Therefore, for one departure the probability is 61 percent, for two successive departures the probability is 7 percent, and we cannot release more than two successive departures between a pair of arrivals while maintaining minimum interarrival separations.

The mixed operation hourly capacity is then from Eq. (12-16)

$$C_m = \frac{3600}{151}[1 + 0.61(1) + 0.07(2)] = 42 \text{ operations}$$

Now let us find required interarrival time if at least one departure is to be released after each arrival and the resulting capacity of the runway system to service mixed operations under this condition. For this to occur, all values of the interarrival matrix must be at least 124 s. Therefore, all values of the interarrival time less than 124 s must be increased to 124 s to release at least one departure between every pair of arrivals. Therefore, the new required interarrival time matrix becomes

$[T_{ij}]$:

		Leading		
		A	B	C
	A	133	124	124
Trailing	B	170	157	124
	C	240	175	146

which results in

$E(\Delta T_{ij}) = 154$ sand the hourly capacity for mixed operations becomes

$$C_m = \frac{3600}{154}[1 + 0.93(1) + 0.07(2)] = 48 \text{ operations}$$

Therefore, by increasing the interarrival separations anytime that a type A aircraft follows a type B or type C aircraft, or anytime type B aircraft follows a type C aircraft, the runway capacity can be increased from 42 to 48 operations per hour. Small increases in capacity can result in significant decreases in delay.

Application of Techniques for Ultimate Hourly Capacity

The hourly capacity of the runway system is defined as the maximum number of aircraft operations that can take place on the runway system in an hour. The maximum number of aircraft operations depends on a number of conditions including, but not limited to, the following:

1. The ceiling and visibility conditions
2. The physical configuration of the runway system
3. The air traffic control system separation rules
4. The runway-use strategy
5. The mix of aircraft using the runway system
6. The ratio of arrivals to departures
7. The number of touch and go operations by general aviation aircraft
8. The number and location of exits from the runway system

It is important to point out that the definition of hourly capacity of runways in this section differs from that which is delay related, since the definition of capacity herein contains no assumptions regarding acceptable levels of delay.

The determination of runway system hourly capacity is normally made through the use of computer programs developed for that purpose [16, 17, 20, 28, 29]. These programs are capable of accommodating virtually any runway-use configuration at an airport and allow for the variation in all the parameters which might affect runway capacity. Based upon the use of these programs and constraining many of the variables to conform to present operating scenarios, an airport capacity handbook has been developed which will allow for the computation of realistic estimates of runway capacity [4, 26]. The material which follows is based upon the FAA's Advisory Circular on estimating airport capacity and delay (AC 150/5360-5) and its included techniques for determining runway hourly capacity [4, 26]

Parameters Required for Runway Capacity

As noted above, to determine the hourly capacity of the runway system it is necessary to ascertain the parameters which will affect capacity. Due to the fact that aircraft separation rules differ in VMC and IMC weather, it is first necessary to determine the ceiling and visibility conditions, or more appropriately, the separation rules applicable to flying conditions when ceiling at the airport is at least 1000 ft and visibility is at least 3 mi. This condition results in VFR flying rules for arriving and departing aircraft. If either or both of these criteria are not met, then IFR flying rules are in effect. Of course, all airports have a period of time when conditions are such that IFR rules apply. Therefore, the hourly capacity of runways is normally specified for each of these conditions.

The physical runway surfaces at an airport can be used in several ways. For example, two parallel runways can be used with arrivals on one runway and departures on the other runway at some point in time. They could also be used with arrivals and departures on one surface and arrivals only on the other surface. These runway-use configurations are called the *runway use strategies* which are dependent on weather conditions, aircraft types, and the spacing between runways. It is necessary to specify the runway use strategies and the percentage of time each strategy is used.

It is also necessary to specify the types of aircraft which can use a given runway as quite often shorter runways are constructed for use by general aviation aircraft only. The aircraft which can use a runway are defined in terms of a *mix index*. This index is simply an indication of the level of operations on the runway by large and heavy aircraft. The mix index is given by Eq. (12-17).

$$MI = C + 3D \qquad (12\text{-}17)$$

where MI = mix index
  C = percentage of aircraft weighing more than 12,500 lb but less than 300,000 lb on the runway
  D = percentage of aircraft with maximum gross weight of 300,000 lb or greater in the mix of aircraft using the runway

The percentage of arrival operations which occur on the runway is also necessary. This is because the spacing rules for arrivals and departures differ. There are three types of operations which can occur, namely, arrivals, departures, and *touch-and-go* operations. A touch-and-go operation is most commonly used by general aviation pilots practicing approaches, landings, and takeoffs. These operations are seldom conducted in poor weather. For the purpose of determining capacity, the parameter called percent arrivals is used to define the proportion of each type of operation which occurs on the runway. In VFR conditions it is also necessary to find the percentage of touch-and-go operations. At times small general aviation airports may have touch-and-go operations which can approach 30 percent of all operations.

The location of runway exits for arriving aircraft must also be known since this affects runway occupancy time. Depending upon the nature of the aircraft using a runway exits should be located at positions which will allow minimum runway occupancy times. If this is not the case, the capacity will be reduced because of excessive runway occupancy times.

As a result of extensive research conducted to determine the capacity of runway systems, the FAA has published a series of charts to determine runway capacity [4, 26]. These charts are used to determine the runway capacity through Eq. (12-18).

$$C = C_b ET \qquad (12\text{-}18)$$

where C = hourly capacity of the runway-use configuration in operations per hour

C_b = ideal or base capacity of the runway-use configuration

E = exit adjustment factor for the number and location of runway exits

T = touch-and-go adjustment factor

The use of this equation and the charts are illustrated by Example Problem 12-7.

Example Problem 12-7   It is required to find the VFR and IFR hourly capacity of the runway system shown in Fig. 12-12. The runway-use strategy is as shown. In VFR weather, the traffic consists of 3 single-engine, 20 light twin-engine, 25 large transport-type, and 2 wide-bodied aircraft. Arrivals constitute 40 percent of the operations and there are approximately three touch–and-go operations.

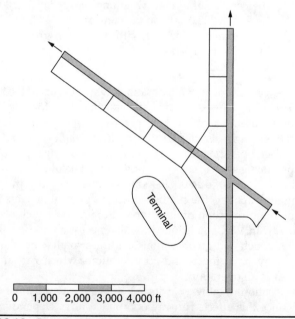

FIGURE 12-12   Runway layout for Example Problem 12-7.

Aircraft	Class	VFR Mix		IFR Mix	
		No.	**%**	**No.**	**%**
Single-engine	A	13	26.0	2	5.9
Twin-engine	B	10	20.0	5	14.7
Transports	C	25	50.0	25	73.5
Wide-bodied	D	2	4.0	2	5.9
Total		50	100.0	34	100.0

TABLE 12-4   Tabulation of Aircraft Mix Index for Example Problem 12-7

In IFR, the small aircraft population count drops to two single-engine and five light twin-engine aircraft. The arrival rate increases to 50 percent and there are no touch-and-go operations.

The capacity of intersecting runways is a function of the location of the intersection from both the arrival and departure threshold. The closer that the intersection is to these thresholds, the greater the capacity. The aircraft are grouped into various classes in VFR and IFR conditions in Table 12-4. The charts used for this configuration in VFR and IFR are taken from references [4, 26] and are given in Figs. 12-13 and 12-14.

From the this tabular data, the mix index can be found for VFR from Eq. (12-17) as

$$MI = C + 3D = 50.0 + 3(4.0) = 62.0$$

and for IFR

$$MI = C + 3D = 73.5 + 3(5.9) = 91.2$$

Using the VFR mix index of 62.0 and the percent arrivals (PA) equal to 40, the base capacity C_b is found from the left side of Fig. 12-13 as about 95 operations per hour. This base value is then adjusted for touch-and-go operations and the location of exits using the right side of this figure. From the given data, the percentage of touch-and-go operations in VFR is equal to 6 percent. Therefore, the touch-and-go adjustment factor T is equal to 1.03. For the mix index of 62.0, only those exits located between 3500 and 6500 ft from the arrival threshold can be counted. There are two such exits, one at 4500 ft and the other at 6000 ft. Therefore, the table then gives an exit factor E of 0.97 for 40 percent arrivals.

Therefore, the hourly capacity of the runway system in VFR is from Eq. (12-18)

$$C = 95(1.03)(0.97) = 95 \text{ operations per hour}$$

The IFR capacity is determined similarly from Fig. 12-14. This will yield, for an IFR mix index of 91.2 and a percent arrivals (PA) of 50 percent, the base capacity, touch-and-go factor, and exit factor as

$$C = 58(1.00)(0.97) = 56 \text{ operations per hour}$$

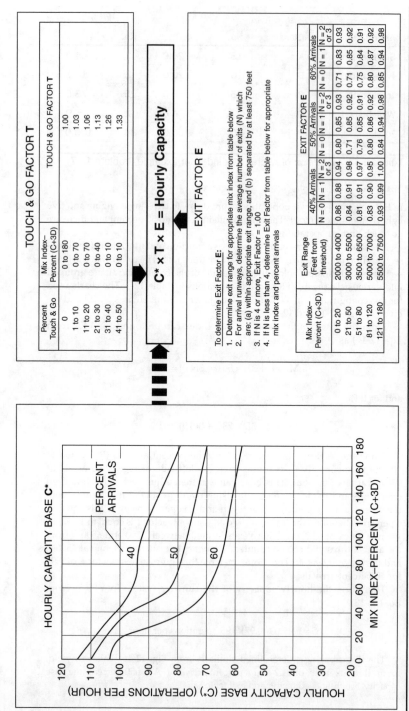

TOUCH & GO FACTOR T

Percent Touch & Go	Mix Index— Percent (C+3D)	TOUCH & GO FACTOR T
0	0 to 180	1.00
1 to 10	0 to 70	1.03
11 to 20	0 to 70	1.06
21 to 30	0 to 40	1.13
31 to 40	0 to 10	1.26
41 to 50	0 to 10	1.33

C* × T × E = Hourly Capacity

EXIT FACTOR E

To determine Exit Factor **E**:

1. Determine exit range for appropriate mix index from table below
2. For arrival runways, determine the average number of exits (N) which are: (a) within appropriate exit range, and (b) separated by at least 750 feet
3. If N is 4 or more, Exit Factor = 1.00
4. If N is less than 4, determine Exit Factor from table below for appropriate mix index and percent arrivals

Mix Index— Percent (C+3D)	Exit Range (Feet from threshold)	40% Arrivals			50% Arrivals			60% Arrivals		
		N = 0	N = 1	N = 2 or 3	N = 0	N = 1	N = 2 or 3	N = 0	N = 1	N = 2 or 3
0 to 20	2000 to 4000	0.86	0.88	0.94	0.80	0.85	0.93	0.71	0.83	0.93
21 to 50	3000 to 5500	0.84	0.91	0.98	0.71	0.85	0.92	0.71	0.85	0.92
51 to 80	3500 to 6500	0.81	0.91	0.97	0.76	0.85	0.91	0.75	0.84	0.91
81 to 120	5000 to 7000	0.83	0.90	0.95	0.80	0.86	0.92	0.80	0.87	0.92
121 to 180	5500 to 7500	0.93	0.99	1.00	0.84	0.94	0.98	0.85	0.94	0.98

Figure 12-13 Hourly capacity in VFR for runway operations in Example Problem 12-7.

518

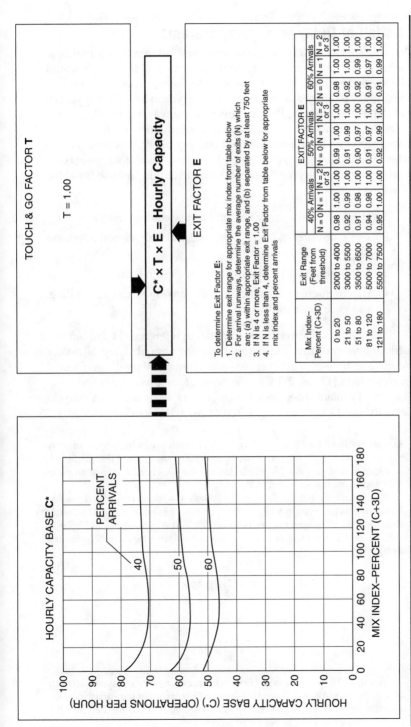

TOUCH & GO FACTOR T

T = 1.00

C* × T × E = Hourly Capacity

EXIT FACTOR E

To determine Exit Factor **E**:
1. Determine exit range for appropriate mix index from table below
2. For arrival runways, determine the average number of exits (N) which are: (a) within appropriate exit range, and (b) separated by at least 750 feet
3. If N is 4 or more, Exit Factor = 1.00
4. If N is less than 4, determine Exit Factor from table below for appropriate mix index and percent arrivals

Mix Index–	Exit Range	EXIT FACTOR E								
Percent (C+3D)	(Feet from threshold)	40% Arrivals			50% Arrivals			60% Arrivals		
		N = 0	N = 1	N = 2 or 3	N = 0	N = 1	N = 2 or 3	N = 0	N = 1	N = 2 or 3
0 to 20	2000 to 4000	0.98	1.00	1.00	0.99	1.00	1.00	0.98	1.00	1.00
21 to 50	3000 to 5500	0.92	0.99	1.00	0.91	0.99	1.00	0.92	1.00	1.00
51 to 80	3500 to 6500	0.91	0.98	1.00	0.90	0.97	1.00	0.92	0.99	1.00
81 to 120	5000 to 7000	0.94	0.98	1.00	0.91	0.97	1.00	0.91	0.97	1.00
121 to 180	5500 to 7500	0.95	1.00	1.00	0.92	0.99	1.00	0.91	0.99	1.00

HOURLY CAPACITY BASE C*

PERCENT ARRIVALS

40

50

60

HOURLY CAPACITY BASE (C*) (OPERATIONS PER HOUR)

MIX INDEX–PERCENT (C+3D)

Figure 12-14 Hourly capacity in IFR for runway operations in Example Problem 12-7.

Computation of Delay on Runway Systems

Delay to aircraft is defined as the difference between the actual time it takes an aircraft to maneuver on the runway and the time it would take the aircraft to maneuver without interference from other aircraft. The runway is defined as the entire runway system including approach and departure airspace [26]. To compute runway system delay, it is necessary to analyze each runway-use configuration for the demand placed upon it. To compute annual runway delay, it is necessary to determine the percentage of time each runway-use configuration is used throughout the year. Normally this will require knowledge of the following factors:

1. The hourly capacity of the runway-use strategy in VFR and IFR
2. The pattern of hourly, daily, and monthly aircraft demand during the design year
3. The peaking of demand during the design hour
4. The frequency of occurrence of runway strategies, ceiling, and visibility conditions

The techniques are outlined in detail in references [4, 26] but it is sufficient to note that the computation of annual delay is a very tedious and time-consuming process, and now is generally performed on computers [17, 28, 29]. The elements of the process are shown in Example Problem 12-8, which uses charts from the FAA Airport Capacity and Delay Advisory Circular AC 150/5360-5.

Example Problem 12-8   The hourly delay to aircraft operating on the runway system in Example Problem 12-7 is to be found for both VFR and IFR conditions. It is known that the peak 15-min demand in the peak hour is 20 operations in VFR and 10 operations in IFR.

The hourly capacity of the runway system was found earlier and yielded 95 operations per hour in VFR and 56 operations per hour in IFR. The hourly demand was 50 operations per hour in VFR and 34 operations per hour in IFR. Therefore, the ratio of hourly demand to hourly capacity is computed as for VFR

$$\frac{D}{C} = \frac{50}{95} = 0.53$$

and for IFR

$$\frac{D}{C} = \frac{34}{56} = 0.61$$

Figure 12-15, which is taken from references for this runway-use strategy [4, 26], the FAA Airport Capacity Advisory Circular, gives the variation of the *arrival delay index* (ADI) and the *departure delay index* (DDI) for VFR and IFR conditions for this runway use for 50 percent arrivals.

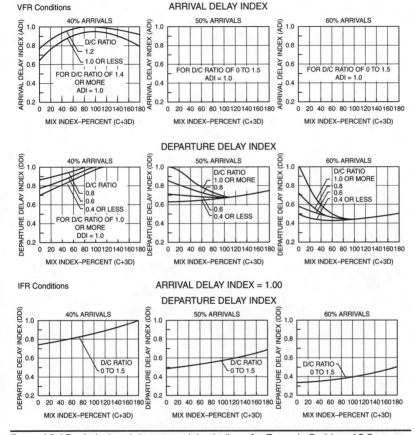

FIGURE 12-15 Arrival and departure delay indices for Example Problem 12-8.

Based upon the respective mix indices for VFR and IFR conditions, this chart gives for VFR

$$ADI = 1.00 \quad \text{and} \quad DDI = 0.65$$

and for IFR

$$ADI = 1.00 \quad \text{and} \quad DDI = 0.57$$

These indices are combined with the respective ratios of demand to capacity to arrive at the *arrival delay factor* (ADF) and the *departure delay factor* (DDF) as follows:

$$ADF = (ADI)\frac{D}{C}$$

and

$$DDF = (DDI)\frac{D}{C}$$

Therefore, for arrivals we have in VFR

$$ADF = (1.00)(0.53) = 0.53$$
$$DDF = (0.65)(0.53) = 0.34$$

and in IFR we have

$$ADF = (1.00)(0.61) = 0.61$$
$$DDF = (0.57)(0.61) = 0.35$$

The average delay for each aircraft is then found from Fig. 12-16 by using the above delay factors and the *demand profile factor*.

The demand profile factor is simply a measure of the peaking of demand in the hour and is defined as the peak 15-min demand divided by the hourly demand. Therefore, the demand profile factors (DPF) are in VFR

$$DPF = \frac{20}{50}(100) = 40$$

and in IFR

$$DPF = \frac{10}{34}(100) = 34$$

From Fig. 12-16, the average delays are found in VFR as

Arrival delay = 1.4 min per aircraft

Departure delay = 0.7 min per aircraft

and in IFR

Arrival delay = 1.0 min per aircraft

Departure delay = 0.3 min per aircraft

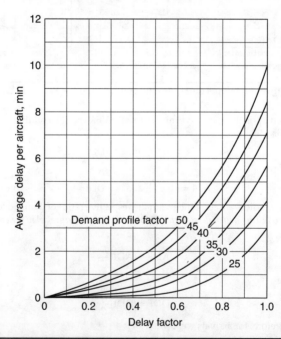

FIGURE 12-16   Variation of average aircraft delay with delay factor (*Federal Aviation Administration*).

Since, there are equal arrivals and departures, the total delay to all aircraft in this hour in both VFR and IFR can be found.

$$\text{Total delay VFR} = 50[1.4(0.5) + 0.7(0.5)] = 52.5 \text{ min}$$
$$\text{Total delay IFR} = 34[1.0(0.5) + 0.3(0.5)] = 22.1 \text{ min}$$

The computation of delay at an airport over various periods of time requires that this procedure be repeated for each hour, day, and month for each runway-use strategy and weather condition and summed over the period of interest. The process is shown in Example Problem 12-9 for the computation of delay on a daily basis.

Example Problem 12-9 To illustrate the airport capacity handbook method [4, 26] of computing the delay throughout a typical day, let us determine the daily delay to aircraft if the runway-use strategy shown in Fig. 12-12 is used throughout the day. The hourly capacity of the runway-use strategy is 80 operations per hour and the number of arrivals equals the number of departures in each hour. The hourly demand throughout the day is given in Table 12-5.

Let us assume that the mix index is equal to 70 and demand profile factor is equal to 25. Both the mix index and the demand profile factor are the same in each hour. Let us also assume that VFR conditions exist throughout the day.

To solve this problem, it is necessary to compute the arrival delay index, the departure delay index, the arrival delay factor, and the departure delay factor for each hour during the day. The computational procedure differs depending upon whether when the hourly demand is less than or equal to the hourly capacity or the hourly demand is greater than the hourly capacity.

For the condition when hourly demand is less than or equal to the hourly capacity the procedure is the same as that shown in Example Problem 12-8.

For example, in hour 1000 the aircraft demand is 60 operations per hour and the runway capacity is 80 operations per hour. Therefore, the ratio of hourly demand to hourly capacity is $60 \div 80 = 0.75$.

From Fig. 12-15, the arrival delay index (ADI) is 1.0 and the departure delay index (DDI) is 0.70. The arrival delay factor is then

$$\text{ADF} = \text{ADI}\frac{D}{C} = 1.0(0.75) = 0.75$$

Hour	Demand	Hour	Demand	Hour	Demand	Hour	Demand
0000	10	0600	20	1200	50	1800	100
0100	10	0700	40	1300	50	1900	70
0200	10	0800	60	1400	40	2000	40
0300	10	0900	50	1500	70	2100	30
0400	10	1000	60	1600	110	2200	20
0500	10	1100	50	1700	120	2300	10

TABLE 12-5 Hourly Aircraft Runway Demand on Typical Day for Example Problem 12-9

and the departure delay factor is

$$DDF = DDI \frac{D}{C} = 0.70(0.75) = 0.53$$

From Fig. 12-16, using these delay factors and a demand profile factor of 25, an average arrival delay of 1.2 min and an average departure delay of 0.2 min are found.

Since the number of arrivals is equal to the number of departures the total delay in hour 1000 is then equal to

$$\text{Delay} = 1.2(0.5)(60) + 0.2(0.5)(60) = 42 \text{ aircraft-minutes}$$

The procedure is the same for hours 0000 through hour 1500.

However, beginning at hour 1600 the hourly demand exceeds the hourly capacity for 3 h. These are called *overloaded hours*. The cumulative demand for these 3 h, 330 operations, exceeds the cumulative capacity available for these 3 h, 240 operations. Therefore, some of these aircraft are not serviced in these 3 h and spill over into later hours. The later hours serve this backlog of demand until the backlog is cleared up. This is shown in Table 12-6. The time from hour 1600 to hour 2000 is called the *saturated period*.

For the overloaded period, hours 1600 through 1800, the total demand is divided by the total capacity to arrive at the average demand to capacity ratio during the overloaded hours. Therefore,

$$\frac{D}{C} = \frac{110 + 120 + 100}{80 + 80 + 80} = 1.38$$

From Fig. 12-15, the arrival delay index is 1.00 and the departure delay index is 0.75 during the overloaded hours. This results in an arrival delay factor of $1.00 \times 1.38 = 1.38$ and a departure delay factor of $0.75 \times 1.38 = 1.04$.

Figure 12-17, which is taken from the FAA airport capacity advisory circular, gives the aircraft delay in the saturated period when the period of overload is 3 h. From this figure, using a demand profile factor of 25, an average arrival delay over this period is found to be 35 min per arrival and an average departure delay over this period is found to be 4 min per departure.

The delay in the saturated period is found by adding the total demand in the saturated period and multiplying by the average delay per operation. The total demand is then the demand from hours 1600 through 2000, or 440 operations.

Hour	Demand	Capacity	Overload	Cumulative Overload
1500	70	80	0	0
1600	110	80	+30	+30
1700	120	80	+40	+70
1800	100	80	+20	+90
1900	70	80	−10	+80
2000	40	80	−40	+40
2100	30	80	−50	0

TABLE 12-6   Overloaded and Saturated Hours Example Problem 12-9

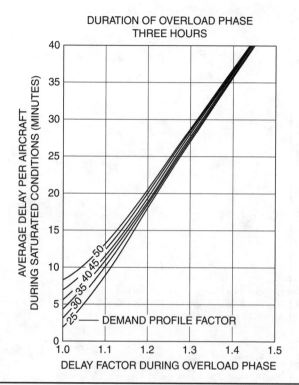

FIGURE 12-17 Average aircraft delay during saturated conditions for an overload period of 3 h (*Federal Aviation Administration*).

Since the number of arrivals is equal to the number of departures in each hour, this yields a total delay in the saturated period of

$$\text{Delay} = 440(0.5)(35) + 440(0.5)(4) = 8580 \text{ aircraft-minutes}$$

In the hours 2100 through 2300 the demand is once again less than the capacity, and the backlog has been cleared up. Therefore, the procedure is the same as used for hours 0000 through 1500.

The results are displayed in tabular format in Table 12-7. The result is that the total delay on this day is equal to 8799 aircraft-hours. The average delay to aircraft on this day is 8799/1050 = 8.4 min.

Graphical Methods for Approximating Delay

A relatively simple technique for estimating delays when demand exceeds capacity has been used in aviation studies [19]. This method is called a deterministic queuing model. In this method, a time scale is established on the X axis to represent the time period being analyzed. On the Y axis, a scale is established for the cumulative number of aircraft which have arrived by some point in time. Therefore, a point on the plot represents the total number of aircraft which have arrived at that point

Hour	Demand	Ratio D/C	Arrivals		Departures per		Minutes of Delay Hourly		
			ADI	ADF	DDI	DDF	Arrival	Departures	Total
0000	10	0.13	1.0	0.13	0.65	0.08	0	0	0
0100	10	0.13	1.0	0.13	0.65	0.08	0	0	0
0200	10	0.13	1.0	0.13	0.65	0.08	0	0	0
0300	10	0.13	1.0	0.13	0.65	0.08	0	0	0
0400	10	0.13	1.0	0.13	0.65	0.08	0	0	0
0500	10	0.13	1.0	0.13	0.65	0.08	0	0	0
0600	20	0.25	1.0	0.25	0.65	0.16	0	0	0
0700	40	0.50	1.0	0.25	0.67	0.34	0	0	0
0800	60	0.75	1.0	0.75	0.70	0.53	1.2	0.2	42
0900	50	0.63	1.0	0.63	0.68	0.43	0.4	0.1	13
1000	60	0.75	1.0	0.75	0.70	0.53	1.2	0.2	42
1100	50	0.63	1.0	0.63	0.68	0.43	0.4	0.1	13
1200	50	0.63	1.0	0.63	0.68	0.43	0.4	0.1	13
1300	50	0.63	1.0	0.63	0.68	0.43	0.4	0.1	13

1400	40	0.50	1.0	0.50	0.67	0.34	0.2	0	4
1500	70	0.88	1.0	0.88	0.73	0.64	1.8	0.4	77
1600	110								
1700	120	1.38	1.0	1.38	0.75	1.04	35	4	8580
1800	100								
1900	70								
2000	40								
2100	30	0.38	1.0	0.38	0.65	0.25	0.1	0	2
2200	20	0.25	1.0	0.25	0.65	0.16	0	0	0
2300	10	0.13	1.0	0.13	0.65	0.08	0	0	0
Daily	1050								8799

TABLE 12-7 Tabulation of Hourly Delay for Example Problem 12-9

in time. The curve which results from plotting the succession of points is actually a representation of the demand $D(t)$. A line of constant, or for that matter variable slope, can be drawn on the same graph to show the service capabilities, or capacity, of a facility. This is the service function $S(t)$. An illustration of such a graph is given in Fig. 12-18.

In this figure, point A represents the time when the demand rate begins to exceed the service rate or capacity. Therefore, delays and queues begin to develop. At point B, the delays and queues which have built since time t_1 will have dissipated, and the demand rate is now less than the service rate. A review of the results displayed on this figure shows that

1. Delays occur from time t_1 to time t_4.

2. The total number of aircraft delayed is the difference between P_4 and P_1.

3. From time t_1 to time t_3 the demand rate exceeds the service rate, and delays and queues increase during this time period.

4. The maximum delay and maximum queue length occur at time t_3 since the demand rate becomes less than the service rate at this time.

5. The delay to any aircraft is given by the magnitude of a horizontal line drawn between the two curves.

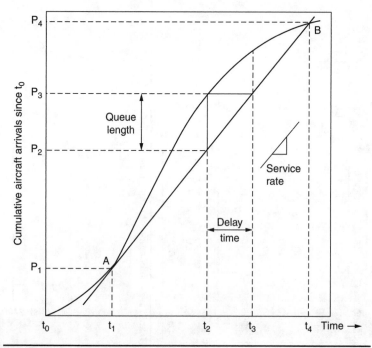

FIGURE 12-18 Deterministic queuing diagram.

6. The length of a queue at any point in time is given by the magnitude of a vertical line drawn between the two curves at that point in time.

7. The area between the two curves represents the total delay to all aircraft which are delayed from time t_1 to time t_4.

This type of analysis is useful in airport planning to estimate the magnitude of delays, number of aircraft delayed, and the cost of delay under assumed operating conditions. It does not, however, give an indication of the delays which occur when average demand is less than the capacity. These are normally calculated by the equations or methods discussed earlier.

Example Problem 12-10 illustrates the application of this technique to a runway system.

Example Problem 12-10 The hourly aircraft demand during a typical day at an airport is given in Table 12-5. The runway system has a capacity to service 80 aircraft per hour without delay.

An analysis of the delay when demand exceeds capacity is to be conducted.

For discussion purposes the hourly pattern of demand and the capacity are plotted in Fig. 12-19. The deterministic model makes the assumption that delay occurs only when demand exceeds capacity. This figure shows that up until hour 1600 the aircraft demand in any hour is less than the runway capacity and therefore delays do not occur. However, beginning at hour 1600 the aircraft demand begins to exceed the runway capacity and therefore delays begin to accrue from this point in time.

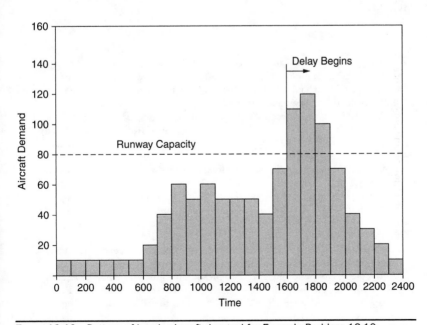

FIGURE 12-19 Pattern of hourly aircraft demand for Example Problem 12-10.

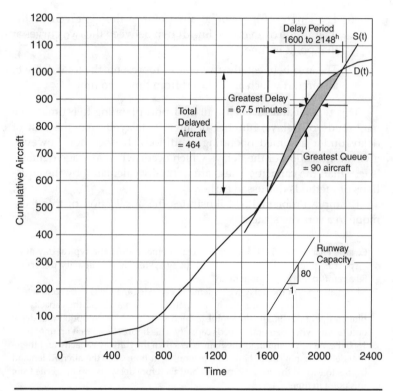

FIGURE 12-20 Plot of cumulative aircraft arrivals versus time for Example Problem 12-10.

The data in Table 12-5 are plotted in Fig. 12-20, where the cumulative hourly demand $D(t)$ and cumulative service rate $S(t)$ are plotted versus time. This figure again shows that at hour 1600 the demand rate begins to exceed the capacity and therefore delay will begin in hour 1600.

The shaded area between the curves represents the period when aircraft delays occur. However, due to the scale on this figure, it is difficult to determine the values of the greatest delay to any aircraft, the greatest X value within the shaded area, the greatest number (queue) of aircraft delayed, the largest Y value within the shaded area, or the total aircraft-hours of delay, the shaded area. Since only the period after hour 1600 contains delay and the Y value within the shaded area represents the difference between demand and capacity, a plot of the cumulative difference between demand and capacity versus time from hour 1600, tabulated in Table 12-8, is shown in Fig. 12-21. This figure effectively expands the scale of the shaded area in Fig. 12-20 and is much easier to use to determine the values of the above delay and queue length parameters.

The greatest Y value on Fig. 12-21 represents the greatest number of aircraft delayed at any point in time, the time period when the curve is above the X axis is the time period during which delay occurs, and the area of the curve above the X axis is the total aircraft-hours of delay.

End Hour	Demand	Demand-Capacity	Cumulative Demand-Capacity
1500			0
1600	110	+30	+30
1700	120	+40	+70
1800	100	+20	+90
1900	70	−10	+80
2000	40	−40	+40
2100	30	−50	−10
2200	20	−60	−70

TABLE 12-8 Cumulative Demand Minus Capacity during Delay Period for Example Problem 12-10

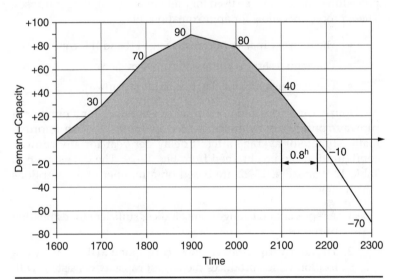

FIGURE 12-21 Plot of demand minus capacity for delay period for Example Problem 12-10.

The largest number of aircraft delayed at any point in time is found to be 90. The service time for an aircraft is the reciprocal of the runway capacity, or the runway will service one aircraft in 0.75 min. The delay time for the aircraft which is delayed the longest is therefore 90(0.75) = 67.5 min. The total number of aircraft hours of delay is the area under this curve or 306 aircraft-hours.

All of the aircraft in hours 1600 to 2000 and 80 percent of the aircraft in hour 2100 are delayed. This amounts to 464 aircraft. Therefore, the average delay to delayed aircraft on this day is 306/464 = 0.66 h = 40 min. The average delay to all aircraft using the runway on this day is 306/1050 = 0.29 h = 17 min.

Application of Techniques for Annual Service Volume

Annual service volume is a level of annual aircraft operations that may be used as a reference in preliminary planning. As annual aircraft operations approach annual service volume, the average delay to each aircraft throughout the year may increase rapidly with relatively small increases in aircraft operations, thereby causing levels of service on the airfield to deteriorate.

When annual aircraft operations on the airfield are equal to annual service volume, average delay to each aircraft throughout the year is on the order of 1 to 4 min. A more precise estimate of actual average delay to aircraft at a particular airport can be obtained using these procedures if this is required in the planning application.

Simplified estimating procedures are available for airport planning purposes to determine the annual service volume and average aircraft delay for runway configurations existing at airports. These procedures should only be used for preliminary estimating purposes. These procedures allow for approximations of

- The hourly capacity of runways in VFR and IFR conditions
- The annual service volume of runways
- The average annual delay to aircraft on runways

Hourly capacities and annual service volumes for a number of runway configurations are presented in Table 12-9. An approximate estimate of average aircraft delay per year for any runway configuration can be obtained from Fig. 12-22. The data shown in Table 12-9 and Fig. 12-22 are based on a number of assumptions which include

1. A representative range of mix indices sufficient for estimating purposes.

2. The hourly capacities are those correspond to the runway utilization which produces the largest capacity consistent with current air traffic control procedures and practices, and this configuration is used 80 percent of the time.

3. One-half of the demand for the use of the runways is by arriving aircraft, and thus, the number of arriving and departing aircraft in a specified period of time is equal.

4. The percentage of touch-and-go operations is a function of the mix index of the airport.

5. Sufficient taxiways exist to permit the capacity of the runways to be fully realized.

6. The impact on capacity of a taxiway crossing an active runway is assumed to be negligible.

Runway Configuration	Mix Index, % (C + 3D)	Hourly Capacity, Operations per Hour		Annual Service Volume, Operations per Year
		VFR	IFR	
A	0–20	98	59	230,000
	21–50	74	57	195,000
	51–80	63	56	205,000
	81–120	55	53	210,000
	121–180	51	50	240,000
B 700' to 2,499'	0–20	197	59	355,000
	21–50	145	57	275,000
	51–80	121	56	260,000
	81–120	105	59	285,000
	121–180	94	60	340,000
C 4,300' or more	0–20	197	119	370,000
	21–50	149	114	320,000
	51–80	126	111	305,000
	81–120	111	105	315,000
	121–180	103	99	370,000
D 700' to 2,499' 2,500' to 3,499'	0–20	295	62	385,000
	21–50	219	63	310,000
	51–80	184	65	290,000
	81–120	161	70	315,000
	121–180	146	75	385,000
E 700' to 2,499' 3,500' or more 700' to 2,499'	0–20	394	119	715,000
	21–50	290	114	550,000
	51–80	242	111	515,000
	81–120	210	117	565,000
	121–180	189	120	675,000
F	0–20	98	59	230,000
	21–50	77	57	200,000
	51–80	77	56	215,000
	81–120	76	59	225,000
	121–180	72	60	265,000
G	0–20	150	59	270,000
	21–50	108	57	225,000
	51–80	85	56	220,000
	81–120	77	59	225,000
	121–180	73	60	265,000
H	0–20	132	59	260,000
	21–50	99	57	220,000
	51–80	82	56	215,000
	81–120	77	59	225,000
	121–180	73	60	265,000

TABLE 12-9 Preliminary Estimates of Hourly and Annual Ultimate Capacities

Runway Configuration	Mix Index, % (C + 3D)	Hourly Capacity, Operations per Hour		Annual Service Volume, Operations per Year
		VFR	IFR	
I	0–20	150	59	270,000
	21–50	108	57	225,000
	51–80	85	56	220,000
	81–120	77	59	225,000
	121–180	73	60	265,000
J	0–20	132	59	260,000
	21–50	99	57	220,000
	51–80	82	56	215,000
	81–120	77	59	225,000
	121–180	73	60	265,000
K 700' to 2,499'	0–20	197	59	355,000
	21–50	145	57	275,000
	51–80	121	56	260,000
	81–120	105	59	285,000
	121–180	94	60	340,000
L 4,300' or more	0–20	197	119	370,000
	21–50	149	114	320,000
	51–80	126	111	305,000
	81–120	111	105	315,000
	121–180	103	99	370,000
M 700' to 2,499'	0–20	295	59	385,000
	21–50	210	57	305,000
	51–80	164	56	275,000
	81–120	146	59	300,000
	121–180	129	60	355,000
N 700' to 2,499'	0–20	295	59	385,000
	21–50	210	57	305,000
	51–80	164	56	275,000
	81–120	146	59	300,000
	121–180	129	60	355,000
O Less than 2,500' / Less than 2,500'	0–20	197	59	355,000
	21–50	147	57	275,000
	51–80	145	56	270,000
	81–120	138	59	295,000
	121–180	125	60	350,000

TABLE 12-9 Preliminary Estimates of Hourly and Annual Ultimate Capacities (*Continued*)

7. There is sufficient airspace to accommodate all aircraft wishing to use the runways and aircraft operations are conducted in a radar environment with at least one runway equipped with an instrument landing system.

8. IFR conditions occur 10 percent of the time.

9. Representative hourly and daily ratios are a function of the mix index.

The order-of-magnitude relationship between average annual delay per aircraft and annual service volume depicted in Fig. 12-22 was derived from historical traffic records and a range of assumptions on likely operating conditions, as itemized above. Typically, the upper portion of the shaded band on Fig. 12-22 is representative of airports primarily serving air carrier operations. Airports serving primarily general aviation operations may typically fall anywhere within the entire shaded band. The dotted curve is the average of the upper and lower limits of the band indicated. Example Problem 12-11 shows the use of these approximate procedures.

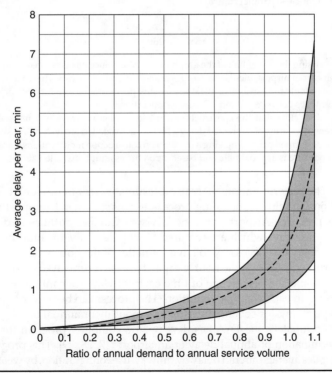

FIGURE 12-22 Relationship between average aircraft delay and ratio of annual demand to annual service volume (*Federal Aviation Administration*).

Example Problem 12-11   An airport has a single runway available to service arrivals and departures. The projected annual demand in a future year is 220,000 operations. The aircraft mix is estimated to consist of 21 percent small single engine, 20 percent small multiengine, 55 percent large commercial aircraft, and 4 percent heavy aircraft. Air carrier operations predominate at the airport and very few touch-and-go operations occur.

It is necessary to determine the annual service volume and average delay to aircraft for the single runway and for closely spaced parallels which may be constructed. The mix index at the airport is computed as

$$MI = 55 + 3(4) = 67$$

From Table 12-9, the annual service volume for each runway is found from runway configuration diagrams A and B with the mix index range of 51 to 80

$$\text{Single-runway ASV} = 205,000 \text{ operations}$$

$$\text{Close parallel-runway ASV} = 260,000 \text{ operations}$$

The ratios of annual demand to annual service volume are then computed for both situations as for a single runway

$$\frac{\text{Demand}}{\text{ASV}} = \frac{220,000}{205,000} = 1.07$$

and for close parallel runways

$$\frac{\text{Demand}}{\text{ASV}} = \frac{220,000}{260,000} = 0.84$$

Using the upper half of the graph in Fig. 12-22, since this is an predominantly air carrier airport, yields the average annual delay per aircraft. These values are for a single runway between 4.0 and 6.0 min per operation and for close parallel runways between 1.2 and 1.7 min per operation.

It is clear that the construction of close parallel runways will represent a benefit in terms of decreased delays. However, a detailed analysis of this should be performed prior to making a decision on construction modifications since this procedure is approximate and based upon the assumptions noted above.

The determination of realistic estimates of aircraft delay is a tedious and time-consuming process, as was shown in Example Problem 12-9. The Airport Capacity Advisory Circular outlines in detail the procedures which are required [4, 26]. Basically, to compute aircraft delay it is necessary to have estimates of the hourly demand on the runway system, the hourly capacity of each runway-use configuration, and the percentage of time each runway-use configuration is utilized in each weather condition. The process can be performed for a typical day, for several days, or on a monthly or annual basis.

Fortunately, the FAA has developed a computer program for the determination of annual delay at an airport [17, 28, 29]. This program compiles average aircraft delay by runway-use configuration, by weather condition, on a daily, weekly, monthly, and annual basis and it presents an annual distribution of the magnitude of delay at the airport.

The annual service volume may be related to specific criteria for the level of average aircraft delay. To do this, the annual demand on the airfield is varied over a range of values which represent upper and lower bounds on the specified average delay criteria. By plotting the average aircraft delay versus the annual demand a curve is developed showing the variation in average aircraft delay as a function of annual demand. By finding the annual demand which corresponds to the delay criteria the annual service volume is defined. It should be noted that the same process may be used on hourly basis to find the practical hourly capacity of a runway-use configuration. The computer model cited above is very useful for this purpose.

Simulation Models

While mathematical models form the fundamental basis for estimating capacity and delay of an airport, such models become extremely complex for most airports. Airports with multiple runways and taxiways, varying use configurations, fleet mixes, and weather conditions render strict analytical methods extremely difficult if not impossible to accurately estimate airfield capacity and delay. Fortunately, superior computing power has become readily available since the late 1990s to apply computer simulation models to accurately and dynamically estimate operating capacity and delays for current, as well as proposed airfield configurations.

The theory and practice of simulation modeling is a highly complex study that exceeds the scope of this text. The Federal Aviation Administration has a team dedicated to development and application of a variety of software tools designed to simulate various elements of the airport and airspace system. These tools simulate and analyze operations from capacity, delay, as well as operational feasibility perspectives. Specific FAA software models include the FAA's airport and airspace simulation model, the airport delay simulation model (ADSIM), and the runway delay simulation model (RDSIM). Each of these tools uses discrete event-based simulations to analyze operations. As with most simulations, these tools apply an infrastructure of nodes and links to define the airfield and airspace configurations, a set of rules for how aircraft are to operate within the infrastructure, and a set of discrete events, in particular the arrival and departure of aircraft within the infrastructure, to simulate the environment.

These programs generate a series of standard reports detailing the flights simulated, along with aggregated statistics, describing the capacity of the system, delays encountered by flights, and any rule violations or conflicts that may have occurred. They also prepare extended reports which compile delay statistics by hour for the various runway-use configurations for arrivals, departures and total operations. These reports can be output in both tabular and graphical format.

There are various other computer simulation models with varying degrees capabilities including ATAC's *SIMMOD*, the Preston Group's total airport and airspace modeler (TAAM) and smaller generic simulation programs such as *ARENA* that may be applied to the airport environment. The reader is encouraged to read these products' promotional materials and/or user guides to gain a greater appreciation of these models.

Gate Capacity

Gate capacity can be defined as the maximum number of aircraft that a fixed number of gates can accommodate during a specified interval of time when there is a continuous demand for service. Gate capacity can be calculated as the inverse of a weighted-average gate occupancy time of all aircraft being served. For example, if an aircraft occupies a gate for an average of 30 min, the capacity of the gate equals two aircraft per hour.

The factors that affect gate capacity are as follows:

1. The number and type of gates available to aircraft.

2. The mix of aircraft demanding apron gates and the gate occupancy time for various aircraft.

3. The percentage of time gates may be used, which reflects the fact that time is required to maneuver aircraft into and out of gate positions and the delay often restricts the amount of time actually available for aircraft gate occupancy.

4. Restrictions in the use of any or all gates.

Type of gate refers to its ability to accommodate a large, medium, or small aircraft. Normally gates at an airport are designated as wide-bodied aircraft gates, narrow-bodied aircraft gates and commuter-aircraft gates. The mix of aircraft refers primarily to size but also the required gate occupancy time. Very large aircraft require certain types of gates in order to process passengers. Time is spent maneuvering at a gate and therefore the gate utilization may not be 100 percent of the time. If the gate occupancy time includes the time to maneuver at gate positions, as well as the normal processing times to load and unload passengers, to fuel and inspect the aircraft, perform cabin service and other routine service, then the utilization will approach 100 percent. Occupancy times vary depending on the size of aircraft and whether it is an originating, turnaround, or through flight. The gate occupancy times expected by aircraft manufacturers are usually given in publications; however, this will vary with each airline and their operating procedures at different stations or airports. Often schedules or lateness of arrivals will result in much greater occupancy times than normally required.

Analytical Models for Gate Capacity

The basis of gate capacity analysis is that the gate time demanded by aircraft should be less than or equal to the gate time available for these aircraft. Two analytical models have been developed for determining the capacity of gates at an airport. One model assumes that all aircraft can use all the gates available at an airport. This is termed an *unrestricted gate-use strategy*. The other model assumes that aircraft of a certain size or airline can only use gates that were specifically designed for these aircraft or airline. This is called a restricted gate use strategy. Both of these models are described and most situations encountered in practice may be approached through one of the two models.

When there are no restrictions on the use of gates, that is, all aircraft can use all the gates, the capacity of the gates C_g can be derived as follows:

Gate time supplied ≥ gate time demanded:

$$\mu_k N_k \geq E(T_g)C_g \qquad (12\text{-}19)$$

where μ_k = gate utilization factor, or the percentage of time in an hour that the gates of type k may be used by aircraft of type i

N_k = number of k type gates available to aircraft of type i

$E(T_g)$ = expected value of the gate occupancy time demanded by aircraft which can use gate type k

C_g = capacity of the gates of type k in aircraft per hour

The expected value of the gate occupancy time $E(T_g)$ is found from the following expression:

$$E(T_g) = \sum m_i T_{gi} \qquad (12\text{-}20)$$

where m_i is the percentage of type i aircraft in the fleet mix using the gates at airport, and T_{gi} is the gate occupancy time required for aircraft of type i at the airport.

The use of these equations is illustrated for unrestricted gate use in Example Problem 12-12.

Example Problem 12-12   An airport has four gates available to all aircraft. The aircraft mix at the airport in the peak hour consists of 30 percent type A, 50 percent type B, and 20 percent type C aircraft. Type A aircraft require a gate occupancy time of 60 min, type B require 45 min, and type C require 30 min. Normally, due to the distribution of demand, the maximum gate utilization which can be expected is 70 percent. It is required to find the capacity of the gates at this airport to process aircraft.

From Eq. (12-19) we have

$$0.70(4)(60) \geq [0.3(60) + 0.50(45) + 0.20(30)]C_g$$

which reduces to

$$C_g = 3.6 \text{ aircraft per hour}$$

It should be observed that, since every aircraft at a gate entails two operations, an arrival and a departure, the hourly capacity of the gates could also be expressed as $2(3.6) = 7.2$ operations per hour. Also if the gate utilization factor is equal to 1, then the ultimate capacity of the gates becomes equal to 5.2 aircraft per hour.

For restricted gate use, the mix of gates and the mix of aircraft using the airport may not be the same. Therefore, it is necessary to find the gate capacity of each type of gate and then determine the overall capacity of the airport based upon gate capabilities as the minimum capacity gate, capacity of any type gate. Mathematically, this becomes

$$C_g = \min(C_{gk}) \tag{12-21}$$

The use of the above analysis for restricted gate use is shown in Example Problem 12-13.

Example Problem 12-13 An airport has 10 gates available for aircraft. These gates are restricted in the types of aircraft which can be accommodated. The five type I gates can accommodate any type of aircraft, the three type II gates cannot accommodate a type A aircraft, and a type III gate can only accommodate a type C aircraft.

The mix and the gate occupancy times of the aircraft using the airport in the peak hour is the same as in Example Problem 12-13. The gate utilization factor is 1.0.

Determine the capacity of the gates to process aircraft at this airport.

The relationship shown in Eq. (12-19) must be solved for each type of gate. There are 5 gates available to type A, 8 for type B, and 10 for type C aircraft. Solving Eq. (12-19) for each gate type yields

$$1.0(5)(60) \geq 0.3(60)C_{gI}$$

$$C_{gI} = 16.67 \text{ aircraft per hour}$$

$$1.0(8)(60) \geq [0.3(60) + 0.5(45)]C_{gII}$$

$$C_{gII} = 11.85 \text{ aircraft per hour}$$

$$1.0(10)(60) \geq [0.3(60) + 0.5(45) + 0.2(30)]C_{gIII}$$

$$C_{gIII} = 12.90 \text{ aircraft per hour}$$

Therefore, the type II gates restrict the aircraft capacity of the airport from Eq. (12-21)

$$C_g = \min(16.67, 11.85, 12.90) = 11.9$$

Therefore, with this mix of aircraft demanding gates at the airport, the gate mix restricts the airport capacity to 11.9 aircraft per hour or 23.8 operations per hour.

It should be noted that only with this capacity is the gate time supplied greater than or equal to the gate time demanded. This is shown as

Gate time supplied ≥ gate time demanded

$$1.0(10)(60) \geq [0.3(60) + 0.5(45) + 0.2(30)](11.9)$$

$$600 \geq 554 \text{ as required}$$

As described in this chapter, airport planning and design with respect to capacity is a critical and complex process. While this chapter focused on the analytical fundamentals of estimating capacity, planners should take careful consideration into estimating capacity from both an analytical and empirical perspective, ideally applying some form of simulation model. It is understood that there is also a balance that must be maintained by the airport planner between the amount of time and budget available to do capacity studies. Some airports with relatively simple airfield configurations and relatively small planning budgets may be sufficiently served with a basic analytical model, or more simply referencing FAA approximation charts. On the other hand, large-scale complex projects should dedicate sufficient resources to properly and comprehensively analyzing capacity and projected delays as part of the planning process. While such involvement may be expensive, the investment made in properly designing for appropriate capacity may very well avoid much higher costs associated with developing infrastructure that cannot accommodate those who use it.

References

1. "A Flexible Model for Runway Capacity Analysis," W. E. Weiss, Master's Thesis, Massachusetts Institute of Technology, Cambridge, Mass., June 1980.
2. *Air and Ground Delay Study: Impact of Increased Operations at Lindbergh Field*, Report No. AOR-89-01, Operations Research Office, Federal Aviation Administration, Washington, D.C., September 1989.
3. *Airport Capacity*, Airborne Instruments Laboratory, Report No. BRD-136, Federal Aviation Agency, Office of Technical Services, Department of Commerce, Washington, D.C., June 1963.
4. *Airport Capacity and Delay*, Advisory Circular AC 150/5060-5, Federal Aviation Administration, Washington, D.C., 1983.
5. *Airport Ground Access*, Report No. FAA-EM-79-4, Federal Aviation Administration, Washington, D.C., October 1978.
6. *Airport System Capacity, Strategic Choices*, Special Report, No. 226, Transportation Research Board, Washington, D.C., 1990.
7. *Air Traffic Control Handbook*, Order No. 7110.65G, Federal Aviation Administration, Washington, D.C., March 1992.
8. "A Methodology for Airport Capacity Analysis," S. L. M. Hockaday and A. Kanafani, *Transportation Research*, Vol. 8, 1974.
9. *Aviation System Capacity Plan 1991–92*, Report No. DOT/FAA/ASC-91-1, U.S. Department of Transportation, Federal Aviation Administration, Washington, D.C., 1991.
10. *Chicago Delay Task Force Technical Report*, Vol. II Model Validation and Existing Airspace/Airfield Performance, Landrum and Brown Aviation Consultants, Chicago, Ill., April 1991.

11. *Critique of the Aircraft Delay Curves in Techniques for Determining Airport Airside Capacity and Delay*, C. T. Ball, Graduate Report No. UCB-ITS-GR-78-1, Institute of Transportation Studies, University of California, Berkeley, Calif., August 1978.
12. *FAA Report on Airport Capacity*, Report No. FAA-EM-74-5, Federal Aviation Administration, Washington, D.C., 1974.
13. *Future Development of the U.S. Airport Network*, Preliminary Report and Recommended Study Plan, Transportation Research Board, Washington, D.C., 1988.
14. *Measuring Airport Landside Capacity*, Special Report, No. 215, Transportation Research Board, Washington, D.C., 1987.
15. *Models for Estimating Runway Landing Capacity with Microwave Landing Systems (MLS)*, V. Tosic and R. Horonjeff, Special Report, National Aeronautics and Space Administration, Ames Research Center, Moffett Field, Calif., September 1975.
16. *Models for Runway Capacity Analysis*, R. M. Harris, MITRE Corporation, Report No. FAA-EM-73-5, Federal Aviation Administration, Washington, D.C., May 1974.
17. *Model Users' Manual for Airfield Capacity and Delay Models*, C. T. Ball, Report No. FAA-RD-76-128, Federal Aviation Administration, Washington, D.C., November 1976.
18. *National Plan of Integrated Airport Systems (NPIAS) 1990–1999*, Federal Aviation Administration, Washington, D.C., 1991.
19. *O'Hare Delay Task Force Study*, Executive Summary and Technical Report, Chicago O'Hare International Airport, Report No. FAA-AGL-76-1, Federal Aviation Administration, Washington, D.C., July 1976.
20. *Procedures for Determination of Airport Capacity*, Report No. FAA-RD-73-111, Vols. 1 and 2, Interim Report, Federal Aviation Administration, Washington, D.C., April 1973.
21. *Reauthorizing Programs of the Federal Aviation Administration, Future Capacity Needs and Proposals to Meet Those Needs*, Subcommittee on Aviation, Committee on Public Works and Transportation, House of Representative Report No. 101-37, US House of Representatives, Washington, D.C., 1990.
22. *SIMMOD: The Airport and Airspace Simulation Model*, Reference Manual, Release 1.1, Federal Aviation Administration, Washington, D.C., October 1990.
23. *South Florida Supplemental Airport Study*, H. True, S. Wolf, and D. Winer, Operations Research Service, Federal Aviation Administration, Washington, D.C., February 1991.
24. *Supporting Documentation for Technical Report on Airport Capacity and Delay Studies*, Report No. FAA-RD-76-153, Federal Aviation Administration, Washington, D.C., June 1976.
25. *Technical Report on Airport Capacity and Delay Studies*, Final Report, Report No. FAA-RD-76-153, Federal Aviation Administration, Washington, D.C., June 1976.
26. *Techniques for Determining Airport Airside Capacity and Delay*, Report No. FAA-RD-74-124, Federal Aviation Administration, Washington, D.C., June 1976.
27. *Terminal Area Forecasts*, Fiscal Years 1991–2005, Report No. FAA-APO-91-5, Federal Aviation Administration, Washington, D.C., July 1991.
28. *Upgraded FAA Airfield Capacity Model*, Vol. I: *Supplemental User's Guide*, W. J. Swedish, The MITRE Corporation, Report No. FAA-EM-81-1, Vol. I, Federal Aviation Administration, Washington, D.C., February 1981.
29. *Upgraded FAA Airfield Capacity Model*, Vol. II: *Technical Descriptions of Revisions*, W. J. Swedish, The MITRE Corporation, Report No. FAA-EM-81-1, Vol. II, Federal Aviation Administration, Washington, D.C., February 1981.
30. *Validation of the SIMMOD Model*, Final Report, J. C. Bobick, ATAC Corporation, Mountain View, Calif., December 1988.
31. National Airspace System: DOT and FAA Actions Will Likes Have a Limited Effect on Reducing Delays during Summer 2008 Travel Season, US GAO Publication GAO-012- 934T, Washington, D.C. 2008.

CHAPTER 13

Finance Strategies for Airport Planning

Introduction

This chapter is designed to provide the airport planner and engineer with some of the fundamental strategies available to finance large scale planning and design projects. Due to the rules associated with many Federal, state, and local programs in the United States, strategies for funding large capital programs are both different and exclusive from funding the day-to-day operations of an airport. Thus, the focus of this chapter is on capital programs, including grants, bond strategies, and private investment, and not on operational revenue strategies, more germane to airport management.

Background

In the very early years of aviation, airport ownership was vested almost entirely in private hands. State and federal financial participation in airport development was virtually nonexistent. The Depression of the 1920s witnessed a collapse in private investments in airports and gave rise to public ownership. As of 2008, the vast majority of airports within the United States are still privately owned and operated, although most of the overall aviation activity, and virtually all commercial airline activity, operates at publicly owned airports. Internationally, many airports are still operated by their respective federal governments, although, many international airports have become owned and operated by for-profit private entities.

Public ownership in airports is vested in a number of different types of government levels, including municipalities, counties, and state ownership. The largest percentage of the 100 busiest airports in the United States is operated by an "authority." An authority is an independent, politically appointed entity, typically comprised of representatives from the municipalities, counties, and/or states in which the airport is located.

Although both the federal and state governments may have provisions which affect the airport, it is the decision of the airport owner, called the airport sponsor, which ultimately determines the development of the airport.

Airport improvements are financed in a variety of ways including federal grants, state grants, airport bonds, and private investment. In addition, capital improvements of a minor nature have been financed from accumulated surpluses from airport revenues.

Federal Funding Programs in the United States

Until 1933, airports for civil use were developed mainly through investments by municipalities and private sources. The Depression was largely responsible for the first substantial federal participation in the development of civil airports. The bulk of the funds provided from 1933 until the beginning of World War II were through work-relief programs. The first program was under the Civil Works Administration (CWA). In the fall of 1933, the CWA provided more than $15 million for airport construction, with most of the money going to smaller communities.

In 1934, the CWA was succeeded by the Federal Emergency Relief Administration (FERA). This agency provided over $17 million for the development of 943 airport projects.

The administration of federal aid for airports was taken over in 1935 by the Works Progress Administration (WPA). The WPA spent $323 million for airport construction in the United States and it was under this program that contributions by municipalities were encouraged and a pattern of cost sharing emerged. The local contributions amounted to about $110 million.

Another federal program contributing to airport development in the 1930s was the Public Works Administration (PWA) which made loans or grants amounting to almost $29 million, primarily to municipalities.

Prior to the start of World War II the federal government spent a total of about $384 million for airport development under the four programs of CWA, FERA, WPA, and PWA, however, it must be recognized that these were primarily work-relief programs and provided no basis for federal support in times of a normal economy.

During World War II the federal government, through the Civil Aeronautics Administration, spent $353 million for the development of landing areas for military use. While the priority in this program was attached to military requirements, the needs of postwar civil aviation were considered in the location and construction of these facilities. During the same period the federal government, through the CAA, spent over $9 million for the development of airports solely for civil use. These two programs are referred to as defense landing area (DLA) and development of civil landing areas (DCLA). The DLA and

DCLA programs were independent of the airports constructed by the war and navy departments. After the war some 500 military airports were declared surplus and turned over to cities, counties, and states. This is the principal reason that today's public ownership is vested in local authorities.

The Federal Aid to Airports Program

At the end of World War II interest was renewed in establishing a federal program for monetary aid for airport development. A resolution was introduced in Congress (H.R. 598, 78th Congress) requiring the Civil Aeronautics Administration to make a survey of airport needs and prepare a report on the subject. These recommendations formed the basis of the Federal Aid to Airports programs, as written in the Federal Airport Act of 1946 (Public Law 79-377). Appropriations of $500 million over a 7-year period were authorized for projects within the United States plus $20 million for projects in Alaska, Hawaii, Puerto Rico, and the Virgin Islands. In 1950 the 7-year period was extended for an additional 5 years (Public Law 81-846). However, annual appropriations approved by Congress were much less than the amounts authorized by the act.

The original act provided that a project shall not be approved for federal aid unless "sufficient funds are available for that portion of the project which is not to be paid by the United States."

Local governments often required 2 to 3 years to make arrangements for raising funds and most of the larger projects were financed locally through the sale of bonds. This method of financing required legislation at the local level and, in some cases, also at the state level. General obligation bonds normally required approval by the electorate. Programs to inform the public on the need for airport improvement must be carefully planned and executed. Thus, after the completion of these events, local governments frequently found that sufficient federal funds were not appropriated to match local funds, and the projects were delayed. Another complaint of local governments had been that Congress failed to fulfill its obligation, since the amount appropriated by Congress fell far short of the amount authorized by the Federal Airport Act. These deficiencies as well as other matters were incorporated in a new bill (S. 1855) and hearings were held before the subcommittee of the Committee on Interstate and Foreign Commerce of the U.S. Senate in 1955. Representatives of the Council of State Governments, the American Municipal Association, the National Association of State Aviation Officials, airport and industry trade associations, and individuals were unanimous in the feeling that air transportation had reached a stage of maturity where many airports were woefully inadequate and greater financial assistance from the federal government would be required to meet the current needs of aviation. After much debate, the bill was approved by the President (Public Law 84-211).

This amending act made no change in the basic policies and purposes expressed in the original act. There were no changes in the requirements with respect to the administration of the grants authorized, such as the distribution and apportionment of funds, eligibility of the various types of airport construction, sponsorship requirements, etc. The primary purpose of the act was to provide provisions granting substantial annual contract authorization in specific amounts over a period of four fiscal years. Airport sponsors were thus furnished assurance that federal funds would be available at the time projects were to be undertaken.

This law provided $40 million for fiscal year 1956 and $60 million for each of fiscal years 1957, 1958, and 1959 for airport construction in the continental United States. It also provided $2.5 million in fiscal year 1956 and $3 million for the three succeeding fiscal years for airport construction in Alaska, Hawaii, Puerto Rico, and the Virgin Islands. Besides the $42.5 million made available in fiscal year 1956 by Public Law 84-211, Congress approved an additional appropriation of $20 million for airport projects.

In 1958, the 85th Congress passed a bill (S. 3502) proposing to extend the Federal Airport Act for 4 years at an annual funding rate of $100 million. This bill was vetoed by the President in September 1958 with a veto statement (S. 3502 Veto Statement), which stated in part:

> I am convinced that the time has come for the federal government to begin an orderly withdrawal from the airport grant program. This conclusion is based, first, on the hard fact that the government must now devote the resources it can make available for the promotion of civil aviation programs which cannot be assumed by others, and second, on the conviction that others should begin to assume the full responsibility for the cost of construction and improvement of civil airports.

In the 86th Congress, much debate involved a significant increase in the federal airport program. Two bills, one introduced in the House ($297 million over a 4-year period) and the other in the Senate ($465 million for a 4-year period), together with the President's recommended bill for a 4-year program of $200 million, were finally merged into a 2-year continuation of the existing aid program at $63 million per year (Public Law 86-72).

Significant changes were the removal of the territorial status from Hawaii and Alaska which had been admitted as states and the exclusion of automobile parking and certain portions of airport building improvement costs as allowable costs.

A 3-year continuation of the federal aid program providing $75 million annually was enacted by the 87th Congress (Public Law 87-255). The provisions of the bill were very similar to those of Public Law 86-72. One new feature of the legislation was that it provided the administrator of the FAA with a discretionary fund of $7 million from the $75 million for developing general aviation airports to relieve congestion at high-density commercial airports. Thus this legislation started the reliever airport program. The Federal Airport Act was again amended in 1964, authorizing the expenditure of $75 million for fiscal years 1965 to 1967 (Public Law 88-280). The final amendment was accepted in 1966, authorizing the continuation of the expenditure of $75 million for fiscal years 1968 to 1970 (Public Law 89-647). This marked the end of the Federal Airport Act, as the Airport and Airway Development Act of 1970 became law in 1970.

During the 24 years of airport funding under the Federal Airport Act, a total of $1.2 billion has been appropriated by the federal government for improvements at 2316 airports involving almost 8000 projects. Much of the capital infrastructure of airports still in existence were funded by, and adhere to the terms of the Federal Aid to Airports program.

The Airport Development Aid Program

Because of the rapidly growing requirements for modernizing the air traffic control system and airport expansion, neither the federal government nor the local authorities were able to fund capital improvements badly needed for the growth of aviation. The FAA needed more money in its budget to accelerate the implementation of a program to modernize the air traffic control system. The $75 million authorized annually by the Federal Aviation Act, together with local matching funds, fell far short of the needs of the local airport authorities to meet the current and projected growth of airport traffic. Cities were unable to raise sufficient funds at the local level to meet the rising costs of airport construction. The Federal Aviation Act of 1946 was supported by general rather than user tax revenues and therefore it had to compete annually with other government programs for scarce federal dollars. The increased competition for fewer dollars resulted in delay and postponement of airport construction throughout the nation. As a result of this situation, aviation organizations representing airport owners, airlines, pilots, and general aviation aircraft owners joined in pressing for more funds for airports and airways and the establishment of a trust fund similar to that for the national highway program. Several bills were introduced in the House and Senate to enact legislation which would remedy the deficiencies in the Federal Airport Act of 1946 (S. 1637, S. 2437, S. 2651). After much debate including the question of whether airport terminal buildings should be included in the legislation (initially they were not), Public Law 91-258 was signed into law in May 1970. As stated in Chap. 1, the law consisted of two parts, referred to as Titles I and II. Title I was known as the Airport and Airway Development Act of 1970, which replaced the

Federal Airport Act of 1946. Title II was known as the Airport and Airway Revenue Act of 1970, and it provided the excise taxes required to furnish the resources necessary to carry out the Title I programs through 1980. These excise taxes provided the revenues to fund the programs under the act and were deposited in the Airport and Airway Trust Fund established by the act.

Title I of the original act provided for $250 million annually for the "acquisition, establishment, and improvement of air navigational facilities" and security equipment required by the sponsor for fiscal years 1971 through 1980. For airport assistance, the Airport Development Aid Program (ADAP) initially authorized a total of $2.5 billion for the 10-year period. The act further specifically authorized $250 million annually through fiscal year 1973 and $275 million each for fiscal years 1974 and 1975 for airports served by air carriers and general aviation airports which relieve high-density air carrier airports. Also authorized were $30 million annually through fiscal year 1973 and $35 million each for fiscal years 1974 and 1975 for all other general aviation airports (Public Law 93-44). Later amendments (Public Law 94-353) raised the program level to range from $500 to $610 million annually through 1980. The Aviation Safety and Noise Abatement Act (Public Law 96-193) further raised the final-year program to $667 million. These amendments provided $435 to $539 million annually for airports serving all segments of aviation and $65 to $95 million annually for general aviation airports. The act also authorized the issuance of planning grants for the preparation of airport system plans and airport master plans. The Planning Grant Program (PGP) was designed to promote the effective location and development of publicly owned airports and to develop a national airport system plan. System plans were prepared by state and regional agencies to formulate air transportation policy, determine facility requirements needed to meet forecast aviation demand, and establish a framework for detailed airport master planning. Airport master plans, which were developed by the airport owner, focused on the nature and extent of the development required to meet the future aviation demand at specific facilities.

The funds that were authorized for airport development for the several classes of airports were apportioned to air carrier and general aviation airports. In the final amended form of the law, two-thirds of the air carrier and commuter service funds were made available to air carrier airports based upon the number of annual enplaned passengers. These were termed *entitlement funds*. The remaining monies were placed in a discretionary fund, of which $15 million annually was apportioned to commuter service airports. Air carrier airports serving aircraft heavier than 12,500 lb were authorized to receive between $150,000 and $10 million annually. Airports serving aircraft weighing less than 12,500 lb were to receive not less than $50,000 annually. Of the funds appropriated for general aviation and general aviation reliever airports, $15 million annually was apportioned to reliever airports. Seventy-five percent of the remaining funds was allocated to the states on the basis of population and area; 1 percent was allocated to Puerto Rico, Guam, the Virgin Islands, American Samoa, and the Trust Territories of the Pacific Islands and was distributed by the Secretary of Transportation.

The maximum federal grant for any specified project varied from 50 to 90 percent of the total eligible project costs over the life of the act, depending upon the type of project being considered for funding. A maximum of 75 percent of the allowable project cost was allowed for airports in areas that enplaned 0.25 percent or more of the total annual passengers enplaned by air carriers certified by the Civil Aeronautics Board. These airports were called the large and medium air traffic hub airports. A maximum of 90 percent of the allowable project cost was allowed for airports in areas that enplaned less than 0.25 percent of the total annual passengers, for small air traffic hub and nonhub airports, and for general aviation airports.

The original act specifically prohibited the use of federal funds for automobile parking facilities or airport buildings except those parts "intended to house facilities or activities directly related to the safety of persons at the airport." However, amendments to the act in 1976, (Public Law 94-353) provided federal funding for the non-revenue-producing public areas of terminal facilities required for the processing of passengers and baggage. In this case the federal share was limited to 50 percent of the project costs, and the airport could not spend more than 60 percent of its enplanement funds on such development.

There was no state apportionment for planning grant funds. The Secretary of Transportation prescribed the regulations governing the award and administration of these grants. When the program first began, the federal government provided up to two-thirds of the cost of planning grant projects. However, amendments to the act in 1976 increased this share to 75 percent of the cost of airport system plans, 90 percent of the cost for master plans at general aviation airports, and a range from 75 to 90 percent of the cost for master plans at air carrier airports, depending upon the number of enplaned passengers.

In administering the Airport and Airways Development Act, the FAA established, in detail, the types of improvements which were eligible for federal aid under the act. In general, items that were eligible included land acquisition, paving and grading, lighting and electrical work, utilities, roads, removal of obstructions to air navigation, fencing, fire and rescue equipment, snow removal equipment, terminal-area development, and physical barriers and landscaping for noise attenuation.

Separate buildings for airport emergency, snow removal, and firefighting equipment were eligible but administration buildings serving air commerce or general aviation were not eligible. Buildings used exclusively for the handling of cargo were also ineligible.

The non-revenue-producing public-use areas of terminal facilities became eligible for funding in 1976 (Public Law 94-353) if these areas were "directly related to the movement of passengers and baggage in air commerce within the boundaries of the airport."

Roads and streets were eligible if they were within the boundary of the airport and were needed for the operation and maintenance of the airport, or were directly related to the movement of passengers and baggage.

Each eligible project was evaluated separately. It was rated on the basis of the aeronautical necessity of the airport, volume and character of traffic, and type of work included in the project. These ratings established a priority score for each increment of work in any one state and were used to program the funds allocated to that state. The ratings were based on such factors as safety, efficiency, and convenience.

A community interested in obtaining federal aid contacted the Airports District Office (ADO) of the FAA in the geographic area in which the airport was located. In states which required that all federal aid be channeled through the state, submission of a request for aid was made through the state aeronautical agency.

If the project qualified for aid, if there were sufficient funds, and if the project was found by the Secretary of Transportation to be acceptable from the standpoint of its economic, social, and environmental effects on the community, then the sponsor was notified that a tentative allocation of funds had been made for part of or all of the items listed in the request. The tentative allocation was an indication that funds had been placed in reserve pending the completion of arrangements for necessary financing, land acquisition, and preparation of plans, specifications, and contract documents.

Upon submitting detailed plans and specifications with a project application and upon approval of the FAA, the sponsor secured bids from contractors and made recommendations for the award of the contracts. At this time the sponsor also formally accepted federal aid and the obligations connected therewith by executing what was known as a "grant agreement." The execution of the grant agreement legally bound the community to fulfill the obligations (sponsors' assurances) set forth in the project application. Some of the sponsor's important obligations included:

1. The sponsor would operate the airport for the use and benefit of the public, on fair and reasonable terms without unjust discrimination.

2. It would keep the airport open to all types, kinds, and classes of aeronautical use without discrimination between such types, kinds, and classes.

3. It would operate and maintain in a safe and serviceable condition the airport and all facilities thereon which are

necessary to serve aeronautical uses other than facilities owned or controlled by the United States.

4. It would make every effort to maintain clear approaches to the runways.

5. It would not charge government owned or military aircraft for the use of runways and taxiways unless the use was substantial.

6. These obligations would remain in effect for not more than 20 years.

The authority to issue grants under this act expired in 1981. During the 11-year period under this legislation, 8809 grants totaling $4.5 billion were approved for airport planning and development at over 1800 airports. This was about four times greater than the total amount provided by the Federal Airport Act of 1946. Over 6700 of these grants were made under the Airport Development Aid Program, and almost 2000 of these grants were made under the Planning Grant Program.

In 1982, Congress enacted the Airport and Airway Improvement Act (Title V of the Tax Equity and Fiscal Responsibility Act of 1982, Public Law 97-248). This act continued to provide funding for airport planning and development under a single program called the Airport Improvement Program (AIP). The act also authorized funding for noise compatibility planning and implementation of noise compatibility programs contained in the Noise Abatement Act of 1979 (Public Law 96-193). It required that to be eligible for a grant, the airport must be included in the National Plan of Integrated Airport Systems (NPIAS). The NPIAS, the successor to the National Airport System Plan (NASP), is prepared by the FAA and published every 2 years and it identifies public-use airports considered necessary to provide a safe, efficient, and integrated system of airports to meet the needs of civil aviation, national defense, and the U.S. Postal Service.

Projects eligible for funding under this legislation were restricted to planning, development, and noise compatibility projects at or associated with public-use airports, including heliports and seaplane bases, which were defined as airports open to the public and publicly owned, or privately owned but designated by the FAA as a reliever airport, or privately owned and having scheduled service and at least 2500 annual enplanements.

Airports were defined in five categories: commercial service airports, primary airports, cargo service airports, reliever airports, and other airports. Commercial service airports are publicly owned airports which enplaned at least 2500 passengers annually and received scheduled service. Primary service airports are commercial service airports which enplaned at least 10,000 passengers annually. Cargo service airports are airports served by aircraft providing air transportation of property only, including mail, with an aggregate annual aircraft landed weight in excess of 100,000,000 lb. Reliever airports are airports in metropolitan areas designated by the FAA as having the function of relieving congestion at large commercial service airports by providing alternative landing areas for general aviation aircraft and which provided more general aviation access to the community. Other airports are the remaining airports, commonly referred to as general aviation airports.

The allocation of funds under the AIP was also defined in the legislation such that these funds are distributed between apportioned and discretionary funds. As amended by the Airport and Airway Safety, Capacity, Noise Improvement and Intermodal Transportation Act of 1992, of the total funds available not more than 44 percent is apportioned as entitlements to primary airports and 3.5 percent is apportioned as entitlements to cargo service airports. Additionally, 12 percent of the total funds is apportioned for states and insular areas. There is a separate apportionment for airports in Alaska, and 2.5 percent is apportioned to the Military Airport Program for current and former military airfields, to enhance the capacity of the national transportation system by enhancement of airport and air traffic control systems in major metropolitan areas. The remaining funds are designated as discretionary funds, which are required to be used so that of the total funds available a minimum of 10 percent is to be used for reliever airports, 12.5 percent for noise compatibility projects, 2.5 percent for nonprimary commercial service airports, and 0.5 percent for integrated system plans for states, regions, or metropolitan areas. Of the remaining discretionary funds 75 percent are to be used for projects to preserve and enhance capacity, safety, and security and projects carrying out noise compatibility planning programs at primary and reliever airports.

The federal share of the costs associated with integrated airport system planning was limited to 90 percent. For individual airports, the federal share of planning and airport development project costs was limited to 75 percent at primary airports and 90 percent at all other airports. The federal share of the costs of noise compatibility projects was limited to 80 percent. The federal share of non-revenue-producing public-area terminal development costs at large, medium, and small hub commercial service airports was limited to 75 percent. The federal share of both revenue-producing and non-revenue-producing public areas in terminal buildings and non-revenue-producing parking lots at nonhub commercial service airports was limited to 85 percent.

The Airport and Airway Improvement Act has been amended several times resulting in significant changes in the provisions of the act and in authorized appropriations. These amendments are included in the Continuing Appropriations Act of 1982, the Surface Transportation Assistance Act, the Airport and Airway Safety and Capacity Expansion Act of 1987, the Airway Safety and Capacity Expansion Act of 1990 the Airport and Airway Safety, Capacity, Noise Improvement and Intermodal Transportation Act of 1992, the Wendell H. Ford Aviation Investment and Reform Act for the Twenty-First Century (AIR-21), and Century of Aviation Reauthorization Act (Vision 100) of 2003. Each of these amendments altered, and often increased, the annual congressional authorized of AIP funding levels and the terms to which they are appropriated. In addition, the terms to which the Airport and Airway Trust Fund is contributed has been modified with the above amendments. Figure 13-1 illustrates the annual authorizations and appropriations of AIP funding since its inception in 1982.

In 2007, the final year of the amendments associated with Vision 100, approximately $3.4 billion in AIP funding was authorized. Appropriated funds from these authorizations were allocated through two primary funding categories: entitlements and discretionary funding (Table 13-1).

AIP entitlements to a primary airport are based on the number of an airport's categorization within the NPIAS and the airport's annual enplanement levels. In 2006, primary airports received annual AIP entitlements ranging from $750,000 to more than $6 million, whereas nonprimary entitlements, offered to nonprimary commercial service, general aviation, and reliever airports, were typically $150,000 per airport.

AIP discretionary funds are grants that may be applied for by airports to fund capital improvement projects, including infrastructure

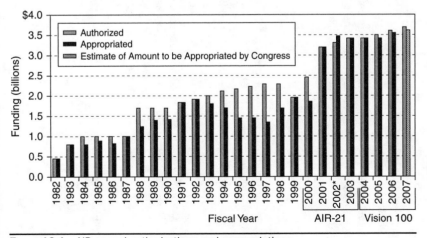

Figure 13-1   AIP annual authorizations and appropriations.

Aviation Taxes	Comment	Tax Rate
		Passengers
Domestic passenger ticket tax	Ad valorem tax	7.5% of ticket price (10/1/99 through 9/30/2007)
Domestic flight segment tax	"Domestic Segment" = a flight leg consisting of one takeoff and one landing by a flight	Rate is indexed by the Consumer Price Index starting 1/1/02 $3.00 per passenger per segment during calendar year (CY) 2003 $3.10 per passenger per segment during CY2004 $3.20 per passenger per segment during CY2005 $3.30 per passenger per segment during CY2006 $3.40 per passenger per segment during CY2007
Passenger ticket tax for rural airports	Assessed on tickets on flights that begin/end at a rural airport. Rural airport: <100K enplanements during 2nd preceding CY, and either 1) not located within 75 miles of another airport with 100K+ enplanements, 2) is receiving essential air service subsides, or 3) is not connected by paved roads to another airport	7.5% of ticket price (same as passenger ticket tax) Flight segment fee does not apply.
International arrival and departure tax	Head tax assessed on pax arriving or departing for foreign destinations (& U.S. territories) that are not subject to pax ticket tax.	Rate is indexed by the Consumer Price Index starting 1/1/99 Rate during CY2003 = $13.40 Rate during CY2004 = $13.70 Rate during CY2005 = $14.10 Rate during CY2006 = $14.50 Rate during CY2007 = $15.10

Flights between continental U.S. and alaska or hawaii	Rate is indexed by the Consumer Price Index starting 1/1/99
	$6.70 international facilities tax + applicable domestic tax rate (during CY03)
	$6.90 international facilities tax + applicable domestic tax rate (during CY04)
	$7.00 international facilities tax + applicable domestic tax rate (during CY05)
	$7.30 international facilities tax + applicable domestic tax rate (during CY06)
	$7.50 international facilities tax + applicable domestic tax rate (during CY07)
Frequent flyer tax	Ad valorem tax assessed on mileage awards (e.g., credit cards)	7.5% of value of miles
Freight/Mail	
Domestic cargo/mail	6.25% of amount paid for the transportation of property by air
Aviation Fuel	
General aviation fuel tax	Aviation gasoline: $0.193/gallon
	Jet fuel: $0.218/gallon
Commercial fuel tax	$0.043/gallon

Source: Federal Aviation Administration.

TABLE 13-1 Airport and Airway Trust Fund Tax Schedule as of 2007

enhancements directly benefiting aviation activity, land acquisition, noise mitigation, and planning studies. These funds are awarded to those projects deemed most important for improving the national airspace system. AIP discretionary funds typically may assume up to 80 percent of a project's capital costs and 100 percent for certain noise mitigation programs.

The Passenger Facility Charge Program

The years following the 1982 Airport and Airway Transportation Act saw significant growth at a few of the nation's airports, due to the hub-and-spoke network strategies adopted by the nation's commercial air carriers. As a result, the vast majority of AIP funding was directed toward investment in these largest hub airports. This allocation of funding left relatively little assistance available to the smaller airports who were in need of capital improvements and other planning activities.

With the goal of providing more funding available to the smaller airports by allowing the larger airports to raise capital funding on an individual basis, the federal government authorized a policy that would allow airports to charge passenger facility charges (PFCs). Through the Aviation Safety and Capacity Expansion Act of 1990, airport operators were permitted to propose collecting a $1, $2, or $3 fee per enplaned passenger. Revenues from PFCs could be used for airport planning and development projects eligible for AIP funding, as well as for the preparation of noise compatibility plans and measures. The provision of PFC has allowed more AIP funds to be allocated to smaller airports with fewer annual enplanements.

The legislation also provided that at those commercial service airports in areas which enplaned at least 0.25 percent of the national annual enplanements in any year, i.e., the large and medium hub airports. The entitlement funds apportioned to the airport based upon passenger enplanements would be reduced by 50 percent of the revenue obtained through the PFCs but these reductions of entitlement funds would not exceed 50 percent of the apportioned funds. The legislation directed that 25 percent of the revenues obtained through a reduction in these apportioned funds are to be placed in the AIP discretionary fund, of which one-half is to be used for small hub airports, and 75 percent is to be used to establish a Small Airports Fund. One-third of the revenues in the Small Airport Fund is to be distributed to general aviation airports and two-thirds to nonhub commercial service airports.

The specific requirements imposed upon airports requesting authority to impose passenger facility charges are contained in FAR Part 158 [15].

In 2000, the passage of the Wendell H. Ford Aviation Investment and Reform Act for the Twenty-First Century (AIR-21) increased the allocation of federal grant programs by increasing the amount of AIP funding that may be released from the Airport and Airway Trust Fund on an annual basis, and allowing PFCs to be set at $4.00 or $4.50. As of 2006, nearly 525 commercial service airports in the United States have participated in the PFC program, generating nearly $2.2 billion in funding for approved capital projects.

The Vision 100 Act of 2003 extended the life of these programs through 2007. As of 2008, a reauthorization of these programs has yet to be passed. Up for debate are a number of issues including how to best fund the Trust Fund and how to best disseminate the funds among the nation's airports, while also making investments in a major modernization of the nation's airspace system.

As airfares have declined, aviation activity outside of airlines has increased, and aviation infrastructure has aged to the point of needing major reinvestment, there has been much debate that the current model of funding the system strictly through airline passenger and landing fees will be insufficient. Much of the debate is focused on the seeking user-based fees from both commercial and general aviation activity to fund the system, much to the consternation of the general aviation community. Also, there has been some debate associated with raising the allowable PFC from $4.50 to $7.00. This has brought concern to commercial air carriers who feel that any additional fees could hurt demand for service.

In addition, debate has surrounded the redistribution of funds from the trust fund. While AIP and PFC funding have traditionally favored the largest of airports, smaller airports that struggle to retain air service, and general aviation airports that have no commercial air service, have received less funding. Potential policies being debated have considered significantly reducing the enplanement-based AIP entitlement funding program, in favor of more discretionary grants. As of publication of this text, these debates continue, leaving current funding programs under continuing resolutions.

State and Local Participation in Financing Airport Improvements

Airport development and aviation planning are a major concern in most states. As of 2003, 30 U.S. states provided some formal financial assistance for airport improvements. Increases in state financial support have been dramatic during the past several years. Data show that expenditures by states for airports have risen from $59 million in 1966 [11] to over $450 billion in 2008 [22].

State sharing in airport development varies depending on whether federal aid is involved and whether this aid is channeled through a state aeronautical agency. Most states require that federal funds be channeled through the state to local sponsors. In those states where this is a requirement, the state normally contributes one-half of the sponsor's share of the project costs, which amounts to approximately one-quarter of the total project cost. If no federal aid is involved, the state often contributes between 50 and 90 percent of the project cost. In other states where revenues are obtained from user taxes, formulas for apportioning the revenues are established.

States obtain the revenues to finance aviation and airport improvement projects from a variety of sources, including the general fund, aviation fuel taxes, aircraft sales and use taxes, and other sources including hangar rents and other property leases, and tax revenue.

Nine U.S. states participate in a federal "State Block Grant" program. This program allows participating states to receive large amounts of AIP funding for distribution among state airports and better manage their aviation system and associated capital improvement plans.

Once state and local funding sources are considered, municipalities provide the remainder of the funds for airport improvements. These funds have come from four principal sources: revenues generated at the airport, taxes for the support of local government as a whole, sale of general obligation bonds, and sale of revenue bonds. In the early years of aviation, the general tax fund was the principal source of local funds, and it still is for small airports. The taxes levied are not earmarked specifically for airport use but are the kind that are normally imposed to operate most of the affairs of local government. As long as the amount of funding is relatively small, this method of financing has not met with much opposition from the citizens in whose political jurisdiction the airport is located.

Bond Financing

As aviation grew and the amounts of funds required became large in relation to other community expenditures, drawing funds from the general tax fund became impractical, and municipalities had to resort to the sale of bonds. Initially these were primarily general obligation type. There were several reasons for resorting to this type of bond financing. First, obligating the entire resources of a local government to back the bonds resulted in much more favorable interest rates than could be obtained by any other form of financing. Second, the projected revenues to be derived from the facility to be developed were often insufficient to utilize any other method of financing.

General Obligation Bonds

As air transport continued to grow and mature, the requirements for airport capital improvements also grew substantially. At the same time, communities were faced with an increased demand for schools, streets, sewage disposal, and other public services. In many cases cities either had reached or were reaching the statutory limit on the amount of general obligation bonds which they could issue, and the cities desired to reserve whatever remaining margin of bonding capacity they had to carry out needed improvements that did not have the revenue potential of airports.

General Airport Revenue Bonds

Airport financing through general obligation bonds still exists for many smaller airports due to fact that these bonds are typically financed at relatively low interest rates, depending on the credit strength of the airport sponsor, lower issuance costs, and little coverage requirements.

Larger airports have moved away from general obligation bonds in recent years, mostly due to the difficulty of receiving large funding and risks associated with obligating the municipality. To overcome the limitation of financing through the sale of general obligation bonds, many communities raise funds through the sale of revenue bonds whenever possible.

General airport revenue bonds (GARBs) are the most common bonds issued for airport capital improvements. While interest rates are generally higher for revenue bonds than they are for general obligation bonds, the differential between the two has decreased considerably since the first issuance of revenue bonds. Revenue bond financing is most successful for those components of the airport which are good revenue producers, such as terminal buildings and parking garages. In addition revenues generated from airline rates and charges, terminal concessions, and other leases contribute to the financing of revenue bonds. GARBs allow for revenues generated from the full spectrum of airport revenue sources to back capital improvement projects, including airfield infrastructure, such as runways or taxiways.

Special Facility Bonds

Some facilities at an airport, such as hangars, hotels, and shopping centers, have been financed through what are known as special facility bonds. These bonds are issued either by the airport sponsor or a single tenant to finance the construction of a single facility, such as a terminal, terminal expansion, maintenance or cargo facilities, or airline ground support facilities. Special facility bonds are backed by the revenue to be generated by the facility or the facility sponsor. The risks associated with these bonds are directly correlated to the financial stability of the sponsor, and as such, specialty facility bonds have become less recommended, particularly for projects backed by the commercial air carriers.

To overcome the risks of a single backer or to build facilities to be used by multiple tenants, such as a consolidated rental car facility, multitenant special facility bonds have been issued. These bonds are backed by more than one sponsor, and typically have greater credit strengths.

PFC Bonds

As the passenger facility charge (PFC) program has grown since its inception in 1990, strategies have been employed to leverage either current or future PFC revenues to fund capital improvements. These strategies are known as PFC leveraging, or the issuance of PFC bonds. There are a number of varieties of PFC leveraging, from the consideration of PFCs as revenue to pay all or part of existing GARB debt service to issuing bonds that will be paid off solely by projected PFC revenue.

CFC Bonds

Similar to PFCs, certain airport tenants assess customer facility charges (CFCs) to generate revenue. This is most common with rental car agencies. Leveraging these CFCs has been a strategy employed to finance the construction of rental car facilities, particularly consolidated rental car facilities, shared by a number of tenants.

The tenant builds the facility on property that is leased from the airport owner. One advantage of this type of airport financing is that it relieves the community of all capital investment in the facility except for utilities and access roads or taxiways. However, it does require the community to commit the land on the airport for 25 to 30 years, the period normally required in a ground lease if the tenant is to secure private financing.

Privatization of Airports

Privatization is a mechanism by which some level of airport management, operation, or ownership is transferred from the public sector to the private sector. Its purpose is to introduce market competition into the operation of airports and to relieve government of the financial burden of providing the large investments required to maintain and operate a system of airports. Proposals exist for the private involvement in airports in several ways including the outright transfer of airport ownership, the leasing of the airport to private sector management firms, and the private development, ownership, and operation of a segment of an airport such as the terminal buildings. Several examples of the various types of airport privatization ventures presently exist [4].

Although in the United States the privatization of airports has been limited, in other countries privatization of airports is a significant issue because of the absence of a specifically designated revenue base dedicated to financing aviation system improvements such as the Airport and Airway Trust Fund. In these countries, investments in aviation facilities compete with other public sector programs for limited funds. Given the significant level of financial investment required to maintain an adequate aviation system and the pressures to limit overall government expenditures, privatization is viewed as a mechanism to finance aviation system improvements and operations with limited public-sector involvement. Recently, despite the slow proliferation, privatization efforts in the United States began again to become more high profile in the early years of the twenty-first century.

Proponents of airport privatization often argue that economic market forces and profit motivation can stimulate private-sector investments, resulting in the development of both new airports and increased capacity at existing airports. Furthermore, it is argued that privatization can lead to cost savings in the management and operation of the airport because of private-sector profit motivation, which tends to lower costs and increase productivity. It is also argued that private sector management and operation of airports can lead to revenue enhancement through market pricing strategies being employed for airport airside and landside services. Market pricing strategies, such as marginal cost pricing, often lead to a more efficient utilization of airport resources and generate revenues necessary to increase capacity in those aspects of airport operations which need capacity enhancement. Furthermore, market pricing can be used to increase revenue from the various commercial enterprises offering services at the airport.

Proposals for the privatization of airports require the careful consideration of several factors, the most fundamental of which is that an airport is essentially a monopoly upon which airport users, i.e., airlines and other aircraft operators, passengers, and shippers, are highly dependent. For this reason regulatory safeguards must be implemented with privatization that maintain freedom of access and nondiscrimination among different groups of airport users, ensure conformity with operating standards and agreements, and satisfy both national and local air transportation policy [3].

Privatization of airports may be categorized as either full privatization or partial privatization. While full privatization is defined as the outright sale of the airport to a private owner/operator, partial privatization is less strictly defined. Partial privatization typically implies the involvement in some manner or other of a private investment or management firm in airport operations and/or capital improvement projects.

Partial privatization in airports typically involves major airport tenants, such as airlines, concessionaires, private companies operating on the airport, or private airport management companies.

Privatized activities range from simple management contract services, to developer financing, to capital investments and the operations of facilities.

Private investment in capital projects may be structured in a number of ways. Two of the more common strategies involve some level of private investment in the capital stage, followed by an operating agreement for a set term. Under a build, operate, and transfer (BOT) agreement, for example, private investment is used to construct and operate a facility for a defined period of time. At the end of the term, the facility becomes the ownership of the airport sponsor.

In a lease, build, and operate (LBO) agreement an airport sponsor is allowed to receive the benefits of privatization without losing control over airport facilities. An LBO agreement allows private investment to build and manage an airport facility, while the property itself is leased by the private company from the airport for a period of time.

Full privatization has seen much of its success internationally, beginning with the privatization of several airports in the United Kingdom in the mid-1980s. Major international private airport operators now own and manage airports in Europe, Asia, Australia, and the Middle East, with limited additional airports owned and operated in North and South America.

As of the publication of this text, the latest large commercial service airport in the United States to undertake privatization efforts is Midway Airport in Chicago.

Financial Planning

The financial plan for the capital improvements at an airport requires a detailed analysis of projected traffic, costs, and revenues. At larger airports significant portions of the capital costs of a project are recovered through revenues from the airlines, concessionaires, and other tenants. The remainder is recovered through capital grants from federal and state sources. Since the airlines become long-term tenants obligated to pay user fees and rents for the facilities utilized in conducting their operations, an evaluation of the conceptual alternatives in the planning phases of a project should only be undertaken with direct input from these users. The financial feasibility of a program is in a large measure determined by the magnitude and reasonableness of the charges and rents paid by airport users and tenants. The financing of general aviation airports continues to be a major problem since the revenue base available is usually insufficient to support significant capital improvements. Therefore, it is imperative that the benefits of such airports to the community be carefully analyzed so that other sources of financing can be demonstrated economically viable.

The determination of the financial feasibility of a project is initiated with an agreement between the airport owners and airport users defining the fiscal policies which will govern the setting of rates and charges for the airport users. The basic process consists of a series of steps, as defined below [21]:

1. Allocation of the capital costs of the project to the various cost centers is established by airport management and airport users. These cost centers are usually categorized as the airfield area, hangar and other operational support building areas, terminal area, concessions area, and other areas of the airport.

2. The net annual costs of the capital construction program are projected, and these costs are assigned to the various cost centers. These costs are amortized over the period specified in the agreement between management and users.

3. Projection and allocation of the net annual administrative, operating, and maintenance costs to each of the cost centers are based upon a knowledge of past cost experience and projections of these anticipated costs for the new facility.

4. Conversion of the total annual capital and administrative, operating, and maintenance costs to a schedule of the fees and rents to be paid by the users of the facilities utilizes available forecasts of aircraft activity, passenger enplanements, parking usage, and other relevant indices of projected airport activity.

Recovery of capital costs requires that the total capital investment in each cost center be determined. Projected costs in the airfield area for runways, taxiways, apron ramps, and land acquisition and improvement and in the terminal area for terminal building construction, land, and terminal support facilities must be ascertained and assigned to the relevant cost center. It is essential that airport support facilities such as access roads, service roads, sanitary and storm sewer systems, electrical and mechanical systems, communication and security services, emergency medical services, and crash, fire, and rescue services be properly apportioned to the appropriate cost centers to eliminate imbalances in the determination of the facility cost-center revenue requirements, which may result in unreasonable rates and charges. For projects where bond issues are utilized to finance portions of the capital costs, the annual cost of debt service (i.e., principal, interest and the required reserve, called coverage) over the recovery period must be included and assigned to the relevant cost center category.

The anticipated costs of airport administration, operations, and maintenance are assigned to each cost center on an annual basis. These costs generally include all direct costs for salaries, materials, supplies, and outside services and related indirect costs.

Terminal costs are divided by the terminal area. Often the location and degree of finishing in ticketing facilities, baggage facilities, office space, and car rental space are taken into consideration to establish rental rates for the terminal building tenants.

Concession and other airport revenues will normally be applied against the appropriate cost center and the net revenue recovery requirements determined. Forecasts of landing weights are used to determine the landing fees charged to the airlines to recover airfield costs. Often the cost of the apron area is isolated as a separate cost center, and ramp fees are established based upon the gate frontage required by the airlines.

Concession area costs are derived from rentals paid and from charging the concessionaire a percentage of the gross receipts. Usually the most significant concession cost center at a large airport is the parking facility. Since the capital costs of a parking structure are considerable, these costs are usually assigned to the terminal-area cost center. The administrative, operating, and maintenance costs of these facilities are recovered through a percentage of the gross receipts charge.

Rate Setting

A commercial service airport is designed to service two distinct groups: the airlines and the commercial entities serving them, and the passengers and those retail enterprises which service them [16]. The airport leases its facilities to the airlines, concessionaires, industries, general aviation, and airport support services. The airlines lease ground for aircraft storage and space for ticket counters, operations, maintenance, and baggage handling, and they pay fees for landing and ramp rights. Cargo and hangar facilities are also utilized by the airlines at specific locations. Concessionaires rent space within the terminal and are charged on the basis of the amount and quality of space rented and a percentage of receipts. Those franchisees outside the terminal, such as taxicab companies, are charged in various ways. One method is based upon the number of passengers enplaned at the airport, and another is a fixed rate based upon the number of times the airport is utilized for the service.

The method for determining rates varies from airport to airport. For the last part of the twentieth century, the most common method was the *residual cost approach*. In such an approach, the total annualized costs of the airport are reduced by the amount of all nonairline revenues, and the remainder is proportioned among the airlines based upon level of activity measures. Those costs apportioned to the terminal area are divided by the gross terminal area to determine space rental charges. Those costs apportioned to the airfield are divided by the total annual gross landing weight of the carriers at the airport to determine landing fees. This type of cost recovery approach essentially guarantees that the airlines will provide the revenues necessary to cover airport costs. It also places the airlines in a unique position relative to airport management in that the airlines have a vested interest in maximizing nonairline revenues to minimize their costs.

Residual cost agreements were commonly longer term agreements, ranging from 20 to 50 or more years. As most of these agreements were signed between 1945 and 1985, many of these agreements are due to expire.

Rather than extend or renew these traditional agreements, airports are exploring more dynamic and flexible rate setting policies. Such approaches attempt to classify airport expenses into distinct cost centers and to apportion the cost of each among users through equitable rates, or to assign the expenses associated with certain cost centers directly to users and to group other expenses to be shared by all users. This type of rate setting is called the *compensatory cost method*.

The test of the validity of any rate setting scheme, however, lies in its ability to reflect rates which are reasonable and justifiable to the airlines, concessionaires, and other tenants. An illustration of the mechanics of rate setting and the determination of measures to assess financial feasibility are contained in Example Problem 13-1.

Example Problem 13-1 Estimate the rates and charges required to support the capital costs of airport development shown in Table 13-2. Use the compensatory cost method of determining rates and charges. The capital costs are financed by issuing 6 percent bonds which are repaid in 20 years. Financial considerations

Airfield	$72,265,000
Apron area and concourses	$46,510,000
Main terminal building	$50,000,000
Parking facilities	$15,400,000
Airport access roads	$4,260,000
Land acquisition	$10,980,000
Total	$199,415,000

TABLE 13-2 Airport Capital Development Costs for Example Problem 13-1

require a 1.25 coverage factor on revenues to support the repayment of airport development bonds. The interest earned at an annual rate of 6 percent on the accumulated balance in the capital recovery fund is to be used to decrease the rates and charges.

The airport is being developed for an initial annual enplaned passenger level of 2 million, and it is expected that the annual demand will increase to 4 million enplaned passengers in 20 years. Air carrier demand is expected to increase from 48,000 to 96,000 annual operations and total aircraft operations are expected to increase from 150,000 to 250,000 annual operations over the 20-year period. Assume that the increases in annual passengers and aircraft operations are constant in each year rather than the growth rate of annual passengers and operations being constant. The air carrier aircraft mix using the airport is expected to consist of 25 percent Boeing 767-200 and 75 percent McDonnell-Douglas MD-87 aircraft. A typical general aviation aircraft will have a maximum certified landing weight of 3000 lb. The airport will be developed with four wide-bodied gates to accommodate the Boeing 767-200 and 12 narrow-bodied gates to accommodate the MD-87. The main terminal building will have an area of 250,000 ft^2 exclusive of the concourses housing the aircraft gates.

The Boeing 767-200 is found to have a maximum certified landing weight of 272,000 lb, a wingspan of 156 ft 1 in, and an average capacity of 236 passengers. The MD-87 has a maximum certified landing weight of 130,000 lb, a wingspan of 107 ft 10 in, and an average capacity of 135 passengers.

The rates and charges to the airport users must recover the capital development costs, including bond interest, over the capital recovery period of the bonds, which is 20 years.

In general, airfield costs are recovered through landing fees, apron area and concourse costs through ramp charges, terminal building costs through square foot rental charges, and parking facility costs through parking rates. The costs for the ground access system and the land acquisition are usually allocated to the other charges to recover these costs. Sometimes the ground access system costs are allocated to either the terminal building or parking charges or to both. In this problem, the ground access system and land acquisition costs total $15,240,000 and represent about 8 percent of the total project cost. Therefore, all rates and charges would be increased by 8 percent to cover these costs.

First, the determination of landing fees is made. Generally, landing fees are charged to only the commercial air carriers at an airport. Since the air carrier demand will double over the 20-year span, the total number of air carrier aircraft landings using the airport over a 20-year period will be about 720,000. Assuming that the air carrier mix over the project period is constant, the average landing weight of the air carrier aircraft is expected to be $0.25 \times 272,000 + 0.75 \times 130,000 = 165,500$ lb. Therefore, the total landed weight of all air carrier aircraft over the project life is found to be $720,000 \times 165,500 = 119,160,000,000$ lb.

If bonds are issued to finance the project, the bond interest payments for 20 years are equal to $72,265,000 \times 0.06 \times 20 = $86,718,000. The required average annual contribution to the capital recovery fund to repay the face value of the bonds, considering the interest earned on the accumulated surplus, is $2,414,500. Therefore, the air carrier landing fee must generate a total revenue of $86,718,000 + 20 \times $2,414,500 = $135,008,000.

To recover the airfield cost through air carrier landing fees, the average rate per 1000 lb of landed weight is found to be $135,008,000 \div 119,160,000 = $1.13. The landing fee that must be assessed to air carrier aircraft to recover the airfield costs is computed for the Boeing 767-200 as $1.13 \times 272 = \$307.36$ and for the MD-87 as $\$1.13 \times 130 = \$146.90$. (The reader should examine the cash flow for each year to determine that the required revenue is attained through the fees calculated. If this is done, it will be found that the landing fee must be increased to about $1.15 per thousand pounds to realize sufficient revenue to pay the bond interest and retire the bonds at the end of 20 years. In any situation in which rates are based on a varying demand, a cash flow analysis should be performed to verify the rates and charges, as average values of demand typically yield either too little or too much revenue.)

It is instructive to look at the impact of a policy which charges general aviation aircraft landing fees on the cost to air carriers. Since there are initially 102,000 annual general aviation operations which increase to 154,000 annual operations in the design year, general aviation aircraft over the project life will conduct about 1,280,000 landings. General aviation aircraft at this airport average about 64 percent of the aircraft fleet. Therefore, the average landed weight of all aircraft using the airport is equal to $0.64 \times 3000 + 0.36 \times 165,500 = 61,500$ lb. The total landed weight of all aircraft over the project life is then found to be $2,000,000 \times 61,500 = 123,000,000,000$ lb.

To recover the airfield cost through landing fees assessed to all aircraft, the average rate per 1000 lb of landed weight is found to be equal to $135,008,000 \div 123,000,000 = \$1.10$. This is not a significant reduction from the case where only air carrier aircraft were charged landing fees. The landing fee that must be assessed to air carrier aircraft to recover the airfield costs is computed for the Boeing 767-200 as $\$1.10 \times 272 = \$299.20$ and for the MD-87 as $\$1.10 \times 130 = \$143.00$. For general aviation aircraft the landing fee is $\$1.10 \times 3 = \$3.30$.

Clearly there is very little benefit to air carrier aircraft from a policy which charges landing fees to all aircraft using the airport. Furthermore, the design of the airfield is significantly impacted by the presence of air carrier aircraft, and these costs are appreciably higher than if the airfield were designed for only general aviation aircraft. Collecting general aviation aircraft landing fees also presents an operational problem for airport management, and the cost of collecting the fee will very likely exceed the fee collected. Airport management must determine the best method of charging general aviation aircraft for airport use if landing fees are not assessed to these users.

Next a determination of ramp fees for air carrier aircraft is made. Normally, ramp fees are charged to the airlines to recover the cost of the apron and concourse system. The ramp fee is based upon the wingspan of the gate design aircraft.

In this problem, the average wingspan of the gate design aircraft is computed as $0.25 \times 156.1 + 0.75 \times 107.84 = 120$ ft. The total wingspan of the aircraft occupying the 16 gates is then $16 \times 120 = 1920$ ft. The cost of the apron and concourse development was $46,510,000. The interest payments on the bonds over 20 years will be $55,812,000. The required average annual contribution to the capital recovery fund to repay the face value of the bonds, considering the interest earned on the accumulated surplus, is $1,551,000. Therefore, the air carrier landing fee must generate a total revenue of $\$55,812,000 + 20 \times \$1,551,000 = \$86,832,000$. Therefore, the cost per foot of gate over the life of the project is $\$86,832,000 \div 1920 = \$45,225$. This means that each of the Boeing 767-200 gates would cost $\$45,225 \times 156.1 = \$7,059,600$ and that each of the MD-87 gates would cost $\$45,225 \times 107.84 = \$4,877,100$. This is a significant cost to the airline leasing the gate. Since there are 720,000 landings over a 20-year period, there will be 45,000 aircraft occupancies at each gate. The cost per aircraft for gate use is then $156.88 for a Boeing 767-200 and $108.38 for an MD-87.

The cost of the main terminal building is usually recovered through square-foot rental charges. However, typically only about 50 percent of the space in a terminal building is rentable. Therefore, the space in the main terminal building which is rentable is $0.50 \times 250,000 = 125,000$ ft^2. The cost of the main terminal building is then recovered based upon this area. The cost of the main terminal building is $50 million. The interest payments on the bonds over 20 years will be $60 million. The required average annual contribution to the capital recovery fund to repay the face value of the bonds, considering the interest earned on the accumulated surplus, is $1,359,300. Therefore, the main terminal building area charges must generate a total revenue of $60,000,000 + $20 \times \$1,359,300 = \$87,186,000$. Therefore, the cost per square foot to recover terminal building costs is then $\$87,186,000 \div 125,000 = \$698$ over the life of the project. Assuming a 20-year project life, this becomes $35 per square foot per year.

Since typically about 40 percent of the terminal building area is rented by the airlines this represents a total annual cost to the airlines of $\$35 \times 0.40 \times 250,000 = \$3.5$ million. This represents a lifetime cost of $70 million to the airlines which is significant since the airlines normally operate from many airports.

Parking charges are used to recover the cost of the parking facility. This requires that one know the number of parkers and average parking duration. If it is assumed that vehicle occupancy rates are 2.5 passengers per vehicle, then 4 million annual enplaned passengers translate to 1.6 million annual vehicles with enplaning passengers in the design year. Since over time the enplaning and deplaning passengers are about the same, this means a total of 3.2 million vehicles on the ground access system during the design year. Typically 70 percent are passenger cars, and of these typically 30 percent park. The number of vehicles parking in the design year is then $0.70 \times 0.30 \times 3,200,000 = 672,000$.

The total number of vehicles parking in the parking facilities over the project life is then about 10,080,000. The parking facility cost is $15,400,000. The interest payments on the bonds over 20 years will be $18,480,000. The required average annual contribution to the capital recovery fund to repay the face value of the bonds, considering the interest earned on the accumulated surplus, is $517,400. Therefore, the terminal area charges must generate a total revenue of $18,480,000 + 20 \times \$517,400 = \$28,828,000$. The average parking rate required is $\$28,828,000 \div 10,080,000 = \$2.86$ per vehicle. This requires that time-related parking rates be established based upon the average vehicle parking time to realize a revenue of $2.86 per parked vehicle.

As noted earlier, one method of recovering land acquisition and ground access system costs is to proportionally increase the other rates for the effect of these costs. In this problem, all the above rates would be increased by 8 percent for this purpose, since these costs are about 8 percent of the total project development costs. Additionally, bonding agencies usually require that the airport demonstrate that its rate structure will realize actual revenues which are 1.25 times the required revenues to retire airport debt. This is called bond *coverage*. Therefore, each of the rates and charges evaluated above must be increased in this problem by $0.08 + 0.25 = 0.33$, or 33 percent, for these purposes.

Based upon the cost recovery and allocation factors discussed above, and to account for the allocation of ground access system and land acquisition costs as well as the impact of bond interest and coverage, the final rates and charges can be determined. A Boeing 767-200 aircraft would be assessed a landing fee of $409 and a ramp fee of $209. An MD-87 aircraft would be assessed a landing fee of $195 and a ramp fee of $144. Tenants would be charged a rental fee of $47 per square foot per year. Parking charges would average about $3.80. The total number of enplaned passengers serviced over this 20-year project life is 60,500,000. The total number of deplaned passengers would also be about 60,500,000. Including the bond coverage requirements, the average cost per passenger for landing fees and ramp charges amounts to $4.92, and the total airport development cost per enplaned passenger is equal to $449,346,000 ÷ 60,500,000 = $7.42. Each of these metrics is an indicator of the financial viability of the development project. In both cases these are very reasonable values.

It should be emphasized that airport rates and charges include not only capital development costs but also the operating and maintenance costs associated with the airport. These charges are usually reevaluated each year by the airport.

The compensatory cost method of determining rates and charges was used in Example Problem 13-1. It is likely that very different rates and charges would be realized if the residual cost method were used. This is shown in Example Problem 13-2.

Example Problem 13-2   Estimate the landing fees required to support the capital costs of the airfield development in Example Problem 13-1 if a passenger facility charge is imposed at the rate of $2 per enplaned passenger for 10 years. This passenger facility charge is dedicated to airfield development. Solve this problem, using the residual cost method of determining rates and charges and assuming that the passenger facility charge is the only additional revenue received by the airport (Table 13-2). As before, the airport is being developed for an initial annual enplaned passenger level of 2 million, and it is expected that the annual demand will increase to 3 million enplaned passengers in 10 years. Therefore, the total revenue gained from the passenger facility charge is $2 × 25,000,000 = $50,000,000. The net cost of airfield development, including interest over the 20-year period, becomes $117,802,100 − $50,000,000 = $67,802,100. The average landing fee to support airfield development over a capital recovery period of 20 years will be determined.

Since the air carrier demand doubles over the 20-year span, the total number of air carrier aircraft landings using the airport over a 20-year period will still be about 720,000. The average landing weight of the air carrier aircraft is still equal to 165,500 lb and the total landed weight of all air carrier aircraft over the project life is still 119,160,000,000 lb.

To recover the net airfield development cost through air carrier landing fees, the average rate per 1000 lb of landed weight is then found to equal $67,802,100 ÷ 119,160,000 = $0.57. Therefore, the landing fee that must be assessed to air carrier aircraft to recover the airfield costs is computed for the Boeing 767-200 as $0.57 × 272 = $155.04 and for the MD-87 as $0.57 × 130 = $74.10. As in Example Problem 13-1, to account for land acquisition and ground access system costs, and coverage, these landing fees must be increased by 33 percent. Therefore, a Boeing 767-200 aircraft would be assessed a landing fee of $206 and an MD-87 aircraft would be assessed a landing fee of $99. The cash flow is shown in Table 13-3.

As may be observed, the residual cost method of determining rates and charges results in a decrease in cost to the airlines.

Year	Aircraft Landing Demand	Enplaned Passenger Demand	Landing Fee Revenue $	PFC Revenue, $	Total Revenue $	Bond Interest $	Capital Recovery Fund	
							Deposit $	Accumul $
1	24000	2000000	2260068	4000000	6260068	4335900	1924168	1924168
2	25263	2111111	2379019	4222222	6601241	4335900	2265341	4304959
3	26526	2222222	2497970	4444444	6942415	4335900	2606515	7169772
4	27789	2333333	2616921	4666667	7283588	4335900	2947688	10547646
5	29053	2444444	2735873	4888889	7624761	4335900	3288861	14469366
6	30316	2555556	2854824	5111111	7965935	4335900	3630035	18967563
7	31579	2666667	2973775	5333333	8307108	4335900	3971208	24076825
8	32842	2777778	3092726	5555556	8648282	4335900	4312382	29833816
9	34105	2888889	3211677	5777778	8989455	4335900	4653555	36277400
10	35368	3000000	3330628	6000000	9330628	4335900	4994728	43448772
11	36632	3100000	3449579	0	3449579	4335900	-886321	45169378
12	37895	3200000	3568531	0	3568531	4335900	-767369	47112172
13	39158	3300000	3687482	0	3687482	4335900	-648418	49290484
14	40421	3400000	3806433	0	3806433	4335900	-529467	51718445
15	41684	3500000	3925384	0	3925384	4335900	-410516	54411036
16	42947	3600000	4044335	0	4044335	4335900	-291565	57384134
17	44211	3700000	4163286	0	4163286	4335900	-172614	60654568
18	45474	3800000	4282237	0	4282237	4335900	-53663	64240180
19	46737	3900000	4401189	0	4401189	4335900	65289	68159879
20	48000	4000000	4520140	0	4520140	4335900	184240	72433711
Total	720000	60500000	$67802100	$50000000	$117802078	$86718000	$31084078	$72433711

TABLE 13-3 Landing Fee Cash Flow Analysis with Passenger Facility Charge for Example Problem 13-2

Evaluation of the Financial Plan

Criteria for measuring the financial effectiveness of an airport plan are usually determined by considering various evaluative measures including these [21]:

1. The effectiveness of functional areas as measured by the ratios of the amount of public space, revenue space, airline exclusive space, and concession space to the total space within the terminal building

2. The relative effectiveness of areas within the terminal building, as indicated by the ratio of airline exclusive space to the number of gates and the ratio of the ramp area to the total building area

3. An evaluation of annual costs and revenues for various items in each of the cost-center categories, as shown by the cost and revenue per enplanement, per operation, per 1000 lb of aircraft landing weight, and per square foot of building space

4. The effectiveness of the schedule plan of the airline, as indicated by the number of departures per gate and enplaned passengers per unit of airline exclusive space

The final determination of the most effective plan is made through the process of discussion and negotiation between airport managers and users. Various assumptions are made concerning the allocation of costs and revenues between cost centers until a consensus is reached. At this point airline lease agreements and concession policies are developed which result in long-term commitments by the airlines and tenants to the airport project. In the final analysis, airport expansion plans must address not only the needs for changes in physical facilities but also the economic, environmental, and financial feasibility associated with such development. It can be expected in this age of limited financial resources, with energy and aircraft equipment needs foremost in the management of airlines, that a clear determination of the feasibility of airport expansion projects will be required before long term commitments for support by the airline will be made.

References

1. *AOCI Uniform Airport Financial Statement*, Airport Operators Council International, Inc., Washington, D.C.
2. *Airport Administration and Management*, John R. Wiley, Eno Foundation for Transportation, Inc., Westport, Conn., 1986.
3. *Airport Economics Manual*, Doc. No. 9562, International Civil Aviation Organization, Montreal, Canada, 1991.
4. *Airport Finance*, Norman Ashford and Clifton A. Moore, Van Nostrand Reinhold, New York, N.Y., 1992.

5. *Airport Planning and Management*, D. F. Smith, J. D. Odegard, and W. Shea, Wadsworth Publishing Company, Belmont, Calif., 1984.
6. *Airport Economic Planning*, Airport Revenues and Expenses, G. P. Howard, Editor, MIT Press, Cambridge, Mass., 1974.
7. *Economics of Airport Operation*, Calendar Year 1972, J. A. Neiss, Federal Aviation Administration, Washington, D.C., April 1974.
8. *Eleventh Annual Report of Operations Under the Airport and Airway Development Act*, Fiscal Year Ended September 30, 1980, U.S. Department of Transportation, Federal Aviation Administration, Washington, D.C., 1981.
9. *Fort Lauderdale-Hollywood International Airport Economic Feasibility*, Preliminary Report, Aviation Planning Associates, Inc., Cincinnati, Ohio, April 1981.
10. *General Aviation and the Airport and Airway System: An Analysis of Cost Allocation and Recovery*, National Business Aircraft Association, Inc., Washington, D.C., April 1981.
11. *Hearings before the Subcommittee on Aviation of the Committee on Commerce*, U.S. Senate, Washington, D.C., July 1969.
12. *National Airport System Plan*, 1978–1987, Federal Aviation Administration, Department of Transportation, Washington, D.C.
13. *National Airspace System Plan*, Federal Aviation Administration, Washington, D.C., 1989.
14. *National Plan of Integrated Airport Systems (NPIAS) 1990–1999*, Federal Aviation Administration, U.S. Department of Transportation, Washington, D.C., 1991.
15. *Passenger Facility Charges*, Part 158, Federal Aviation Regulations, Federal Aviation Administration, Washington, D.C., June 1991.
16. *Planning for Airport Access: An Analysis of the San Francisco Bay Area*, Conference Publication 2044, National Aeronautics and Space Administration, Ames Research Center, Moffett Field, Calif., May 1978.
17. *Reauthorizing Programs of the Federal Aviation Administration, Future Capacity Needs and Proposals to Meet Those Needs*, Subcommittee on Aviation, Committee on Public Works and Transportation, House of Representative Report No. 101-37, U.S. House of Representatives, Washington, D.C., 1990.
18. Senate Report No. 97–97, *Airport and Airway System Act of 1981*, Committee on Commerce, Science, and Transportation, U.S. Senate, Washington, D.C., May 15, 1981.
19. *State Funding of Airport and Aviation Programs*, National Association of State Aviation Officials, Washington, D.C., 1981.
20. *Tenth Annual Report of Accomplishments Under the Airport Improvement Program*, Fiscal Year 1991, Federal Aviation Administration, Washington, D.C., 1992.
21. *The Apron-Terminal Complex*, The Ralph M. Parsons Company, Federal Aviation Administration, Washington, D.C., September 1973.
22. *The States and Air Transportation: Expenditures and Tax Revenues*, Center for Aviation Research and Education, National Association of State Aviation Officials, Silver Spring, Md., 1991.
23. *Airport Planning and Management*, 5th ed., Alex Wells and Seth Young, McGraw Hill Publishing, New York, N.Y., 2003.
24. *Innovative Finance and Alternative Sources of Revenue for Airports*, Cindy Nichol, Transportation Research Board, ACRP Synthesis 1, Washington, D.C., 2007.

CHAPTER 14
Environmental Planning

Introduction

The current concern for an assessment and understanding of the environmental, ecological, and sociological consequences of development actions has resulted in the emergence of a holistic approach to planning. This approach views all actions as being undertaking in a single system and examines the consequences of these actions in terms of the entire system. Traditionally, proposals for transportation facilities have been evaluated in terms of sound engineering and technological principles, economic criteria, and benefits to the users and community. However, policy decisions today are being made with a more complete awareness of the impacts of these decisions on both users and nonusers from economic, social, environmental, and ecological viewpoints.

Airports must be planned in a manner which ensures their compatibility with the environs in which they exist. There are many serious compatibility problems which presently exist in the vicinity of airports which represent a serious confrontation between two important characteristics of urban economics, the need for airports to meet transportation needs and the continuing demand for community expansion. Airport planning must be conducted within the context of a comprehensive regional plan. The location, size, and configuration of an airport must be coordinated with the existing and planned patterns of development in a community, considering the effect of airport operations on people, ecological systems, water resources, air quality, and the other areas of community concern [9].

This chapter presents an overview of the factors which must be considered to assess and evaluate the impact of airport development decisions in the context of a system's approach to planning.

Policy Considerations

In the United States, the overall basis for policies related to the consideration of the environmental, ecological, and social impacts of airport development is rooted in the National Environmental Policy Act of 1969 (Public Law 91-190). The policy of the Department of Transportation (DOT) is

To integrate national environmental objectives into the missions and programs of the department and to:

1. Avoid or minimize adverse environmental effects wherever possible;
2. Restore or enhance environmental quality to the fullest extent practicable;
3. Preserve the natural beauty of the countryside and public park and recreational lands, wildlife and waterfowl refuges, and historic sites;
4. Preserve, restore, and improve wetlands;
5. Improve the urban physical, social and economic environment;
6. Increase access to opportunities for disadvantaged persons; and
7. Utilize a systematic, interdisciplinary approach in planning and decision making which may have an impact on the environment. [43]

To implement this policy the FAA has established an environmental assessment and consultation process which provides the relevant officials, policy makers, and the public with an understanding of the potential environmental consequences of proposed actions and ensures that the decision-making process includes environmental assessments as well as economic, technological, and other factors relevant to the decision. It requires that environmental impact statements and negative declarations serve to document and record compliance with this policy and reflect a thorough study of all relevant environmental factors using a systematic, comprehensive, and interdisciplinary approach. The National Environmental Policy Act (NEPA) requires

All agencies of the Federal government to include in every recommendation or report on proposals for legislation and other major Federal actions affecting the quality of the human environment, a detailed statement on:

1. The environmental impact of the proposed action;
2. Any adverse environmental effects which cannot be avoided should the proposal be implemented;
3. Alternatives to the proposed action;
4. The relationship between local short-term uses of man's environment and the maintenance and enhancement of long-term productivity; and
5. Any irreversible and irretrievable commitments of resources which would be involved in the proposed action should it be implemented.

Complementing this overall policy statement, the FAA also established an Aviation Noise Abatement Policy (Public Law 96-193 and Public Law 101-508) to significantly reduce the adverse impacts of aviation noise on existing land uses and to achieve a substantial degree of noise compatibility between airports and their environs. It has endorsed coordinated actions between aircraft operators and owners, the FAA, the airport owners and sponsors, and the community. It proposed several actions to achieve airport noise control and land-use compatibility including source noise reductions through aircraft retrofit and replacement, modifications of landing and takeoff procedures, and compatibility plans which have the objective of containing severe noise impacts within airport controlled areas.

Noise is the most apparent impact of an airport upon the community but due consideration is required for all of those social, economic, environmental, and ecological factors which are influenced by airport activity. These factors may be grouped into four categories which can be identified as pollution factors, social factors, ecological factors, and engineering and economic factors [22]. The pollution factors include air and water quality, noise, and construction impacts. The social factors include land development, the displacement and relocation of businesses and residences, parks and recreational areas, historic places and archeological resources which may be impacted, areas which are unique because of natural or scenic beauty, and the consistency of the proposed development with local planning. The ecological factors include the impact on wildlife and waterfowl, flora and fauna, endangered species, and wetlands or coastal zones. The engineering and economic factors include a consideration of flood hazards, costs of construction and operation, benefits of implementation, and energy and natural resource use.

The FAA has identified the requirements for environmental impact assessment (EA) reports, environmental impact statements (EIS), and findings of no significant impact (FONSI) for various types of projects, and has also categorically excluded certain types of projects from the requirements of a formal environmental assessment [2, 43]. Table 14-1 lists a breakdown of the type of environmental study required for some common airport planning actions.

The general format for an environmental study consists of a statement of need for the proposal, an inventory of problems and issues, an identification of constraints and opportunities, an identification of the improvement components including physical and non-physical entities, measures to increase benefits and reduce harm, a discussion of the alternatives and their impacts, and the manner and degree of community and public agency involvement in the process [2, 36, 43].

Typical actions normally requiring environmental assessment:
Airport location
New runway
Major runway extension
Runway strengthening to permit use by noisier aircraft
Major expansion of terminal or parking facilities
Establishment or relocation of instrument landing system
Land acquisition
    Required for facility modifications
    Relocation of business or residences
    Affecting historical, recreational, or archaeological resources
    Affecting wetlands, coastal zones, or floodplains
    Affecting endangered or threatened species
Typical actions normally requiring environmental impact statement:
Adoption of a new airport system plan if criteria are
    substantially different from former plan
First-time airport location or airport layout planned
New runways capable of serving air carrier traffic in metropolitan areas

*See Federal Aviation Administration [43].
Source: Federal Aviation Administration [2].

TABLE 14-1 Environmental Study Requirements of Airport Development Project*

Pollution Factors

Air Quality

Many of the larger, more densely populated urban areas are facing serious difficulties associated with the emission of dangerous gaseous and particulate matter into the atmosphere due to industrial processes, combustion, and transportation. Air pollution affects the public welfare including the personal comfort and health of man, causes damage to soil, water, vegetation, wildlife, animals, deterioration of property and the erosion of property values, and a reduction in visibility resulting in losses of aesthetic appeal and increased hazards in transportation. Air pollution is defined as the introduction of foreign substances or compounds into the air or the alteration of the concentrations of naturally occurring elements. Hub airports with a considerable volume of commercial jet aircraft traffic may contribute substantially to this problem.

Air quality is defined by the concentration level of six pollutants for which standards have been adopted, namely, carbon monoxide, hydrocarbons, nitrogen oxides, sulfur dioxide, suspended particulates, and photochemical oxidants. The standards are specified in the Clean Air Act and consist of two categories, primary standards related to health and secondary standards related to welfare.

The amount of a particular pollutant produced by an aircraft is a function of the type of engines and the mode of operation of the aircraft [17]. An analysis must include a consideration of aircraft idling at the gate and runway threshold, engine power run-ups, taxiing, takeoff, climb-out, approach, and landing. The dispersion of the pollutants is studied through the use of either emission models or diffusion models. The emission model assumes a uniform dispersion of the pollutants within the atmosphere of concern, whereas the diffusion model uses emissions or emission rates together with physical and meteorological conditions to determine concentrations of pollutants. A study of the air quality impacts for an airport project requires a determination of ambient air quality, local meteorological conditions, the mix, number, and paths of aircraft using the airport, and the emission rate of the aircraft in different operating modes. It also requires a knowledge of the operating characteristics and volume of ground transportation modes providing access to and services at the airport, and the point sources of pollution occasioned by the normal operation of an airport. A flow chart of the interaction of those factors which are normally considered in an air quality study at an airport is given in Fig. 14-1. The results of an air quality study are typically displayed on maps which show the before and after concentration of pollutants in the area of the airport, together with charts indicating the level of compliance with air quality standards.

Water Quality

Water is one of the most valuable resources on earth. Not only is it essential for the maintenance of life itself but it is also used by man in nearly all daily activities. As the population has grown, so has the demand for water, and today, that need is so great that in many areas of the world the need has outpaced the supply. The construction and operation of airport facilities can contribute to the degradation of the quality and reduction of the quantity of groundwaters or surface waters. Water quality can be affected by the addition of soluble or insoluble organic or inorganic materials into rivers, streams, and aquifers resulting in a water source which is inadequate to support aquatic life and other uses such as fishing, swimming, and water supply needs. Changes in the cover, composition, and topography of the ground in the vicinity of airport sites can cause changes in the amount, peaking, routing, and filtration of runoff and the recharge area of aquifers. Construction-related activities may cause the introduction of materials and wastes into streams and water sources, increases in the volumes of sanitary wastes and water supply demand, and increases in storm water management systems.

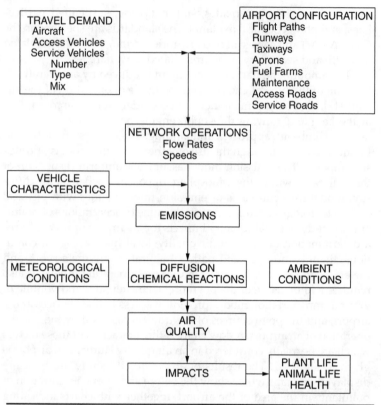

FIGURE 14-1   Flow chart illustrating air quality study process for airports.

A water quality study for an airport facility should address both the direct and indirect effects of the project on water quality [8, 21]. The direct effects include soil erosion, the amount and composition of runoff from the facility, infiltration, spills, turbidity, and the quantities of water supply and sewage disposal needs. Indirect effects include the accelerated weathering of exposed geologic and construction materials, disruption of nutrient cycles for the support of life, and the extraction of construction materials which may alter natural filtering, the degree of imperviousness of soils, and water storage capacity. Typically, a water quality study will identify the source and receptors of pollutants, and the amount of degradation which the introduction of pollutants will cause. It will also address the impact on the quantity of water sources through a determination of flow rates, flow and recharge areas, permeability, infiltration, and flow interruptions. Construction measures utilized to minimize degradation of water quality and supply include the construction of check dams, sediment traps, berms, dikes, channels, and slope drains, sodding and seeding, brush barriers, and paving. Wastewater management plans are usually prepared integral with a review of this impact area [8, 21, 30].

Aircraft and Airport Noise

The effects that noise from aircraft have on communities surrounding airports present a serious problem to aviation. Since commercial jet transport operations began in 1958, the public reaction to aircraft noise has been vigorous. Because of these reactions much has been learned about the generation and propagation of noise and about human reactions to noise. On the basis of this knowledge, procedures have been developed which permit the planner to estimate the magnitude and extent of noise from airport operations and to predict community response. Several of these procedures are outlined here. The impact of aircraft noise on a community is dependent upon several factors including the magnitude of the sound, the duration of the sound, the flight paths used during takeoff and landing, the number and types of operations, the operating procedures, the aircraft mix, the runway system utilization, the time of day and season, and meteorological conditions. The response of communities to exposure to aircraft noise is a function of the land and building use, the type of building construction, the distance from the airport, the ambient noise level, and community attitudes [9, 45].

Quantifying Aircraft Noise

Like most other environmental issues, aircraft noise has many dimensions. Most of these dimensions relate to the reaction of people to aircraft noise. These reactions relate to the sound level, the varying sensitivity of the human ear to different frequencies or pitches of sound, the frequency of occurrence of aircraft noise intrusions, the time of day these intrusions occur, and the number of intrusions that occur over a period of time such as a day. Given this range of dimensions, it is not surprising that several metrics of aircraft noise have been developed over the years. While some were built into the first sound meters over a half century ago, most have been developed since the introduction of the first transport category turbojet aircraft in the late 1950s.

Many metrics have been developed over the years to describe aircraft noise. Some of the more common ones are presented in the following paragraphs. The goal of these metrics is to quantify aircraft noise in a manner which relates the physical aspects of sound to human assessments of loudness and noisiness. These metrics are the basis of most noise analyses conducted at airports throughout the United States and elsewhere. In addition, there are other similar noise metrics with specialized purposes and these are also discussed. These are effective perceive noise level (EPNL), composite noise rating (CNR), and noise exposure forecast (NEF).

Sound Pressure and Sound Pressure Level

All sounds come from a sound source such as a musical instrument, a voice speaking, or an airplane passing overhead. Sound energy radiated by such sources is transmitted through the air in sound waves which are tiny pressure fluctuations just above and below atmospheric pressure. These pressure fluctuations, called *sound pressures*, impinge on the ear, creating audible sound. Sound pressures are quantified by the root-mean-square (RMS) value, that is, the square root of the average squared pressure fluctuation over some brief period of time (about 1 s for aircraft noise purposes) as shown in Eq. (14-1).

$$p_{rms} = \left(\frac{1}{T} \int_{t=0}^{T} p(t)^2 \, dt \right)^{1/2} \tag{14-1}$$

where p_{rms} = root mean square sound pressure
$p(t)$ = deviation from atmospheric pressure at time t
T = averaging time, 1 s for airport noise purposes

The human auditory system is sensitive to a very wide range of RMS sound pressures. The loudest sounds people can hear without pain have about 1 million times the RMS sound pressure as the faintest sounds people can hear. Equally remarkable is the way the auditory system perceives *changes* in loudness. To a first approximation, equal *percentage* changes in RMS sound pressure are perceived as equal changes in loudness. Hence, at higher RMS sound pressures, larger absolute changes in RMS sound pressure are required to make a noticeable difference in loudness than at lower RMS sound pressures. The smallest difference in RMS sound pressure the human auditory system can detect is about 10 percent.

For these reasons a logarithmic, or *decibel scale*, is well suited for quantifying sound in a manner which relates to human perception. In its logarithmic form, RMS sound pressure is called the RMS *sound pressure level* (SPL). Sound pressure level is the logarithm of the ratio of two squared pressures, the numerator containing the pressure of the sound source of interest and the denominator containing a reference pressure, as shown in Eq. (14-2). The units of sound pressure level are decibels (dB).

$$L_p = 10 \log \left(\frac{p_{rms}^2}{p_0^2} \right) \tag{14-2}$$

where L_p = RMS sound pressure level
p_{rms} = RMS sound pressure
p_0 = reference pressure of 20×10^{-6} newtons per square meter or 2.90×10^{-9} pounds per square inch
$\log$ = logarithm to the base 10

The value of p_0 has been chosen to approximate the lowest RMS sound pressure a healthy young adult can hear. Substituting this barely audible RMS sound pressure for p_{rms} in Eq. (14-2) produces a sound pressure level of 0 dB. In contrast, an RMS sound pressure 1 million times greater produces a sound pressure level of 120 dB. Most sounds in our day-to-day environment have sound pressure levels on the order of 30 to 100 dB. Two useful rules of thumb for comparing sound pressure levels are that, on an average, people perceive a 6 to 10 dB increase in the sound pressure level as a doubling of subjective loudness and changes of less than 2 or 3 dB are not readily detectable outside of a laboratory environment.

The A-Weighted Sound Level

Another important attribute of sound is its frequency, or *pitch*. For a pure tone this is the number of times per second the sound pressure oscillates back and forth about atmospheric pressure. The unit of frequency is hertz (Hz) but may also be referred to as cycles per second in references predating the adoption of hertz as an international standard. Virtually all sounds contain energy across a broad range of frequencies. Even a single note of a musical instrument contains a fundamental frequency plus a number of overtones.

The normal frequency range of hearing for a young adult extends from a low of 16 Hz to a high of about 16,000 Hz. However, the human auditory system is not equally sensitive across this entire range. Frequencies in the range of 2000 to 4000 Hz sound louder than lower or higher frequencies when heard at the same RMS sound pressure level. Thus, it is possible for two different sounds with the same sound pressure level to sound different in loudness.

For this reason the *A-weighted sound level* (A-level) was developed. Incorporated in almost every commercially available sound level meter, a standardized A-weighting filter adds gain or attenuation to different frequencies in a manner approximating the sensitivity of the human ear. The frequency response of the filter has a ±3 dB effect in the midfrequency range between 500 and 10,000 Hz and increasing attenuation outside this range. Although the A-weighting filter is only an approximation to a complex physiological process, one sound judged louder than another will generally have a higher A-weighted sound level. Similarly, two sounds judged equally loud will generally have nearly the same A-weighted sound levels. A range of commonly encountered A-weighted sound levels is shown in Fig. 14-2.

For environmental assessment purposes, the A-weighted sound level represents a significant improvement over the overall (unweighted) sound pressure level. Unweighted sound pressure levels are rarely, if ever, used in environmental analyses. All federal agencies dealing with community noise, including transportation, have adopted the A-weighted sound level as the basic unit for analysis of environmental impacts.

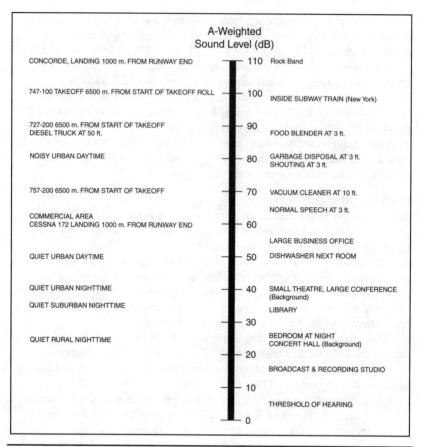

A-Weighted Sound Level (dB)

Left	dB	Right
CONCORDE, LANDING 1000 m. FROM RUNWAY END	110	Rock Band
747-100 TAKEOFF 6500 m. FROM START OF TAKEOFF ROLL	100	INSIDE SUBWAY TRAIN (New York)
727-200 6500 m. FROM START OF TAKEOFF DIESEL TRUCK AT 50 ft.	90	FOOD BLENDER AT 3 ft.
NOISY URBAN DAYTIME	80	GARBAGE DISPOSAL AT 3 ft. SHOUTING AT 3 ft.
757-200 6500 m. FROM START OF TAKEOFF	70	VACUUM CLEANER AT 10 ft.
COMMERCIAL AREA CESSNA 172 LANDING 1000 m. FROM RUNWAY END	60	NORMAL SPEECH AT 3 ft.
QUIET URBAN DAYTIME	50	LARGE BUSINESS OFFICE DISHWASHER NEXT ROOM
QUIET URBAN NIGHTTIME	40	SMALL THEATRE, LARGE CONFERENCE (Background)
QUIET SUBURBAN NIGHTTIME		LIBRARY
	30	
QUIET RURAL NIGHTTIME		BEDROOM AT NIGHT CONCERT HALL (Background)
	20	
		BROADCAST & RECORDING STUDIO
	10	
		THRESHOLD OF HEARING
	0	

FIGURE 14-2   Common environmental A-weighted sound levels in decibels (*Harris Miller Miller Hanson*).

A-weighted sound levels are measured in decibels as are unweighted sound pressure levels and several other metrics discussed in this chapter. Thus, the measurement units themselves do not identify the quantity being reported. To avoid ambiguity the quantity, in this case the A-weighted sound level, should always be reported along with the units. An example would be an A-weighted sound level of 85 dB. [Although not meeting current acoustical terminology standards, A-weighted sound levels may be reported in the literature as dBA, dB(A), or simply *A-weighted*.]

Maximum A-Weighted Sound Level

In addition to sound *level*, another important dimension to environmental sound is its variation over time. For example, a distant highway with relatively steady traffic produces a fairly continuous background sound level with moment-to-moment variations of only

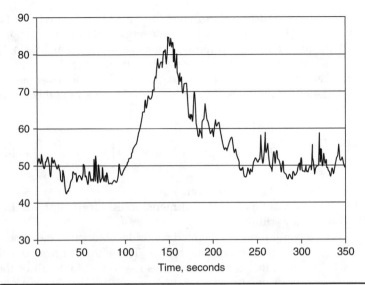

FIGURE 14-3   Typical A-weighted sound level time history of an aircraft pass-by (*Harris Miller Miller Hanson*).

a few decibels. In contrast, an aircraft pass-by produces a distinct, transient noise event. During an aircraft pass-by, the sound level emerges out of the fluctuating background environment, continues to increase until the aircraft passes the observer, and then decreases to blend in with the background as the aircraft recedes into the distance. Figure 14-3 illustrates this phenomenon.

For reporting as well as comparison purposes it is desirable to use a single number for describing the sound level of such a noise event. A convenient metric is the maximum A-weighted sound level. This value is convenient to measure as it requires an observer to simply note the maximum reading on a sound level meter. It is also convenient to describe since most people can relate to the loudest part of a noise event. In Fig. 14-3 the maximum A-weighted sound level is 85 dBA.

Sound Exposure Level

While being a very useful metric of aircraft noise events, the maximum level does not address the time element, or *duration*, of the event. During the late 1960s and early 1970s several psychoacoustic listening studies were conducted to investigate how people assessed the relative noisiness of noise events with differing durations. All other things being equal, it was found that increased duration resulted in greater perceived noisiness. On average, the studies determined that people were willing to trade a doubling of duration for a 3-dB reduction in maximum A-level sound. This finding supported a simple model for subjective noisiness, noise events with equal time-integrated A-weighted sound energy are rated as equally noisy.

Developed to address this finding, the *sound exposure level* (SEL) is defined as the total A-weighted sound energy contained in the noise event. Its description as a continuous integral is shown in Eq. (14-3). The units are decibels. Theoretically, this integral could approach infinity as T becomes large. If the integration takes place over the top 20 dB, the computation will be only 0.1 dB less than the theoretical maximum.

$$L_{AE} = 10 \log\left(\frac{1}{T_0}\int_{t=0}^{T} 10^{L_{At}/10}\,dt\right) \tag{14-3}$$

where L_{AE} = sound exposure level

$T_0 = 1$ s to maintain a dimensionless argument for the logarithm

L_{At} = continuous A-weighted sound level function describing the noise event time history. The limits of t from 0 to T are sufficient to encompass the top 10 to 20 dB of the noise event.

For measurement purposes, the continuous integral presents two difficulties, namely, a continuous, mathematical function for the A-weighted sound level time history is never known, and the time limits of integration are nebulous since there is no precisely defined beginning or end to an aircraft noise event which slowly emerges from, and then blends back, into a time-varying background. These difficulties are circumvented using discrete samples of A-weighted sound level, the summation approximation of Eq. (14-4) and empirically derived guidelines for the limits of i from 1 to N.

$$L_{AE} = 10 \log\left(\frac{1}{T_0}\sum_{i=1}^{N} 10^{L_{A,i}/10}\,\Delta t\right) \tag{14-4}$$

where $L_{A,i}$ is the instantaneous, ith A-weighted sound level measured every 0.5 s and Δt is 0.5 s. The limits of i from 1 to N are sufficient to perform the summation over at least the top 10 dB of the noise event.

An accepted sampling interval Δt is 0.5 s. If the summation starts, i equal to 1, with the first sample to come within 20 dB of the maximum value and continues until the last sample, i equal to N, is within 20 dB of the maximum, this approximation will be about 0.1 dB lower than the theoretical value. If the summation is performed only over the top 10 dB of the time history, the discrepancy will be less than 1 dB. This discrete summation approximation to the integral is used in all sound level meters and monitoring devices. The use of Eq. (14-4) in computing SEL is illustrated in Example Problem 14-1.

Example Problem 14-1   The following sample of A-weighted sound levels was measured at 0.5-s intervals during an aircraft flyover: 64.5, 66.7, 67.1, 69.2, 71.3, 73.2, 74.1, 75.6, 77.8, 79.1, 78.6, 77.2, 75.7, 74.5, 72.6, 71.1, 69.7, 68.6, 68.0, and 66.4 dB.

To determine the SEL of this noise event, we must substitute into Eq. (14-4) obtaining

$$L_{AE} = 10 \log\left[\frac{1}{1}(10^{64.5/10} + 10^{66.7/10} + \cdots + 10^{66.4/10})0.5]\right]$$

$$= 10 \log(253,734,091) = 84.0 \text{ dB}$$

It will be observed that even though none of the individual sound events had an A-weighted sound level in excess of 79.1 dB the effect of the duration of the noise event leads to a numerically higher value for the SEL.

Because of the sound level durations involved with typical aircraft pass-bys, the SEL will always be numerically larger than the maximum A-weighted sound level of the event. For most aircraft overflights, the difference is on the order of 7 to 12 dB. Factors affecting this difference are aircraft speed (the greater the speed, the smaller the difference) and the distance to the aircraft at its closest point of approach to the observer (the greater the distance, the greater the difference).

Equivalent Steady Sound Level

The preceding discussion focused on measures of sound associated with individual events. However, it is frequently necessary to quantify sound levels over longer periods of time, such as an hour, several hours, or even a day. Such needs arise when tracking diurnal patterns, describing cumulative exposure over intermediate exposure periods (such as during school or office hours), or cumulative exposure over a 24-h day. In contrast to the energy *summation* metrics used for individual noise events, energy *average* metrics are used for longer time periods. One such metric is the *equivalent steady sound level* (QL). Although not meeting current acoustical terminology standards, the equivalent steady sound level reported in the literature as *energy average sound level*, L_{eq} or LEQ and the units are decibels. Mathematically, the equivalent steady sound level is the sound pressure level shown in Eq. (14-2) calculated using a long-term RMS sound pressure (T in Eq. (14-1) equals to the time period of interest). As a practical matter, the equivalent steady sound level is almost always calculated from a time series of A-weighted sound levels acquired with a sound level meter with readings taken at 0.5 s intervals or less. Equation (14-5) shows the manner in which the equivalent steady sound level is computed from a discrete time series of data.

$$L_{eq} = 10 \log\left(\frac{1}{T}\sum_{i=1}^{N} 10^{L_{A,i}/10}\Delta t\right) \qquad (14\text{-}5)$$

where L_{eq} = equivalent steady sound level
$L_{A,I}$ = instantaneous ith A-weighted sound level measured every 0.5 s
T = time period of interest (e.g., 1 h)
Δt = typically 0.5 s or less
$N = T/\Delta t$, where T and Δt must be in the same units

An equivalent and computationally more efficient manner of expressing Eq. (14-5) is

$$L_{eq} = 10 \log\left(\frac{1}{N}\sum_{i=0}^{N}10^{L_{A,i}/10}\right) \tag{14-6}$$

where $L_{A,i}$ is the instantaneous ith A-weighted sound level measured every 0.5 s and N is total number of sound level samples.

The computational process described in Eqs. (14-5) and (14-6) does not make any distinction between sources, that is, it accumulates sound levels produced by both aircraft and nonaircraft sources. When QL is computed in this manner it is called *total* QL. However, it is often useful to know only the aircraft component. The aircraft component can be calculated from the sound exposure levels of individual events using Eq. (14-7).

$$L_{eq} = 10 \log\left(\frac{1}{T}\sum_{j=1}^{M}10^{L_{AE,j}/10}\right) \tag{14-7}$$

where $L_{AE,j}$ = sound exposure level produced by the jth aircraft pass-by during the time period
T = time period of interest (e.g., 1 h) measured in seconds
M = number of aircraft noise events during the period T

Functionally, this equation accumulates all of the aircraft sound energy from multiple events, then spreads it out uniformly over the time period by dividing by the length of the period (not just the length of time that aircraft were present).

The computation of hourly average sound level is illustrated in Example Problem 14-2.

Example Problem 14-2 The following sound exposure levels for four aircraft flyovers were measured in a 1-h period: 84.0, 89.1, 90.2, and 86.6 dB.

To compute the hourly average sound level, we must substitute into Eq. (14-7) obtaining

$$L_{eq} = 10 \log\left[\frac{1}{3600}(10^{84.0/10} + 10^{89.1/10} + 10^{90.2/10} + 10^{86.6/10})\right]$$

$$= 10 \log 713.339 \approx 58.5 \text{ dB}$$

It will be observed that even though the sound exposure level of each aircraft flyover was greater than 58.5 dB the averaging process reduces the hourly average sound exposure level.

Experience has shown the concept of an *average* sound level is often misinterpreted by the affected public as an underreporting or understatement of their noise environment. Their concern is that the metric does not report the *total* noise energy over the time period. As can be seen in Eqs. (14-5), (14-6), and (14-7), this metric, as well as other *average* metrics, does indeed include all of the noise energy. Each and every noise event, no matter how high or low the sound level, increases the value of the metric. Viewed another way, the average value is the total noise energy adjusted by a constant, 10 log *T*.

The local community component of the equivalent steady sound level is also a frequently reported statistic which serves as a basis of comparison for the aircraft component. It may be estimated using a variant of Eq. (14-6) which accumulates sound levels only during subintervals of the total period when no aircraft are present. It is an *estimate* because there is no way of knowing the community sound level contribution during periods when aircraft are present. Equation (14-8) shows the basic summation process. Each summation in the equation represents a nonaircraft subinterval.

$$L_{eq} = 10 \log \left[\frac{1}{N} \left(\sum_{i=1}^{N_1} 10^{L_{A,i}/10} + \sum_{i=1}^{N_2} 10^{L_{A,i}/10} + \cdots + \sum_{i=1}^{N_n} 10^{L_{A,i}/10} \right) \right] \quad (14\text{-}8)$$

where $L_{A,i}$ = instantaneous ith A-weighted sound level measured every 0.5 s

N_1, N_2, N_n = number of sound level samples in each subinterval containing no aircraft noise

N = total number of samples which is equal to ($N_1 + N_2 + \cdots + N_n$)

This metric is referred to as the hourly average sound level when 1 h of averaging time is used. (Although not meeting current acoustical terminology standards, the hourly average sound level may be reported in the literature as *the hourly noise level, HNL, hourly L_{eq}* or *1-h L_{eq}*.) In airport applications, hourly average sound levels may be used for plotting and visualizing diurnal trends. Eight and twenty-four hour periods are referred to as 8-h and 24-h average sound levels. The symbol L_{eq} is generic referring to any arbitrary period of time. To avoid ambiguity, the subscript is replaced by the appropriate time frame. Thus, the symbols L_{1h}, L_{8h}, and L_{24h}, are used for 1-, 8-, and 24-h periods, respectively.

Day-Night Average Sound Level

As the name implies, the day-night average sound level, DNL, is a metric used to describe sound exposure over a 24-h period and the units are decibels. Computationally it is identical to the 24-h average sound level with one important difference. The DNL incorporates a time-of-day weighting which adds 10 dB to sound levels occurring between 10 p.m. and 7 a.m. While the magnitude of the weighting periodically becomes a topic of discussion within the scientific community, the intent is to account for a presumed increase in human sensitivity to noise during nighttime hours. While the formal definition is a continuous integral, Eq. (14-9) shows the formula for computing the total (aircraft plus community sources) DNL from discrete samples of the A-weighted sound level.

$$L_{dn} = 10 \log\left(\frac{1}{86,400} \sum_{i=1}^{N} 10^{(L_{A,i}+W_i)/10} \Delta t \right) \tag{14-9}$$

where L_{dn} = day-night average sound level for 1 day
   $L_{A,i}$ = instantaneous ith A-weighted sound level measured every 0.5 s
   86,400 = number of seconds in a day
   W_i = time-of-day weighting for the ith A-weighted sound level (0 dB if it occurred between 7 a.m. and 10 p.m., 10 dB if it occurred between 10 p.m. and 7 a.m.)
   Δt = typically 0.5 s or less and the units must be in seconds
   N = equal to 86,400/Δt

The aircraft component of DNL may be computed from sound exposure levels of individual events using Eq. (14-10).

$$L_{dn} = 10 \log\left(\frac{1}{86,400} \sum_{j=1}^{M} 10^{(L_{AE,j}+W_j)/10} \right) \tag{14-10}$$

where $L_{AE,j}$ = sound exposure level produced by the jth aircraft pass-by during the day
   W_j = time-of-day weighting for the jth aircraft pass-by (0 dB if it occurred between 7 a.m. and 10 p.m., 10 dB if it occurred between 10 p.m. and 7 a.m.)
   M = number of aircraft noise events during 24-h period

The application of this equation to determine the DNL of several aircraft flyovers at various times during the day is illustrated by Example Problem 14-3.

Example Problem 14-3   The following sound exposure levels of five aircraft flyovers were measured over the course of a 24-h period: 81.2 dB at 6:03 a.m., 95.1 dB at 10:32 a.m., 79.2 dB at 2:15 p.m., 88.8 dB at 7:33 p.m., and 71.2 dB at 10:05 p.m.

To compute the DNL, we must substitute into Eq. (14-10). The sound exposure levels at 6:03 a.m. and 10:05 p.m. must be increased by the time of day weighting of 10 dB since these flyovers occurred between 10:00 p.m. and 7:00 a.m.

$$L_{dn} = 10 \log \left[\frac{1}{86,400} (10^{91.2/10} + 10^{95.1/10} + \cdots + 10^{81.2/10}) \right]$$

$$= 10 \log 63,978.85 = 48.1 \text{ dB}$$

To find the aircraft which has the greatest and least contribution to the day-night average sound level the $(L_{AE,j} + W_j)/10$ value of the quantity 10 must be evaluated for each aircraft. Clearly, by adding the time of day weighting, we see that the aircraft flyover at 10:32 a.m. is the greatest contributor and the aircraft flyover at 2:15 is least contributor to the day-night average sound exposure level.

A useful rule of thumb for estimating the contribution of DNL to a single daytime (7 a.m. to 10 p.m.) noise event may be obtained by simplifying Eq. (14-10) for the condition, where M is equal to 1. The approximation shown in Eq. (14-11) is accurate to within 0.5 dB.

$$L_{dn} \approx L_{AE} - 50 \tag{14-11}$$

where L_{AE} is the sound exposure level of a single aircraft pass-by.

The use of Eqs. (14-10) and (14-11) to compute the DNL of a single daytime noise event is illustrated by Example Problem 14-4.

Example Problem 14-4 Let us determine the DNL produced by a single daytime noise event with a sound exposure level of 105 dB.

Using Eq. (14-10), we have

$$L_{dn} = 10 \log \left[\frac{1}{86,400} (10^{105.0/10}) \right]$$

$$= 10 \log 366,004 = 55.6 \text{ dB}$$

Using Eq. (14-11), we have

$$L_{dn} \approx 105 - 50 \approx 55 \text{ dB}$$

If this noise event were added to the noise events in Example Problem 14-3, we would find that the DNL was increased to 56.3 dB or there would be an increase of 0.7 dB.

Environmental reporting criteria often require annual average values of DNL. Both airport and atmospheric factors contribute to day-to-day variability in the DNL observed at a particular location near the airport. In cases where average values must be computed from measurements, the averaging must be done on a sound energy basis. Equation (14-12) shows the formula for computing the annual average value.

$$L_{dn,\text{annual}} = 10 \log\left(\frac{1}{365}\sum_{i=1}^{365}10^{L_{dn,i}/10}\right) \qquad (14\text{-}12)$$

where $L_{dn,i}$ is the DNL for the ith day of the year.

This equation assumes 365 individual DNL values are to be used in the averaging process. For conditions where the number of days differs from 365 (leap years, missing data, etc.) the available number of data points should be used in the summation and the number 365 replaced by the actual number of data points used.

Representative values of DNL range from a low of 40 to 45 dB in extremely quiet isolated locations to highs of 80 or 85 dB immediately adjacent to a busy truck route or just off the end of a runway at an active military air base. The U.S. Environmental Protection Agency (EPA) identified this measure as the most appropriate means of evaluating community (including aircraft) noise in 1974 [28]. Most other public agencies dealing with noise exposure, including the FAA, the Department of Defense, and the Department of Housing and Urban Development (HUD), have also adopted DNL in their guidelines and regulations.

Time Above Threshold Level

The preceding metrics quantify noise exposure in terms of sound level or sound energy. An alternate descriptor uses duration, or time, as the basic metric. The metric is *time above* (TA), defined as the length of time that the A-weighted sound level exceeds a specified *threshold* level over a given period of time. Typically TA is reported as the numbers of minutes per day that the A-weighted sound level exceeds values of 55, 65, 75, 85, 95, and 105 dB. The historical appeal of TA has generally been one of simplicity, that is, TAs are arithmetically additive. TAs for single noise events can be arithmetically added to compute hourly TAs, and hourly TAs can be arithmetically added to form 24-h values. Proponents of TA argue that the arithmetic addition process allows easy-to-understand assessments of major contributors to 24-h totals. TA may be required for some environmental analyses. However, at the present time there are no accepted criteria or land-use compatibility guidelines using TA.

Other Single-Event Sound-Level Metrics

The *perceived noise level* (PNL) and *effective perceived noise level* (EPNL) are quantities similar to A level and sound exposure level, respectively. They were developed specifically to correlate with subjective response to aircraft sound. The perceived noise level, in units of PNdB, is a quantity which varies from moment to moment, just like the A-weighted sound level. As a general rule, the perceived noise level is approximately 13 dB greater than the A-weighted sound level.

The effective perceived noise level, in units of EPNdb, is a single event metric which sums the perceived noise level in a manner similar to the way SEL sums A-level. EPNL, however, also incorporates a tone correction adjustment to account for the increased subjective noisiness of sounds containing discrete frequency tones (like those produced by turbofan engine compressor blades). The formula for computing EPNL is shown in Eq. (14-13) [39].

$$L_{EPN} = 10 \log \left(\frac{1}{T_0} \sum_{i=1}^{N} 10^{(L_{PN,i} + TC_i)/10} \Delta t \right) \qquad (14\text{-}13)$$

where L_{EPN} = effective perceived noise level
   $L_{PN,i}$ = instantaneous ith perceived noise level measured every 0.5 s
   TC_i = instantaneous ith tone correction
   Δt = 0.5 s
   T_0 = 10 s. The limits of i from 1 to N are sufficient to perform the summation over the top 10 dB of the noise event.

Both the PNL and the tone correction are computed from sound pressure levels measured in individual one-third octave bands from 50 to 10,000 Hz. The magnitude of the tone correction ranges from 0 to 6 dB depending upon the frequency where the tone occurs and the sound pressure level of the tone relative to the broadband noise in the same frequency range. As a general rule, the EPNL is about 3 dB greater than SEL but can be more if very noticeable pure tones are present, or less at very large distances.

Because of the complexity involved in measurement, sophisticated frequency analyses and nonlinear amplitude adjustments are required, they are not used in the United States for routine environmental analyses. Their current use is limited to aircraft airworthiness certification under Federal Aviation Regulations, Part 36 [39].

Other 24-h Sound-Level Metrics

The *community noise equivalent level* (CNEL) adopted in California airport noise standards was actually a forerunner of DNL. The computation procedure is virtually identical to DNL. Equations (14-9) and (14-10) can be used to compute CNEL, the only difference is the use of 3 weighting periods instead of 2. For 7 a.m. to 7 p.m. the weighting is 0 dB, for 7 p.m. to 10 p.m. the weighting is 4.77 dB (the actual weighting is a factor of 3 in sound energy and 10 log 3 or 4.77 dB), and for 10 p.m. to 7 a.m. the weighting is 10 dB. The only difference between the two metrics is the approximately 5 dB weighting during the three evening hours from 7 p.m. to 10 p.m. Numerically, CNEL is always greater than DNL but from a practical standpoint this difference is rarely more than 1 dB.

Before the adoption of DNL, two other descriptors of daily noise exposure were used to quantify noise impacts around airports. Neither is still in active use in the United States today. The *composite noise rating* (CNR) was one of the first 24-h metrics to embody individual aircraft sound levels, their frequency of occurrence, and their time-of-day in a single number rating. Predating the development and use of personal computers by more than two decades, CNR calculations were performed using a handbook procedure published in 1963 under a joint effort by the U.S. Air Force and the FAA. CNR used the maximum perceived noise level as the single event sound level descriptor.

The *noise exposure forecast* (NEF) was developed in 1967 and quickly replaced CNR. The NEF uses EPNL as the single event sound level descriptor and a sound energy summation process similar to DNL. Equation (14-14) shows the formula for computing NEF. Because of the computational complexities involved in their underlying single event metrics, both CNR and NEF fell into disuse with the adoption of DNL.

$$\text{NEF} = 10 \log \left(\sum_{j=1}^{M} 10^{(L_{\text{EPN},j} + W_j)/10} \right) - 88 \qquad (14\text{-}14)$$

where $L_{\text{EPN},j}$ = EPNL produced by the jth aircraft pass-by during the day

W_j = time-of-day weighting for the jth aircraft pass-by (0 dB if it occurred between 7 a.m. and 10 p.m., 12 dB if it occurred between 10 p.m. and 7 a.m.)

88 = adjustment factor designed to shift the metric to a lower numeric range not occupied by any other then-current 24-h metric

Because of differences in frequency weightings, differences in accounting for the durations of individual events, and differences in the evening and nighttime weightings, there is no exact functional relationship between these three metrics. Within ±3 dB, however, the relationship shown in Eq. (14-15) has been found to be valid. Thus, for DNL and CNEL values of 65, an NEF value of 30, and a CNR value of 100, all indicate approximately the same degree of noise exposure, within ±3 dB.

$$L_{dn} \approx \text{NEF} + 35 \approx \text{CNR} - 35 \qquad (14\text{-}15)$$

Aircraft Noise Effects and Land-Use Compatibility

The effects of noise on people can be classified into one of two categories, namely, behavioral effects and health or physiological effects. Behavioral effects are those that are associated with activity interference. These effects include annoyance, interference with communication, mental activity, rest, and sleep. Health effects are those that produce hearing loss or nonauditory effects such as cardiovascular disease and hypertension.

Various federal agencies have developed guidelines for assessing the compatibility of noise with land uses, including the EPA, HUD, and the FAA. All of the guidelines are based on the day-night average sound level (DNL) and were designed to protect public health and welfare, but also take into account the feasibility of controlling noise [28, 44].

Speech Interference

One of the primary effects of aircraft noise is its tendency to drown out or *mask* speech, making it difficult or impossible to carry on a normal conversation without interruption. The sound level of speech decreases as distance between a talker and listener increases. As the level of speech decreases in the presence of background noise, it becomes harder and harder to hear. Figure 14-4 presents typical distances between a talker and listener for satisfactory outdoor conversations in the presence of different steady A-weighted background sound levels for three degrees of vocal effort, namely, raised, normal, and relaxed. As the background level increases, the individuals must either talk louder or must get closer together to continue their conversation.

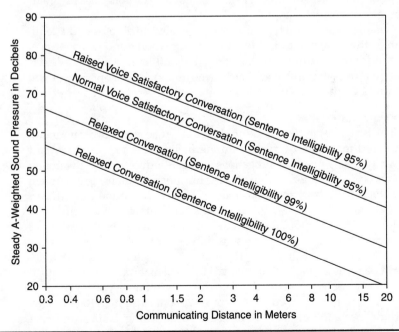

Figure 14-4 Maximum distances outdoors over which conversation is satisfactorily intelligible outdoors in a steady noise environment (*Adapted from Environmental Protection Agency [28]*).

As indicated in the figure, *satisfactory conversation* does not always require hearing every word, 95 percent intelligibility is acceptable for many conversations. This is because a few unheard words can be inferred when they occur in a familiar context. However, for relaxed conversation, people have higher expectations of hearing speech and require complete 100 percent intelligibility. Any combination of talker-listener distances and background noise that falls below the bottom line in Fig. 14-4, thus ensuring 100 percent intelligibility, represents an ideal environment for outdoor speech communication and is considered necessary for acceptable indoor conversation as well.

One implication of the relationships in Fig. 14-4 is that for typical communication distances of 3 or 4 ft (1 to 1.5 m), acceptable outdoor conversations where 95 percent intelligibility is acceptable can be carried on in a normal voice as long as the background A-weighted sound level is less than about 65 dB. In other situations, where greater speech intelligibility is required, background levels must be lower. For example, indoors, where 100 percent intelligibility is desired, the background A-weighted sound level must be less than about 45 dB. If the noise exceeds either of these levels, as might occur when an aircraft passes overhead, intelligibility is lost unless vocal effort is increased or communication distance decreased.

A second implication of these relationships is that an acceptable A-weighted background level of 60 to 65 dB outdoors does not guarantee an acceptable background level indoors. This is because most housing construction typically provides about 15 dB of sound attenuation from outside to inside the building when windows are open. Thus, only if the outdoor A-weighted sound level is 60 dB or less is there a reasonable chance that the resulting indoor sound level will afford acceptable conversation inside.

Sleep Interference

The disruptive effects of noise on sleep can be of concern in communities exposed to aircraft overflights during nighttime hours. Over the past two decades, many investigations of noise-induced sleep disruption have been conducted worldwide. The functional relationship most often evaluated has been the probability of a sleep disruption created by a single noise intrusion of a given sound level. Four major research review studies [10, 32, 37, 46] all support the same general finding that increased sound exposure level (SEL) results in higher probabilities of sleep disruption. Currently, however, there are no guidelines or acceptability criteria for assessing the cumulative impact of many aircraft noise intrusions of various SELs over the course of a nighttime sleeping period.

The aforementioned reviews provide some insight as to why such guidelines and criteria have not been forthcoming. Data acquisition methodologies, the treatment of mitigating variables, and the choice of both dose and disruption metrics are far from standardized, even for studies limited to short-term, transient noise events similar to those produced by aircraft overflights.

Despite all of this variability, first order dose-response curves have been developed which attempt to relate the probability of an arousal, either a sleep stage change or an actual awakening, to the sound exposure level of a single noise event [11, 26]. At the present time caution should be exercised in drawing inferences from such curves. Among other things, the preponderance of underlying data generally represents laboratory listening conditions. Thus the curves could be expected to better predict reaction to new and unfamiliar sounds rather than older, more familiar ones. When compared with in-laboratory studies, the limited data available from in-home investigations using familiar sound sources suggests that arousal probabilities may be on the order of only one-eighth those observed from unfamiliar sources. In addition to these adaptation issues, uncertainty still remains on important questions such as the cumulative effects of multiple noise intrusions, the effects of noise on falling asleep as opposed to awakening, and the extent to which sleep deprivation represents a quantifiable physiological problem.

Community Annoyance

Social survey data have long made it clear that individual reactions to noise vary widely for a given 24-h average sound level. As a group, however, the aggregate response of people to factors such as speech and sleep interference and desire for an acceptable environment is predictable and relates well to measures of cumulative noise exposure such as DNL. Figure 14-5 shows the most widely recognized

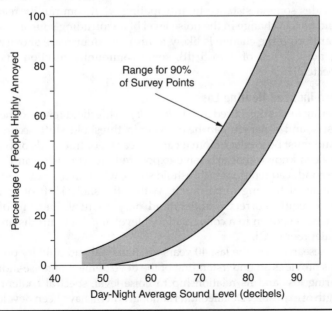

FIGURE 14-5 Percentage of people highly annoyed as a function of day-night average sound level (*Adapted from Schultz, Journal of the Acoustical Society of America [49]*).

relationship between day-night average sound level and the percent-age of people highly annoyed, regardless of the noise source. Based on data from 18 studies of the attitudes of people toward noise conducted worldwide, the curve indicates that the relationship between group reaction and 24-h exposure is quantifiable. The curve shows that at DNLs as low as 55 dB, approximately 5 percent of the people will still be highly annoyed with their noise environment. The percentage increases more rapidly as the DNL increases above 65 dB [49].

Separate work by the EPA suggests that overall community reaction to a noise environment is also dependent on the level of the intruding noise as compared with the level of the existing noise. Research was conducted to determine the relationship between intruding noise level and community reaction for 55 cases of community noise intrusion where reactions were known [16]. The data were normalized to the same set of conditions so that the cases were comparable. In particular, the conditions were adjusted to an existing noise environment, without the intruding noise, of about 60 dB. The data show that sporadic complaints occurred when the intruding equivalent noise level was between about 59 to 65 dB, and wide-spread complaints occurred when the intruding noise fell between about 63 and 75 dB.

The implication of this research is that complaints may begin to occur when aircraft DNL is approximately equal to the background DNL, and that widespread complaints start to occur when the aircraft DNL exceeds the background DNL by 3 to 5 dB. Such a conclusion provides some assistance in anticipating what community reaction could be to a change in the noise level of an intruding source, such as an airport. If the change is likely to result in an increase above existing noise levels of 3 to 5 dB, some community reaction may be expected [16].

Noise-Induced Hearing Loss

Hearing loss is measured as *threshold shift*. Threshold refers to the quietest sound a person can hear. When a threshold shift occurs, the sound must be louder before it can be heard. For hundreds of years it has been known that excessive exposure to loud noises can lead to noise-induced temporary threshold shifts, which in time can result in permanent hearing impairment. With a threshold shift of 25 dB a person could correctly understand only about 90 percent of the sentences spoken in a conversational level at a 3 ft (1 m) distance in a quiet room [42].

Research over the last 40 years on industrial and military populations provides an understanding of the development of noise-induced hearing loss and its relationship to noise level, spectral content and length of exposure. Detailed international criteria have been developed that identify maximum noise exposures that do not produce noise-induced hearing loss in any segment of the population exposed [1].

The U.S. Occupational Safety and Health Administration (OSHA) regulation [41] identifies the maximum permissible A-weighted sound exposure of 90 dB for 8 h.

It is extremely unlikely that aircraft noise around airports could ever produce hearing loss. For example, it would take continuous exposure to more than 1000 overflights per day with an SEL of 100 dB each to produce a time-weighted average sound level of 85 dB. If this occurred 5 days a week for 40 years, and if people were exposed to this outdoors without any attenuation from buildings, the resultant noise exposure would start to produce a *noise-induced permanent threshold shift* (NIPTS) of less than 10 dB in the most sensitive 10 percent of the population.

Nonauditory Health Effects

Concern is often raised that noise has adverse effects on human health other than hearing. In spite of considerable worldwide research, however, no unambiguous scientific evidence to relate quantitatively any noise environment with the origin of or contribution to any clinical nonauditory disease. Even the most recent research, conducted at levels above the limits for conservation of hearing, failed to give consistent results. Most authoritative reviews, such as the World Health Organization Environmental Health Criteria Document on noise [33], agree that "research on this subject has not yielded any positive evidence, so far, that disease is caused or aggravated by noise exposure, insufficient to cause hearing impairment." For practical noise control considerations, the present status of our knowledge means that by using criteria that prevent noise induced hearing loss, minimize speech and sleep disruption, and minimize community reactions and annoyance, any effects on health will also be prevented. In general, these guidelines should not be regarded as identifying levels of exposure that are desirable but rather as a balancing of what is desirable with what is feasible.

Noise and Land-Use Compatibility Guidelines

Based on the relationships between noise and the collective response of people to their environment, DNL has become accepted as a standard for evaluating community noise exposure and as an aid in decision-making regarding the compatibility of alternative land uses.

In their application to airport noise in particular, DNL projections have two principal functions:

1. To provide a means for comparing existing noise conditions with those that might result from the implementation of noise abatement procedures or from forecast changes in airport activity

2. To provide a quantitative basis for identifying and judging potential noise impacts

Both of these functions require the application of objective criteria. Government agencies dealing with environmental noise have devoted significant attention to this issue and have developed noise and land-use compatibility guidelines to help federal, state, and local officials with the noise evaluation process.

In FAR Part 150 [6], which defines procedures for developing airport noise compatibility programs, the FAA has established DNL as the official cumulative noise exposure metric for use in airport noise analyses, and has developed guidelines for noise and land-use compatibility evaluation. Table 14-2 presents these guidelines.

The guidelines represent a compilation of extensive scientific research into noise-related activity interference and attitudinal response. However, reviewers of DNL contours should recognize the highly subjective nature of response to noise and the special circumstances that can either increase or decrease the tolerance of an individual. For example, a high nonaircraft background or ambient noise level, such as from ground traffic, can reduce the significance of aircraft noise. Alternatively, residents of areas with unusually low background levels may find relatively low levels of aircraft noise very annoying. Response may also be affected by expectation and experience. People often get used to a level of noise exposure that guidelines suggest may be unacceptable, and similarly, changes in exposure may generate a response that is far greater than that which the guidelines might suggest.

Finally, the cumulative nature of DNL means that the same level of noise exposure can be achieved in an essentially infinite number of ways. For example, a large increase in relatively quiet flights can counter balance a smaller reduction of relatively noisy operations, with no net change in DNL. The increased frequency of operations can annoy residents, despite the apparent unchanged status quo of the noise. With these cautions in mind, the guidelines of the FAA for compatible land use can be combined with DNL contours indicating points of equal exposure to identify the potential types and locations of land uses and the degree of their incompatibility. Note that, by these guidelines, all land uses are considered compatible with aircraft day-night average sound levels below 65 dB. This does not mean that people will not complain or otherwise be disturbed by aircraft noise at lower levels, as has been shown earlier, nor does it preclude individual communities or other jurisdictions from adopting lower standards to meet local needs.

Determining the Extent of the Problem

The extent of a potential or ongoing airport noise problem is generally quantified in one of two ways:

1. Prediction using computer-based simulation models

2. Measurement through portable or permanent monitoring systems

| | Yearly Day-Night Average Sound Level, DNL, dB | | | | | |
	Below 65	65–70	70–75	75–80	80–85	Over 85
Residential Use						
Residential other than mobile homes and transient lodgings	Y	N	N	N	N	N
Mobile home park		Y	N	N	N	N	N
Transient lodgings		Y	N	N	N	N	N
Public Use						
Schools		Y	N	N	N	N	N
Hospitals and nursing homes	Y	25	30	N	N	N
Churches, auditoriums, and concert halls	Y	25	30	N	N	N
Governmental services	Y	Y	25	30	N	N
Transportation		Y	Y	Y	Y	Y	Y
Parking		Y	Y	Y	Y	Y	N
Commercial Use						
Offices, business and professional	Y	Y	25	30	N	N
Wholesale and retail—building materials, hardware, and farm equipment	Y	Y	Y	Y	Y	N
Retail trade—general	Y	Y	25	30	N	N
Utilities		Y	Y	Y	Y	Y	N
Communication		Y	Y	25	30	N	N
Manufacturing and Production						
Manufacturing general	Y	Y	Y	Y	Y	N
Photographic and optical	Y	Y	25	30	N	N
Agriculture (except livestock) and forestry	Y	Y	Y	Y	Y	Y
Livestock farming and breeding	Y	Y	Y	N	N	N
Mining and fishing, resource production and extraction		Y	Y	Y	Y	Y	Y

Table 14-2   FAA Noise and Land-Use Compatibility Guidelines

	Yearly Day-Night Average Sound Level, DNL, dB					
	Below 65	65–70	70–75	75–80	80–85	Over 85
Recreational						
Outdoor sports arenas and spectator sports		Y	Y	Y	N	N	N
Outdoor music shells, amphitheaters	Y	N	N	N	N	N
Nature exhibits and zoos	Y	Y	N	N	N	N
Amusements, parks, resorts and camps	Y	Y	Y	Y	Y	Y
Golf courses, riding stables and water recreation		Y	Y	25	30	N	N

Notes: Y(Yes) Land use and related structures compatible without restrictions
N(No) Land use and related structures are not compatible and should be prohibited
25, 30, or 35 Land use and related structures generally compatible; measures to achieve outdoor-to-indoor Noise Level Reduction of 25, 30, or 35 dB must be incorporated into design and construction of structure.
There are special provisions pertaining to many of the compatibility designations that are not included here; refer to FAR Part 150 [6] for details.
Source: Federal Aviation Administration [36]

TABLE 14-2   FAA Noise and Land-Use Compatibility Guidelines (*Continued*)

The simulation models produce maps depicting contours of equal sound level such as DNL. Measurements are used to provide or confirm input to the simulation models as well as to confirm model predictions at specific ground locations.

The Integrated Noise Model (INM) and NOISEMAP

Two computer-based simulation models are currently used in the United States. Both produce maps showing contours of equal day-night average sound level. Developed by the FAA, the integrated noise model (INM) is most often used for civil airports [29]. NOISEMAP, developed by the U.S. Air Force, is generally used for military airbases but is also used for civil and joint-use airports. The FAA has approved both models for use in airport noise studies. The two models require the same basic input parameters but formats differ.

Use of either model requires inputs in two principal categories, namely, aircraft noise and performance data, and aircraft operational data. The major difference between the two categories of input is that the first is generally not airport dependent while the second is airport specific and must be individually developed for each airport.

Aircraft Noise and Performance Data

The INM uses a standard, internal noise and performance data base containing a large number of aircraft types. The model uses the noise data to determine the SEL of specific aircraft types as a function of thrust and distance from the observer. The performance data used by the model define the length of the takeoff roll, climb rate, speed, and thrust management for both departures and arrivals.

Aircraft Operational Data

The INM also requires operational input data specific to the airport under study. These data are often difficult to obtain as they are not routinely collected by either the airport or the FAA. To address this problem, airports are beginning to develop specific data collection procedures for this specific purpose. Operational inputs describe activity at the airport using average values during the period of interest and include the following:

1. Physical description of the airport runways, including any displaced takeoff or landing thresholds

2. Runway utilization percentages

3. Number of aircraft operations by aircraft type for all noise-significant aircraft types in the fleet mix

4. Day-night split of operations by aircraft type

5. Flight corridor descriptions

6. Flight corridor utilization percentages

Noise Model Output

Both the INM and NOISEMAP produce output in two forms, namely, contour maps of equal day-night average sound level, and detailed tabular analyses for user specified ground locations. Figure 14-6 shows an example of a DNL contour map. A typical map shows contours from 60 to 80 dB at 5-dB intervals. For presentation purposes, these contours are superimposed graphically on a good quality base map or aerial photograph.

In addition to DNL contours, SEL contours can also be helpful in addressing issues of sleep and speech interference and for analyzing the effects of noise abatement procedures, such as proposed noise abatement flight tracks. Graphical comparisons of SEL contours of various aircraft types can also provide powerful images for comparing noise emissions of differing aircraft types.

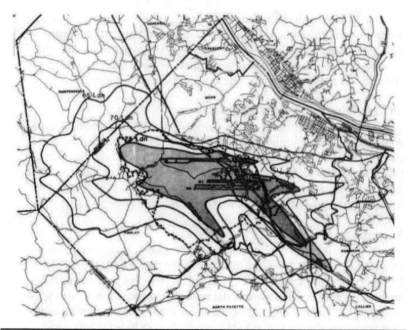

Figure 14-6 Sound exposure level contour map for Greater Pittsburgh International Airport (*Aviation Planning Associates, Inc. [7]*).

Tabular listings for user-specified ground locations show not only the predicted DNL but also the SEL and DNL contribution of individual aircraft by runway and flight corridor. This information is invaluable for understanding the major contributors to the total DNL. It can also be used to compare the model predictions with data from noise monitoring locations. Such comparisons often provide the basis for fine-tuning model inputs as well as promoting public confidence in the computer model and the contours it produces.

Aircraft Noise and Operations Monitoring

Many civil aviation airports in the United States have installed aircraft noise monitoring systems to assist in managing airport-community relations. The first systems, installed 20 or more years ago, performed strictly sound level monitoring. Current technology systems have evolved into complete noise monitoring systems capable of providing information on both aircraft sound levels and aircraft operations. The primary uses of airport noise and operations monitor systems are to help establish and monitor compliance with noise abatement procedures, verify trends in overall fleet noise, and provide input and validation data for computer-based airport noise simulation models.

When people complain about aircraft noise the complaint is often followed by a reference to some operational characteristic of the aircraft which differed from their expectations, For example, "they're not supposed to fly directly over my house," "that aircraft flew too low," or "they never used to use that runway so often." While admittedly anecdotal in nature, such informative complaints can be extremely helpful in pinpointing the operational source of the complaint and in starting a process of noise mitigation and community education.

The operations side of the monitoring system provides airport managers with the tools to verify the underlying cause of complaints, determine the extent of identified problems, and provide an objective basis for seeking solutions. The primary source of information for modern systems is data routinely collected by the FAA with their Automated Terminal Radar System (ARTS). The ARTS retains information sufficient to reconstruct the three-dimensional flight trajectory, aircraft type, airline, flight number and type of operation (departure or arrival) for every commercial aircraft movement. Modern, computer-based operations monitoring systems access these data, provide extensive on-line data storage capacity, and embody sophisticated data base management systems for retrieving, sorting, and reporting the enormous volumes of data they acquire.

Useful, long-term summary statistics from operations monitoring systems include the percentage runway utilization, with breakdowns by departures and arrivals and by aircraft type, and overall traffic counts, with breakdowns by aircraft type and by time-of-day. Detailed presentations of actual aircraft flight tracks, such as those shown in Fig. 14-7, are extremely helpful for examining noise abatement

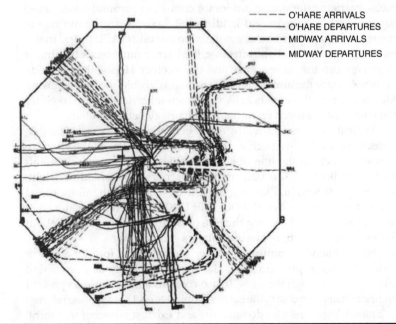

– – – – O'HARE ARRIVALS
——— O'HARE DEPARTURES
– ▪ – ▪ MIDWAY ARRIVALS
——— MIDWAY DEPARTURES

FIGURE 14-7   Radar derived air carrier departure flight tracks in Chicago area terminal airspace (*Landrum and Brown Aviation Consultants [14]*).

alternatives. The data from operations monitoring systems are also one of the few objective data sources for preparing accurate and defensible airport noise contours.

The sound level monitoring side of the system consists of a number of remote microphones located in the community surrounding the airport and a central processing site usually located at airport administrative offices. Microphones are located on top of 7-m-high poles and the microphone signal processed in real time *at the pole* to compute and store most all of the sound level metrics of interest. Data are then transmitted digitally from the microphone site to the central station using a modem and voice-grade telephone lines.

Finding Solutions

Table 14-3 presents a matrix of aircraft-related noise problems and potential solutions. In general, solutions to mitigate noise impacts seek to increase distance between the aircraft and noise-sensitive elements of the community, reduce noise levels at the source, or reduce the numbers of noise events in noise-sensitive areas. Some specific solutions require FAA expertise and approval and hence, the involvement of the agency should be sought at the earliest possible opportunity. Details of some solutions are discussed in the following paragraphs.

Noise Barriers

Noise barriers offer opportunities for controlling ground-based noise sources such as takeoff and landing roll, taxiway and apron movements, aircraft power-backs, auxiliary power units (APUs), and maintenance engine runs. To be effective, the barrier must break the line of sight between the noise source and the receiver. Hence, they provide no benefit once the aircraft is airborne and is visible above the barrier. Maximum effectiveness is achieved when a barrier is close to either the source or the receiver, rather than halfway between them.

Typical barriers are walls, earth berms or wall-berm combinations. Long buildings, such as the terminal itself, also make effective barriers. Blocking the line of sight to APUs and low engine aircraft such as the Boeing 737 usually requires barriers of only modest height, assuming flat terrain. Blocking the line of sight to high tail-mounted engines, such as those on the DC-10 or L-1011, presents a greater challenge. Barriers just blocking the line of sight generally provide about 5 dB of noise reduction. Higher barriers provide more.

For maintenance runups, a barrier is often in the form of a pen or series of walls. A pen surrounds the aircraft as closely as possible but allows entry through the front. It also contains a blast shield to prevent engine exhaust damage to the barrier. Complete enclosures, often referred to as *hush houses*, feature doors, roofs and exhaust silencing treatment. They are used where large amounts of noise reduction are required.

Consider these actions		Taxiing	Departure	Approach	Landing roll	Training flights	Maintenance	Ground equipment
Airport plan	Changes in runway location, length, or strength	•	•	•	•	•		
	Displaced thresholds			•		•		
	High-speed exit taxiways	•			•			
	Relocated terminals	•					•	•
	Isolating maintenance run-ups or use of test stand noise suppressors and barriers	•					•	•
Airport and airspace use	Preferential or rotational runway use*	•	•	•	•	•		
	Preferential flight track use or modification to approach and departure procedures.*		•	•		•		
	Restrictions on ground movement of aircraft*	•						
	Restrictions on engine run-ups or use of ground equipment						•	•
	Limitations on number or types of operations on types of aircraft	•	•	•	•	•	•	•
	Use restrictions Rescheduling Move flights to another airport	•	•	•	•	•	•	•
Aircraft operation	Raise glide slope angle or intercept*			•		•		
	Power and flap management*		•	•		•		
	Limited use of reverse thrust*				•			
Land use	Land or easement acquisition	•	•	•	•	•	•	•
	Joint development of airport property	•	•	•	•	•	•	•
	Compatible-use zoning	•	•	•	•	•	•	•
	Building code provisions and sound insulation of buildings	•	•	•	•	•	•	•
	Real property noise notices		•	•	•	•	•	•
	Purchase assurance		•	•	•	•	•	•
Noise program management	Noise-related landing fees	•	•	•	•	•		
	Noise monitoring		•	•			•	•
	Establish citizen complaint mechanism Establish community participation program	•	•	•	•	•	•	•

*These are examples of restrictions that involve the FAA's responsibility for safe implementation. They should not be set in place unilaterally by the airport operator.
Source: Federal Aviation Administration [36].

Table 14-3   Matrix of Noise Control Actions

Along the runway sideline, especially in the vicinity of start-of-takeoff roll, barriers are most effectively placed near the residences they are meant to protect. Obstruction height clearance requirements usually preclude placing barriers close enough to the runway to be effective in these locations.

Barrier performance can be degraded by temperature inversions and winds with a component blowing in the direction of source to receiver. This is especially true if the barrier cannot be located as close as desired to the source or receiver. Under these atmospheric conditions, refracted sound travels a higher curved path from the source to receiver, and sound attenuation is reduced or eliminated under extreme conditions such as in high wind.

Sound Insulation

Sound insulation of structures, such as residences, seeks to improve the environment indoors through treatment of the structure itself. FAA funding criteria for sound insulation projects seek a 5-dB transmission loss improvement and a day-night average sound level (DNL) goal of 45 dB indoors. Windows are usually the weak link in the sound attenuation properties of structures. With windows open the noise reduction properties of other parts of the structure are largely irrelevant and a noise reduction up to 14 dB is all that can be expected. With windows closed noise reduction is greater, but the additional reduction is dependent on the extent of

1. Any remaining air gaps such as around windows and doors, and through attic and basement vents

2. The thickness and number of panes of glazing

3. The weight of exterior doors

4. The weight of roofing and walls

Cost-effective sound insulation programs can achieve 25 to 35 dB of noise reduction through attention to air gaps (caulking around door and window frames, insulation of walls and attics, sound absorbing material around attic vents and soffits), window treatment (replacement of jalousie or poorly fitting windows, and use of double strength or double pane glass in the form of special acoustical windows or storm windows), and doors (replacement of hollow core with solid core units). In order to be effective during the summer months, central air conditioning must also be part of a basic noise insulation package so proper ventilation can be achieved with windows closed.

Enhancing roof and wall weight can provide additional benefit once the aforementioned items are no longer the weak link. However, the cost of ensuring that other elements are not the weak link, such as installing triple instead of double glazing and sophisticated air duct treatment, added to the cost of the structural enhancements themselves generally increases the cost significantly.

Preferential Runway System

The preferential runway concept is based on optimizing runway utilization under wind, weather, demand, and airport layout constraints to minimize population impacts by taking advantage of uneven population distribution around the airport. Preference is given, weather permitting, to those runways for which arrivals or departures affect the fewest people. Considerable effort can be devoted to determining which runway flight track combinations create the least noise impact and to developing with the FAA a workable plan that can be implemented. Monitoring the effectiveness of any preferential runway use program is important and can also require significant effort to develop and implement.

Noise Abatement Departure Procedures and Flight Tracks

The FAA has developed a recommended noise abatement takeoff procedure involving power settings and profile characteristics for turbojet-powered aircraft with maximum certificated gross takeoff weights in excess of 75,000 lb [34]. Most domestic airlines have incorporated this procedure or an equivalent in their flight manuals. The National Business Aircraft Association (NBAA) has also developed and recommended noise abatement procedures for turbojet business aircraft. The objectives of the NBAA program are to ensure that jet aircraft noise abatement procedures are safe, standardized and uncomplicated while at the same time being effective at reducing noise levels in the community. Noise abatement departure procedures can also include use of specific headings and turns to avoid populated areas. The INM may be used to assess the effectiveness of such procedures.

Noise abatement flight paths can offer significant opportunities for noise abatement where distribution of incompatible land uses is uneven. Typically, noise abatement flight paths are designed to avoid the noise sensitive areas and route air traffic over less sensitive areas. Implementation of these kinds of flight paths will also require extensive interaction with the FAA. Again, the INM may be useful in assessing the noise impacts of various flight tracks.

Airport Use Restrictions

Noise-based airport use restrictions address noise control through reductions in the average noisiness of the aircraft that use the airport. Use restrictions have come under court challenge, especially by the FAA, as unduly restrictive of interstate commerce. In general, the courts have found restriction of an airport to be legal if they are

1. Reasonable in the circumstances of the particular airport
2. Carefully tailored to the local needs and to community expectations

3. Based upon data which support the need and rationale for the restriction

4. Not unduly restrictive of interstate commerce

Several types of restrictions may be considered, particularly if they are considered "voluntary." Curfews or other nighttime use restrictions are designed to reduce or eliminate noisy operations during late-night hours when people may be particularly sensitive to noise. Such restrictions can have large DNL benefits relative to the number of aircraft operations affected because of the 10-dB penalty added to noise between 10:00 p.m. and 7:00 a.m. when computing DNL. Aircraft operators may react by canceling operations by restricted aircraft types, switching to quieter aircraft types, or rescheduling.

Full curfews, such as eliminating all nighttime flights, have been found to be *overbroad* and to impose *undue burden* on interstate commerce, and are often viewed as *arbitrary and capricious*. The overbroad issue has to do with the fact that a full curfew may deny access to the airport by users who, in fact, could operate quietly at night without significant disruption to sleep. A full curfew might have interstate commerce implications because of nighttime activity to and from out-of-state destinations. The arbitrary and capricious test has to do with whether or not a use restriction can be justified in terms of its noise benefits. Perhaps the most important point to be made is that a detailed quantitative noise analysis should be developed to provide justification for any noise-based use restriction.

Use restrictions can also be based on FAA noise certification categories. These categories are identified in FAR Part 36 and discussed later. These restrictions limit the use of the airport based on the noise certification stage of the aircraft. For example, an airport may adopt a restriction that limits the use of the airport to stage 3 aircraft at night.

As part of the certification process specific noise levels are measured for each aircraft [38] and use restrictions can be based on these specific levels. For example, an airport could prohibit nighttime departures by aircraft with certified noise levels exceeding 108 EPNdb.

Use restrictions do not have to be based on certified noise levels. Certified noise levels may not be available for some older transport category aircraft or for many general aviation aircraft. Certified noise levels may also be deemed unrepresentative of the sound levels produced under actual local operating conditions. In such cases, it may be preferable to set limits based on other published data [25] or on the noise levels measured at the airport itself.

Noise-based landing fees provide an economic incentive to discourage the operation of noisier aircraft, especially during noise-sensitive times of the day. Noise-based landing fees are proportional in some way to the noise produced by the aircraft. For example, an operator may be charged more for a takeoff by a stage 2 aircraft than for the same operation by a stage 3 aircraft. Alternatively, the fee may be higher for a night departure than for a day departure by the same aircraft. To be effective, however, the fee structure must be set high enough to affect airport user decision making.

Noise Regulations

Federal aviation noise regulations are identified in a number of forms. The highest form of regulations are those set forth in various parts of Title 14 of the Code of Federal Regulations (14 CFR). This section of the federal code is called the Federal Aviation Regulations (FAR). The FAA also publishes orders and advisory circulars. Orders are procedures to which FAA staff must adhere in performing their responsibilities. To the extent that FAA approves actions by others in the aviation industry (airports, airlines, etc.), the orders apply to them as well. Advisory Circulars are printed documents which provide useful guidance and information often related to the FAR or FAA orders.

FAR Part 36

FAR Part 36 sets noise standards that aircraft must meet to obtain type and airworthiness certificates for operation in the United States [39]. First promulgated in 1969 for application to civil subsonic turbojets and large (over 12,500 lb) propeller-driven aircraft, the government subsequently amended the regulation to address civil supersonic aircraft, small (not over 12,500 lb) propeller aircraft, and rotary-wing aircraft such as helicopters. FAR Part 36 also prescribes the procedures for aircraft manufacturers and others to use in measuring aircraft noise for certification purposes. The FAA publishes companion Advisory Circulars which present measurement results [25, 31, 38].

In 1977, the certification limits were made more stringent, leading to the classification of aircraft into three groups known as *stages. Stage 1* aircraft are those that were flying before the regulation was initially adopted and were never required to meet level limits when they were first issued. *Stage 2* aircraft are those that met the original (1969) noise emission limits but not the revised (1977) limits. *Stage 3* aircraft are those newest, quietest, types that must meet the revised limits.

The regulation requires that aircraft meet gross weight based noise limits at three locations. Figure 14-8 shows the required measurement locations for turbojet and large propeller aircraft. These are under the takeoff path 6500 m from brake release, under the approach path 2000 m from runway threshold, and along the flight track sideline 450 m from the runway centerline (650 m for older turbojet aircraft). The sideline measurement is at the point of maximum sideline noise. In practice this is normally to the side of takeoff, not landing.

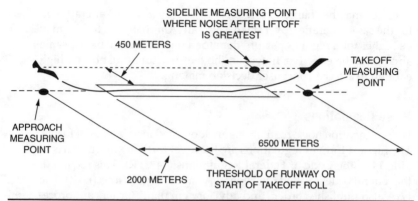

FIGURE 14-8   FAR Part 36 noise measurement locations (*Federal Aviation Administration [39]*).

Figures 14-9 through 14-11 show the original Stage 2 noise limits and the lower stage 3 limits for each of these three locations. Shown with the limits are several examples of the actual certificated levels for a variety of different aircraft. Note that some of the quieter types include the McDonnell-Douglas DC-8-70 series, DC-9-80s (also known as the MD-80), and the Boeing 757-200 and 767-200. FAR Part 36 certification noise levels are published and regularly updated in Advisory Circulars [25, 31, 38].

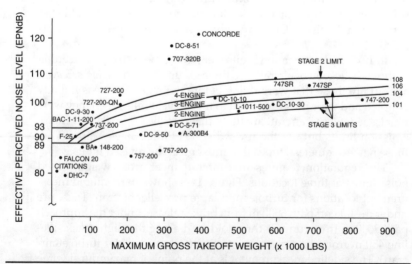

FIGURE 14-9   FAR Part 36 certification levels for takeoff noise (*Federal Aviation Administration [38]*).

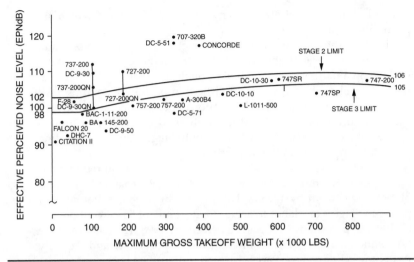

FIGURE 14-10   FAR Part 36 certification levels for approach noise (*Federal Aviation Administration [38]*).

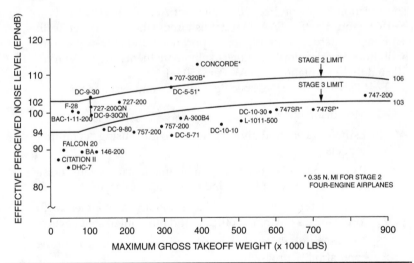

FIGURE 14-11   FAR Part 36 certification levels for sideline noise (*Federal Aviation Administration [38]*).

FAR Part 91

FAR Part 91 limits civil aircraft operations in the United States based on FAR Part 36 certification status. The noise elements of FAR Part 91 were first adopted in 1977. This regulation prohibits operation of civil subsonic turbojet aircraft with maximum weights over 75,000 lb unless they were certificated under FAR Part 36 Stage 2 or 3 limits. FAR Part 91 has led to the elimination and ongoing prohibition of all Stage 1 operations in the United States in civil subsonic turbojets over

75,000 lb but does not affect aircraft of 75,000 lb or less which are primarily corporate aircraft.

In 1990, the federal government enacted the Airport Noise and Capacity Act of 1990 (Public Law 101-508). This act called for the FAA to develop a national aviation noise policy and regulations to implement the policy. This was accomplished, in part, by amending FAR Part 91 to require the phased elimination of Stage 2 operations in civil subsonic turbojets over 75,000 lb by the end of 1999 with limited waivers through 2003. This will leave only Stage 3 aircraft operating in the air carrier fleet in the near future.

The interim compliance schedule of FAR Part 91 required that aircraft operators remove 25 percent of their Stage 2 airplanes by the end of 1994, 50 percent by 1996 and 75 percent by 1998 or, alternatively, phase-in Stage 3 airplanes to achieve a fleet mix of 55 percent Stage 3 by 1994, 65 percent by 1996 and 75 percent by 1998. The balance of the FAA national noise policy required by the Airport Noise and Capacity Act is embodied in FAR Part 161 which is discussed later.

FAR Part 150

The Aviation Safety and Noise Abatement Act of 1979 (Public Law 96-193) required the FAA to establish regulations that set forth national standards for identifying airport noise and land-use incompatibilities and develop programs to eliminate them. The FAA promulgated these regulations as FAR Part 150 [6].

FAR Part 150 prescribes specific standards and systems for

1. Measuring noise

2. Estimating cumulative noise exposure using computer models

3. Describing noise exposure including instantaneous noise levels, single event levels and cumulative exposure

4. Coordinating noise compatibility program development with local land-use planning officials and other interested parties

5. Documenting the analytical process and development of the compatibility program

6. Submitting documentation to the FAA

7. FAA and public review processes

8. FAA approval or disapproval of the submission

A full FAR Part 150 submission consists of two basic elements, namely, a *noise exposure map* (NEM) and its associated documentation, and a *noise compatibility program* (NCP). It is possible, however, to submit only the NEM. In addition to these elements, a critical ingredient to a successful FAR Part 150 program is a thorough and effective public involvement program.

Noise Exposure Map

The Noise Exposure Map (NEM) document describes the airport layout and operation, aircraft-related noise exposure, land uses in the airport environs, and the resulting noise-related land-use compatibility situation. It addresses the year of submission and five years into the future. It includes graphic depiction of existing and future noise exposure resulting from aircraft operations, and of land uses in the airport environs. Documentation must accompany the noise exposure map that describes the data collection and analysis undertaken in its development. The basic output of the map development is identification of existing and potential future noise and land-use incompatibilities. FAR Part 150 includes a table presenting noise and land-use compatibility guidelines, shown earlier in Table 14-2.

Noise Compatibility Program

Following development of a NEM, which essentially defines the extent of noise and land-use incompatibility, the airport proprietor may elect to develop a Noise Compatibility Program (NCP). In developing a noise compatibility program, the airport proprietor must consider all potential compatibility measures, including the airport layout, operational and use alternatives, and land use alternatives. FAR Part 161, discussed next, further regulates the evaluation and adoption of airport use restrictions. The ultimately developed program is essentially a list of the actions the airport proprietor proposes to undertake to minimize existing and future noise and land-use incompatibilities. The noise compatibility program documentation must recount the development of the program, including a description of all measures considered, the reasons that individual measures were accepted or rejected, how measures will be implemented and funded, and predicted effectiveness of individual measures and the overall program.

Following FAA acceptance of the NEM and approval of the NCP program submissions, the airport operator may apply for FAA funding of program implementation. Official FAA approval of the NCP does not eliminate the requirements for a formal environmental assessment of any proposed actions pursuant to requirements of the National Environmental Policy Act (NEPA) [2, 43], however, acceptance of the submission is a prerequisite to application for funding of implementation actions.

Public Involvement

At every stage of the FAR Part 150 planning process opportunities exist to apprise airport neighbors, user groups, and local officials of project alternatives, and to solicit their comments, criticism, and support. A key objective is to utilize the broadest possible definition of the *airport public* which includes more than just the residents of areas around the airport. A balanced discussion of issues must include representatives of the aviation industry including local, state, regional, and national agencies with jurisdiction, the business community, and individual airport users. Successful public information programs include the following elements:

1. Regular advisory committee meetings
2. Technical and other subcommittee meetings as required
3. Informational newsletters tailored for public distribution
4. Informal workshops open to the general public
5. A final public hearing
6. Briefings to local public officials

FAR Part 161

The second major element of the national noise policy enacted through the Airport Noise and Capacity Act of 1990 is FAR Part 161. It establishes requirements that an airport operator must meet prior to promulgating any airport noise or access restriction on the use of Stage 2 or Stage 3 aircraft [40].

Under FAR Part 161 an airport proprietor may impose restrictions on Stage 2 aircraft operations as long as two conditions are met. First, the proprietor must prepare an analysis of the anticipated costs and benefits of the proposed restriction. Second, the proprietor must provide proper notice of the restriction to the public and to affected parties. The cost-benefit analysis required by FAR Part 161 must include

1. An analysis of the anticipated or actual costs and benefits of the proposed restriction
2. A description of alternative restrictions which were considered
3. A description of the alternative measures considered that do not involve airport restrictions
4. A comparison of the costs and benefits of the alternative measures which were considered

FAR Part 161 imposes substantial impediments to local restrictions on Stage 3 aircraft. No local Stage 3 restriction may become effective unless it has been submitted to and approved by the FAA. The process for FAA review and approval has three principal elements [5]:

1. The collection and analysis of data to justify the restriction and to explain its environmental and economic impact
2. The notification of the public and allowance of adequate time for comment on the proposed restriction
3. The submission of the restriction for FAA review and approval

An airport cannot implement a stage 3 aircraft restriction unless it complies with each of the above elements.

FAR Part 161 stipulates that both types of restrictions may be developed through the FAR Part 150 process. The benefit of using FAR Part 150 as a mechanism for developing a rule restricting airport access is the availability of federal funding. Since such federal funding may be used for the preparation of a FAR Part 150 NCP, this funding can be used for the preparation of any noise or access restriction that is included in the FAR Part 150 study. The major disadvantage of submitting Stage 2 restrictions to the FAA as part of the FAR Part 150 submission is that a formal submission will invoke the approval process which is otherwise not necessary under FAR Part 161. Stage 3 restrictions, on the other hand, require FAA review whether or not they are developed during a FAR Part 150 process.

Construction Impacts

The construction of facilities at airports can result in temporary and long-term impacts on the community and travelers. Those factors which are of primary concern during the construction process include soil erosion, water and air quality, noise due to construction equipment and methods, the source and quantity of construction materials, disruption and relocation of businesses and residences, the continued operation of existing facilities both on and off-airport during the construction process, and the interference with other construction projects [23].

A review of the environmentally sensitive areas and facilities should be undertaken to identify those which will be subject to impact and the likely duration and consequences of these activities. The location and quantity of cut and fill must be identified and methods to minimize the effect of construction activities on soil erosion during construction should be identified. Procedures for the handling of construction materials and wastes to minimize the introduction of particulate matter into the air and water resources are required. Those land uses which will be adversely affected by noise from construction activities should be identified and the optimal routing of construction vehicles and timing of activities chosen to minimize damage. Obvious positive impacts of a major construction activity are the increases in employment and payroll for personnel associated with the project and the purchase of materials and supplies from local firms which support the local economy. However, certain businesses and residences in the vicinity of the project may be subject to disruption due to the possible rerouting and congestion of vehicular traffic and restrictions on access to land uses.

The study of the direct and indirect socioeconomic impacts on the community should include an identification of the location, timing, and amount of impact throughout the entire construction period.

Social Factors

Land Development

The development of an airport can result in a change in the pattern of land use activity both in the vicinity of the airport and in the geographic region. These land development activities may result in changes in the level of economic activity and population growth and demography. The location of an airport generally results in the inducement of land development in the vicinity of the of the airport which may be measured in terms of the density of industrial, commercial, retail, agricultural and residential use. It usually impacts the nature, magnitude, and operating patterns of other transportation modes providing access to the airport and those land uses associated with its development. The presence of the airport may induce industrial or commercial activities to move into or expand within the region due to increased accessibility to markets and materials.

An airport will affect land development as a function of its direct economic impact on the region. The following land development activities should be studied [2, 43]:

1. Plant relocation from outside the region which will require construction activities

2. Increases in the production or sales of existing business enterprises requiring new or expanded capital facilities

3. Increased expenditures in tourist and recreational facilities resulting in requirements for new or expanded facilities and increases in retail sales

4. Expansion of agricultural markets resulting in increased productivity and resource utilization

5. Increased demand for specialized facilities for such activities as business or convention centers

6. Expansion of commercial and financial markets resulting in a demand for additional facilities

The study of the land development impacts requires an analysis of those factors which influence industrial, commercial, and residential location decisions including accessibility to raw materials, labor, and markets, the costs of production and transportation, and those quality of life and community factors associated with such decisions. The analysis is usually conducted through an examination of historical trends for the area and similar locations and use surveys and economic models to predict land development. The study should identify the nature and extent of existing zoning ordinances in the vicinity of the airport and recommend changes necessary to accommodate the likely development in a compatible manner. The study should also identify those requirements for regional policy decisions needed to stimulate overall land development in accordance with the stated objectives of the communities affected.

Displacement and Relocation

The construction or expansion of an airport often creates a need for additional land, the relocation of residences, businesses, and community facilities, a disruption in business activity and community character and cohesiveness, the impairment of community service functions, and an increased demand for public services. The assessment is directed toward determining the type, extent, characteristics, and effects of displacement and relocation and mitigation measures to minimize adverse consequences.

The boundaries of the areas affected are determined from area maps. The community structure is usually defined in terms of population demography, growth, and density, housing and business characteristics such as the type, distribution, condition, value, occupancy and vacancy levels, open land, recreational resources, and community services. The availability of relocation resources for those land uses required for airport and ancillary activities are identified and the changes in the demand for public services are quantified. Comparisons of the relative impact of project alternatives are usually made which attempt to address the following items [43]:

1. The nature, location, and extent of the displacement of homes, businesses, and community facilities

2. The creation of physical barriers or divisions

3. The impairment of mobility, accessibility, community services, and community facilities

4. The disruption of homes, businesses, and community facilities during construction

5. The nature, availability, adequacy, and compatibility of relocation resources

6. The nature, location, and extent of land use changes

7. The aesthetic appeal of the design for the facility and surrounding environment

Parks, Recreational Areas, Historical Places, Archeological Resources, and Natural and Scenic Beauty

Particular attention is required to determine the impact upon parks, recreational areas, open spaces, cultural and historic places, archeological resources, and natural and scenic beauty. The analyst must identify the type of facilities which will be impacted, their size and use, and those measures which can be implemented to preserve the nature, character, compatibility, and accessibility of the facilities. It is particularly important to document the relative impact of project alternatives on these types of facilities and to avoid the acquisition of such lands for the implementation of a project alternative.

The assessment can be performed on the basis of a participatory evaluation with those groups which possess expertise or interest in this impact area. Suggested generalized evaluation criteria might include [2, 43]:

1. The existence, nature, and extent of any physical alteration to the facilities

2. The degree of conformity of the planned facilities with the existing environment

3. The disruption of access

4. The disruption of the ambient environment

5. The compatibility of access induced development with the facilities

Consistency with Local Planning

The planning and design of airports can have significant effects on the economy, land use, infrastructure, and nature of community development. The planning effort must be carried on in an environment which is compatible and coordinated with other local planning efforts and guidelines. Care must be exercised from the inception of planning to assess the impact of project alternatives and operations on the goals and objectives of communities and to identify those facets of the project which may present conflicts between existing plans or community goals. Modifications to airport project proposals which will be in conformity with local policies and plans must be explicitly examined. Policies required to preserve the overall development objectives of the community should be identified and mechanisms to implement such policies proposed.

The assessment of impacts in this area requires an identification of and coordination with those federal, state, and local agencies which have a concern or jurisdiction in matters related to airport and community development actions, the delineation of clear statements of goals and objectives, the integration of airport plans with local comprehensive land use, economic, and transportation development plans, and the establishment of a continuing dialogue on issues related to these plans. The presentation of the results of planning efforts and development recommendations in public forums, with mechanisms for citizen participation in the planning and review process, together with timely and well documented responses to community concerns is essential.

Ecological Factors

Wildlife, Waterfowl, Flora, Fauna, Endangered Species

The consideration of the impact of airport development on changes in the natural state of land and waterways is essential to protect ecosystems. Living and nonliving elements, plants, and animals all interact on land and in water to produce a highly interdependent system of aquatic and terrestrial ecosystems. The relationship between species and the ecosystem is essential to maintain the life support system for wildlife, waterfowl, flora, fauna, and endangered species [20, 51, 52]. Of particular importance is vegetation, plant and animal life. The principal impacts which may occur are the loss of or injury to the organisms or the loss or degradation of the ecosystem.

The use of land for airport development creates disturbances and disruptions to flora and fauna. The specific project elements often include the clearing and grubbing of land areas, changes in the composition and nature of the topography, and interferences with water shed patterns. Thus airports can destroy the natural habitat and feeding grounds of wildlife and eliminate or reduce flora essential to the maintenance of the ecological balance in the area. Particular hazards may be presented to birds and aircraft due to striking birds, and care must be exercised in choosing airport sites to avoid land which attracts birds and natural migration routes. The protection for endangered species in the United States is legislated through the contents of the Endangered Species Act of 1973 (Public Law 93-205) and lists of endangered species are published [20, 51, 52]. Reference should also be made to state and location regulations in this regard.

The assessment techniques used include the identification of the important aquatic and terrestrial organisms present in the area and a determination of the life support systems required for the different species. An analysis is performed to determine the impacts on vegetation requirements, food chains, and habitats of these species, as well as their tolerance to air and water pollution. Care must be taken in the case of aquatic species to examine the effects of soil erosion, flooding, and sedimentation on stream beds where food chains, spawning grounds, and habitats exist.

Wetlands and Coastal Zones

Improperly planned or operated drainage systems at airports can cause contaminants to enter streams, lakes, and waterways. The normal operation of an airport results in contamination potential through aircraft and ground vehicle washing, servicing, and fueling, airport and aircraft maintenance, and terminal services. In the construction phase of a project there is a high potential for contamination through clearing, grubbing, pest control, and changes in topography. Changes in the natural drainage patterns of the area are very common due to the nature of airport development projects. Preservation of recharge areas and stream flows, the elimination of flooding and sedimentation problems, and the preservation of the quality and routing of water resources are all vitally essential to the maintenance of water quality and the protection of ecosystems.

Flood Hazards

The flood hazard potential of any development is a necessary consideration since alterations in the topography, cover, and soil characteristics on the property are inevitable. The storage capacity of local rivers, streams, canals, and groundwater areas can be exceeded due to changes in the magnitude and paths of runoff from storms and high rainfall or thawing events. The analysis of the potential for flooding is conducted by evaluating the characteristics of the ground surface, soil materials, topography, and floodplains, the historical frequency and intensity of storms, and storm water drainage and retention facilities. The methods for conducting such analyses are discussed in Chap. 9.

If it is found that the project design increases the potential for on or off-site flooding, those areas subject to these effects are identified and the mechanisms required to alleviate the hazards are incorporated into the project design. The construction of new or increased capacity storm sewers and impounding areas, channels, and dikes are most commonly indicated. Changes in the elevation of facilities and the slope and cover of the ground surface at the site can also be of considerable benefit in reducing flood hazards.

Engineering and Economic Factors

Costs of Construction and Operation

All engineering planning studies consider the capital, maintenance, and operating costs of all feasible alternatives as an integral part of the planning process. For airport projects these costs include land acquisition, purchases of land leases and covenants to protect aircraft operations and environmental quality, facility construction, operating, maintenance, and administrative costs. Typically, construction costs are derived from quantity takeoffs of materials which are related to locally appropriate cost indices for the various construction items [13, 18, 19]. The relationship between the overall cost factor is related by a concept of bench mark cost indices for various components of the terminal building [50]. A tabulation of these indices is given in Table 14-4. The capital costs usually include materials, supplies, labor, and engineering.

Location	Area Category	Area Type	Cost Unit	Benchmark Index per Unit		
				Shell	Tenant	Total Cost
Terminal	Type A: Passenger-handling facilities	1. Lobbies	ft²	1.00	NA*	
		2. Waiting rooms				
		3. Circulation				
		4. Rest rooms				
		5. Counter areas			2.00	3.00
		6. Baggage claim facilities, including claim device			0.50	1.50
		7. Service and storage areas			NA	
Terminal	Type B: Airline/tenant operations space, partly finished	1. Customer service offices	ft²	0.65	0.40	1.05
		2. Agent supervisor offices, checkout, and agent lounge			0.50	1.15
		3. Toilets			1.55	2.20
		4. VIP/IPR rooms			1.10	1.75
		5. Lost and found			0.72	1.37
Terminal or connector	Type C: Airline operations space, lower level unfinished	1. Offices	ft²	0.60	0.55	1.15
		2. Tire shop (including equipment)			0.62	1.22
		3. Storerooms			0.30	0.90
		4. Ready and lunch rooms			0.43	1.03
		5. Lockers			0.20	0.80
		6. Toilets			1.80	2.40
		7. Planning center and load planning			0.85	1.45

TABLE 14-4 Unit-Cost Indices for Space and Special Equipment at Airport Terminal Buildings

Location	Area Category	Area Type	Cost Unit	Benchmark Index per Unit		
				Shell	Tenant	Total Cost
Connector	Type D: Passenger-handling	1. Corridors 2. Rest rooms 3. Service and storage area 4. Boarding areas	ft²	0.80	NA 0.26	 1.06

*NA = not applicable.

These cost data have been compiled for new terminal and concourse construction, and they may not be applicable for remodeling or small additions to existing facilities. They do not include installed equipment, such as loading bridges, claim devices, ramp utilities, and so on.

Source: Federal Aviation Administration [50].

TABLE 14-4 Unit-Cost Indices for Space and Special Equipment at Airport Terminal Buildings (*Continued*)

The construction costs for the various items are normally made at different points in time and, therefore, for comparative and evaluative purposes these are brought back to some common point in time in order that value may be properly attributed to the construction needs. The operating, maintenance and administrative costs are usually annualized. These costs are normally estimated through comparisons with similar installations, historical trends, and economic influences. The costs of capital and the required coverage are also included to arrive at the total program cost.

An overview of the categories and items typically found in airport projects for which capital, operating, maintenance, and administrative cost estimates are required includes:

1. Airfield facilities

   a. Runways, taxiways, and aprons

   b. Fueling and fixed power systems, crash, fire and rescue units

   c. Air traffic control facilities, lighting, and navigational aids

2. Terminal building facilities

   a. Terminal buildings and connectors

   b. General aviation servicing buildings and hanger areas

   c. Boarding devices, mechanical and electrical systems

   d. Communications and security systems

   e. Air cargo buildings

   f. Maintenance and administrative buildings

   g. Furnishings

3. Access facilities

   a. Roadways, drives, and curb frontage

   b. Rental car, limousine, and transit areas

   c. Parking lots and garages

   d. Graphics, signage, and lighting

4. Infrastructure facilities

   a. Landscaping and drainage

   b. Utilities including water supply, sewage disposal, power supply systems

   c. Land acquisition

For the purposes of evaluation these costs are usually related to passenger and aircraft traffic characteristics such as the cost per enplaned passenger or cost per air carrier operation. The costs are also allocated among the various users and nonusers of the project, normally tenants, concessionaires, airlines, general aviation, cargo businesses, federal, and state, and local governmental agencies as appropriate.

Economic Benefits and Fiscal Requirements

The evaluation of the economic and financial feasibility of the project requires and identification of both the level and allocation of the benefits and costs of the project, as well as a revenue analysis performed for the various cost centers. Both direct and indirect benefits can accrue to the users of the airport and to the community in which the airport is located. Generally user benefits include reductions in delay, fuel consumption, time, and other operating and maintenance costs. These can usually be derived relative to dollar value. Nonuser and community benefits take the form of increased economic activity, rises in employment, and purchases of goods and services. These can also be evaluated through classical economic techniques [2, 8, 21].

Although it may be possible to justify expenditures from an economic standpoint, it may not be possible to generate or capture the value of these benefits from a revenue viewpoint. A revenue analysis seeks to identify the revenue requirements by cost center and the level of revenue required to cover project costs. Normally, the costs are allocated to facility users and rents, rate and charges, and concession agreements are negotiated on the principle that all users should pay their fair and proportionate share of the costs of providing, maintaining, operating, and administering the facilities they use.

Various indices are used to determine the reasonableness of revenue requirements including the percentage of revenue generated which is paid for terminal rents or landing fees, and the revenue required per enplaned passenger. A comparison of the bonding capacity of the governmental units concerned with the project is vital in order to determine the influence of the project on other public revenue requirements.

Energy and Natural Resources

The use of new technology in power generation systems at airports, the efficient layout of apron areas and taxiing routes, improvements in the capacity of runway systems, the installation of navigational aids, and the effective uses of construction materials can substantially reduce the costs and resource use of energy. A detailed examination of the impact of airport design elements on energy consumption should be performed. Typically comparisons between existing and planned systems, and between alternative systems for proposed facility modifications relative to fuel consumption of aircraft and terminal systems may yield essential information concerning their feasibility and merit.

Summary

Though the incentive for the study of environmental, sociological, and ecological factors in the evaluation of engineering projects was initially provided through national, state, and local legislation, the state of the art has evolved to the point that a better and more complete understanding of the short- and long-term implications of these projects is leading to more efficient engineering designs. True the costs of planning have increased because of the need to study several criteria in the evaluation of planning proposals, but the potential for the overall reduction in the real *costs* of these proposals on the long-term requirements of society through the use of comprehensive planning approaches has also been increased.

References

1. *Acoustic Determination of Occupational Noise Exposure and Estimation of Noise-Induced Hearing Impairment*, Publication 1999, International Organization for Standardization, Geneva, Switzerland, 1990.
2. *Airport Environmental Handbook*, Order No. 5050.4A, Federal Aviation Administration, Washington, D.C., 1985.
3. *Airport Landscaping for Noise Control Purposes*, Advisory Circular AC 150/5320-14, Federal Aviation Administration, Washington, D.C., 1978.
4. *Airport Master Plans*, Advisory Circular AC 150/ 5070-6A, Federal Aviation Administration, Washington, D.C., 1985.
5. *Airport Noise: A Guide to the FAA Regulations under the Airport Noise and Capacity Act*, Cutler and Stanfield and Harris Miller Miller and Hanson, Inc., Lexington, Mass., January 1992.
6. *Airport Noise Compatibility Planning*, Part 150, Federal Aviation Regulations, Federal Aviation Administration, Washington, D.C., 1991.
7. *Airport Noise Compatibility Study*, Greater Pittsburgh International Airport, Aviation Planning Associates, Cincinnati, Ohio, 1986.
8. *Airport Planning and Environmental Assessment*, Notebook Series, 4 Vols., DOT P 5600.5, Department of Transportation, Washington, D.C., 1978.
9. *Airport Planning Manual, Part 2*, Land Use and Environmental Control, 2d ed., Document No. 9184-AN/902, International Civil Aviation Organization, Montreal, Canada, 1985.
10. *Analysis of the Predictability of Noise-Induced Sleep Disturbance*, K. S. Pearsons, D. S. Barber, and B. G. Tabachnick, Report No. HSD-TR-89-029, U.S. Air Force, Washington, D.C., 1989.
11. "Applied Acoustical Report: Criteria for Assessment of Noise Impacts on People," L. S. Finegold, C. S. Harris, and H. E. Von Gierke, submitted to *Journal of Acoustical Society of America*, New York, N.Y., 1992.
12. *Aviation Noise Effects*, S. J. Newman, Report No. FAA-EM-85-2, Federal Aviation Administration, Washington, D.C., 1985.
13. *Building Construction Cost Data*, R. S. Means Company, Inc., Duxbury, Mass.
14. *Chicago Delay Task Force Technical Report*, Vol. I: Chicago Airport/Airspace Operating Environment, Landrum and Brown Aviation Consultants, Chicago, Ill., April 1991.
15. *Citizen Participation in Airport Planning*, Advisory Circular AC 150/5050-4, Federal Aviation Administration, Washington, D.C., 1975.
16. *Community Noise*, Wyle Laboratories, Report No. DOT-NTID300.3, Office of Noise Abatement and Control, U.S. Environmental Protection Agency, Washington, D.C., 1971.

17. *Compilation of Air Pollution Emission Factors*, Report No. AP-42, Environmental Protection Agency, Washington, D.C., Periodically Revised.
18. *Dodge Guide for Estimating Public Works Construction Costs*, McGraw-Hill Information Systems Company, New York, N.Y.
19. *Dodge Manual for Building Construction Pricing and Scheduling*, McGraw-Hill Information Systems Company, New York, N.Y.
20. *Endangered and Threatened Wildlife and Plants, Fish and Wildlife Service*, Department of the Interior, Washington, D.C.
21. *Environmental Assessment Notebook Series*, Report No. DOT P5600.4, 7 Volumes, Department of Transportation, Washington, D.C., 1975.
22. "Environmental Considerations in Airport Planning," C. V. Robart, Course Notes for Airport Planning and Design Short Course, University of California, University Extension, Berkeley, Calif., June 1977.
23. Environmental Impact Assessment Report for the Expansion of the Existing Terminal Complex at the Fort Lauderdale-Hollywood International Airport, Aviation Division, Broward County Department of Transportation, Fort Lauderdale, Fla., 1980.
24. *Environmental Protection*, Annex 16 to the Convention on International Civil Aviation, Vol. 1: Aircraft Noise, 2d ed., International Civil Aviation Organization, Montreal, Canada, 1988.
25. *Estimated Airplane Noise Levels in A-Weighted Decibels*, Advisory Circular AC 36-3F, Federal Aviation Administration, Washington, D.C., 1990.
26. *Federal Agency Review of Selected Noise Analysis Issues*, Federal Interagency Committee on Noise (FICON), Washington, D.C., 1992.
27. *General Operating and Flight Rules*, Part 91, Federal Aviation Regulations, Federal Aviation Administration, Washington, D.C., 1992.
28. Information on Levels of Environmental Noise Requisite to Protect Public Health and Welfare with an Adequate Margin of Safety, U.S. Environmental Protection Agency, Arlington, Va., 1974.
29. *INM Integrated Noise Model Version 3*, User's Guide, Report No. FAA-EE-81-17, Office of Environment and Energy, Federal Aviation Administration, Washington, D.C., 1982.
30. *Management of Airport Industrial Waste*, Advisory Circular AC 150/5320-15, Federal Aviation Administration, Washington, D.C., 1991.
31. *Measured or Estimated (Uncertified) Airplane Noise Levels*, Advisory Circular AC 36-2C, Federal Aviation Administration, Washington, D.C., 1986.
32. *Measures of Noise Level: Their Relative Accuracy in Predicting Objective and Subjective Responses to Noise During Sleep*, J. S. Lucas, Report No. EPA- 600/1-77-010, Environmental Protection Agency, Washington, D.C., 1977.
33. *Noise, Environmental Health Series No. 12*, World Health Organization, Geneva, Switzerland, 1980.
34. *Noise Abatement Departure Profiles*, Advisory Circular AC 91-53, Federal Aviation Administration, Washington, D.C., 1978.
35. *Noise Certification Handbook*, AC 36-4B, Federal Aviation Administration, Washington, D.C., 1988.
36. *Noise Control and Compatibility Planning for Airports*, Advisory Circular AC 150/5020-1, Federal Aviation Administration, Washington, D.C., 1983.
37. "Noise-Induced Sleep Disturbance and Their Effect on Health," B. Griefahn and A. Muzet, *Journal of Sound and Vibration*, Vol. 59, No. 1, New York, N.Y., 1978.
38. *Noise Levels for Certified and Foreign Aircraft*, Advisory Circular AC 36-1F, Federal Aviation Administration, Washington, D.C., 1992.
39. *Noise Standards: Aircraft Type and Airworthiness Certification*, Part 36, Federal Aviation Regulations, Including Changes 1 to 21, Federal Aviation Administration, Washington, D.C., 1991.
40. *Notice and Approval of Airport Noise and Access Restrictions*, Part 161, Federal Aviation Regulations, Federal Aviation Administration, Washington, D.C., 1991.
41. "Occupational Noise Exposure; Hearing Conservation Amendment," Federal Register 48(46), Occupational Safety and Health Administration, Washington, D.C., 1983.

42. *Physiological, Psychological and Social Effects of Noise*, K. D. Kryter, Reference Publication 1115, National Aeronautics and Space Administration, Washington, D.C., 1984.
43. *Policies and Procedures for Considering Environmental Impacts*, Order No. 1050.1D, Federal Aviation Administration, Washington, D.C., 1986.
44. *Public Health and Welfare Criteria for Noise*, U.S. Environmental Protection Agency, Arlington, Va., 1973.
45. *Recommended Method for Computing Noise Contours Around Airports*, Circular-205 AN1/25, International Civil Aviation Organization, Montreal, Canada, 1988.
46. "Research on Noise-Disturbed Sleep Since 1973," B. Griefahn, Proceedings, The Third International Congress on Noise as a Public Health Problem, ASHA Report No. 10, Frieburg, West Germany, 1980.
47. *Special Report: Summary of Significant Provisions of the Final FAA Noise Rule*, Airports Association Council International-North America, Washington, D.C., 1991.
48. *Study of Soundproofing Public Buildings Near Airports*, Wyle Laboratories, Report No. DOT-FAA- AEQ-77-9, Federal Aviation Administration, Washington, D.C., 1977.
49. "Synthesis of Social Surveys on Noise Annoyance," T. J. Schultz, *Journal of the Acoustical Society of America*, Vol. 64, No. 2, New York, N.Y., 1978.
50. *The Apron and Terminal Building Planning Report*, Report No. FAA-RD-75-191, Federal Aviation Administration, Washington, D.C., July 1975.
51. *Threatened Wildlife of the United States*, Fish and Wildlife Service, Department of the Interior, Washington, D.C.
52. *United States List of Endangered Fauna*, Fish and Wildlife Service, Department of the Interior, Washington, D.C.

CHAPTER 15

Heliports

Introduction

A rotorcraft is a rotary winged aircraft that can lift vertically and sustain forward flight by power-driven rotor blades turning on a vertical axis. The helicopter is a vehicle which essentially can take off from and land in a nearly vertical direction. This is known as vertical takeoff and landing (VTOL). Since the helicopter is by far the most advanced and utilized of the vertical takeoff aircraft, the emphasis on this chapter is on ground facilities for helicopters and other rotary wing type aircraft, referred to as heliports.

Heliports

A *heliport* is defined as an identifiable area on land, water, or structures, including buildings or facilities, used or intended to be used for the landing and takeoff of helicopters or other rotary wing type aircraft [10]. A *helideck* is a heliport located on a floating or an off-shore structure. A *helistop* is defined as an area developed and used for helicopter landings and takeoffs to drop-off or pickup passengers or cargo. A *helipad* is defined as a paved or other surface used for parking helicopters at a heliport.

The Nature of Helicopter Transportation

Helicopters are used for a variety of aviation activities including aerial observation, sightseeing, agricultural application, law enforcement, fire fighting, emergency medical services, transporting personnel and supplies to offshore oil rigs, traffic and news reporting, corporate and business transportation, personal transportation, and heavy lifting. Helicopters are also extensively used in military operations throughout the world. Transportation by helicopter can generally be classified into two general categories, namely, private operations and commercial operations. Private operations are of the same nature as general aviation and commercial operations are similar to scheduled air carrier activity.

Most of the helicopters used in private operations have a capacity of 1 to 5 persons and have maximum gross weights between 3000 and 6000 lb. Helicopters in commercial operations are of greater capacity, typically between 10 and 50 passengers, and have maximum gross weights between 10,000 and 50,000 lb. Primarily because of the difference in size, heliport facilities for private operations are normally much smaller than those for commercial operations.

Private operations include construction, forest and police patrol, crop dusting, advertising, emergency medical service and rescue, and transport to off-shore oil well locations. Commercial operations may be classified into two types. One is transportation in large metropolitan areas between several airports in the region and between airports and the business center in the region. The second is intercity transportation between cities not necessarily in the same metropolitan area.

Experience to date shows that the helicopter has achieved its greatest success in the first type of service, the operating service area being within a radius of about 50 mi. The continued development of the second type of service will depend on the ability of helicopters to compete favorably with fixed-wing aircraft with respect to speed and economics over stage lengths up to about 300 mi.

Despite the growth of transport by helicopter, it only accounts for a small percentage of the total number of persons traveling by air. Helicopter operating costs have gradually been reduced but they are still considerably higher than those for fixed-wing aircraft.

Characteristics of Helicopters

A helicopter is a powered aircraft which gains its lift from the rotary motion of airfoil surfaces. The distinctive characteristic of a helicopter is its ability to hover through application of power to the rotating airfoils. The practical consequences of this characteristic are a much greater range of flight speeds and flight attitudes than is the case with conventional aircraft and the ability to land on and takeoff from comparatively small areas. When on the ground, helicopters have the ability of taxiing under their own power. Helicopters in private operation typically have cruise speeds between 90 and 130 kn, ranges between 300 and 400 nm, and passenger capacities between 2 and 10. Helicopters in the transport category typically have cruise speeds from 100 to 150 kn, ranges between 300 and 700 nm, and passenger capacities between 10 and 50.

While helicopters can ascend vertically from the ground, prolonged vertical ascents severely restrict load-carrying capacities. The usual procedure is to employ vertical ascent only to initiate the takeoff. As with other aircraft, takeoffs are usually made into the wind. The initial vertical rise for takeoff is aided by a ground cushion built up by the pressure of the air directed against the ground by the revolving rotors. After a few feet of vertical ascent, horizontal acceleration is begun until climb-out speed is reached. Prior to reaching climb-out speed, the helicopter can be flown in a horizontal path or in a slightly ascending path. Climb-out and descent speeds vary from 30 to 60 kn. Just before touchdown, the helicopter hovers momentarily 5 to 10 ft above the landing pad.

From a safety standpoint, the operation of single-engine helicopters requires that emergency landing areas be available along the entire flight path. In case of engine failure, safe landings using autorotation can be made if space is available. Autorotation is the continuation of rotor rotation in flight after cessation of power. Sufficient height must be reached by the helicopter in order to utilize the principle of autorotation. Twin-engine helicopters, on the other hand, are designed to permit continuation of flight and even a moderate rate of climb in the event once the engine fails. For these types of helicopters, it is not necessary from a safety standpoint to have space available for emergency landing along the entire route. Virtually all small helicopters in private operation are single engine. Most of the helicopters, with the exception of a few models, are also single engine. Twin-engine helicopters are not, however, designed to hover with only one engine in operation. On takeoff, therefore, in the event an engine fails before the helicopter has reached a one-engine-out flight speed, a landing must be made. To take care of this eventuality, sufficient space must be provided ahead of the landing area for an emergency landing. The single-engine helicopter also needs this area, but in addition requires space for emergency landings all along the flight path. If helicopters are designed to hover with one engine out, space ahead of the landing area is not required.

Because it is not economically practical for a helicopter to ascend and descend vertically, unobstructed approach-departure paths leading to the heliport are required. To protect the approach-departure path, obstructions are not permitted to extend above a prescribed inclined plane, an approach surface, beginning at the heliport and extending to specified distance from the heliport. The obstruction clearance requirements specified by the FAA are discussed later and ICAO has adopted similar recommendations.

Helicopters can be either single-rotor or tandem rotor and powered by one or two engines. The landing gear can consist of pontoons for landing on water, skids, or wheels equipped with rubber tires. When wheels are used, the landing gear normally consists of two main wheels and a single nose or tail wheel, or four wheels. Figure 15-1 shows several small utility helicopters at the Port Authority of New York and New Jersey heliport on the East River in New York City. The principal dimensions of representative helicopters used for private and commercial operations are shown in Fig. 15-2 and tabulated in Table 15-1.

Factors Related to Heliport Site Selection

The selection of a heliport site in an urban area requires the consideration of many factors, the most important of which are:

1. The best locations to serve potential traffic
2. The provision of minimum obstructions in the approach and departure areas

FIGURE 15-1   Port Authority of New York and New Jersey heliport (*Bell Helicopter Textron, Inc.*).

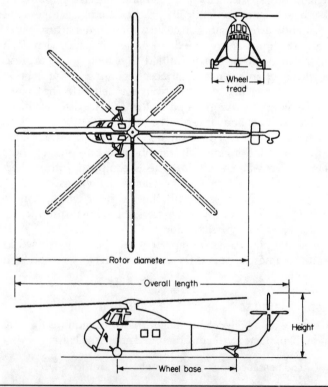

FIGURE 15-2   Dimensional definitions for helicopters.

Aircraft	Rotor Diameter, ft	Overall Length, ft	Height, ft	Wheelbase, ft	Wheel Tread, ft	Gross Weight, lb	Maximum Passengers
Aerosp 330J	49.5	59.8	16.9	13.2	7.9	16,315	19
Aerosp 332L	51.2	61.4	16.2	17.2	9.8	18,410	24
Bell-212	48.0	47.3	13.0	7.6	8.3	11,200	14
Bell-214ST	52.0	62.2	13.2	8.1	8.3	17,500	18
B-Vertol 107II	50.0	83.3	16.9	24.9	12.9	20,000	25
B-Vertol 234	60.0	99.0	18.7	25.8	10.5	48,500	44
B-Vertol 360	83.7	49.7	19.4	32.7	11.4	36,160	30
Sikorsky S-61N	62.0	73.0	18.9	23.5	14.0	20,500	28
Sikorsky S-64	72.0	88.5	25.4	24.4	19.8	42,000	45
Sikorsky S-76B	44.0	52.5	14.5	16.4	8.0	11,400	12
Westland 30300	42.5	52.1	16.3	17.8	9.3	16,000	19

Sources: International Civil Aviation Organization and Federal Aviation Administration.

TABLE 15-1 Dimensions of Typical Commercial Helicopters

3. The provision of minimum disturbance from noise and desirable location with respect to adjacent land use

4. The provision of adequate access to surface transportation and parking

5. The cost to acquire and develop

6. The provision of two approach paths separated by at least 90° and oriented with respect to prevailing winds

7. The avoidance of traffic conflicts between helicopters and other air traffic

8. The consideration of turbulence and visibility restrictions presented by nearby buildings

9. The provision of emergency landing areas along the entire route for single-engine helicopters

Final selection of a heliport site will usually require a compromise among these various factors. The most severe problems can be expected in large, highly developed metropolitan areas. In large urban areas heliports should be planned on a regional basis. The first step is to prepare an estimate of the demand for helicopter services and the origins and destinations of this demand. The second step is to select a heliport site or sites which can reasonably satisfy the demand and yet meet the requirements cited above.

The principal market for commercial helicopter transportation has been in large urban areas between one or more airports and the central business district. Therefore, it is essential that the downtown heliport be centrally located near the hotel area and the business district. Likewise adequate provision for helicopters should be made at airports. In extremely large urban complexes, there may be outlying smaller centers, and secondary heliports are needed so that the benefits of air transportation can extend to these centers. A heliport must have good access to streets, highways, and public transit facilities so that passengers using buses, personal vehicles, or mass transit can easily reach the facility.

Noise

The noise caused by helicopter operations within or adjacent to built-up urban areas is and will continue to be an extremely important factor in planning for helicopter transport, as it has been with fixed-wing aircraft. Manufacturers are aware of this problem and continue to study ways in which noise can be minimized.

A heliport should be located so that the noise generated by helicopters will not cause excessive disturbance to surrounding developments. The noise factor is most critical underneath the flight path on takeoff and landing. The amount of sound that can be tolerated by the average person is dependent upon a number of factors, including the overall noise level, its frequency, and its duration, the type of development (residential, industrial, etc.) surrounding the source of the noise, and the ambient sound level in the area. A greater amount of noise can be tolerated in industrial areas than in residential areas. Docks and other waterfront sites offer some of the best possibilities for heliport location in large, congested urban centers. Approach and noise problems can usually be overcome by making the use of water areas for heliport location. The downtown heliport in New York City is an example of such a facility.

Noise generated by small two- and three-seat helicopters can be tolerated in business and industrial areas, but the noise generated by large multiengine helicopters powered by turbine engines can exceed tolerable levels even in business and industrial areas. It is well to check with the manufacturers concerning the latest information on the levels of noise generated by the several transport type helicopters.

To minimize the noise, it is desirable to orient the landing pad so that landings and takeoffs are made over areas where noise would be least objectionable. Considerably more latitude can be exercised in this respect for helicopters than with fixed-wing aircraft.

Protection of Approach and Departure Paths

Zoning is necessary both to control the location of heliport sites for maximum benefit to the community and to provide safety in helicopter operations by protection of the surrounding airspace. The dimensions of the approach-departure paths for various types of heliport operations are discussed below.

Turbulence and Visibility

Another factor which must be considered in the selection of a site for a heliport is the effect of turbulence over roof surfaces and downdrafts near buildings. This factor is of particular importance for rooftop heliports. If there is doubt in the planner's mind, the site should be flight-checked with a helicopter.

Poor visibility can be an important factor to consider for sites on tall buildings, that is, those of 100 ft or more in height. The cloud deck seldom reaches the ground, but at higher levels the heliport might find itself enveloped in fog when the ground is clear.

Physical Characteristics of a Heliport

A heliport is defined as a facility which is intended to be used for the landing and takeoff of helicopters, and may include space for helicopter parking, buildings, servicing facilities, and vehicular parking. The *final approach and takeoff* (FATO) *area* is a defined area over which the final phase of the approach maneuver to a hover or a landing is completed and from which the takeoff maneuver is commenced. The *touchdown and liftoff* (TLOF) *area* is a hard surfaced load bearing area typically located within the final approach and takeoff area on which a helicopter may touch down or lift off. Functionally, the terminal area requirements for the parking, servicing, and fueling of helicopters and the processing of passengers and ground vehicles are no different from the requirements for fixed-wing aircraft.

Heliports are usually classified according to use as follows:

Military heliport: Facilities operated by one of the branches of the armed services. The design criteria are specified by the branch of the service and usually prohibit nonmilitary uses.

Federal heliport: Facilities operated by a nonmilitary agency or department of the federal government. They are used to carry out the functions appropriate to the agency.

Private-use heliport: Facilities which are restricted in use by the owner. These may be publicly owned but their use is restricted, as in police or fire department use.

Public-use heliport: Facilities which are open to the general public and do not require the prior permission of the owner to land. The extent of the facilities available may limit operations to helicopters of specified sizes or weights.

Commercial service heliport: Public use and public owned facilities which are designed for the use of helicopters in commercial passenger or cargo service which enplane 2500 passengers annually and receive scheduled passenger service with helicopters.

Personal-use heliport: Facilities which are used exclusively by the owner.

The principal components of a heliport are the final approach and touchdown area, the touchdown and liftoff area, and, for large heliports, taxiways, helicopter parking areas, and the terminal building area. The relationships between these components are shown in Fig. 15-3.

Final Approach and Takeoff Area

The final approach and takeoff area (FATO) is a surface from which the helicopter can land or take off. The FAA allows the FATO to be any shape as long as it is enclosed by a square of the dimensions indicated in Table 15-2. ICAO specifies that the FATO is a circle. Its size depends primarily on the overall length of the largest helicopter to be accommodated by the heliport. Because of the dust that can be created by the rotor of a helicopter, it is necessary to prepare the surface of the landing and takeoff area so that it will be free of dust

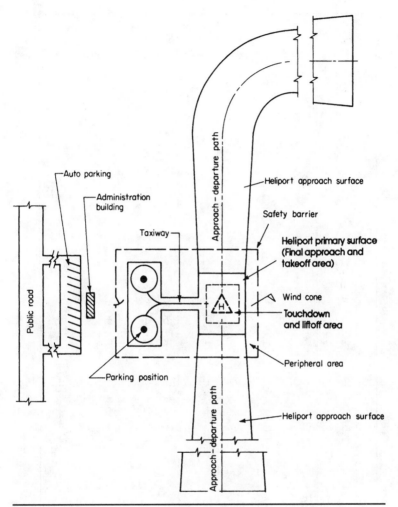

FIGURE 15-3 Typical heliport layout (*Federal Aviation Administration*).

(e.g., turf or pavement). It is recommended that paving or stabilizing the soil of the takeoff and landing area to improve the load carrying ability of the surface, minimize the erosive effects of rotor downwash, and to facilitate surface runoff due to rain or snow.

The recommended dimensions of the final approach and takeoff area are given in Table 15-2. For precision instrument operations the final approach and takeoff area is 300 ft wide by 1225 ft long and incorporates a *final approach reference area* (FARA) which is an obstruction-free area 150 ft by 150 ft located at the far end of the final approach and takeoff area.

	FAA			ICAO	
	Private	Public Utility	Commercial Transport	Land	Water
Final approach and takeoff area					
Length	$1.5L^b$	$1.5L^b$	200 ft^b	$1.5D^h$	$1.5D + 10\%^i$
Width	$1.5L^b$	$1.5L^b$	$2R^d$	$1.5D^h$	$1.5D + 10\%^i$
Clearance	$1/3R^c$	$1/3R^c$	30 ft	$0.25D^i$	$0.25D^i$
Touchdown and liftoff area					
Length and width	$1.5U$	$1.0L$	$1.0L^e$	$1.5U$	
Parking area					
Clearance^f	$1/3R^g$	30 ft	30 ft	—	
Minimum width		$1.5U$	$1.5U$	—	
One way taxiway route width					
Hover operations		$R + 60$ ft	$R + 60$ ft	$2R$	
Ground operations		$R + 40$ ft	$R + 40$ ft	7.5–20 m^j	
Parallel taxiway route width					
Hover operations		$R + 90$ ft	$R + 90$ ft		
Ground operations		$R + 70$ ft	$R + 70$ ft		
Taxiway pavement width		$2T$	$2T$		
Air transit route				$7R^k$	

[a] L is overall length of design helicopter; R is rotor diameter of design helicopter; U is maximum of undercarriage length or width of design helicopter; D is the overall length or width of the design helicopter, whichever is greater; T is wheel tread of design helicopter.

[b] May need to be adjusted for elevation; see AC 150/5390-2B.

[c] Minimum of 10 ft for private; minimum of 20 ft for public.

[d] 100 ft for public owned.

[e] Position on major axis of FATO with its center at least 50 ft from end or edge of FATO.

[f] Cannot lie under approach or climb path.

[g] Minimum of 20 ft.

[h] The overall length or width of the design helicopter, whichever is greater; for class 2 or class 3 helicopters on water heliports this is $2D$; the FATO described is circular with this diameter.

[i] For VFR; for IFR the clearance should be at least 45 m on each side of the centerline and 60 m beyond the ends.

[j] Depending upon the main gear span; separation between parallel taxiways should be 60 m on the side and 90 m on the ends of the final approach and takeoff area.

[k] For daytime operations; $10D$ for nighttime operations

[l] The same as for an aircraft parking area; see Chap. 6.

Sources: Federal Aviation Administration [10] and International Civil Aviation Organization [3].

TABLE 15-2 Geometric Design Standards for Heliports[a]

Touchdown and Liftoff Area

Within the final approach and takeoff area, an area is designated for the normal everyday landing of helicopters. The TLOF area is usually defined by a solid border painted on the pavement surface. The recommended dimensions of the touchdown and liftoff area are given in Table 15-2.

Peripheral Area

A peripheral or clearance area surrounding the final approach and takeoff area is recommended as an obstruction free safety zone. The area should be kept free of objects hazardous to the operation of helicopters. The clearance from the edges of the final approach and takeoff area required for this area is also given in Table 15-2.

Effect of Wind

Although helicopters can maneuver in much higher crosswinds than fixed-wing aircraft, the takeoff and landing area should preferably be oriented as nearly as possible to permit operation into the wind. At the present time it appears that the crosswind characteristics of the helicopter will be such that for a majority of cases a rectangular takeoff and landing area need be oriented in one direction only.

Terminal Area

At heliports where the volume of traffic is relatively small, the loading and unloading of passengers can be accomplished within the final approach and takeoff area. As traffic increases, it becomes necessary to provide additional space for the parking of helicopters and passenger processing. This is usually accomplished on a helipad which is an area adjacent to the terminal building for processing passengers. This area provides for one or more parking spaces for helicopters and is similar in nature and function to the gate or ramp area provided on the apron adjacent to airport terminal facilities. Clearances between adjacent helicopter parking positions are provided so that a separation is provided between the rotor planes of helicopters as shown in Table 15-2.

The helipad is connected to the final approach and takeoff area by taxiways and taxilanes. Helicopters may traverse taxiways in a hover or ground mode on single or parallel taxiway routes. The width of these taxiway routes and the taxilanes adjacent to the helipad are given in Table 15-2. The taxilane widths are the same as the taxiway routes in a ground mode of operation.

Approach and Takeoff Climb Path

There is a requirement that heliports have at least one approach and takeoff climb path which is free of obstructions that should be established on the basis of the direction of prevailing winds and the access route that has the fewest obstacles in the flight path. As conditions permit, additional approach and takeoff climb paths should be established to facilitate operations at times when winds are from other directions. At private use heliports, it is recommended that these paths flare out in the horizontal plane from the final approach and takeoff area at the rate of 1:20 and slope upward at the rate of 8:1. These paths terminate when the design helicopter attains a safe en route altitude. These paths may curve to avoid objects or noise sensitive areas when necessary. The FAA is in the process of developing design criteria for curved visual approaches and recommends that the FAR Part 77 specifications be applied until these criteria are developed. For a commercial service airport it is recommended that at least one instrument approach and takeoff climb path be established. For public use heliports the dimensions of the final approach and climb paths correspond to those specified under FAR Part 77 as discussed below.

Obstruction Clearance Requirements

Imaginary obstruction clearance surfaces are established for each class of heliport in FAR Part 77 [10]. For heliports, the principal surfaces are the approach and departure surfaces, the transitional surfaces, and the heliport protection zone. The *heliport protection zone* is the area on the ground below the approach surface from the edge of the final approach and takeoff area to the point where the approach surface is 35 ft above the elevation of the final approach and takeoff area. The horizontal surface required for airports is not necessary for heliports. The approach surface requirements for visual, nonprecision instrument and precision instrument operations specified by the FAA are given in Table 15-3. Similar requirement are specified

	Types of Approach		
	Visual	**Nonprecision**	**Precision**
Length	4,000	10,000	25,000*
Inner width	†	500	1,000
Outer width	500	5,000	6,000
Slope	8:1	20:1	34:1‡
Transitional			
Inner width	†	†	600
Outer width	250	600	1,500
Slope	2:1	4:1	7:1

*Begins 1225 ft from the far end of the final approach and takeoff area.
†Width of final approach and takeoff area.
‡For a 3° glide slope; 22.7:1 for a 4.5° glide slope; 17:1 for a 6° glide slope; glide slope can be increased in 0.1° increments with corresponding corrections to approach slope.
Source: Federal Aviation Administration [10].

TABLE 15-3 FAR Part 77 Approach Surface Dimensions for Heliports, ft

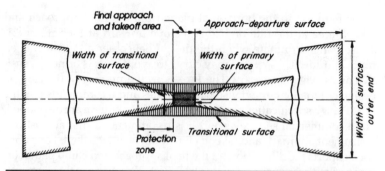

FIGURE 15-4   Obstruction clearance requirements for heliports (*Federal Aviation Administration [10]*).

by ICAO [3]. Precision instrument operations by commercial helicopters are very limited and thus visual or nonprecision instrument operation criteria will suffice for most heliports. The various FAR Part 77 surfaces are shown in Fig. 15-4. It should be noted that curved-path approaches and departures are currently not permitted under IFR conditions but with the implementation of microwave landing systems (MLS) this is subject to change. The FAA has developed specifications for the MLS critical areas and siting requirements which should be consulted if the installation of an MLS is contemplated. The specifications for IFR are a function of the nature of the navigational aids and references [10] should be consulted prior to establishing landing and takeoff paths.

The various dimensions specified by the FAA for a commercial service heliport are illustrated in Example Problem 15-1.

Example Problem 15-1   Let us design the layout of a commercial service heliport for operations with a Boeing-Vertol 234 design helicopter. Let us assume that the helipad or parking apron adjacent to the passenger terminal building will require space for four helicopters. Let us assume that the heliport elevation is 800 ft above mean sea level. A ground taxiway route is to be provided from the touchdown and liftoff area to the helipad.

From Table 15-1 the design helicopter has a rotor diameter D of 60 ft, an overall length L of 99 ft, a height of 18.7 ft, a wheelbase of 25.8 ft, a wheel tread of 10.5 ft, a maximum gross weight of 48,500 lb, and a maximum capacity 44 passengers.

For a commercial service heliport, from Table 15-2 the final approach and takeoff area is required to have a minimum length of 200 ft and a minimum width of 2 times the rotor diameter or 2(60) or 120 ft. An object-free clearance width of at least 30 ft from the edges of the final approach and takeoff area is also required.

The length and width of the liftoff and touchdown area is equal to the overall length of the design helicopter or 99 ft. Let us use 100 ft for these dimensions. To provide for the length of the helicopter and the minimum object-free area distance from the final approach and takeoff area, the minimum length of the taxiway leading to and from the touchdown and liftoff area is 100 + 30 or 130 ft.

Parking positions must have a minimum width of 1.5 times the undercarriage length or undercarriage width whichever is greater. The greater of these two dimensions is the wheelbase and therefore, the required minimum width of a parking position is 1.5 (25.8) or 39 ft. Let us provide 40 ft. However, since the rotor diameter of the design helicopter is 60 ft, the minimum width of the parking position must be 60 ft of which 40 ft will be paved. There must also be a clearance between the edges of adjacent parking positions 30 ft. This results in the minimum distance between the centerlines of adjacent parking positions being 90 ft.

A paved ground taxiway will be provided and this must have a minimum width of 2 times the wheel tread or 2(10.5) or 21 ft. The ground taxiway route is also required to have a safety area width of the rotor diameter plus 40 ft or 60 + 40 or 100 ft. Taxiways or taxilanes in the vicinity of the terminal building must also have this safety area to provide clearances between the building and parked helicopters. Therefore, the minimum distance from the centerline of the taxiway or taxilanes is 50 ft.

The layout of the heliport with the corresponding dimensions is shown in Fig. 15-5.

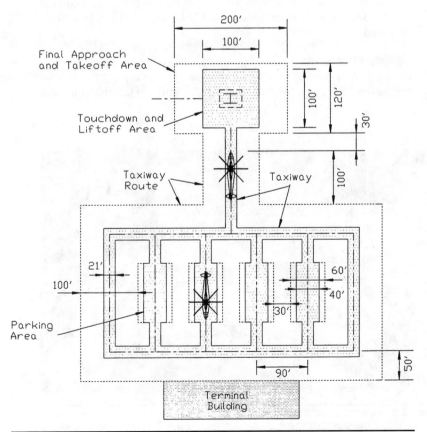

Figure 15-5 Commercial service heliport layout for Example Problem 15-1.

Marking of Heliports

The primary purpose for marking heliports is to identify the area clearly as a facility for the use of helicopters. The requirements for marking heliports are specified by the FAA and the ICAO. Essentially these requirements consists of painting an equilateral square with an "H" in the center of the touchdown and liftoff area. For hospital heliports a white cross is also inscribed within the square along with the letter "H" as shown in Fig. 15-6 for the heliport located at the Alexian Brothers Medical Center in Dade County, Florida. Marking delineating the edges of the final approach and takeoff area should be broken white lines whereas on the touchdown and liftoff area the edges should be delineated by continuous white lines. Taxiway centerlines are delineated by solid yellow lines and taxi route centerline and apron edge markings should be solid yellow lines. Taxiway edges should be marked by a double solid yellow line. A painted yellow line is also recommended to define the centerline of parking positions and when these positions vary in the amount of clearance provided a number enclosed by a circle should be painted on the entrance to the parking position to indicate the largest helicopter that can be accommodated.

Lighting of Heliports

For operation during hours of darkness various types of lights are suggested [10]. The amount of lighting depends on the character and

Figure 15-6 Heliport located at Alexian Brothers Medical Center in Dade County, Florida (*Howard Needles Tammen & Bergendoff and Alexian Brothers Medical Center*).

volume of operations. More lighting is required for scheduled air carrier operations than for private heliports with occasional use.

The minimum recommendations for private use and public use heliports consist of lighting of the perimeter of either the final approach and takeoff area or the touchdown and liftoff area with lights with yellow lenses uniformly spaced at 25-ft intervals. At public use heliports green lights are used to define the taxiway and taxilane centerlines. For commercial service heliports the touchdown and liftoff area is delineated by yellow lights located 10 ft from the outside edge. At such heliports in-pavement green lights are recommended for taxiway and taxilane centerlines. Blue retroreflective markers are also used at these airports to identify taxiway entrance and exit points and to define taxiway edges. Perimeter lighting defining the touchdown and liftoff area of a commercial service heliport is shown in Fig. 15-7. All objects that penetrate the obstruction clearance surfaces should be lighted with red colored lights.

Other useful lighting aids are the landing direction lights, visual glide path indicators, and heliport identification beacons. The landing direction lights are miniature approach lights since they extend only 75 ft. The color is yellow. Visual glide path indicators are also recommended for visual operations at commercial service heliports. The lowest on-course signal should provide a 1° clearance over any object in the approach path within 10° horizontally on either side of the approach path centerline. The optimal location is on the extended runway centerline of the approach path such that it will bring the

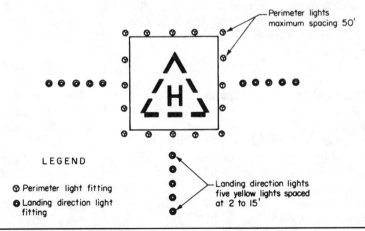

Figure 15-7   Heliport lighting configuration (*Federal Aviation Administration [10]*).

helicopter to between a 3- and 8-ft hover distance over the touch-down and liftoff area. Heliport identification beacons, alternate flashing white-green-yellow lights should be located with one quarter of a mile from a commercial service heliport.

Elevated Heliports

When ground level sites are not available or are unsuitable, an elevated site may be practical. Elevated heliports may be located on piers or other structures over water, as well as on buildings. The dimensions of the touchdown and liftoff area are the same as for heliports on the ground, but the final approach and takeoff area can be smaller and there is no need for peripheral areas. When planning a rooftop heliport, a thorough study should be made of the air currents caused by the presence of adjacent buildings. Roof areas make it possible to locate the heliport closer to the center of business activities in a city, provided the facility is environmentally acceptable. Another advantage is that the land cost is partially absorbed by the tenancy of the lower floors of the building. However, it should be realized that of the operations of any size, space on the floors below the takeoff and landing area may have to be devoted to uses such as lobby, freight, and baggage handing. The possible disadvantage of height with respect to visibility was mentioned earlier. A heliport 100 ft or more above the ground would require a higher cloud base than a ground heliport to provide the same operating safety. A downtown commercial heliport would require, in addition to lobby space, car parking facilities relatively close by.

Where heliports are built on elevated structures, the strength of the floor should be greater than the strength of the landing gear of the helicopter. The loads imposed by helicopters and recommendations concerning the structural design of elevated structures are discussed in the next section.

Structural Design of Heliports

Helicopters using facilities on land are usually supported on tubular skids or wheels equipped with rubber tires. Helicopters equipped with conventional landing gear wheels are normally supported by two main wheels and one tail or nose wheel. For larger helicopters each main landing gear consists of two wheels. Each main gear typically supports 40 to 45 percent of the weight of the helicopter and the tail or nose wheel supports the remainder of the weight, approximately 10 to 20 percent. If the helicopter is supported by tubular skids, 50 percent of the weight is supported by each skid.

The strength requirements for the touchdown and liftoff area are determined by considering the static load, dynamic load, and downwash load of the helicopter. Both the static load and the dynamic load are applied through the landing gear contact area whereas the downwash load is applied over a contact area defined by the diameter of the rotors. The FAA recommends that for design purposes the touchdown and liftoff area should be capable of supporting 150 percent of the maximum takeoff weight of the design helicopter.

Heliports at Airports

A large number of helicopters will operate into airports to serve traffic from the downtown area and surrounding communities. Accordingly, provisions should be made at an airport for the landing and takeoff of helicopters. The takeoff and landing area should be located to

1. Provide maximum separation from fixed-wing aircraft traffic patterns so as to avoid creating a conflict in takeoff and landing operations.

2. Be as close as possible to passenger check-in areas for fixed-wing aircraft to avoid long walking distances for passengers.

3. Avoid as much as possible the mixing of taxiing fixed-wing aircraft and helicopters, since helicopters taxi at relatively low speeds.

It is recommended that if simultaneous same direction diverging helicopter and fixed-wing aircraft operations are to be conducted under visual flight rule conditions, a runway centerline to a final approach and takeoff area and a touchdown and liftoff area centerline separation of 700 ft be provided. A 2500-ft minimum separation is required for radar departures under instrument flight rule conditions [10]. Helicopter parking apron areas should meet the same runway clearance standards as those required for fixed-wing aircraft parking [5].

Alternative locations for heliports at an airport are the roof of the terminal building, the apron adjacent to the terminal building used by fixed-wing aircraft, and the area adjacent to the terminal building separate from the fixed-wing aircraft apron. There are advantages and disadvantages to all three locations. Normally, a ground-level site is preferred. The most convenient and least expensive method for accomplishing this is to reserve a part of the fixed-wing aircraft apron for the takeoff and landing of helicopters. If this is not convenient, a special pad for helicopter operations on the aircraft side of the terminal building should be provided.

Figure 15-8 shows the heliport at Miami International Airport.

Figure 15-8   Heliport at Miami International Airport (*Howard Needles Tammen and Bergendoff & Miami International Airport*).

References

1. *A Canadian STOL Air Transport System—A Major Program*, Report No. 11, Canadian Science Council, Ottawa, Canada, 1970.
2. *Aerodrome Design Manual*, Part 2: *Taxiways, Aprons, and Holding Bays*, 2d ed., International Civil Aviation Organization, Montreal, Canada, 1983.
3. *Aerodromes, Annex 14 to the Convention on International Civil Aviation*, Vol. II, *Heliports*, International Civil Aviation Organization, Montreal, Canada, 1990.
4. *A Guide to STOL Transportation System Planning*, The DeHavilland Aircraft of Canada, Limited, Ottawa, Canada, January 1970.
5. *Airport Design*, Advisory Circular AC 150/5300-13 Federal Aviation Administration, Washington, D.C., 1989.
6. *Canadian STOL Demonstration Service Montreal STOL Port Master Plan*, ST-71-8, Canadian Air Transportation Administration, Ottawa, Canada, March 1972.
7. *Certification and Operations of Scheduled Air Carriers with Helicopters*, Part 127, Federal Aviation Regulations, Federal Aviation Administration, Washington, D.C., 1974.
8. *Guide for the Planning of Small Airports*, Roads and Transportation Association of Canada, Ottawa, Canada, 1980.
9. *Helicopter Annual*, Helicopter Association International, Alexandria, Va., 1992.
10. *Heliport Design*, Advisory Circular AC 150/5390-2B, Federal Aviation Administration, Washington, D.C., September 2004.
11. *Objects Affecting Navigable Airspace*, Part 77: Federal Aviation Regulations, Federal Aviation Administration, Washington, D.C., 1989.
12. *Planning and Design Criteria for Metropolitan STOL Ports*, Advisory Circular AC 150/5300-8, Federal Aviation Administration, Washington, D.C., April 1975.
13. "Planning STOL Facilities," L. Schaefer, Paper No. 690421, *Proceedings*, Society of Automotive Engineers, New York, 1969.

14. *Provisional Criteria for STOL Port Zoning*, Canadian Air Transportation Administration, Ottawa, Canada, August 1973.
15. *Quiet Turbofan STOL Aircraft for Short-Haul Transportation*, Contractor Report Nos. NASA CR 114612 and CR 114613, National Aeronautics and Space Administration, Washington, D.C., June 1973.
16. *Rotorcraft Master Plan*, Federal Aviation Administration, Washington, D.C., November 1990.
17. *STOL Aircraft Future Trends*, Transport Aircraft Council, Aerospace Industries Association of America, Inc., Washington, D.C., May 1971.
18. *STOL Port Manual*, 1st ed., Doc 9150-AN/899, with Amendment 1, International Civil Aviation Organization, Montreal, Canada, 1988.
19. *STOL-VTOL Air Transportation Systems*, C. Hintz, Jr., Civil Aeronautics Board, Washington, D.C., 1970.
20. *Studies in Short Haul Air Transportation—Effects of Design Runway Length, Community Acceptance, Impact on Return on Investment and Fuel Cost Increases*, R. S. Shevell and D. W. Jones, Jr., Stanford University, National Aeronautics and Space Administration, Ames Research Center, Moffett Field, Calif., July 1973.
21. *Study of Aircraft in Intraurban Transportation Systems, San Francisco Bay Area*, Boeing Company, Contractor Report No. NASA CR-114347, National Aeronautics and Space Administration, Ames Research Center, Moffett Field, Calif., 1970.
22. *Study of Quiet Turbofan STOL Aircraft for Short-Haul Transportation*, McDonnell-Douglas Corporation Report No. MDC-J4371, for National Aeronautics and Space Administration, Moffett Field, Calif., June 1973.
23. *Study of Short-Haul Aircraft Operating Economies*, Contractor Report No. CR-137685, Ames Research Center, National Aeronautics and Space Administration, Moffett Field, Calif., September 1975.
24. *United States Standard for Terminal Instrument Procedures (TERPS)*, FAA Order 8260.3B, with Changes 1 through 12, Federal Aviation Administration, Washington, D.C., 1992.
25. *Vertiports*, Advisory Circular AC 150/5390-3, Federal Aviation Administration, Washington, D.C., 1991.
26. "V/STOL Aircraft: The Future Role in Urban Transportation as a Pickup and Distribution System," R. H. Miller, Proceedings of Symposium on Transportation and the Prospects for Improved Efficiency, National Academy of Engineering, Washington, D.C., October 12, 1972.

Index